Endocytosis in Plants

Jozef Šamaj
Editor

Endocytosis in Plants

Editor
Jozef Šamaj
Centre of the Region Haná for Biotechnological
and Agricultural Research
Faculty of Science
Palacký University in Olomouc
Olomouc
Czech Republic

ISBN 978-3-642-32462-8 ISBN 978-3-642-32463-5 (eBook)
DOI 10.1007/978-3-642-32463-5
Springer Heidelberg New York Dordrecht London

Library of Congress Control Number: 2012948295

Printed on acid-free paper

Springer is part of Springer Science+Business Media (www.springer.com)

Editor Biography

Jozef Šamaj received his Ph.D. degree in Plant Physiology from Comenius University in Bratislava, Slovakia. He completed three post-doctoral stays in France, Germany, and Austria, supported by Eurosilva, the Alexander von Humboldt Foundation, and the Marie Curie Programme. He co-edited four books and co-authored more than 115 research papers, reviews, and book chapters. He received his habilitation degree from the University of Bonn, Institute of Cellular and Molecular Botany in 2004, where he was a senior lecturer and group leader. Since 2010, he has been a full professor and department leader at the Palacký University Olomouc, Centre of the Region Haná for Biotechnological and Agricultural Research. His research is focused on the functional characterization of mitogen-activated protein kinases as well as cytoskeletal and vesicular trafficking proteins during plant development and stress responses. His laboratory is using integrated cell-biological, genetic, and functional proteomic approaches.

Preface

During the past 6 years plant endocytosis developed into a flourishing research field. The role of clathrin-mediated endocytosis in the internalization of some plasma membrane proteins was firmly established while alternative clathrin-independent endocytic routes such as fluid-phase and flotillin-dependent endocytosis were also described. Plant endosomes turned from enigmatic organelles to subcellular compartments with partially defined molecular topology and function. Chapter 1 of this book provides an overview of diverse methods recently introduced into plant endocytosis research. Up-to-date methodological approaches such as proteomics and advanced microscopy including light sheet microscopy or fluorescence recovery after photobleaching (FRAP) combined with super resolution microscopy start to be applied in plant endocytic research. Chapter 2 is devoted to chemical genomics, providing a new generation of more specific chemical inhibitors which in combination with automated quantitative microscopy (cellomics) provide another very powerful tool to study endocytosis in plants. Chapters 3 and 4 are focused on the crucial role of endocytosis in the establishment and maintenance of polarity in diverse types of plant cells. Chapter 5 describes fluid phase endocytosis in specialized storage plant cells while Chap. 6 provides a very useful overview of physical factors which have some impact on endocytosis. The next three chapters are focused on plasma membrane proteins such as receptors, auxin transporters and water channels, nicely demonstrating biologically relevant roles of endocytosis in the regulation of signalling proteins as well as auxin and water transport in plant cells and tissues. Chapters 10–14 deal with crucial molecular players regulating different steps of endocytosis, namely Rab GTPases, SNAREs, SCAMP, sorting nexins, retromer, and ESCRT proteins. The next two chapters summarize the importance of endocytosis in plant cell interaction with pathogens and symbiotic microbes. The final chapter provides an overview of the role of the cytoskeleton in the different types of endocytosis in plants and other organisms.

In total, the present book summarizes the latest advances in the field of plant endocytosis. Moreover, it also provides several excellent examples of biological relevance of endocytosis in physiological processes controlling cell polarity, shape, water and nutrition uptake, or biotic interactions of plant cells with pathogens and symbionts. These surely belong to important fields of plant biology.

I would like to thank all authors for their excellent contributions to this book. I hope that reader will enjoy and appreciate it.

This book is dedicated to my family.

Olomouc, June 2012 Jozef Šamaj

Contents

Update on Methods and Techniques to Study Endocytosis in Plants

Olga Šamajová, Tomáš Takáč, Daniel von Wangenheim, Ernst Stelzer and Jozef Šamaj

Abstract The growing interest in the investigation of endocytosis, vesicular transport routes, and corresponding regulatory mechanisms resulted in the exploitation of cell biological, genetic, biochemical, and proteomic approaches. Methods and techniques such as site-directed and T-DNA insertional mutagenesis, RNAi, classical inhibitor treatments, and recombinant GFP technology combined with confocal laser scanning microscopy (CLSM) and electron and immune-electron microscopy were routinely employed for investigation of endocytosis in plant cells. However, new approaches such as high-throughput confocal microscopy screens on mutants and proteomic analyses on isolated vesicular compartments and root cells treated with vesicular trafficking inhibitors (both focused on the identification of new endosomal proteins), together with chemical genomics and advanced microscopy approaches such as Förster resonance energy transfer (FRET), fluorescence recovery after photobleaching (FRAP), light sheet-based fluorescence microscopy, and super-resolution microscopy provided a significant amount of new data and these new methods appear as extremely promising tools in this field.

O. Šamajová · T. Takáč · J. Šamaj (✉)
Centre of the Region Haná for Biotechnological and Agricultural Research,
Faculty of Science, Department of Cell Biology, Palacký University, Šlechtitelů 11
CZ-783 71 Olomouc, Czech Republic
e-mail: jozef.samaj@upol.cz

D. von Wangenheim · E. Stelzer
Physical Biology Group, Frankfurt Institute for Molecular Life Sciences (FMLS),
Goethe Universität Frankfurt am Main, Max-von-Laue-Street 15
60438 Frankfurt am Main, Germany

J. Šamaj (ed.), *Endocytosis in Plants*,
DOI: 10.1007/978-3-642-32463-5_1,

1 Introduction

Endocytosis is a dynamic process of intracellular uptake of plasma membrane and extracellular cargos, which is controlled by a network of regulatory proteins. In addition, endocytosis of some plasma membrane proteins is modulated by their posttranslational modifications (PTMs) such as mono-ubiquitination and/or phosphorylation. Endocytosis is highly sensitive to the changes in external and internal physical (see chapter by Baluška and Wan in this volume) and chemical conditions (see chapter by Li et al. in this volume). Diverse molecules including proteins, lipids, and carbohydrates (Müller et al. 2007; Ovečka et al. 2010) are internalized by endocytosis. Spatial and temporal regulation and variability of endocytosis in diverse plant cell types, tissues, and organs emphasize the careful choice of methodological approaches used to study this highly dynamic biological process. This chapter provides an overview of methods and techniques, which were used to study plant endocytosis in recent years.

2 Chemical and Biochemical Methods

2.1 Using Conventional Chemical Inhibitors to Study Endocytosis

Well-established inhibitory compounds such as brefeldin A (BFA) and wortmannin are useful to study endocytosis in plant cells (Müller et al. 2007). The sensitivity of certain proteins to these inhibitors may indicate their subcellular localization and participation in specific steps of endocytic vesicular transport. During the last years, inhibitors were broadly employed for microscopic observations (Müller et al. 2007) and also in biochemical and proteomic studies (Luczak et al. 2008; Takáč et al. 2011a; Takáč et al. 2012).

BFA, a macrocyclic lactone, targets some BFA-sensitive adenosine diphosphate ribosylation factor-guanine nucleotide exchange factors (ARF GEFs), thus inhibiting secretory and recycling vesicular trafficking pathways in yeast, mammalian, and plant cells (Nebenführ et al. 2002; Geldner et al. 2003; Takáč et al. 2011a). A latest proteomic study revealed that profilin 2, an actin binding protein, is involved in the formation of BFA-induced compartments in Arabidopsis roots (Takáč et al. 2011a).

Concanamycin A, an inhibitor of V-ATPase was used to prove importance of V-ATPase activity for Golgi ultrastructure as well as to study *trans*-Golgi network (TGN) and multivesicular body (MVB) structural integrity (Dettmer et al. 2006; Viotti et al. 2010; Scheuring et al. 2011). Application of concanamycin A caused colocalization of TGN and MVB molecular markers, thus substantially contributing to the new finding that MVBs origin from TGN in Arabidopsis (Scheuring et al. 2011).

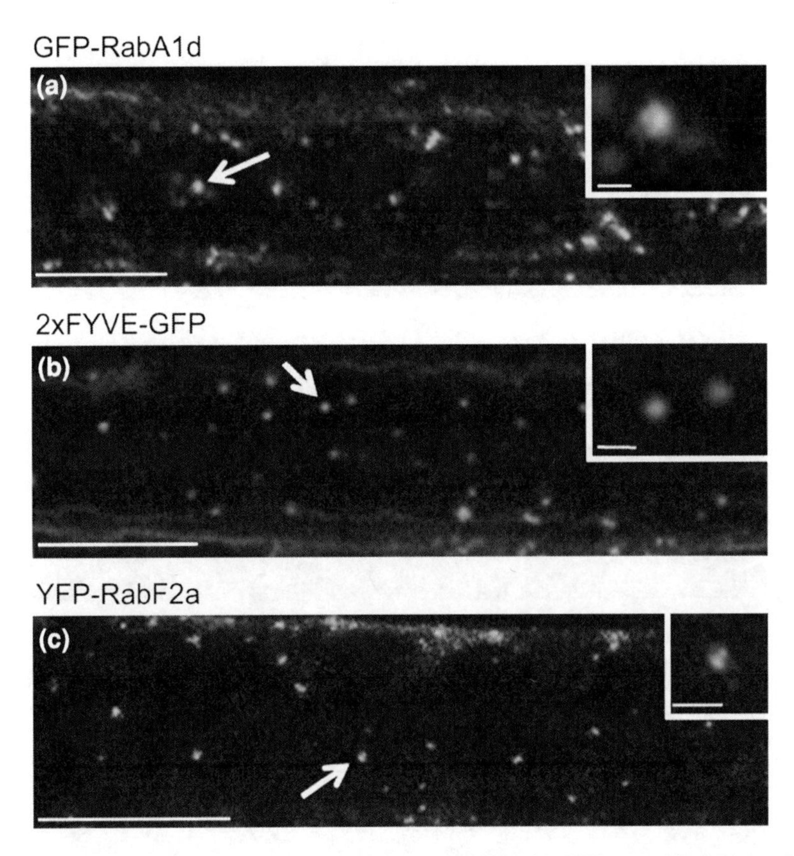

Fig. 1 Visualization of early endocytic compartment/TGN with GFP-tagged RabA1d and late endocytic compartments with 2xFYVE-GFP and YFP-tagged RabF2a using CLSM. Typical early and late endocytic compartments are indicated with Ï arrows. *Bars* represent 10 μm for panels **a**, **b**, and **c** and 1 μm for insets

Furthermore, there are several chemical compounds inhibiting clathrin-mediated endocytosis. Wortmannin inhibits phosphatidyl-inositol-3-kinase (PI3 K) and phosphatidyl-inositol-4-kinase (PI4 K) in a dose-dependent manner and also blocks clathrin-mediated endocytosis consequently to the aggregation and stabilization of clathrin-coated pits at the plasma membrane (Matsuoka et al. 1995; Ito et al. 2012; Figs. 1 and 2). The same effect was shown for LY294002 (2-(4-morpholinyl)-8-phenyl-4H-1-benzopyran-4-one) which is a synthetic compound selectively inhibiting PI3 K (Etxeberria et al. 2005; Baroja-Fernandez et al. 2006; Lee et al. 2008a). These two compounds also block vacuolar transport due to the fusion and swelling of prevacuolar compartments (PVCs), which are identical with MVBs (Matsuoka et al. 1995; Wang et al. 2009; Takáč et al. 2012). However, wortmannin also caused depletion of TGN compartments which probably fused with MVBs in wortmannin-treated Arabidopsis root cells (Takáč et al. 2012). Other inhibitors

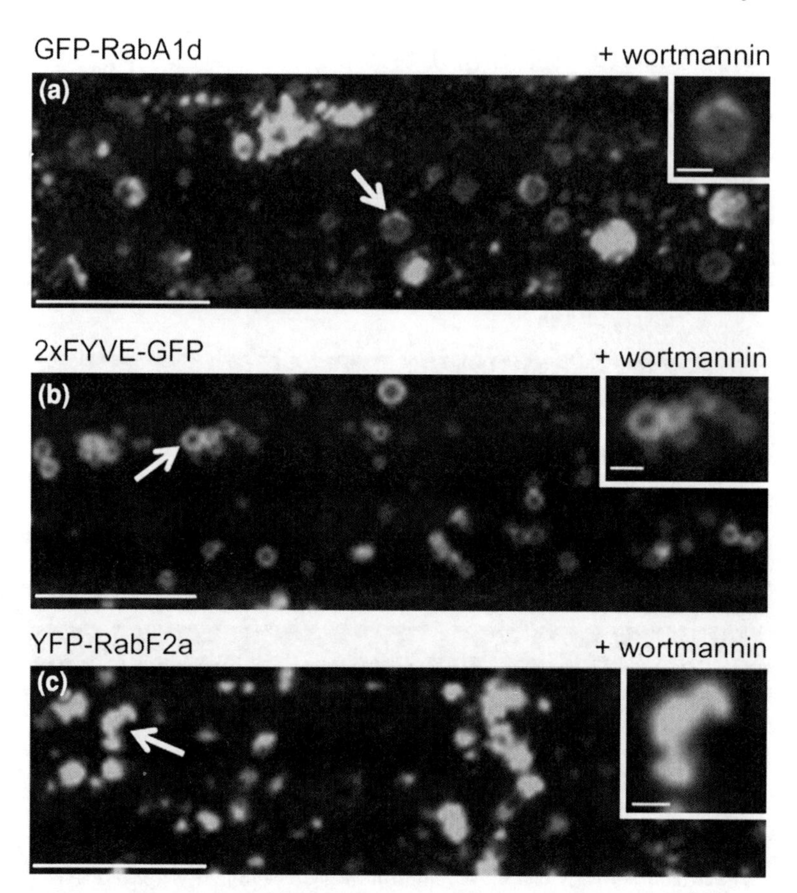

Fig. 2 Effect of wortmannin on early endocytic compartment/*TGN* (GFP-RabA1d) and late endocytic compartments (2xFYVE-GFP and YFP-RabF2a). The wortmannin-induced swelling and partial aggregation of *TGN* compartments as well as formation of larger clustered structures designated as wortmannin-induced compartments are indicated by *arrows*. Bars represent 10 μm for panels **a**, **b** and **c** and 1 μm for insets

such as tyrphostin A23 inhibited cargo sorting into clathrin-coated vesicles during clathrin-mediated endocytosis (Banbury et al. 2003) while ikarugamycin was used to inhibit clathrin-dependent endocytosis in tobacco protoplast culture (Onelli et al. 2008; Bandmann and Homann 2012; Nagawa et al. 2012).

Additionally, several routinely used auxin transport inhibitors such as 2,3, 5-triiodobenzoic acid (TIBA) and 2-(1-pyrenoyl) benzoic acid (PBA) inhibit vesicle trafficking in plant, yeast, and mammalian cells affecting actin dynamics (Dhonukshe et al. 2008) while nystatin and filipin inhibit lipid raft-mediated endocytosis (Kale et al. 2010; Ovečka et al. 2010; Luu et al. 2012).

2.2 Chemical Genomics Identifies New Inhibitors of Endocytosis and Vesicular Trafficking

More specific and targeted inhibitors are necessary to study complex vesicular trafficking pathways including the endocytosis. Significant effort has been devoted to the identification of new vesicular trafficking inhibitors through large-scale chemical genomics approaches (see chapter by Li et al. in this volume). In a chemical library screening study, altered pollen germination and pollen tube polar growth were used as selection criteria to find compounds altering vesicular trafficking (Robert et al. 2008). Using microscopic studies it was shown that endosidin 1 (ES1) interfered with endocytosis, causing the selective accumulation of PIN2, AUX1, and the brassinosteroid receptor BRI1 in distinct endomembrane compartments termed "endosidin bodies". ES1 also altered the circadian system of Arabidopsis, and it was proposed that the ability of ES1 to modify endosome trafficking is connected with an actin-stabilizing effect (Tóth et al. 2012). In a complementary study, a chemical library of 46418 compounds was screened for potential effects on vesicular transport by means of high-throughput CLSM (Drakakaki et al. 2011). For this purpose, distinct subcellular chemical-induced relocalizations of plasma membrane marker proteins such as PIN1-GFP, PIN2-GFP, and BRI1-GFP (all under the control of the respective native promoters) were tested in Arabidopsis root tips. A new compound named endosidin 3 was shown to target TGN but not Golgi, while it also altered the plasma membrane localization of small GTPase ROP6. Finally, another compound named endosidin 5 enhanced trafficking of plasma membrane proteins toward the vacuole.

2.3 Cell Fractionation and Isolation of Endosomes

The isolation of organelles involved in endocytosis and vesicular transport in combination with immunolocalization, biochemical, and proteomic analyses is one of the most effective ways to dissect the intracellular transport processes. Recently, vesicular TGN compartments containing SYP61 were isolated from Arabidopsis plants expressing *proSYP61::SYP61-CFP* using sucrose gradient fractionation, followed by immunopurification with antibodies against GFP (Drakakaki et al. 2012). SYP61 is a TGN-localized member of a Q-SNARE complex formed by SYP41/SYP61/VTI12 (Zouhar et al. 2009). Immunoprecipitation was performed using protein A agarose beads coupled to GFP antibody and rabbit IgG. Plant vesicle extracts were pretreated with unconjugated beads to remove non-specific binding proteins and then incubated with the antibody-coated beads to collect vesicles containing SYP61-CFP (Drakakaki et al. 2012). Sucrose density gradient fractionation was also used to isolate the MVB fraction and to prove the presence of vacuolar sorting receptors (VSR) in MVBs of germinating mung bean seeds.

First, the VSR-enriched fraction in sucrose gradients was detected by using VSR-1 antibody, and subsequently immunogold negative staining of isolated MVBs was performed with VSR-1 and BP-80 CT antibodies (Wang et al. 2007).

Sucrose gradient fractionation combined with immunoblots also represents a widely used approach to localize proteins related to endosomal transport. For example, the retromer-like protein complex is associated with PVC and with high-density sedimenting membranes. Immunogold negative staining identified these membranes as coated microvesicles with 90 nm diameter (Oliviusson et al. 2006). GRV2/RME-8 protein was identified as a protein controlling the transport between MVB and vacuole. The sucrose gradient fractionation showed that this protein may be distributed along an array of diverse but interconnected endomembrane compartments (Silady et al. 2007). Membrane fractions separated in sucrose gradients were either probed against RabA4b, or spotted onto nitrocellulose membranes and immunoblotted with monoclonal antibodies recognizing cell wall polysaccharides (Kang et al. 2011). The pattern of terminally fucosylated xyloglucan distribution nearly followed the pattern of RabA4b distribution, suggesting TGN localization. To simplify the detection of the protein of interest, microsomal membranes of BY-2 cells overexpressing SCAMP2-YFP were separated by sucrose density gradient fractionation and subsequently separated by SDS-PAGE. SCAMP2-YFP was found in plasma membrane and TGN fractions suggesting that it is involved in the transport between these two subcellular compartments (Toyooka et al. 2009). Recently, signal transducing mitogen-activated protein kinase MPK6 was localized to the plasma membrane and TGN using subcellular fractionation combined with immunoblotting, and these results were supported by diverse microscopic methods (Müller et al. 2010).

2.4 Isolation of Plasma Membrane Lipid Rafts

Lipid rafts are sterol/sphingolipid-enriched liquid crystalline-ordered membrane microdomains important for protein and lipid subcellular trafficking and for intracellular signalling (Simon-Plas et al. 2011). They are usually isolated from the plasma membrane fraction prepared by using an aqueous two-phase polyethylene glycol (PEG)-dextran partitioning system. Detergent-resistant membrane (DRM) proteins are isolated by the solubilization of plasma membrane by the addition of Triton X-114, followed by the addition of sucrose in certain, tissue-dependent final concentration, overlaid with successive concentration steps of sucrose in TBS buffer, and then centrifuged at 200,000 g for 16 h. DRMs could be recovered above the 30–35 % interface as an opaque band (Borner et al. 2005; Morel et al. 2006; Lefebvre et al. 2007). Alternatively, an Optiprep step gradient can be used (40/30/0 % OptiPrep). In this case, DRMs are isolated on the top of the gradient simplifying the procedure (Carmona-Salazar et al. 2011).

In the search of lipid raft origins, DRMs were isolated also from endoplasmic reticulum (ER) and Golgi fractions again on the basis of Triton X-100 insolubility (Laloi et al. 2007). The lipid composition of Golgi DRMs showed a marked similarity to plasma membrane-derived DRMs suggesting that plasma membrane lipid rafts originate in Golgi apparatus (Laloi et al. 2007).

2.5 Proteomic Approaches to Study Endocytosis

Proteomics is a valuable tool for the investigation of various aspects of plant development (Takáč et al. 2011b; Weckwerth 2011) and also represents a challenging strategy to investigate endocytosis. With the help of proteomic techniques it is possible to identify a spectrum of proteins in certain tissues at certain time. However, the capability of proteomics to investigate endocytosis is limited due to the dynamic character and spatial variability of endocytotic proteins.

A significant advancement of the knowledge on organelle proteomes was achieved during the last few years (Dunkley et al. 2006; Lilley and Dupree 2007; Agrawal et al. 2010).

The organelle proteomic analysis, like all other biochemical subcellular localization studies, requires absolute purity of the respective organelle fractions. One of the most powerful approaches to overcome this problem is membrane density gradient fractionation combined with isotope tags for relative and absolute quantitation (iTRAQ) (Dunkley et al. 2006). In brief, proteins of membrane fractions are solubilized, digested, and the resulting peptides are labeled by different iTRAQ reagents. Such peptides are subjected to 2D LC MS/MS analysis in conjunction with 2D liquid chromatography, which enables the quantification of protein distributions within a gradient. Proteins with similar density gradient distributions and therefore, localizations, are clustered (Lilley and Dupree 2007).

Very recently, the SYP61 compartment was isolated using two-step immunopurification combined with sucrose gradient fractionation (Drakakaki et al. 2012). Isolated proteins were identified by nano-liquid chromatography coupled to tandem mass spectrometry and two-dimensional nano-ultra-performance liquid chromatography (UPLC). Proteins detected in the SYP61 sample but not in IgG control were considered specific to TGN compartment possessing SYP61. About 60 % of the identified proteins were known or predicted to be associated with the endomembrane system and 32 % were predicted as a putative cargo transported by vesicular transport. Microscopic observations were employed to validate the localization of individual proteins. This study detected all members of SYP61 SNARE complex in the proteome of SYP61 compartment while SYP43 was identified as a putative SYP61 interactor. In addition, the proteome of SYP61 compartment encompasses important regulators of membrane trafficking such as RabD2b and YIP1, as well as the VSR3, VSR4, and VSR7 (Drakakaki et al. 2012).

Since endocytosis and vesicular trafficking are dynamic processes affected by chemical compounds in certain defined steps, differential proteomics using such compounds can significantly contribute to our better understanding of these processes. For example, the differential proteomic analysis of Arabidopsis roots treated by BFA was recently published (Takáč et al. 2011a). Using comprehensive, gel-free, and gel-based proteomic approaches it was shown that proteins known to accumulate in BFA compartments are upregulated in the whole cell protein fraction. A combination of proteomic approaches with microscopic observations revealed an important role of cytoskeletal protein profilin 2 in the interplay of vesicular transport and the cytoskeleton. Additionally, the BFA-induced accumulation of ER resident proteins revealed by proteomics was confirmed by independent CLSM and electron microscopy approaches (Takáč et al. 2011a). In a similar way, the changes in the Arabidopsis root proteome were analyzed after wortmannin treatment. This study revealed that TGN-localized RabA1d, a small Rab GTPase is downregulated by wortmannin (Takáč et al. 2012). Wortmannin also affected post-Golgi compartments leading to depletion of TGN and fusion/swelling of MVBs, indicating the possible consumption of TGN by MVBs, as revealed by CLSM and electron microscopy analyses (Wang et al. 2009; Takáč et al. 2012).

Several proteomic studies employing Triton X-100-based isolation procedures aimed to analyze lipid rafts in Arabidopsis (Borner et al. 2005), BY-2 tobacco cells (Morel et al. 2006), *Medicago truncatula* (Lefebvre et al. 2007) and the two monocotyledonous species oat and rye (Takahashi et al. 2012). The solubilization of DRMs using buffers containing high concentrations of detergents may be incompatible with reverse phase separation and substantially decrease the efficiency of trypsin to digest proteins. Therefore, the separation of proteins either by 1D SDS-PAGE followed by excision of protein lanes (Morel et al. 2006; Lefebvre et al. 2007) or 2D electrophoresis (Borner et al. 2005) overcomes these problems. Proteins involved in signalling and response to biotic and abiotic stresses, cellular trafficking, and cell wall metabolism are more abundant in DRMs compared to the rest of the plasma membrane (Morel et al. 2006; Takahashi et al. 2012).

3 Molecular Biological Methods

3.1 Cloning and Fluorescent Tagging of Endocytic Proteins for Visualization and Colocalization Studies

Recently, the most common method of subcellular visualization of endocytic proteins is based on their cloning and tagging with fluorescent proteins such as GFP and its structural analogs such as YFP and CFP. Occasionally, other fluorescent proteins such as mCherry and mRFP were used. The number of fluorescently tagged endocytic proteins for cell biological studies is continuously increasing. Here, we provide a brief overview and highlight the most important examples of these proteins.

In most cases, these proteins were cloned under the control of the strong constitutive cauliflower mosaic virus promoter 35S. In order to minimize problems with silencing, variable expression, and expression bias, the ubiquitin promoter UBQ10 was also used for the constitutive expression of some endocytic marker proteins (Geldner et al. 2009; Herberth et al. 2012). Nevertheless, there are also examples of gene cloning under their own native promoters as it was in the case of *GNOM* (Geldner et al. 2003; Miyazawa et al. 2009), *GNL1* and *GNL2* (Richter et al. 2011), *VAN3* (Naramoto et al. 2010), *BRI1* (Wang et al. 2005), *BAK1* (Chinchilla et al. 2007), *FLS2* (Robatzek et al. 2006), *SNARE* (Ebine et al. 2008; Zhang et al. 2011), *AP-3 δ* and *AP-3 β* (Zwiewka et al. 2011), *PIN1* (Benková et al. 2003), *PIN2* (Xu and Scheres 2005), *PIN3* (Friml et al. 2002a), *PIN7* (Blilou et al. 2005), *AUX1* (Swarup et al. 2004), *BOR1* (Takano et al. 2010), *DRP1A* (Konopka and Bednarek 2008a) and *SNX1* (Jaillais et al. 2006) genes. Importantly, some work focused on the preparation of molecular markers which proved to be very useful for visualization of endocytic compartments including GFP-tagged VHA1, RabA1d, RabA1f, RabA2, RabA3 for early endosomes/TGN and RabF2a, RabF2b, ESCRT for late endosomes/MVBs (Ueda et al. 2001, 2004; Voigt et al. 2005; Chow et al. 2008; Spitzer et al. 2009; Ovečka et al. 2010). Some of above-mentioned molecular markers for endocytic compartments were tagged also with YFP (YFP-RabA1e, YFP-RabF2a, YFP-RabD1 and YFP-RabD2) or with red fluorescent proteins as in the case of DsRed-FYVE, VHA1-RFP, ARA6-mRFP, mRFP-Ara7, VAN3-mRFP, DRP1A-mRFP1, mRFP-Rha1, and mCherry-RabF2a/Rha1 (Voigt et al. 2005; Dettmer et al. 2006; Miao et al. 2008; Geldner et al. 2009; Naramoto et al. 2009; Takano et al. 2010; Mravec et al. 2011) and were subsequently used for successful colocalization studies with other endocytic proteins tagged with GFP (e.g. Ueda et al. 2004; Voigt et al. 2005; Miao et al. 2008; Pinheiro et al. 2009).

In the last years, several plasma membrane proteins have been identified to be internalized to plant cells by endocytic pathways including clathrin-dependent and lipid raft-dependent endocytosis (Lam et al. 2007; Mayor and Pagano 2007; Müller et al. 2007; Doherty and McMahon 2009; Bassil et al. 2011; Chen et al. 2011; Kitakura et al. 2011; Reyes et al. 2011; Li et al. 2012; see also chapters by Li et al. and Bassil and Blumwald in this volume). Some of these proteins are crucial for the control of plant development, water, and ion homeostasis and for interactions of plants with pathogens.

Polar auxin transport is a fundamental process shaping the plant body and it is controlled primarily by plasma membrane localized auxin transporters such as PIN-FORMED1 (PINs) and AUXIN-RESISTANT1 (AUX1). These transporters are polarly organized in the plasma membranes of opposite cellular poles. Their turnover is regulated via endocytosis as revealed by dynamic studies using GFP/YFP-tagged PIN1, PIN2, PIN3, and AUX1 in diverse cell types (Swarup et al. 2004; Dhonukshe et al. 2007, 2008; Kleine-Vehn et al. 2008; Jelínková et al. 2010; Kitakura et al. 2011; see also chapter by Nodzinski et al. in this volume). PIN proteins and their polar subcellular localization at the plasma membrane determine the direction and rate of cellular export and intercellular transport of auxin

(Petrášek et al. 2006; Wisniewska et al. 2006). PIN proteins are constitutively recycled between the plasma membrane and endosomes (Geldner et al. 2003; Dhonukshe et al. 2007), aiming at polarity establishment and rapid polarity alterations during plant development and organogenesis (Kleine-Vehn et al. 2010). Endocytosis of PIN proteins is clathrin-dependent (Dhonukshe et al. 2007) while their vesicular recycling depends on an ARF-GEF protein called GNOM (Geldner et al. 2003). Moreover, the clathrin-dependent endocytosis of PINs is inhibited by auxin (Paciorek et al. 2005; Robert et al. 2010). AUX1-YFP polar localization and subcellular trafficking, unlike PIN1 dynamics, is independent of GNOM but it is sensitive to sterol disruption. Thus, AUX1 and PINs seem to use different trafficking pathways in plants (Kleine-Vehn et al. 2006). Further, it was shown that PIN transcytosis occurs by endocytic recycling and alternative recruitment of the same cargo molecules by apical and basal sides of polarized cells (Kleine-Vehn et al. 2008).

Fluorescently tagged (with GFP) plasma membrane aquaporin PIP2.1 can be internalized to plant cells by two alternative pathways, including a classical clathrin-dependent pathway, or a lipid raft-associated pathway employed during salt stress, as recently shown by GFP-tagged PIP2.1 in combination with FRAP technology (Mongrand et al. 2004; Morel et al. 2006; Li et al. 2011; Luu et al. 2012). Thus, PIP2.1 might be involved in the multiple modes of regulating water permeability through the dynamic heterogeneous distribution in the plasma membrane and recycling pathways (Li et al. 2011).

Plant brassinosteroids (BR) are recognized at the plasma membrane by the receptor called BRASSINOSTEROID INSENSITIVE 1 (BRI1) belonging to transmembrane serine/threonine protein kinases. Interestingly, GFP-tagged BRI1 localizes to the plasma membrane and to the endosomes of Arabidopsis root cells independently of brassinosteroid treatment (Geldner et al. 2007; Irani et al. 2012). Another plasma membrane localized receptor kinase involved in perception and signalling of bacterial flagellin during plant–pathogen interactions is FLAGELLIN SENSITIVE2 (FLS2) (Robatzek et al. 2006). Ligand-activated GFP-FLS2 moves to endosomes and it is further sorted to the vacuole for lytic degradation, depending on ubiquitination and phosphorylation of this receptor (Robatzek et al. 2006). It is not clear yet, whether signalling of flg22-activated FLS2 is associated with plasma membrane or rather with endosomes.

Other molecular components essential for internalization of plasma membrane proteins, such as clathrin and dynamin, were studied in plant cells using GFP recombinant technology. The large GTPase dynamin is required for scission of CCVs. Dynamics of clathrin foci as revealed by fluorescently tagged clathrin light chain CLC-GFP at the plasma membrane was studied by using variable-angle epifluorescence microscopy (VAEM) (Konopka and Bednarek, 2008b). Furthermore, it was correlated with the dynamics of two dynamins, DRP1A and DRP1C, by dual color labeling and live cell imaging. DRP1C-GFP colocalized with a clathrin light chain fluorescent fusion protein (CLC-FFP) in dynamic foci which depend on functional clathrin-mediated endocytosis, cytoplasmic streaming and cytoskeleton (Konopka et al. 2008). Recently, it was shown that clathrin plays a fundamental role in plant cell polarity, growth, patterning, and organogenesis (Kitakura et al. 2011).

Several proteins such as epsin, auxilin, synaptojanin, synaptotagmin and annexin are positioned at the interface between clathrin coats and the cytoskeleton (Šamaj et al. 2004) during clathrin mediated endocytosis in yeast, animals and likely also in plants. So far, epsin-homology domain proteins EHD1 and EHD2 were identified in plants. They were visualized as GFP-tags and implicated in endocytosis (Bar et al. 2008).

Boron and iron transporters as well as Na^{+}/H^{+} antiporters were also tagged with GFP and their internalization via endocytosis was visualized using CLSM (Takano et al. 2010; Bassil et al. 2011; Yoshinari et al. 2012).

Except for the aforementioned plasma membrane localized and associated proteins, also proteins regulating vesicular trafficking, associated with early and late endosomes represented by TGN and MVB/PVC compartments, were extensively studied in plant cells using recombinant GFP technology. Among these proteins several small Rab GTPases tagged with GFP, YFP, or mRFP such as RabA4b, RabA1d, RabA1e, RabA1f, RabA2 were localized to TGN (Preuss et al. 2004, 2006; Chow et al. 2008; Ovečka et al. 2010) while others such as RabF2a and RabF2b were rather localized to MVBs (Haas et al. 2007).

Further, secretory carrier membrane proteins (SCAMPs) tagged with GFP, YFP, or RFP are localized to the plasma membrane and TGN/early endosome, whereas VSRs tagged with GFP mediate the sorting of soluble vacuolar cargo molecules, and they are localized to the MVBs and vacuole of pollen tubes (Lam et al. 2007; Wang et al. 2010).

The retromer, a multiprotein complex, is involved in the recycling of transmembrane VSRs from late endosomes/MVBs to the TGN. VSRs mediate the transport of vacuolar/lysosomal hydrolases from TGN to the lytic compartments (lysosome/vacuole). It was shown that GFP-tagged VSRs are localized preferentially to MVBs in tobacco BY-2 cells and in Arabidopsis suspension culture cells (Miao et al. 2006, 2008; Oliviusson et al. 2006). In addition to retromer and VSRs, sorting nexins have also been studied in plant cells with fluorescent-tagging technology (Jaillais et al. 2006, 2008).

Another protein complex essential for the formation of internal vesicles in MVBs is the endosomal sorting complex required for transport (ESCRT). Using GFP technology ESCRT components ELC and VPS23 were shown to colocalize with MVB markers (Spitzer et al. 2006, 2009; Richardson et al. 2011; see also chapter by Otegui et al. in this volume).

An alternative method to visualize endosomes is to use GFP/RFP-tagged specific peptide domains such as FYVE which binds to phosphoinositol 3-phosphate enriched in the endosomal membranes (Gaullier et al. 1998; Voigt et al. 2005; Veermer et al. 2006). It is likely that this molecular marker binds to two populations of endosomes, likely representing early endosomes/TGN and late endosomes/MVBs in different cell types (Bar et al. 2008; Salomon et al. 2010).

3.2 Molecular Interactions Between Endocytic Proteins

Endocytosis is a complex process depending on highly regulated interactions between diverse types of molecules, e.g. ligand-receptor, protein–protein, and protein-lipid interactions. Increased internalization of brassinosteroid receptor BRI1 was observed in protoplasts as a consequence of coexpression of BRI1 together with its co-receptor BAK1. It was proposed that BAK1 regulates BRI1 endocytosis and trafficking (Russinova et al. 2004). In vivo, BAK1 can form a complex with FLS2 in a ligand-dependent manner (Chinchilla et al. 2007). It was shown in Arabidopsis protoplasts that epsin1, a homolog of epsin, binds and interacts with clathrin, AP-1, vacuolar sorting receptor1 (VSR1) and VTI11, and it is involved in the vacuolar trafficking of soluble proteins at the TGN (Song et al. 2006). Further, EHD2 is essential for endocytosis of the plasma membrane receptor LeEix2 tagged with GFP because binding of the coiled-coil region of EHD2 to the cytoplasmic domain of the LeEix2 receptor is required for inhibition of receptor internalization and signalling (Bar and Avni 2009; Bar et al. 2009). PIN polarity and transcytosis are regulated by Ser/Thr protein kinase PINOID and protein phosphatase 2A in the pathway which is independent of GNOM (Kleine-Vehn et al. 2009).

Clathrin and adaptor proteins (AP) form complexes, which are associated with clathrin-coated pits (CCPs) at the plasma membrane and with clathrin-coated vesicles (CCVs) during their budding. APs are recruited to CCPs and CCVs from the cytosol and they are necessary for further recruitment of clathrin and cargos to CCPs and CCVs. AP2 complex has a role in trafficking from the plasma membrane to TGN while AP1 complex takes part on clathrin-dependent endosomal sorting at the level of TGN/early endosome (Drake et al. 2000). Recently, it was proposed that AP3 complex likely functions as a clathrin adaptor complex and plays a role in protein sorting at the TGN and/or endosomes in plants (Zwiewka et al. 2011).

One major class of proteins associated with vesicular trafficking are SNARE proteins. Recently, RabF1/Ara6 was found to interact with SNARE and to regulate complex formation at the plasma membrane (Ebine et al. 2011).

4 Genetic Methods

Several genetic approaches such as forward genetics on ethyl methane sulfonate (EMS)-induced mutants, reverse genetics on knock-out T-DNA insertional mutants as well as site-directed mutagenesis of functional sites in proteins undergoing or regulating endocytosis was used for functional studies on these proteins. Here, we selected some examples of these genetic approaches.

Most functional analyses on these proteins were likely performed on knockout T-DNA insertional mutants. For example, single, double, and triple knock-out *pin1*, *pin2*, *pin3*, *pin4*, *pin7* mutants as well as single *aux1* mutant in auxin transporters were used to study their function and complex regulation of auxin distribution in the Arabidopsis root (Friml et al. 2002a, b; Benková et al. 2003; Friml et al. 2003; Swarup et al. 2004; Blilou et al. 2005). Knockout mutants *ara6, ara7, rha1*, (Haas et al. 2007; Ebine et al. 2011) were used to study biological functions of small Rab GTPases and together with *skd1* mutant also to study biogenesis of MVBs (Haas et al. 2007). Other mutants such as *vamp72* and *syp22* were used to study biological functions of plant SNARE proteins in endocytosis, seed development as well as their functional link with Rab GTPase Ara6 (Ebine et al. 2008, 2011). Next, *bak1* (Chinchilla et al. 2007), *epsin1* (Song et al. 2006), *snx2b* (Phan et al. 2008), *van3* (Naramoto et al. 2010), and *trs120* (Thelmann et al. 2010) carrying T-DNA insertions disrupting functions of corresponding genes helped to determine roles of these key endocytic protein players. The double *chmp1a chmp1b* mutant in ESCRT-related Charged MVB Proteins show MVBs with less lumenal vesicles, mislocalization of auxin transporters, and defects in embryo polarity establishment (Spitzer et al. 2009). Recently, the *chc1*, *chc2* knockout mutants together with dominant-negative CHC1 (HUB) transgenic lines, all defective in clathrin heavy chain, showed aberrant endocytosis of PINs leading to auxin transport-related phenotypes (Kitakura et al. 2011).

Another set of mutants were point mutants obtained by classical EMS-induced mutagenesis. Among these mutants were $gnom^{R5}$ (Geldner et al. 2003), *aux1* (Swarup et al. 2004), and *bor1-1* (boron transporter, Noguchi et al. 1997), again revealing the crucial role of mutated proteins in the plant endocytosis. Additionally, fluorescence imaging-based screen of an EMS-mutated plant population identified the *ben1* mutant, shown by forward genetics to be an ARF-GEF protein called BIG, which is involved in early endocytosis of plasma membrane proteins (Tanaka et al. 2009). A similar approach led to the identification of *pat2* and *pat4* mutants, defective in AP necessary for biogenesis of vacuoles (Feraru et al. 2010; Zwiewka et al. 2011).

Site-directed mutagenesis of functional amino acid residues in endocytic proteins was also used to study function of these proteins in plant endocytosis. Point mutations generating the amino acid substitutions Ser-26 to Asn (S26N), Gln-71 to Leu (Q71L), and Asn-125 to Ile (N125I) were created to study biological functions of RabA2 and RabA3 localizing to TGN during cell plate formation (Chow et al. 2008). Similar artificial locking of Rab and Arf GTPases in either GTP (constitutive active membrane-bound form), or GDP (inactive/dominant-negative cytoplasmic form) state represents a powerful and widely used tool to study biological function of these proteins (Xu and Scheres 2005; Dhonukshe et al. 2006; Chow et al. 2008; Nielsen et al. 2008; Böhlenius et al. 2010; Ebine et al. 2011). Site-directed mutagenesis was also used for elucidation of the ubiquitination of plasma membrane proteins for internalization and vacuolar trafficking (Herberth et al. 2012) as well as to study functional relevance of some domains in auxin influx carrier AUX1 (Yang et al. 2006), brassinosteroid receptor BRI1 (Wang et al. 2005), inositol transporters INT1 and INT4 (Wolfenstetter et al. 2012), and epsin homology domain protein EHD2 (Bar et al. 2009).

5 Electrophysiological Methods

Electrophysiological approaches to study exo- and endocytosis were mostly based on patch-clamp capacitance measurements. Capacitance measurements take advantage of the fact that exo- and endocytosis are associated with changes in plasma membrane area leading to proportional changes in the electrical membrane capacitance (Bandmann et al. 2011). As the membrane capacitance is proportional to the membrane area, the surface area and thus the diameter of the vesicle can be determined from the vesicle capacitance. The observed different kinetics of single-vesicle membrane capacitance could be grouped into four different categories, representing a variable behavior with respect to transient or permanent fusion and fission of vesicles. Using capacitance measurements, it is also possible to estimate the diameter of the vesicles. Real-time patch-clamp recordings were undertaken to monitor single-vesicle fusion and fission in order to resolve the kinetic properties of the release and incorporation of the secretory cargo to the plasma membrane (Thiel et al. 2009). The data show that single vesicles can, in a rhythmic fashion, make and break contact with the plasma membrane of plant protoplasts. Such oscillations are only possible if the two processes are linked by a distinct feedback system.

Recently, patch-clamp capacitance measurements were applied for the investigation of glucose uptake in BY-2 cells (Bandmann and Homann 2012). In the presence of glucose a strong decrease in the number of fusion events and transient fission events was recorded, while the frequency of permanent endocytic events increased fourfold (Bandmann and Homann 2012). The inhibition of clathrin coat formation by ikarugamycin did not prevent the stimulatory effect of glucose on endocytosis indicating the clathrin-independent endocytosis of glucose in BY-2 cells.

6 Vital Fluorescent Markers for Endocytosis

6.1 Fluorescently Labeled Endocytic Cargo

The existence of distinct fluid phase endocytosis (FPE) in plant was proved using Na-dependent fluorescent marker Coro-Na (membrane-impermeable form) and membrane marker FM 4-64 (Etxeberria et al. 2009). Recently it was shown that FPE is clathrin-independent (Onelli et al. 2008; Bandmann and Homann 2012). Using inhibitors of CME, uptake of fluorescent-labeled glucose derivative 2-NBDG [2(N-(7-Nitrobenz-2-oxa-1,3-diazol-4-yl)amino)-2-deoxyglucose] was not affected (Bandmann and Homann 2012). Further evidence for FPE independence on CME is the demonstration that GFP-Flot1 does not colocalize with CLC-mOrange and also TyrA23 (tyrphostin A23, an inhibitor of endocytotic cargo recruitment in clathrin-mediated endocytosis) did not affect GFP-Flot1 movement (Li et al. 2012). Very recently, the fluorescently labeled endocytotic marker FITC-BSA was used to discover and describe the endocytic route of nutrient uptake in diverse carnivorous plants (Adlassnig et al. 2012).

6.2 Vital FM Styryl Dyes

Fluorescent FM dyes, water-soluble lipophilic styryl compounds, are virtually nonfluorescent in aqueous media but turn to be fluorescent following incorporation to the outer leaflet of the plasma membrane. The plasma membrane incorporation of FM dyes and their subsequent endocytic internalization and distribution among endomembranes appointed them a crucial role in endocytic trafficking visualization in living plant cells. The most widely used FM dyes in plant cell biology are FM 4-64 and FM 1-43 which were successfully used as live endocytic tracers not only in different fungal and plant cells including fungal hyphae, *Chara* internodal cells, pollen tubes, root epidermal cells and root hairs, stomata, leaf epidermal cells, but also in protoplasts and tobacco BY-2 suspension cells (Ovečka et al. 2005; Voigt et al. 2005; Xu and Scheres 2005; Sousa et al. 2008; Zonia and Munnik 2008; Scheuring et al. 2011). Studies based on FM4-64 uptake experiments combined with colocalization of this dye with TGN markers revealed that TGN acts as an early endosome, receiving first endocytosed material from the plasma membrane (Dettmer et al. 2006; Lam et al. 2007, 2009; Jaillais et al. 2008). FM 4-64 was also microinjected directly into the cytosol of BY-2 cells and *Tradescantia virginiana* stamen hair cells (van Gisbergen et al. 2008). Additionally, advanced CLSM technique based on FRET measurements comparing membrane-bound FM 4-64 with cytoplasmic GFP also showed that the dye is taken up into cells preferentially by endocytosis (Griffing 2008). Recently, it was shown that FM 4-64 and FM 5-95 in tobacco BY-2 and Arabidopsis cell suspensions initialize transient re-localization of auxin carriers, and FM 1-43 affects their activity. Re-localization was not blocked by inhibitors of endocytosis or cytoskeletal drugs. However, no changes in localization of auxin carriers were observed in Arabidopsis root epidermis and cortex cells labeled with FM dyes (Jelínková et al. 2010).

6.3 Filipin and di-4-ANEPPDHQ

The antibiotic filipin is a polyene fluorochrome that binds to structural sterols. Since filipin has fluorescent properties it can serve as a vital probe for in vivo visualization of structural sterols in the plasma membrane and endosomes (Grebe et al. 2003; Kleine-Vehn et al. 2006; Liu et al. 2009; Ovečka et al. 2010; Boutté et al. 2011). Additionally, sterols complexed by filipin can be visualized at the ultrastructural level (Ovečka et al. 2010). Filipin was used to label structural sterols in the plasma membrane and to study their internalization and endosomal trafficking in epidermal cells of intact Arabidopsis roots (Grebe et al. 2003) and in root hairs (Ovečka et al. 2010). Early endocytic trafficking of structural sterols seems to be actin dependent and BFA-sensitive, and might involve endosomes enriched with RabF1/Ara6 (Grebe et al. 2003). Another fluorescent dye visualizing structural sterols is di-4-ANEPPDHQ (Liu et al. 2009). Recently, it was shown that phosphatidyl-inositol-3-phosphate (PI3P) is abundant on the outer surface of plant cell plasma membranes and it mediated pathogen effector proteins entry involving lipid raft-mediated endocytosis (Kale et al. 2010). The inhibition of the accumulation of effector proteins was achieved using wortmannin (inhibitor of PI3P biosynthesis and clathrin-mediated endocytosis) and filipin and nystatin (inhibitors of lipid raft-mediated endocytosis). This observation suggested that the disruption of lipid rafts impaired the distribution of PI3P on the outer surface of plasma membrane (Kale et al. 2010).

6.4 Labeled Ligands for Signalling Receptors

Very recently, a fluorescently labelled brassinosteroid analog (castasterone coupled to Alexa Fluor 647) was used as a ligand for GFP-tagged BRI1 (Irani et al. 2012). This study, in contrary to a previous report regarding BRI signalling from endosomes (Geldner et al. 2007), revealed that BRI1 signalling is actually restricted to the plasma membrane, while BRI1 internalization to endosomes leads to the attenuation of castasterone-induced signalling (see also chapter by Di Rubbo and Russinova in this volume).

7 Immunolocalization Methods for In Situ Localization of Endosomal Proteins

Immunolocalization with specific antibodies raised against endocytic proteins is a very reliable technique which can considerably support and strengthen results obtained by fluorescent GFP/YFP/RFP-tagging of these proteins. For more robust species such as maize, a technique based on embedding in Steedman wax and subsequent immunolocalization on dewaxed semithin sections is a common and useful method of choice (Vitha et al. 2000a, b; Baluška et al. 2004). In Arabidopsis seedlings and suspension BY-2 cells, the whole mount immunolocalization was the technique mostly used so far (Hejátko et al. 2006; Szechyńska-Hebda et al. 2006). The main advantage of immunolocalization techniques is that immunofluorescence data obtained on lower resolution can be correlated with high-resolution localization using immunogold electron microscopy (EM). The latter technique allows the unambiguous identification of proteins associated with endocytotic uptake as well as with early and late endosomes. Thus, clathrin was localized at the plasma membrane (Dhonukshe et al. 2007; Müller et al. 2010), VHA and Rab GTPases were localized to TGN (Nielsen et al. 2008) while RabF2a and RabF2b were localized to MVBs (Haase et al. 2007) by immunogold EM.

Specific antibodies against PIN1 (Gälweiler et al. 1998; Benková et al. 2003), PIN2 (Müller et al. 1998), Ara6 and Ara7 (Ueda et al. 2001) were used in Arabidopsis and anti-VSR (Miao et al. 2006, 2008) in BY-2 cells for subcellular localization of corresponding proteins by using immunolocalization methods. Moreover, immunolocalization with specific antibodies against SCAMP1 and VSR revealed that SCAMP is localized to apical endocytic vesicles, while VSRs are localized to the MVB and vacuole in lily pollen tubes (Wang et al. 2010).

Very recently, immunogold EM helped to identify fusions of the PVC/MVB (including internal MVB vesicles) with vacuole. Thus, cargo ubiquitination-independent and PVC-mediated degradation of plasma membrane proteins in the vacuole was proposed for plant cells (Cai et al. 2012).

8 Microscopic Methods

The progress in microscopic imaging in the past few years, involving for example high-throughput analyses, or sophisticated new instrumentation, such as light sheet and super-resolution microscopy, brought novel analytic tools, also for the investigation of proteins undergoing or regulating endocytosis.

8.1 Identification of New Endocytic Proteins by High-Throughput Microscopy Screen

A high-throughput fluorescence imaging uses epifluorescence or confocal microscopy to screen tissues or cells for specific phenotype. The exact targeting (visualization) of specific vesicular transport processes requires fluorescence-tagged proteins, which selectively localize to certain organelles, or take part in specific processes. A fluorescence imaging-based forward genetic screen, was performed on EMS-mutagenized *PIN1pro::PIN1-GFP* Arabidopsis transgenic plants treated with BFA by using epifluorescence microscopy (Tanaka et al. 2009). Arabidopsis mutants that do not efficiently internalize and/or accumulate PIN1-GFP in the BFA compartments were identified in this search. Using this approach, three mutant loci, *BFA-visualized endocytic trafficking defective1* (*ben1*), *ben2* and *ben3* were identified. Fine mapping revealed that *BEN1* encodes an ARF-GEF vesicle trafficking regulator belonging to the BIG class. Further detailed study suggested that this ARF-GEF is involved in endocytosis of plasma membrane proteins and localizes to early endocytic compartments distinct from GNOM-positive recycling endosomes.

In a similar study, EMS mutagenized *PIN1pro::PIN1-GFP* Arabidopsis transgenic plants were screened for aberrant PIN1-GFP distribution resulting in the detection of several *protein-affected trafficking* (*pat*) mutants (Feraru et al. 2010). It was found that *pat2* is specifically defective in the biogenesis, identity, and function of lytic vacuoles but shows normal sorting of proteins to storage vacuoles. *PAT2* encodes a putative b-subunit of adaptor protein complex 3 (AP-3) (Feraru et al. 2010). High-throughput confocal microscopy was carried out also to analyze the quantitative differences between distinct endomembrane vesicles in leaf epidermal tissue and to find differences in the quantity of GFP-2xFYVE in endosomal compartments upon exposure of Arabidopsis plants to biotic or abiotic stresses (Salomon et al. 2010). Toward this goal, a spinning-disk microscope enabling the observation of samples in multi-well plates was employed and marked differences in the quantity of endomembrane compartments were found in leaf cells. Moreover, substantial increase in the number of GFP-2xFYVE compartments after biotic and cold stress was observed, whereas dark caused decreased numbers of these endosomes.

8.2 Confocal Laser Scanning Microscopy

CLSM became the most common and valuable method in endocytic investigations being a platform for visualization of fluorescently labeled endocytotic markers or proteins fused with fluorescent tags. For example it was widely used for defining the colocalization of endocytic proteins with the membrane tracker FM4-64 (Ito et al. 2012) and for colocalizations of clathrin, SNXs, VSR2, AAA ATPase Vps4p/SKD1, AtVSR reporters, and storage vacuolar cargo with endocytic markers (Haas et al. 2007; Miao et al. 2008; Foresti et al. 2010; Niemes et al. 2010; Pourcher et al. 2010; Ito et al. 2012). CLSM was also used for the detection of the sensitivity of PIN1, GNOM, SCAMP1, SNX, selected Rab GTPases, and other proteins to vesicular trafficking drugs such as BFA, concanavalin A and wortmannin (Geldner et al. 2003; Dettmer et al. 2006; Lam et al. 2007; Jaillais et al. 2008; Takáč et al. 2011a; Takáč et al. 2012; Fig. 2). The dynamics of vesicles and visualization of zones of exocytosis and endocytosis in tobacco pollen tubes was performed by pulse-chase observations of endocytic tracker FM4-64 and FM1-43 using CLSM (Zonia and Munnik 2008). For the observations of highly dynamic endosomal systems, a spinning-disk confocal time-lapse microscopy proved to be a promising tool (Nakano 2002). It is capable for high-speed sequential acquisition and significantly minimizes sample bleaching. For example, EYFP-RabA4d in growing pollen tubes was monitored by this imaging system (Szumlanski and Nielsen 2009). It also helped to identify TGN/EE as a dynamic and independent compartment which only temporarily moves together with Golgi (Viotti et al. 2010). It was also used for high-throughput CLSM screen using FYVE-labeled endosomes (Salomon et al. 2010).

In conventional CLSM, the laser beam is focused through the tissue whereas in total internal reflection fluorescence microscopy (TIRFM) the laser beam is reflected off the surface at a critical angle of surface plasmon generation (Sparkes et al. 2011).

It is valuable mostly for the visualization of processes in the close proximity of plasma membrane as it was shown to monitor the behavior of FM-labeled vesicles in growing pollen tubes (Wang et al. 2009), GFP-tagged clathrin, dynamin, and flotillin at the plasma membrane (Konopka and Bednarek 2008a; Li et al. 2012) or GFP and GDP locked versions of small RabGTPase Ara6 (Ebine et al. 2011).

8.3 Light Sheet-Based Fluorescence Microscopy

Live microscopic imaging is one of the major tools in experiments targeted at understanding the development of organisms. The primary constraint in all such experiments is to maintain the specimen at near-physiological conditions. This requires minimally invasive imaging methods with sufficient spatial resolution and the temporal capability to capture the biological processes on site. Processes such as ion dynamics, cytoskeletal reorganization, cell division, cytokinesis, and intracellular trafficking are completed within seconds to minutes, and, therefore, require short imaging intervals but a high spatial resolution. Capturing and tracking endosomes that move at a speed of several μm/sec is one of the most challenging efforts. Since plants scatter and absorb light well, they have to be exposed to relatively strong light fluencies to ensure a proper illumination and the detection of signals across their entire volume. This potentially results in heating and photo-toxicity problems, which can induce the malfunction and finally the death of the plant. Therefore, if the experimental measurements are to represent a normal development, the specimen must be maintained at an appropriate physiological condition, and must remain accessible for microscopic observation at high image acquisition rates.

Conventional microscopy and CLSM usually employ the same objective lens for both fluorescence excitation and detection. Confocal theta fluorescence microscopy (Stelzer and Lindek 1994) introduced the systematic use of at least two separate lenses for illumination and detection. In such a system the optical axes are arranged orthogonally, improving the axial resolution and resulting in an almost spherical point spread function. The wide-field implementation of the theta principle, illuminates the specimen throughout an entire plane with a light sheet and collects the emitted light at a perpendicular axis. The use of light sheets for imaging purposes has been known for more than one hundred years and their use has been suggested for macroscopic imaging (e.g. Voie et al. 1993), but the applicability for high-resolution light microscopy was not realized until a few years ago (Huisken et al. 2004). Almost all theoretical aspects (single/multiple photons, circular and annular apertures) have been covered in a series of papers published by Stelzer and Lindek (Lindek and Stelzer 1996; Lindek et al. 1996a, b) and have been picked up and further developed by many other authors (e.g., Sätzler and Eils 1997).

However, in light sheet-based fluorescence microscopy (LSFM), optical sectioning arises from the overlap between the focal plane of the detection system and the central plane of a light sheet. The thickness of the light sheet is similar to, and in many instances even thinner than the depth of field of the detection system,

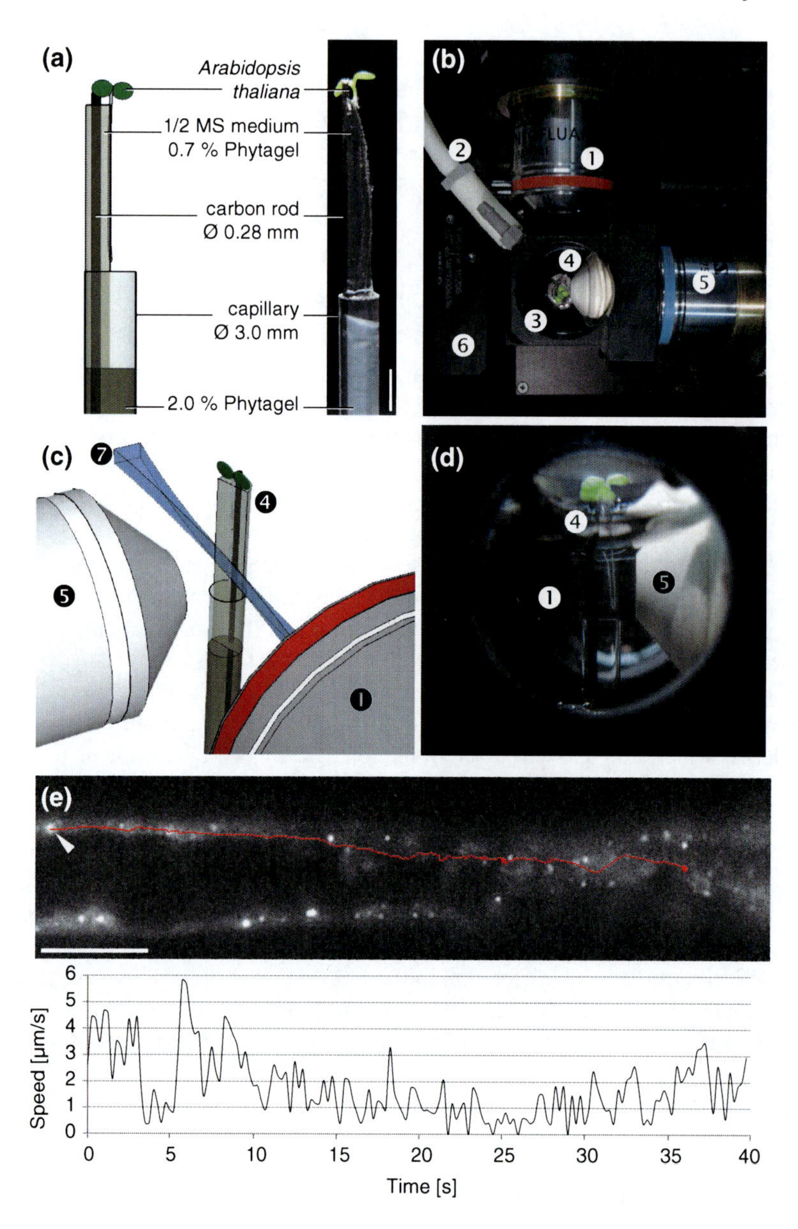

which means that only fluorophores close to the focal plane of the detection system are excited and contribute to the image. It also means that illumination causes no photo damage either in front or behind the focal plane. Therefore, LSFM exposes a specimen to 200 times less energy than a conventional and up to 5,000 times less energy than a confocal fluorescence microscope (Keller and Stelzer 2008). Importantly, LSFM takes advantage of state-of-the-art scientific cameras (e.g.,

◀ **Fig. 3** *Arabidopsis thaliana* in the monolithic digital scanned laser light sheet-based fluorescence microscope (mDSLM). The plant grows in an upright position in the specimen chamber. While the leaves remain in the air, the root system is perfused with liquid half-strength MS medium. Only the fluorophores in a thin planar volume that overlaps with the focal plane of the detection lens are excited. Thus, fluorophores outside this volume do not blur the image and are not subject to photo bleaching. Four motors below the chamber move the plant along x/y/z and rotate it around the vertical axis, which is orthogonal to the detection axis. **a** Sample holder for microscopy of upright plants. *A. thaliana* seedlings grow in a capillary on the surface of a vertically positioned ½ MS medium containing 0.7 % Phytagel. For the image acquisition process, the phytagel cylinder is extruded from the capillary, which is rigidified by an embedded carbon rod. **b** Top view of the specimen chamber. **c** 3D model of the sample in front of the objective lens. **d** Side-view into the chamber with the root in front of the detection lens. Legend: *1* illumination objective lens, *2* perfusion system, *3* specimen chamber, *4 A.thaliana* in capillary, *5* detection objective lens, *6* microscope stage, *7* laser light sheet. **e** The initial image of a single plane time-lapse recording of an *A. thaliana* root hair. The 6-day-old plant expresses the late endosome marker YFP-RabF2a (Geldner et al. 2009). The *red line* indicates the track of a single endosome during a 40-second-long time frame. **d** The endosome speed over time. Images were recorded at a rate of 4 fps for a 40 s stretch with the pair EC Plan-Neofluar 5×/0.16 and N-Apochromat 63×/1.0 W. Scale bar: **a** 3 mm, **e** 10 μm

CCD, EM-CCD, sCMOS, time-of-flight sensitive cameras) and easily records 100 frames per second (fps) with a dynamic range of 12–14 bits and image sizes exceeding 2,000 pixels by 2,000 lines. It typically achieves an isotropic resolution between 250 and 300 nm.

The first results of live imaging of growing multicellular plant structures with LSFM were published recently (Maizel et al. 2011; Sena et al. 2011; Fig. 3). Dynamics of the movement of subcellular organelles is a typical and fast event. We recorded the movements of endosomal compartments labeled by an YFP-RabF2a reporter (Geldner et al. 2009) and successfully tracked single endosomes.

The plant is placed vertically in a medium-filled chamber close to the common focal point of the two objective lenses. The root grows on the surface of a phytagel cylinder cast in a glass capillary. For the image acquisition process, the phytagel cylinder is extruded from the capillary, which is rigidified by an embedded carbon rod. A perfusion system ensures that the entire medium in the plant chamber is exchanged every 15 min. The plant is inserted from above but it is held from below. Thus, the opening remains accessible for diurnal illumination provided by a standard lamp and can be spectrally adjusted (Fig. 3).

8.4 Fluorescence Recovery After Photobleaching and Super-Resolution Microscopy

FRAP has many important applications, mainly for the investigation of the dynamics of plasma membrane located proteins and highly motile compartments such as endosome. It was also used to monitor vesicle fusion to the plasma membrane during exocytosis in growing pollen tubes (Lee et al. 2008b). The apical region of tobacco pollen tubes expressing RLK-GFP (receptor-like kinase) was photobleached and the recovery of RLK-GFP fluorescence was monitored. This experiment, together with BFA treatment showed that RLK-GFP is targeted to plasma membrane during exocytosis (Lee et al. 2008b). FRAP analysis was also used to investigate correlations between cell plate formation and vesicular transport (Dhonukshe et al. 2006). It was found that the endocytotic tracker FM4-64 is typically delivered to non-attached cell plates by endocytotic vesicles and not by a direct connection with the parental plasma membrane (Dhonukshe et al. 2006). To investigate the determination of PIN protein distribution, the recovery of PIN1-GFP at the plasma membrane of stele cells was monitored by FRAP in Arabidopsis control cells, BFA-treated cells, and in *vps29* mutant cells. This experiment revealed that PIN1-GFP recovery in the plasma membrane is conditioned by endocytotic recycling and involves the retromer protein VPS29 (Jaillais et al. 2007). Dynamics of PIN3-GFP was followed in response to gravistimulation on both lateral sides of individual columella cells. A pronounced recovery at the new bottom, but not at the new top side of the cell was detected by using FRAP. PIN3 is translocated to the bottom cell side from the preexisting pool of proteins in the cell (Kleine-Vehn et al. 2010). FRAP experiments were carried out also on membranes labeled with YFP-GDAP1 (the plant golgin GRIP-related ARF-binding domain-containing Arabidopsis protein 1) in cells expressing GTP locked ARF1 (ARF1Q71L). The cycling of the golgin was affected in the *arf1* mutant suggesting that the activation of ARF1 is necessary for the binding of GDAP1 to membranes (Matheson et al. 2007). The membrane localization of ARF1 protein is conditioned by the presence of AGD5 (ARF-GAP domain) protein as revealed by FRAP experiments in tobacco leaf epidermal cells co-expressing *ARF1* with mutant *YFP-AGD5R59Q* (Stefano et al. 2010). Recently, FRAP was used to study dynamics of plant aquaporin PIP2.1 (Li et al. 2011; Luu et al. 2012) and PIN proteins (Kleine-Vehn et al. 2010; Furutani et al. 2011).

A recent study employing super-resolution STED microscopy combined with FRAP revealed that PINs are delivered to the center of polar plasma membrane domains by super-polar recycling. They are accumulating in non-mobile membrane clusters showing reduced lateral diffusion which results in longer polar retention of PINs in the plasma membrane. Clathrin-dependent endocytosis of PINs takes place especially at membrane edges (Kleine-Vehn et al. 2011).

8.5 Förster Resonance Energy Transfer

FRET technology is used to investigate protein–protein interactions based on energy non-radiative transfer from the donor to the acceptor fluorophore, leading to quenching of donor fluorescence and an increase in acceptor fluorescence (Padilla-Parra and Tramier 2012). The interaction between YFP-AtRabH1b and the *trans*-Golgi coiled-coil protein AtGRIP-GFP was investigated using the combination of FRET with fluorescence lifetime imaging microscopy (FLIM) technology (Osterrieder et al. 2009). A fluorophore has a characteristic lifetime that can be influenced by changes in temperature, environment, calcium ion concentration, and the occurrence of FRET. Tobacco leaves transiently transformed with either GDP or GTP locked AtRabH1b proteins were used in this study. The co-expression of AtGRIP-GFP and mRFP-AtRab-H1b resulted in quenching of AtGRIP-GFP in the FRET experiment indicating direct protein–protein interaction.

8.6 Electron Microscopy and Electron Tomography

Transmission electron microscopy (TEM) provided crucial information representing milestones in plant endocytosis research. First of all, TEM studies on high-pressure frozen freeze-substituted probes helped to unambiguously identify endocytotic compartments such as TGNs/early endosomes and MVBs/late endosomes in plant cells (Tse et al. 2004; Dettmer et al. 2006; Hause et al. 2006; Müller et al. 2007; Fig. 4). Several lines of evidence provided by TEM contributed to the identification of TGN compartment as both endocytotic and secretory/exocytotic organelle (Viotti et al. 2010). TEM also revealed homotypic fusions of TGN vesicles in BFA-treated cells (Hause et al. 2006; Takáč et al. 2011a) and heterotypic fusions of TGN with MVB in BY-2 cells (Lam et al. 2007) or mung bean cotyledons (Wang et al. 2009) and also in Arabidopsis roots after wortmannin treatment (Takáč et al. 2012).

Electron tomography on cryofixed, freeze-substituted, and plastic-embedded samples allows three-dimensional visualization and display of dynamic, pleiomorphic structures at a resolution of 7 nm in cell volumes up to 25 μm^3 including TGN/early endosomes and MVBs/late endosomes (Donohoe et al. 2006; Otegui et al. 2006; Kang and Staehelin 2008; Limbach et al. 2008; Staehelin and Kang 2008). It was proved to be a valuable approach to investigate the structural features of Golgi-associated and detached, free TGN cisternae/compartments (Kang et al. 2011). By combining electron tomography analysis and immunogold EM labeling of serial thin sections, it was shown that the transformation of a *trans*-Golgi cisterna into TGN coincides with the binding of RabA4b and PI-4 Kβ1 proteins to subdomains of the TGN cisternae that give rise to secretory vesicles (SVs) but not to CCVs. TGN membranes are clearly resolved and a majority of the immunogold particles are seen close to SV-type budding domains that possess an $\sim$5 nm thick, darkly stained coat. The early endosomal marker proteins VHA-a1-GFP and SYP61-CFP colocalize with RabA4b on TGN cisternae. Moreover, the loss of PI-4 Kβ1 and PI-4 Kβ2 activities resulted in pronounced morphological changes of TGN morphology and in an apparent loss of control over SV size (Kang et al. 2011).

Immunogold EM labeling detected GFP-KNOLLE (a cytokinesis-specific syntaxin) in internal vesicles of MVBs in late-mitotic cells and in vacuoles after cytokinesis. The V-ATPase inhibition by concanamycin A caused the accumulation of unfused membrane vesicles. Therefore, traffic from the TGN to the plane of

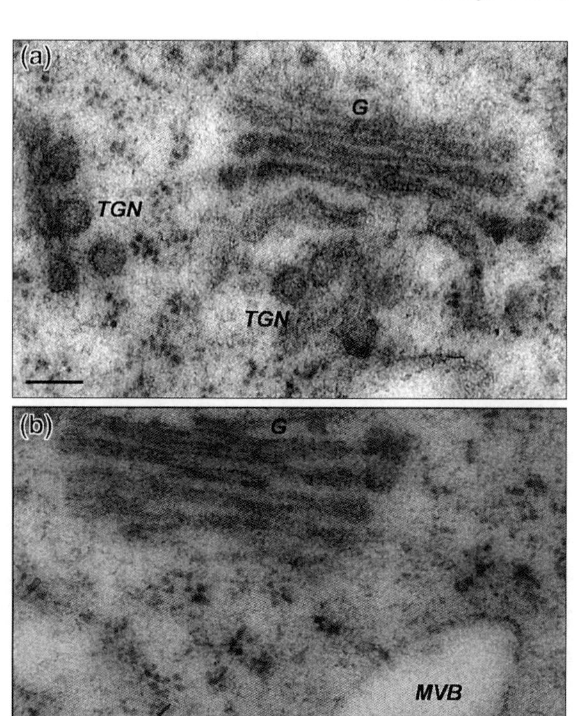

Fig. 4 Visualization of *trans*-Golgi networks (*TGNs*) representing early endosomes (**a**) and multivesicular body (*MVB*) representing late endosome (**b**) using electron microscopy on high-pressure frozen freeze-substituted samples of Arabidopsis root epidermal cells. *G* Golgi apparatus. *Bars* represent 200 nm

cell division is essential for cytokinesis (Reichardt et al. 2007). Immunogold EM labeling of clathrin using an antibody detecting a plant clathrin heavy-chain peptide revealed the localization of clathrin not only on internalizing vesicles at the plasma membrane, but also at the TGN (Dhonukshe et al. 2007; Müller et al. 2010; Fig. 5). The stages of clathrin-coated pit formation, vesicle fission, and the loss of clathrin coat from internalized vesicles were visualized using this technique (Dhonukshe et al. 2007). EM considerably supported importance of V-ATPase for endosome identity. Application of concanamycin A altered the identity of TGN (Viotti et al. 2010). The inhibition of V-ATPase caused substantial changes of TGN and Golgi stack morphology and interferes with BFA action (Dettmer et al. 2006) The altered Golgi morphology was also found in V-ATPase RNAi plants (Brüx et al. 2008).

It was also shown that post-TGN transport toward lytic vacuoles occurs independent of clathrin. Using TEM, a clear connection between MVBs and TGN-like structures was shown after the short recovery after concanamycin A treatment. This study provided evidence that MVBs are derived from the TGN through maturation (Scheuring et al. 2011). In a recent study, the immunogold EM labeling localized flotillin 1 (Flot1) to clusters at the plasma membrane and in endosomal

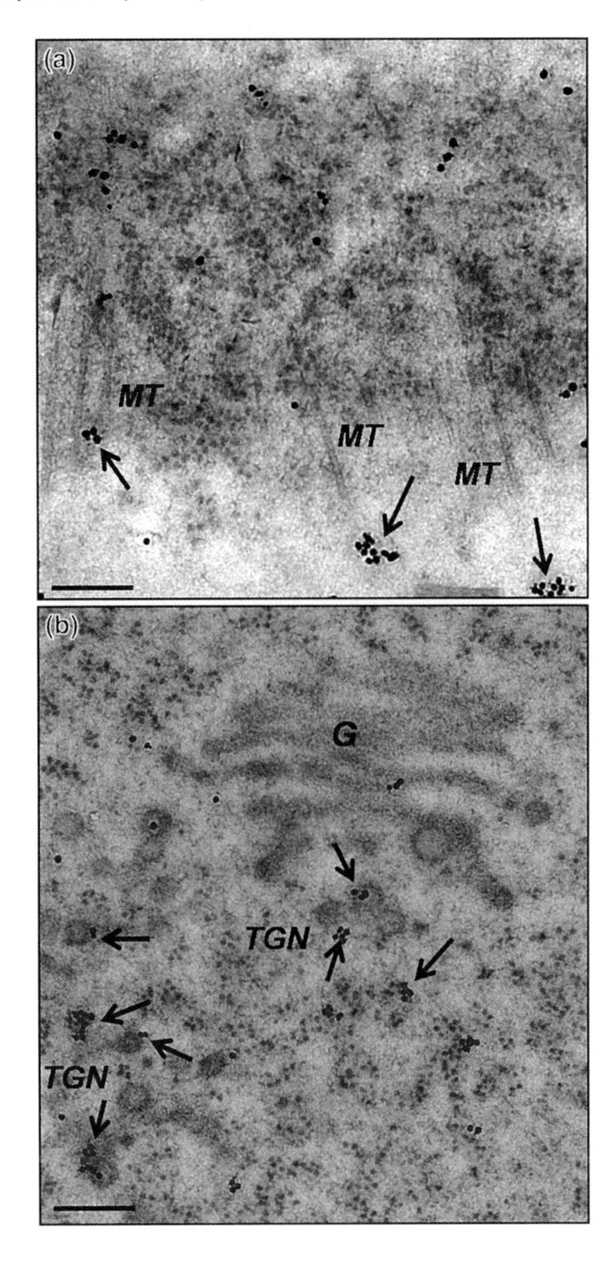

Fig. 5 Immunogold electron microscopy localization of clathrin (*arrows*) associated with plasma membrane on the cortical section underneath of the plasma membrane containing cortical microtubules (*MTs* in image **a**) and clathrin localization on *trans*-Golgi network (*TGN*) vesicles (**b**, *arrows*) in Arabidopsis root epidermal cells prepared by high-pressure freezing and freeze substitution. *G* Golgi apparatus. Gold particles have 15 nm. Bars represent 250 nm

structures. This study showed that Flot1 is involved in a clathrin-independent endocytic pathway in Arabidopsis (Li et al. 2012). Immunogold EM also provided valuable data about the localization of crucial regulators of vesicular trafficking. For example, VHA-E was found to be localized in the tonoplast and TGN using immunogold labeling (Dettmer et al. 2006). *Arabidopsis thaliana* homologs of the three yeast members of the retromer complex (Vps35, Vps29, and Vps26) were localized at the surface of the MVB (Oliviusson et al. 2006). Further, AAA ATPase VPS4/SKD1 homolog was found to be localized to the cytoplasm and to multivesicular endosomes (Haas et al. 2007). Immunogold EM of high-pressure frozen/freeze-substituted samples identified the secretory carrier membrane proteins (SCAMP1)-positive organelles as tubular-vesicular structures at the TGN with clathrin coats (Lam et al. 2007).

9 Conclusions and Future Prospects

This chapter provides an overview of selected conventional and modern technological approaches, which were used in the research field of plant endocytosis during last years. Conventional genetic, biochemical, and microscopic techniques such as confocal and EM provided valuable but still preliminary information about organization and regulation of endocytosis in plant cells. Particulary, electron tomography and immunogold EM substantially contributed to the knowledge about the spatial organization of membraneous subcellular compartments including TGNs and MVBs in plant cells. Obtaining deeper knowledge in this field necessitates the development of novel approaches and techniques. Proteomics, one of quickly developing approaches from omics technologies, was effectively used in the endocytosis research in the past few years. In combination with biochemical approaches such as co-immunoprecipitation it might provide a challenging tool to study protein interactions regulating endocytic transport pathways. New methods such as chemical genetics combined with high-throughput CLSM screening or proteomic screens combined with molecular cell biology provide powerful methods to study plant endocytosis in more broad context. Together with molecular biology, these high-throughput methods will help to develop either novel genetic and protein markers or specific inhibitors for different stages of endocytic transport. Spatial resolution and dynamics of endocytic proteins might be further resolved by super-resolution microscopy while new generation of light sheet and spinning disk microscopes will provide spatio-temporal information about plant endosomes in living cells under almost environmental conditions. In general, present research achievements provide convincing evidence about the effectivity of comprehensive methodological approaches in the investigation of plant endocytosis.

Acknowledgments We would like to thank George Komis for critical reading of the manuscript. This work was supported by structural research grants from EU and the Czech Republic to the Centre of the Region Haná for Biotechnological and Agricultural Research, Faculty of Science, Palacký University, Olomouc, Czech Republic (grant No. ED0007/01/01) and by grant No. P501/11/1764 from the Czech Science Foundation (GAČR).

References

Adlassnig W, Koller-Peroutka M, Bauer S, Koshkin E, Lendl T, Lichtscheidl IK (2012) Endocytotic uptake of nutrients in carnivorous plants. Plant J: Cell Mol Biol. doi: 10.1111/j.1365-313X.2012.04997.x

Agrawal GK, Bourguignon J, Rolland N et al (2010) Plant organelle proteomics: collaborating for optimal cell function. Mass Spectrom Rev. doi:10.1002/mas.20301

Baluška F, Šamaj J, Hlavačka A, Kendrick-Jones J, Volkmann D (2004) Actin-dependent fluid-phase endocytosis in inner cortex cells of maize root apices. J Exp Bot 55:463–473

Banbury DN, Oakley JD, Sessions RB, Banting G (2003) Tyrphostin A23 inhibits internalization of the transferrin receptor by perturbing the interaction between tyrosine motifs and the medium chain subunit of the AP-2 adaptor complex. J Biol Chem 278:12022–12028

Bandmann V, Kreft M, Homann U (2011) Modes of exocytotic and endocytotic events in tobacco BY-2 protoplasts. Mol Plant 4:241–251

Bandmann V, Homann U (2012) Clathrin-independent endocytosis contributes to uptake of glucose into BY-2 protoplasts. Plant J 70:578–584. doi:10.1111/j.1365-313X.2011.04892.x

Bar M, Aharon M, Benjamin S, Rotblat B, Horowitz M, Avni A (2008) AtEHDs, novel Arabidopsis EH-domain-containing proteins involved in endocytosis. Plant J 55:1025–1038

Bar M, Avni A (2009) EHD2 inhibits ligand-induced endocytosis and signaling of the leucine-rich repeat receptor-like protein LeEix2. Plant J 59:600–611

Bar M, Sharfman M, Schuster S, Avni A (2009) The Coiled-coil domain of EHD2 mediates inhibition of LeEix2 endocytosis and signaling. PLoS One 4(11):e7973. doi:10.1371/journal.pone.0007973

Baroja-Fernandez E, Etxeberria E, Muñoz FJ, Morán-Zorzano MT, Alonso-Casajús N, Gonzalez P, Pozueta-Romero J (2006) An important pool of sucrose linked to starch biosynthesis is taken up by endocytosis in heterotrophic cells. Plant Cell Physiol 47:447–456

Bassil E, Ohto MA, Esumi T, Tajima H, Zhu Z, Cagnac O, Belmonte M, Peleg Z, Yamaguchi T, Blumwald E (2011) The Arabidopsis intracellular Na^+/H^+ antiporters NHX5 and NHX6 are endosome associated and necessary for plant growth and development. Plant Cell 23:224–239

Benková E, Michniewicz M, Sauer M, Teichmann T, Seifertová D, Jürgens G, Friml J (2003) Local, efflux-dependent auxin gradients as a common module for plant organ formation. Cell 115:591–602

Blilou I, Xu J, Wildwater M, Willemsen V, Paponov I, Friml J, Heidstra R, Aida M, Palme K, Scheres B (2005) The PIN auxin efflux facilitator network controls growth and patterning in Arabidopsis roots. Nature 433:39–44

Böhlenius H, Mørch SM, Godfrey D, Nielsen ME, Thordal-Christensen H (2010) The multivesicular body-localized GTPase ARFA1b/1c is important for callose deposition and ROR2 syntaxin-dependent preinvasive basal defense in barley. Plant Cell 22:3831–3844

Borner GHH, Sherrier DJ, Weimar T et al (2005) Analysis of detergent-resistant membranes in Arabidopsis. Evidence for plasma membrane lipid rafts. Plant Physiol 137:104–116

Boutté Y, Men S, Grebe M (2011) Fluorescent in situ visualization of sterols in Arabidopsis roots. Nat Protocols 6(4):446–456

Brüx A, Liu TY, Krebs M, Stierhof YD, Lohmann JU, Miersch O, Wasternack C, Schumacher K (2008) Reduced V-ATPase activity in the trans-golgi network causes oxylipin-dependent hypocotyl growth inhibition in Arabidopsis. Plant Cell 20:1088–1100

Cai Y, Zhuang X, Wang J, Wang H, Lam SK, Gao C, Wang X, Jiang L (2012) Vacuolar degradation of two integral plasma membrane proteins, AtLRR84A and OsSCAMP1, is cargo ubiquitination-independent and prevacuolar compartment-mediated in plant cells. Traffic 13:1023–1040. doi:10.1111/j.1600-0854.2012.01360.x

Carmona-Salazar L, El Hafidi M, Enríquez-Arredondo C et al (2011) Isolation of detergent-resistant membranes from plant photosynthetic and non-photosynthetic tissues. Anal Biochem 417:220–227

Chen X, Irani NG, Friml J (2011) Clathrin-mediated endocytosis: the gateway into plant cells. Curr Opin Plant Biol 14:674–682

Chinchilla D, Zipfel C, Robatzek S, Kemmerling B, Nürnberger T, Jones JD, Felix G, Boller T (2007) Aflagellin-induced complex of the receptor FLS2 and BAK1 initiates plant defence. Nature 448:497–500

Chow CM, Neto H, Foucart C, Moore I (2008) Rab-A2 and Rab-A3 GTPases define a trans-golgi endosomal membrane domain in Arabidopsis that contributes substantially to the cell plate. Plant Cell 20:101–123

Dettmer J, Hong-Hermesdorf A, Stierhof Y-D, Schumacher K (2006) Vacuolar H^+-ATPase activity is required for endocytic and secretory trafficking in Arabidopsis. Plant Cell 18:715–730

Doherty GJ, McMahon HT (2009) Mechanisms of endocytosis. Annu Rev Biochem 78:857–902

Dhonukshe P, Baluška F, Schlicht M, Hlavačka A, Šamaj J, Friml J, Gadella TW Jr (2006) Endocytosis of cell surface material mediates cell plate formation during plant cytokinesis. Dev Cell 10:137–150

Dhonukshe P, Aniento F, Hwang I, Robinson DG, Mravec J, Stierhof YD, Friml J (2007) Clathrin-mediated constitutive endocytosis of PIN auxin efflux carriers in Arabidopsis. Curr Biol 17:520–527

Dhonukshe P, Tanaka H, Goh T, Ebine K, Mähönen AP, Prasad K, Blilou I, Geldner N, Xu J, Uemura T, Chory J, Ueda T, Nakano A, Scheres B, Friml J (2008) Generation of cell polarity in plants links endocytosis, auxin distribution and cell fate decisions. Nature 456:962–966

Donohoe BS, Mogelsvang S, Staehelin LA (2006) Electron tomography of ER, Golgi and related membrane systems. Methods 39:154–162

Drakakaki G, Robert S, Szatmari AM, Brown MQ, Nagawa S, Van Damme D, Leonard M, Yang Z, Girke T, Schmid SL, Russinova E, Friml J, Raikhel NV, Hicks GR (2011) Clusters of bioactive compounds target dynamic endomembrane networks in vivo. Proc Natl Acad Sci U S A. doi:10.1073/pnas.1108581108

Drakakaki G, van de Ven W, Pan S, Miao Y, Wang J, Keinath NF, Weatherly B, Jiang L, Schumacher K, Hicks G, Raikhel N (2012) Isolation and proteomic analysis of the SYP61 compartment reveal its role in exocytic trafficking in Arabidopsis. Cell Res 22:413–424. doi:10.1038/cr.2011.129

Drake MT, Zhu Y, Kornfeld S (2000) The assembly of AP-3 adaptorcomplex-containing clathrin-coated vesicles on synthetic liposomes. Mol Biol Cell 11:3723–3736

Dunkley TPJ, Hester S, Shadforth IP et al (2006) Mapping the Arabidopsis organelle proteome. Proc Natl Acad Sci U S A 103:6518–6523

Ebine K, Okatani Y, Uemura T, Goh T, Shoda K, Niihama M, Morita MT, Spitzer C, Otegui MS, Nakano A, Ueda T (2008) A SNARE complex unique to seed plants is required for protein storage vacuole biogenesis and seed development of *Arabidopsis thaliana*. Plant Cell 20:3006–3021

Ebine K, Fujimoto M, Okatani Y, Nishiyama T, Goh T, Ito E, Dainobu T, Nishitani A, Uemura T, Sato MH, Thordal-Christensen H, Tsutsumi N, Nakano A, Ueda T (2011) A membrane trafficking pathway regulated by the plant-specific RAB GTPase ARA6. Nat Cell Biol 13:853–859

Etxeberria E, Baroja-Fernandez E, Muñoz FJ, Pozueta-Romero J (2005) Sucrose-inducible endocytosis as a mechanism for nutrient uptake in heterotrophic plant cells. Plant Cell Physiol 46:474–481

Etxeberria E, Gonzalez P, Pozueta J (2009) Evidence for two endocytic pathways in plant cells. Plant Sci 177:341–348

Feraru E, Paciorek T, Feraru MI, Zwiewka M, De Groodt R, De Rycke R, Kleine-Vehn J, Friml J (2010) The AP-3 β adaptin mediates the biogenesis and function of the lytic vacuoles in Arabidopsis. Plant Cell 22:2812–2824

Foresti O, Gershlick DC, Bottanelli F, Hummel E, Hawes C, Denecke J (2010) A recycling-defective vacuolar sorting receptor reveals an intermediate compartment situated between prevacuoles and vacuoles in tobacco. Plant Cell 22:3992–4008

Friml J, Wisniewska J, Benková E, Mendgen K, Palme K (2002a) Lateral relocation of auxin efflux regulator PIN3 mediates tropism in Arabidopsis. Nature 415:806–809
Friml J, Benková E, Blilou I, Wisniewska J, Hamann T, Ljung K, Woody S, Sandberg G, Scheres B, Jürgens G, Palme K (2002b) AtPIN4 mediates sink-driven auxin gradients and root patterning in Arabidopsis. Cell 108:661–673
Friml J, Vieten A, Sauer M, Weijers D, Schwarz H, Hamann T, Offringa R, Jürgens G (2003) Efflux-dependent auxin gradients establish the apical-basal axis of Arabidopsis. Nature 426:147–153
Furutani M, Sakamoto N, Yoshida S, Kajiwara T, Robert HS, Friml J, Tasaka M (2011) Polar-localized NPH3-like proteins regulate polarity and endocytosis of PIN-FORMED auxin efflux carriers. Development 138:2069–2078
Gaullier JM, Simonsen A, D'Arrigo A, Bremnes B, Stenmark H, Aasland R (1998) FYVE fingers bind PtdIns(3)P. Nature 394:432–433
Gälweiler L, Guan C, Müller A, Wisman E, Mendgen K, Yephremov A, Palme K (1998) Regulation of polar auxin transport by AtPIN1 in Arabidopsis vascular tissue. Science 282:2226–2230
Geldner N, Anders N, Wolters H, Keicher J, Kornberger W, Muller P, Delbarre A, Ueda T, Nakano A, Jürgens G (2003) The Arabidopsis GNOM ARF-GEF mediates endosomal recycling, auxin transport, and auxin-dependent plant growth. Cell 112:219–230
Geldner N, Hyman DL, Wang X, Schumacher K, Chory J (2007) Endosomal signaling of plant steroid receptor kinase BRI1. Genes Dev 21:1598–1602
Geldner N, Dénervaud-Tendon V, Hyman DL, Mayer U, Stierhof YD, Chory J (2009) Rapid, combinatorial analysis of membrane compartments in intact plants with a multicolor marker set. Plant J 59:169–178
van Gisbergen P, Esseling-Ozdoba A, Vos J (2008) Microinjecting FM4-64 validates it as a marker of the endocytic pathway in plants. J Microsc 231:284–290
Grebe M, Xu J, Möbius W, Ueda T, Nakano A, Geuze HJ, Rook MB, Scheres B (2003) Arabidopsis sterol endocytosis involves actin-mediated trafficking via ARA6-positive early endosomes. Curr Biol 13:1378–1387
Griffing LR (2008) FRET analysis of transmembrane flipping of FM4-64 in plant cells: is FM4-64 a robust marker for endocytosis? J Microsc 231:291–298
Haas TJ, Sliwinski MK, Martínez DE, Preuss M, Ebine K, Ueda T, Nielsen E, Odorizzi G, Otegui MS (2007) The Arabidopsis AAA ATPase SKD1 is involved in multivesicular endosome function and interacts with its positive regulator LYST-interacting protein 5. Plant Cell 19:1295–1312
Hause G, Šamaj J, Menzel D, Baluška F (2006) Fine structural analysis of brefeldin a-induced compartment formation after high-pressure freeze fixation of maize root epidermis: compound exocytosis resembling cell plate formation during cytokinesis. Plant Signal Behav 1:134–139
Hejátko J, Blilou I, Brewer PB, Friml J, Scheres B, Benková E (2006) In situ hybridization technique for mRNA detection in whole mount Arabidopsis samples. Nat Protoc 1:1939–1946
Herberth S, Shahriari M, Bruderek M et al (2012) Artificial ubiquitylation is sufficient for sorting of a plasma membrane ATPase to the vacuolar lumen of Arabidopsis cells. Planta. doi:10.1007/s00425-012-1587-0
Huisken J, Swoger J, Del Bene F, Wittbrodt J, Stelzer EH (2004) Optical sectioning deep inside live embryos by selective plane illumination microscopy. Science 305:1007–1009
Irani NG, Di Rubbo S, Mylle E, Van den Begin J, Schneider-Pizoń J, Hniliková J, Šíša M, Buyst D, Vilarrasa-Blasi J, Szatmári AM, Van Damme D, Mishev K, Codreanu MC, Kohout L, Strnad M, Caño-Delgado AI, Friml J, Madder A, Russinova E (2012) Fluorescent castasterone reveals BRI1 signaling from the plasma membrane. Nat Chem Biol 8:583–589. doi:10.1038/nchembio.958
Ito E, Fujimoto M, Ebine K, Uemura T, Ueda T, Nakano A (2012) Dynamic behavior of clathrin in *Arabidopsis thaliana* unveiled by live imaging. Plant J 69:204–216. doi:10.1111/j.1365-313X.2011.04782.x

Jaillais Y, Fobis-Loisy I, Miege C, Rollin C, Gaude T (2006) AtSNX1 defines an endosome for auxin-carrier trafficking in Arabidopsis. Nature 443:106–109

Jaillais Y, Santambrogio M, Rozier F et al (2007) The retromer protein VPS29 links cell polarity and organ initiation in plants. Cell 130:1057–1070

Jaillais Y, Fobis-Loisy I, Miège C, Gaude T (2008) Evidence for a sorting endosome in Arabidopsis root cells. Plant J 53:237–247

Jelínková A, Malínská K, Simon S, Kleine-Vehn J, Parezová M, Pejchar P, Kubes M, Martinec J, Friml J, Zazímalová E, Petrásek J (2010) Probing plant membranes with FM dyes: tracking, dragging or blocking? Plant J 61:883–892

Kale SD, Gu B, Capelluto DG, Dou D, Feldman E, Rumore A, Arredondo FD, Hanlon R, Fudal I, Rouxel T, Lawrence CB, Shan W, Tyler BM (2010) External lipid PI3P mediates entry of eukaryotic pathogen effectors into plant and animal host cells. Cell 142:284–295

Kang B-H, Staehelin LA (2008) ER-to-Golgi transport by COPII vesicles in Arabidopsis involves a ribosome-excluding scaffold that is transferred with the vesicles to the Golgi matrix. Protoplasma 234:51–64

Kang BH, Nielsen E, Preuss ML, Mastronarde D, Staehelin LA (2011) Electron tomography of RabA4b- and PI-4 Kβ1-labeled trans Golgi network compartments in Arabidopsis. Traffic 12:313–329

Keller PJ, Stelzer HK (2008) Quantitative in vivo imaging of entire embryos with digital scanned laser light sheet fluorescence microscopy. Curr Opin Neurobiol 18:624–632

Kitakura S, Vanneste S, Robert S, Löfke C, Teichmann T, Tanaka H, Friml J (2011) Clathrin mediates endocytosis and polar distribution of PIN auxin transporters in Arabidopsis. Plant Cell 23:1920–1931

Kleine-Vehn J, DhonuksheP Swarup R, Bennett M, Friml J (2006) Subcellular trafficking of the Arabidopsis auxin influx carrier AUX1 uses a novel pathway distinct from PIN1. Plant Cell 18:3171–3181

Kleine-Vehn J, Dhonukshe P, Sauer M, Brewer PB, Wiśniewska J, Paciorek T, Benková E, Friml J (2008) ARF GEF-dependent transcytosis and polar delivery of PIN auxin carriers in Arabidopsis. Curr Biol 18:526–531

Kleine-Vehn J, Huang F, Naramoto S, Zhang J, Michniewicz M, Offringa R, Friml J (2009) PIN auxin efflux carrier polarity is regulated by PINOID kinase-mediated recruitment into GNOM-independent trafficking in Arabidopsis. Plant Cell 21:3839–3849

Kleine-Vehn J, Ding Z, Jones AR, Tasaka M, Morita MT, Friml J (2010) Gravity-induced PIN transcytosis for polarization of auxin fluxes in gravity-sensing root cells. Proc Natl Acad Sci U S A 107:22344–22349

Kleine-Vehn J, Wabnik K, Martinière A, Łangowski Ł, Willig K, Naramoto S, Leitner J, Tanaka H, Jakobs S, Robert S, Luschnig C, Govaerts W, Hell SW, Runions J, Friml J (2011) Recycling, clustering, and endocytosis jointly maintain PIN auxin carrier polarity at the plasma membrane. Mol Syst Biol 7:540. doi:10.1038/msb.2011.72

Konopka CA, Bednarek SY (2008a) Comparison of the dynamics and functional redundancy of the Arabidopsis dynamin-related isoforms DRP1A and DRP1C during plant development. Plant Physiol 147:1590–1602

Konopka CA, Bednarek SY (2008b) Variable-angle epifluorescence microscopy: a new way to look at protein dynamics in the plant cell cortex. Plant J 53:186–196

Konopka CA, Backues SK, Bednarek SY (2008) Dynamics of Arabidopsis dynamin-related protein 1C and a clathrin light chain at the plasma membrane. Plant Cell 20:1363–1380

Lam SK, Siu CL, Hillmer S, Jang S, An G, Robinson DG, Jiang L (2007) Rice SCAMP1 defines clathrin-coated, *trans*-Golgi-located tubular-vesicular structures as an early endosome in tobacco BY-2 cells. Plant Cell 19:296–319

Lam SK, Cai Y, Tse YC, Wang J, Law AH, Pimpl P, Chan HY, Xia J, Jiang L (2009) BFA-induced compartments from the Golgi apparatus and *trans*-Golgi network/early endosome are distinct in plant cells. Plant J 60:865–881

Laloi M, Perret AM, Chatre L, Melser S, Cantrel C, Vaultier MN, Zachowski A, Bathany K, Schmitter JM, Vallet M, Lessire R, Hartmann MA, Moreau P (2007) Insights into the role of specific lipids in the formation and delivery of lipid microdomains to the plasma membrane of plant cells. Plant Physiol 143:461–472

Lee Y, Bak G, Choi Y, Chuang WI, Cho HT, Lee Y (2008a) Roles of phosphatidylinositol 3-kinase in root hair growth. Plant Physiol 147:624–635

Lee YJ, Szumlanski A, Nielsen E, Yang Z (2008b) Rho-GTPase-dependent filamentous actin dynamics coordinate vesicle targeting and exocytosis during tip growth. J Cell Biol 181:1155–1168

Lefebvre B, Furt F, Hartmann MA, Michaelson LV, Carde JP, Sargueil-Boiron F, Rossignol M, Napier JA, Cullimore J, Bessoule JJ, Mongrand S (2007) Characterization of lipid rafts from *Medicago truncatula* root plasma membranes: a proteomic study reveals the presence of a raft-associated redox system. Plant Physiol 144:402–418

Li X, Wang X, Yang Y, Li R, He Q, Fang X, Luu DT, Maurel C, Lin J (2011) Single-molecule analysis of PIP2;1 dynamics and partitioning reveals multiple modes of Arabidopsis plasma membrane aquaporin regulation. Plant Cell 23:3780–3797

Li R, Liu P, Wan Y, Chen T, Wang Q, Mettbach U, Baluška F, Šamaj J, Fang X, Lucas WJ, Lin J (2012) A membrane microdomain-associated protein, Arabidopsis flot1, is involved in a clathrin-independent endocytic pathway and is required for seedling development. Plant Cell 24:2105–2122. doi:10.1105/tpc.112.095695

Lilley KS, Dupree P (2007) Plant organelle proteomics. Curr Opin Plant Biol 10:594–599

Limbach C, Staehelin LA, Sievers A, Braun M (2008) Electron tomographic characterization of a vacuolar reticulum and of six vesicle types that occupy different cytoplasmic domains in the apex of tip-growing chara rhizoids. Planta 227:1101–1114

Lindek S, Cremer C, Stelzer EH (1996a) Confocal theta fluorescence microscopy using two-photon absorption and annular apertures. Optik 102:131–134

Lindek S, Cremer C, Stelzer EH (1996b) Confocal theta fluorescence microscopy with annular apertures. Appl Opt 35:126–130

Lindek S, Stelzer EH (1996) Optical transfer functions for confocal theta fluorescence microscopy. J Opt Soc Am A 13:479–482

Liu P, Li RL, Zhang L, Wang QL, Niehaus K, Baluška F, Šamaj J, Lin J (2009) Lipid microdomain polarization is required for NADPH oxidase-dependent ROS signaling in *Picea meyeri* pollen tube tip growth. Plant J 60:303–313

Luczak M, Bugajewska A, Wojtaszek P (2008) Inhibitors of protein glycosylation or secretion change the pattern of extracellular proteins in suspension-cultured cells of *Arabidopsis thaliana*. Plant Physiol Biochem 46:962–969

Luu DT, Martinière A, Sorieul M, Runions J, Maurel C (2012) Fluorescence recovery after photobleaching reveals high cycling dynamics of plasma membrane aquaporins in Arabidopsis roots under salt stress. Plant J 69:894–905

Maizel A, von Wangenheim D, Federici F, Haseloff J, Stelzer EH (2011) High-resolution live imaging of plant growth in near physiological bright conditions using light sheet fluorescence microscopy. Plant J 68:377–385

Matheson LA, Hanton SL, Rossi M, Latijnhouwers M, Stefano G, Renna L, Brandizzi F (2007) Multiple roles of ADP-ribosylation factor 1 in plant cells include spatially regulated recruitment of coatomer and elements of the Golgi matrix. Plant Physiol 143:1615–1627

Matsuoka K, Bassham DC, Raikhel NV, Nakamura K (1995) Different sensitivity to wortmannin of two vacuolar sorting signals indicates the presence of distinct sorting machineries in tobacco cells. J Cell Biol 130:1307–1318

Mayor S, Pagano RE (2007) Pathways of clathrin-independent endocytosis. Nat Rev Mol Cell Biol 8:603–612

Miao Y, Yan PK, Kim H, Hwang I, Jiang L (2006) Localization of green fluorescent protein fusions with the seven Arabidopsis vacuolar sorting receptors to prevacuolar compartments in tobacco BY-2 cells. Plant Physiol 142:945–962

Miao Y, Li KY, Li HY, Yao X, Jiang L (2008) The vacuolar transport of aleurain-GFP and 2S albumin-GFP fusions is mediated by the same prevacuolar compartments in tobacco BY-2 and Arabidopsis suspension cultured cells. Plant J 56:824–839

Miyazawa Y, Ito Y, Moriwaki T, Kobayashi A, Fujii N, Takahashi T (2009) A molecular mechanism unique to hydrotropism in roots. Plant Sci 177:297–301

Mongrand S, Morel J, Laroche J, Claverol S, Carde JP, Hartmann MA, Bonneu M, Simon-Plas F, Lessire R, Bessoule JJ (2004) Lipid rafts in higher plant cells: purification and characterization of Triton X-100-insoluble microdomains from tobacco plasma membrane. J Biol Chem 279:36277–36286

Morel J, Claverol S, Mongrand S, Furt F, Fromentin J, Bessoule JJ, Blein JP, Simon-Plas F (2006) Proteomics of plant detergent-resistant membranes. Mol Cell Proteomics 5:1396–1411

Mravec J, Petrášek J, Li N, Boeren S, Karlova R, Kitakura S, Pařezová M, Naramoto S, Nodzyński T, Dhonukshe P, Bednarek SY, Zažímalová E, de Vries S, Friml J (2011) Cell plate restricted association of DRP1A and PIN proteins is required for cell polarity establishment in Arabidopsis. Curr Biol 21:1055–1060

Müller A, Guan C, Gälweiler L, Tänzler P, Huijser P, Marchant A, Parry G, Bennett M, Wisman E, Palme K (1998) AtPIN2 defines a locus of Arabidopsis for root gravitropism control. EMBO J 17:6903–6911

Müller J, Mettbach U, Menzel D, Šamaj J (2007) Molecular dissection of endosomal compartments in plants. Plant Physiol 145:293–304

Müller J, Beck M, Mettbach U, Komis G, Hause G, Menzel D, Šamaj J (2010) Arabidopsis MPK6 is involved in cell division plane control during early root development, and localizes to the pre-prophase band, phragmoplast, *trans*-Golgi network and plasma membrane. Plant J 61:234–248

Nagawa S, Xu T, Lin D, Dhonukshe P, Zhang X, Friml J, Scheres B, Fu Y, Yang Z (2012) ROP GTPase-dependent actin microfilaments promote PIN1 polarization by localized inhibition of clathrin-dependent endocytosis. PLoS Biol 10:e1001299. doi:10.1371/journal.pbio.1001299

Nakano A (2002) Spinning-disk confocal microscopy-a cutting-edge tool for imaging of membrane traffic. Cell Struct Funct 27:349–355

Naramoto S, Sawa S, Koizumi K, Uemura T, Ueda T, Friml J, Nakano A, Fukuda H (2009) Phosphoinositide-dependent regulation of VAN3 ARF-GAP localization and activity essential for vascular tissue continuity in plants. Development 136:1529–1538

Naramoto S, Kleine-Vehn J, Robert S, Fujimoto M, Dainobu T, Paciorek T, Ueda T, Nakano A, Van Montagu MC, Fukuda H, Friml J (2010) ADP-ribosylation factor machinery mediates endocytosis in plant cells. Proc Natl Acad Sci U S A 107:21890–21895

Nebenführ A, Ritzenthaler C, Robinson DG (2002) Brefeldin A: deciphering an enigmatic inhibitor of secretion. Plant Physiol 130:1102–1108

Nielsen E, Cheung AY, Ueda T (2008) The regulatory RAB and ARF GTPases for vesicular trafficking. Plant Physiol 147:1516–1526

Niemes S, Labs M, Scheuring D, Krueger F, Langhans M, Jesenofsky B, Robinson DG, Pimpl P (2010) Sorting of plant vacuolar proteins is initiated in the ER. Plant J 62:601–614

Noguchi K, Yasumori M, Imai T, Naito S, Matsunaga T, Oda H, Hayashi H, Chino M, Fujiwara T (1997) *bor1-1*, an *Arabidopsis thaliana* mutant that requires a high level of boron. Plant Physiol 115:901–906

Oliviusson P, Heinzerling O, Hillmer S, Hinz G, Tse YC, Jiang L, Robinson DG (2006) Plant retromer, localized to the prevacuolar compartment and microvesicles in Arabidopsis, may interact with vacuolar sorting receptors. Plant Cell 18:1239–1252

Onelli E, Prescianotto-Baschong C, Caccianiga M, Moscatelli A (2008) Clathrin-dependent and independent endocytosis pathways in tobacco protoplasts revealed by labeling with charged nanogold. J Exp Bot 59:3051–3068

Osterrieder A, Carvalho CM, Latijnhouwers M, Johansen JN, Stubbs C, Botchway S, Hawes C (2009) Fluorescence lifetime imaging of interactions between Golgi tethering factors and small GTPases in plants. Traffic 10:1034–1046

Otegui MS, Herder R, Schulze J, Jung R, Staehelin LA (2006) The proteolytic processing of seed storage proteins in Arabidopsis embryo cells starts in the multivesicular bodies. Plant Cell 18:2567–2581

Ovečka M, Lang I, Baluška F, Ismail A, Illeš P, Lichtscheidl IK (2005) Endocytosis and vesicle trafficking during tip growth of root hairs. Protoplasma 226:39–54
Ovečka M, Berson T, Beck M, Derksen J, Šamaj J, Baluška F, Lichtscheidl IK (2010) Structural sterols are involved in both the initiation and tip growth of root hairs in *Arabidopsis thaliana*. Plant Cell 22:2999–3019
Paciorek T, Zažímalová E, Ruthardt N, Petrášek J, Stierhof YD, Kleine-Vehn J, Morris DA, Emans N, Jürgens G, Geldner N, Friml J (2005) Auxin inhibits endocytosis and promotes its own efflux from cells. Nature 435:1251–1256
Padilla-Parra S, Tramier M (2012) FRET microscopy in the living cell: different approaches, strengths and weaknesses. BioEssays 34:369–376. doi:10.1002/bies.201100086
Phan NQ, Kim SJ, Bassham DC (2008) Overexpression of Arabidopsis sorting nexin AtSNX2b inhibits endocytic trafficking to the vacuole. Mol Plant 1:961–976
Petrášek J, Mravec J, Bouchard R, Blakeslee JJ, Abas M, Seifertová D, Wisniewska J, Tadele Z, Kubes M, Covanová M, Dhonukshe P, Skupa P, Benková E, Perry L, Krecek P, Lee OR, Fink GR, Geisler M, Murphy AS, Luschnig C, Zažímalová E, Friml J (2006) PIN proteins perform a rate-limiting function in cellular auxin efflux. Science 312:914–918
Pinheiro H, Samalova M, Geldner N, Chory J, Martinez A, Moore I (2009) Genetic evidence that the higher plant Rab-D1 and Rab-D2 GTPases exhibit distinct but overlapping interactions in the early secretory pathway. J Cell Sci 122:3749–3758
Pourcher M, Santambrogio M, Thazar N, Thierry AM, Fobis-Loisy I, Miège C, Jaillais Y, Gaude T (2010) Analyses of SORTING NEXINs reveal distinct retromer-subcomplex functions in development and protein sorting in *Arabidopsis thaliana*. Plant Cell 22:3980–3991
Preuss ML, Serna J, Falbel TG, Bednarek SY, Nielsen E (2004) The Arabidopsis Rab GTPase RabA4b localizes to the tips of growing root hair cells. Plant Cell 16:1589–1603
Preuss ML, Schmitz AJ, Thole JM, Bonner HK, Otegui MS, Nielsen E (2006) A role for the RabA4b effector protein PI-4Kbeta1 in polarized expansion of root hair cells in *Arabidopsis thaliana*. J Cell Biol 172:991–998
Reichardt I, Stierhof YD, Mayer U, Richter S, Schwarz H, Schumacher K, Jürgens G (2007) Plant cytokinesis requires de novo secretory trafficking but not endocytosis. Curr Biol 17:2047–2053
Reyes FC, Buono R, Otegui MS (2011) Plant endosomal trafficking pathways. Curr Opin Plant Biol 14:666–673
Richardson LGL, Howard ASM, Khuu N, Gidda SK, McCartney A, Morphy BJ, Mullen RT (2011) Protein–protein interaction network and subcellular localization of the *Arabidopsis thaliana* ESCRT machinery. Front. Plant Sci. 2:20
Richter S, Müller LM, Stierhof YD, Mayer U, Takada N, Kost B, Vieten A, Geldner N, Koncz C, Jürgens G (2011) Polarized cell growth in Arabidopsis requires endosomal recycling mediated by GBF1-related ARF exchange factors. Nat Cell Biol 4:80–86
Robatzek S, Chinchilla D, Boller T (2006) Ligand-induced endocytosis of the pattern recognition receptor FLS2 in Arabidopsis. Genes Dev 20:537–542
Robert S, Chary SN, Drakakaki G, Li S, Yang Z, Raikhel NV, Hicks GR (2008) Endosidin1 defines a compartment involved in endocytosis of the brassinosteroid receptor BRI1 and the auxin transporters PIN2 and AUX1. Proc Natl Acad Sci U S A 105:8464–8469
Robert S, Kleine-Vehn J, Barbez E, Sauer M, Paciorek T, Baster P, Vanneste S, Zhang J, Simon S, Čovanová M, Hayashi K, Dhonukshe P, Yang Z, Bednarek SY, Jones AM, Luschnig C, Aniento F, Zažímalová E, Friml J (2010) ABP1 mediates auxin inhibition of clathrin-dependent endocytosis in Arabidopsis. Cell 143:111–121
Russinova E, Borst JW, Kwaaitaal M, Yanhai Yin Y, Caño-Delgado A, Chory J, de Vries SC (2004) Heterodimerization and endocytosis of Arabidopsis brassinosteroid receptors BRI1 and AtSERK3 (BAK1). Plant Cell 16:3216–3229
Salomon S, Grunewald D, Stüber K, Schaaf S, MacLean D, Schulze-Lefert P, Robatzek S (2010) High-throughput confocal imaging of intact live tissue enables quantification of membrane trafficking in Arabidopsis. Plant Physiol 154:1096–1104

Sätzler K, Eils R (1997) Resolution improvement by 3-D reconstructions from tilted views in axial tomography and confocal theta microscopy. Bioimaging 5:171–182

Šamaj J, Baluška F, Voigt B, Schlicht M, Volkmann D, Menzel D (2004) Endocytosis, actin cytoskeleton, and signaling. Plant Physiol 135:1150–1161

Scheuring D, Viotti C, Krüger F, Künzl F, Sturm S, Bubeck J, Hillmer S, Frigerio L, Robinson DG, Pimpl P, Schumacher K (2011) Multivesicular bodies mature from the trans-Golgi network/early endosome in Arabidopsis. Plant Cell 23:3463–3481

Sena G, Frentz Z, Birnbaum KD, Leibler S (2011) Quantitation of cellular dynamics in growing Arabidopsis roots with light sheet microscopy. PLoS One 6(6):e21303. doi:10.1371/journal.pone.0021303

Silady RA, Ehrhardt DW, Jackson K, Faulkner C, Oparka K, Somerville CR (2007) The GRV2/RME-8 protein of Arabidopsis functions in the late endocytic pathway and is required for vacuolar membrane flow. Plant J 53:29–41

Simon-Plas F, Perraki A, Bayer E, Gerbeau-Pissot P, Mongrand S (2011) An update on plant membrane rafts. Curr Opin Plant Biol 14:642–649

Song J, Lee MH, Lee GJ, Yoo CM, Hwang I (2006) Arabidopsis EPSIN1 plays an important role in vacuolar trafficking of soluble cargo proteins in plant cells via interactions with clathrin, AP-1, VTI11, and VSR1. Plant Cell 18:2258–2274

Sousa E, Kost B, Malhó R (2008) Arabidopsis phosphatidylinositol-4-monophosphate 5-kinase 4 regulates pollen tube growth and polarity by modulating membrane recycling. Plant Cell 20:3050–3064

Sparkes IA, Graumann K, Martinière A, Schoberer J, Wang P, Osterrieder A (2011) Bleach it, switch it, bounce it, pull it: using lasers to reveal plant cell dynamics. J Exp Bot 62:1–7

Spitzer C, Schellmann S, Sabovljevic A, Shahriari M, Keshavaiah C, Bechtold N, Herzog M, Müller S, Hanisch FG, Hülskamp M (2006) The Arabidopsis *elch* mutant reveals functions of an ESCRT component in cytokinesis. Development 133:4679–4689

Spitzer C, Reyes FC, Buono R, Sliwinski MK, Haas TJ, Otegui MS (2009) The ESCRT-related CHMP1A and B proteins mediate multivesicular body sorting of auxin carriers in Arabidopsis and are required for plant development. Plant Cell 21:749–766

Staehelin LA, Kang BH (2008) Nanoscale architecture of endoplasmic reticulum export sites and of Golgi membranes as determined by electron tomography. Plant Physiol 147:1454–1468

Stefano G, Renna L, Rossi M, Azzarello E, Pollastri S, Brandizzi F, Baluska F, Mancuso S (2010) AGD5 is a GTPase-activating protein at the *trans*-Golgi network. Plant J 64:790–799

Stelzer EH, Lindek S (1994) Fundamental reduction of the observation volume in far-field light microscopy by detection orthogonal to the illumination axis: confocal theta microscopy. Opt Commun 111:536–547

Swarup R, Kargul J, Marchant A, Zadik D, Rahman A, Mills R, Yemm A, May S, Williams L, Millner P, Tsurumi S, Moore I, Napier R, Kerr ID, Bennett MJ (2004) Structure-function analysis of the presumptive Arabidopsis auxin permease AUX1. Plant Cell 16:3069–3083

Szechyńska-Hebda M, Wedzony M, Dubas E, Kieft H, van Lammeren A (2006) Visualisation of microtubules and actin filaments in fixed BY-2 suspension cells using an optimised whole mount immunolabelling protocol. Plant Cell Rep 25:758–766

Szumlanski AL, Nielsen E (2009) The Rab GTPase RabA4d regulates pollen tube tip growth in *Arabidopsis thaliana*. Plant Cell 21:526–544

Takáč T, Pechan T, Richter H, Müller J, Eck C, Böhm N, Obert B, Ren H, Niehaus K, Šamaj J (2011a) Proteomics on brefeldin A-treated Arabidopsis roots reveals profilin 2 as a new protein involved in the cross-talk between vesicular trafficking and the actin cytoskeleton. J Proteome Res 10:488–501

Takáč T, Pechan T, Šamaj J (2011b) Differential proteomics of plant development. J Proteomics 74:577–588

Takáč T, Pechan T, Šamajová O, Ovečka M, Richter H, Eck C, Niehaus K, Šamaj J (2012) Wortmannin treatment induces changes in Arabidopsis root proteome and post-golgi compartments. J Proteome Res 11:3127–3142. doi:10.1021/pr201111n

Takahashi D, Kawamura Y, Yamashita T, Uemura M (2012) Detergent-resistant plasma membrane proteome in oat and rye: similarities and dissimilarities between two monocotyledonous plants. J Proteome Res 11:1654–1665. doi:10.1021/pr200849v

Takano J, Tanaka M, Toyoda A, Miwa K, Kasai K, Fuji K, Onouchi H, Naito S, Fujiwara T (2010) Polar localization and degradation of Arabidopsis boron transporters through distinct trafficking pathways. Proc Natl Acad Sci U S A 107:5220–5225

Tanaka H, Kitakura S, De Rycke R, De Groodt R, Friml J (2009) Fluorescence imaging-based screen identifies ARF GEF component of early endosomal trafficking. Curr Biol 19:391–397

Thellmann M, Rybak K, Thiele K, Wanner G, Assaad FF (2010) Tethering factors required for cytokinesis in Arabidopsis. Plant Physiol 154:720–732

Thiel G, Kreft M, Zorec R (2009) Rhythmic kinetics of single fusion and fission in a plant cell protoplast. Ann NY Acad Sci 1152:1–6

Tóth R, Gerding-Reimers C, Deeks MJ, Menninger S, Gallegos RM, Tonaco IA, Hübel K, Hussey PJ, Waldmann H, Coupland G (2012) Prieurianin/endosidin1 is an actin stabilizing small molecule identified from a chemical genetic screen for circadian clock effectors in *Arabidopsis thaliana*. Plant J doi. doi:10.1111/j.1365-313X.2012.04991.x

Toyooka K, Goto Y, Asatsuma S, Koizumi M, Mitsui T, Matsuoka K (2009) A mobile secretory vesicle cluster involved in mass transport from the Golgi to the plant cell exterior. Plant Cell 21:1212–1229

Tse YC, Mo B, Hillmer S, Zhao M, Lo SW, Robinson DG, Jiang L (2004) Identification of multivesicular bodies as prevacuolar compartments in *Nicotiana tabacum* BY-2 cells. Plant Cell 16:672–693

Ueda T, Yamaguchi M, Uchimiya H, Nakano A (2001) Ara6, a plant-unique novel type RabGTPase, functions in the endocytic pathway of *Arabidopsis thaliana*. EMBO J 20:4730–4741

Ueda T, Uemura T, Sato MH, Nakano A (2004) Functional differentiation of endosomes in Arabidopsis cells. Plant J 40:783–789

Vermeer JE, van Leeuwen W, Tobeña-Santamaria R, Laxalt AM, Jones DR, Divecha N, Gadella TW Jr, Munnik T (2006) Visualization of PtdIns3P dynamics in living plant cells. Plant J 47:687–700

Viotti C, Bubeck J, Stierhof YD, Krebs M, Langhans M, van den Berg W, van Dongen W, Richter S, Geldner N, Takano J, Jürgens G, de Vries SC, Robinson DG, Schumacher K (2010) Endocytic and secretory traffic in Arabidopsis merge in the *trans*-Golgi network/early endosome, an independent and highly dynamic organelle. Plant Cell 22:1344–1357

Vitha S, Baluška F, Braun M, Šamaj J, Volkmann D, Barlow PW (2000a) Comparison of cryofixation and aldehyde fixation for plant actin immunocytochemistry: aldehydes do not destroy F-actin. Histochem J 32:457–466

Vitha S, Baluška F, Jásik J, Volkmann D, Barlow PW (2000b) Steedman's wax for F-actin visualization. In: Staiger CJ, Baluška F, Volkmann D, Barlow PW (eds) Actin: a dynamic framework for multiple plant cell functions. Kluwer Academic Publishers, Dordrecht, pp 619–636

Voie AH, Burns DH, Spelman FA (1993) Orthogonal-plane fluorescence optical sectioning: three-dimensional imaging of macroscopic biological specimens. J Microsc 170:229–236

Voigt B, Timmers AC, Šamaj J, Hlavačka A, Ueda T, Preuss M, Nielsen E, Mathur J, Emans N, Stenmark H, Nakano A, Baluška F, Menzel D (2005) Actin-based motility of endosomes is linked to the polar tip growth of root hairs. Eur J Cell Biol 84:609–621

Wang X, Goshe MB, Soderblom EJ, Phinney BS, Kuchar JA, Li J, Asami T, Yoshida S, Huber SC, Clouse SD (2005) Identification and functional analysis of in vivo phosphorylation sites of the Arabidopsis BRASSINOSTEROID-INSENSITIVE1 receptor kinase. Plant Cell 17:1685–1703

Wang J, Li Y, Lo SW, Hillmer S, Sun SS, Robinson DG, Jiang L (2007) Protein mobilization in germinating mung bean seeds involves vacuolar sorting receptors and multivesicular bodies. Plant Physiol 143:1628–1639

Wang J, Cai Y, Miao Y, Lam SK, Jiang L (2009) Wortmannin induces homotypic fusion of plant prevacuolar compartments. J Exp Bot 60:3075–3083

Wang H, Tse YC, Law AH, Sun SS, Sun YB, Xu ZF, Hillmer S, Robinson DG, Jiang L (2010) Vacuolar sorting receptors (VSRs) and secretory carrier membrane proteins (SCAMPs) are essential for pollen tube growth. Plant J 61:826–838

Weckwerth W (2011) Green systems biology-from single genomes, proteomes and metabolomes to ecosystems research and biotechnology. J Proteomics 75:284–305

Wisniewska J, Xu J, Seifertová D, Brewer PB, Ruzicka K, Blilou I, Rouquié D, Benková E, Scheres B, Friml J (2006) Polar PIN localization directs auxin flow in plants. Science 312:883

Wolfenstetter S, Wirsching P, Dotzauer D, Schneider S, Sauer N (2012) Routes to the tonoplast: the sorting of tonoplast transporters in Arabidopsis mesophyll protoplasts. Plant Cell 24:215–232

Xu J, Scheres B (2005) Cell polarity: ROPing the ends together. Curr Opin Plant Biol 8:613–618

Yang Y, Hammes UZ, Taylor CG, Schachtman DP, Nielsen E (2006) High-affinity auxin transport by the AUX1 influxcarrier protein. Curr Biol 16:1123–1127

Yoshinari A, Kasai K, Fujiwara T, Naito S, Takano J (2012) Polar localization and endocytic degradation of a boron transporter, BOR1, is dependent on specific tyrosine residues. Plant Signal Behav 7:46–49

Zhang L, Zhang H, Liu P, Hao H, Jin JB, Lin J (2011) Arabidopsis R-SNARE proteins VAMP721 and VAMP722 are required for cell plate formation. PLoS One 6(10):e26129. doi:10.1371/journal.pone.0026129

Zonia L, Munnik T (2008) Vesicle trafficking dynamics and visualization of zones of exocytosis and endocytosis in tobacco pollen tubes. J Exp Bot 59:861–873

Zouhar J, Rojo E, Bassham DC (2009) AtVPS45 is a positive regulator of the SYP41/SYP61/VTI12 SNARE complex involved in trafficking of vacuolar cargo. Plant Physiol 149:1668–1678

Zwiewka M, Feraru E, Möller B, Hwang I, Feraru MI, Kleine-Vehn J, Weijers D, Friml J (2011) The AP-3 adaptor complex is required for vacuolar function in Arabidopsis. Cell Res 21:1711–1722

Chemical Effectors of Plant Endocytosis and Endomembrane Trafficking

Ruixi Li, Natasha V. Raikhel and Glenn R. Hicks

Abstract Plant endocytosis and endomembrane trafficking relies on the coordination of a highly organized and dynamic network of intracellular organelles. Membrane trafficking and associated signal transduction pathways provide critical cellular regulation of plant development and response to environmental stimuli. However, the efficiency of studies on this complex network has been hampered due to the rapid and dynamic nature of endomembrane trafficking as well as gene redundancy and embryonic lethality in mutagenesis-based strategies. Chemical genomics emerged in recent years as a complementary approach to illuminate biological functions through the integration of organic chemistry, biology, and bioinformatics to overcome gene redundancy. The approach presents significant advantages in dosage dependence and reversibility, which offers the ideal ability to study dynamic endomembrane trafficking processes in real time. In this chapter, several successful examples of chemical screening focused on the endomembrane system is presented to illuminate the efficiency and power of chemical genomics in dissecting endomembrane trafficking and its regulation of plant development and environmental responses. Perspectives are also presented to suggest directions for future development of this field.

Keywords Endocytosis · Chemical biology · Chemical genomics · Endomembrane trafficking · Bioactive compounds · High-throughput screening

R. Li · N. V. Raikhel · G. R. Hicks (✉)
Department of Botany and Plant Sciences, Center for Plant Cell Biology,
University of California, Riverside, CA 92521, USA
e-mail: ghicks@ucr.edu

J. Šamaj (ed.), *Endocytosis in Plants*,
DOI: 10.1007/978-3-642-32463-5_2, © Springer-Verlag Berlin Heidelberg 2012

1 Endomembrane Trafficking in Plants Cells

Endocytosis is a complicated and dynamic cellular process for the uptake of extracellular molecules or internalization of plasma membrane proteins and lipids through the formation of closed vesicles (Low and Chandra 1994; Robinson et al. 2008). After internalization, vesicles are either fused with lytic compartments for degradation or recycled back to the plasma membrane. Endocytosis is facilitated by the highly organized endomembrane system which provides the functional compartments necessary for the exchange of proteins, lipids, and polysaccharides via transport intermediates (Jürgens 2004; Šamaj et al. 2005). Several endomembrane trafficking pathways converge in the continuous endomembrane system. The secretory route starts from the endoplasmic reticulum (ER), operates progressively through the *cis*, medial and *trans*-Golgi network (TGN), and finally delivers cargoes to the PM or vacuole (Bassham et al. 2008). While anterograde transport serves as the primary route for newly synthesized proteins from ER to the Golgi and endosomes, retrograde transport mainly functions to maintain the localization of ER or Golgi-resident proteins or components of the trafficking machinery (Hicks and Raikhel 2010) (Fig. 1). Endomembrane trafficking is well studied in animal systems for its importance in the medical field, such as nutrient uptake, cholesterol clearance from blood, and down regulation of many receptor-mediated signalling processes (Murphy et al. 2005; Robinson et al. 2008).

Multiple endomembrane trafficking pathways have been elucidated in animal systems of which clathrin-dependent endocytosis is most widely studied. Clathrin-independent pathways, including caveolae/lipid raft-mediated endocytosis, fluid-phase endocytosis, and phagocytosis have also been investigated in different cells, although the mechanisms are not as clear as the clathrin-dependent pathway due to the lack of specific markers (Šamaj et al. 2004; Murphy et al. 2005). Research on plant endocytosis has lagged for many years due to controversy about the possibility of endocytosis in plant cells in which membrane invagination could be hindered because of the turgor pressure from the cell wall (Cram 1980). In the past three decades, endocytosis in plant cells began to be gradually accepted with the utilization of electron dense tracers, styryl dyes, filipin-labeled plant sterols, the fluid-phase marker Lucifer Yellow, and fluorescent protein-tagged markers in different endomembrane compartments (Hillmer et al. 1986; Geldner et al. 2001; Baluška et al. 2002, 2004; Geldner et al. 2003; Grebe et al. 2003; Hurst et al. 2004). With the help of these techniques, it is now accepted that turgidity is not an obstacle for endocytosis in plant cells (Gradmann and Robison 1989). Even in highly turgid cells such as guard cells, endocytosis can be detected, as indicated by the recent investigation of the potassium channel KAT1 which undergoes constitutive endocytosis and recycling in guard cells (Sutter et al. 2007).

Consistent with animal systems, endocytosis in plant cells is mediated by clathrin-dependent and clathrin-independent pathways, although the existence of caveolae in plant cells is still not clear (Murphy et al. 2005). Multiple endomembrane compartments and the corresponding marker proteins are recognized

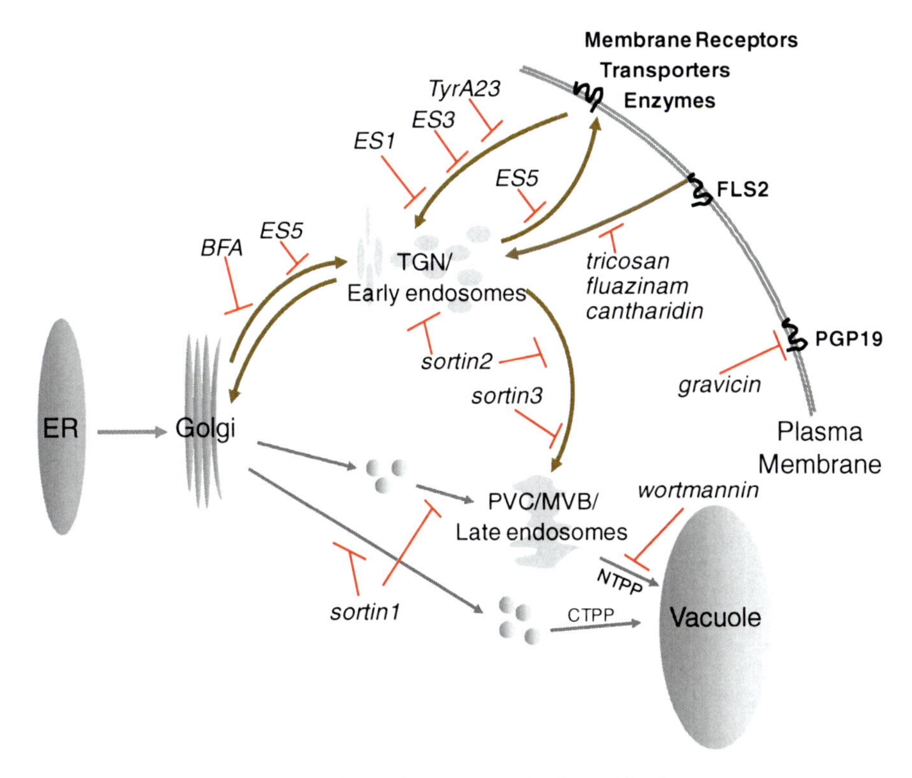

Fig. 1 A basic overview of the endomembrane system in plants. Membrane compartments (*gray color*) are ER, Golgi, TGN/early endosomes, PVC/multivesicular bodies (MVB)/late endosomes or other intermediate compartments, and vacuole. In plants, endosomes are contiguous with the TGN and PVC/MVB. These schematic symbols represent a continuum of intermediate compartments defined by distinct cargoes and vesicle transport components. Potential routes of vesicle transport are indicated (*brown arrows*) and include endocytosis, recycling to the plasma membrane, and transport toward the vacuole. Vacuole transport can utilize the N-terminal pro-peptide (NTPP), C-terminal pro-peptide (CTPP), and other routes. Also indicated are endomembrane processes perturbed by novel compounds including sortin1 (NTPP and CTPP targeting and vacuole biogenesis) (Zouhar et al. 2004; Rosado et al. 2011), sortin2 (endosome involved in trafficking to the vacuole) (Norambuena et al. 2008), sortin3 (possible late endosome compartment) (Chanda et al. 2009), gravicin (cognate target is the auxin-related multidrug P-glycoprotein 19, PGP19) (Rojas-Pierce et al. 2007), TyrA 23 (blocks cargo recruitment into CCVs from PM) (Chanda et al. 2009), endosidin1 (ES1) (perturbs an early SYP61/VHA-a1 endosome compartment) (Robert et al. 2008), ES3 (blocks protein trafficking from the PM by affecting TGN/EE compartment), ES5 (blocks recycling), ES7 (perturbation of secretory pathways) (Drakakaki et al. 2011), and several inhibitors (triclosan, fluazinam and cantharidin) of flg22-induced FLS2 endocytosis (Serrano et al. 2007). Several well-established bioactive molecules are BFA that causes agglomeration of Golgi and TGN and wortmannin (Wort) that affects late trafficking to the vacuole. Recent high-throughput screens in plants (see text) will provide a suite of compounds affecting other endomembrane trafficking processes. Figure modified from Hicks and Raikhel (2010) with permission

within endocytic routes. By the observation of sterol tracers, the first station for cargo delivery is the early endosome (EE). In animal systems, Rab5 GTPase has been established as the convenient marker gene for EE (Tonkin et al. 2006). In Arabidopsis, two Rab5 homologs-RabF1 (Ara6) and RabF2b (Ara7) were identified and co-localized with the sterol tracer FM4-64 after short time uptake (Merigout et al. 2002). In plants, the TGN is expected to be functionally equivalent to the EE, which is labeled by markers including the membrane-integral V-ATPase subunit, VHA-a1 (Dettmer et al. 2006), and the plant-specific syntaxin SYP61 (Robert et al. 2008). The TGN also functions as a sorting compartment through which endocytic vesicles are either sorted to late multivesicular bodies (MVBs) or pre-vacuolar compartment (PVC) for targeting to the lytic vacuole or recycled to the plasma membrane. In plants, the vacuolar sorting receptor BP-80 and pre-vacuolar syntaxin SYP21 (PEP12) are recognized as marker genes for PVC (Bowers and Stevens 2005; Rosado et al. 2011), whereas the ARF-GEF GNOM is widely accepted as the marker for putative recycling endosome (Marles et al. 2003).

As evidenced by proliferating research articles and reviews in the last few years, plant endocytosis impacts many critical events during the plant life cycle, including embryo patterning, lateral organ differentiation, root hair formation, hormone signal transduction, and defense responses (Morita et al. 2002; Carter et al. 2004; Voigt et al. 2005; Robatzek et al. 2006; Geldner et al. 2007; Dhonukshe et al. 2008a). Endosomes are essential compartments for protein targeting, recycling, and degradation, and they are of vital importance for processes including auxin signalling and transport (Dhonukshe et al. 2006, 2007a; Sauer and Kleine-Vehn 2011), BR signalling (Geldner et al. 2007; Robert et al. 2008), boron uptake (Takano et al. 2008), blue light responses (Christie 2007), and plant immunity (Robatzek et al. 2006; Sharfman et al. 2011) among others.

Among the well-characterized roles of endosomes, the most widely studied one is the maintenance of polar localization of auxin transporters, namely the PIN proteins. Unlike their animal counterparts, plant cells have no tight junction-like barriers to restrict lateral diffusion between the apical and basolateral membrane domains. Constitutive endocytosis functions as a powerful mechanism to counteract lateral diffusion of PIN proteins within the plasma membrane, thus the maintenance of polar distribution. The establishment of polar localization of PIN proteins consists of three steps: non-polar secretion, clathrin-dependent endocytosis, and subsequent polar recycling, similar to transcytosis in animal cells (Dhonukshe et al. 2008a, see also Chap. 8 by Nodzynski et al. in this volume). This mechanism ensures the dynamic regulation of PIN polarity in response to developmental and environmental stimulus, for example, in the rapid reorientation of PIN3 in columella cells during gravitropic response (Norambuena et al. 2009; Kleine-Vehn et al. 2010) and switch of PIN polarity during embryogenesis (Benková et al. 2003; Beutler et al. 2009). Mutations of Rab5 or clathrin subunits result in disruption of endocytosis and auxin-related developmental defects (Dhonukshe et al. 2007a, 2008a). Endocytosis also plays a key role in signal transduction processes. Although the brassinosteroid (BR) receptor BRI1

(BRASSINOSTEROID-INSENSITIVE 1) undergoes constitutive endocytosis independent of ligand binding, endosomal localization of this receptor and other endosome components are probably important in signal transduction, as evidenced by the positive effects on BRI1 signalling of fungal toxin, brefeldin A (BFA) (Geldner et al. 2007). Unlike BRI1, endocytosis of flagellin receptor FLAGELLIN-SENSITIVE 2 (FLS2) is induced by ligand (Chinchilla et al. 2007). After internalization, FLS2 is targeted for degradation by ubiquitination which subsequently terminates the signal transduction process (Gohre et al. 2008). So in this case, endocytosis works as a negative feedback of the signalling process. Thus, the dynamic trafficking machinery and balance of cargo localization in the endomembrane system needs to be well maintained for proper development and growth regulation during the plant life cycle.

2 Chemicals Known to Perturb Endomembrane Trafficking

Our knowledge of the mechanisms regulating plant endomembrane trafficking has been enriched with the help of technological advancement. Due to high quality confocal laser scanning microscopy (CLSM) and transmission electron microscopy (TEM), together with the development of histological dyes, contrast-enhancing techniques, and innovative imaging techniques, the ability to image endomembrane structure has been greatly enhanced (Serrano et al. 2007, see also Chap. 1 by Šamajová et al. in this volume). In recent years, the introduction of genetically encoded fluorophores, such as GFP and its mutated isoforms (Chalfie et al. 1994), have further facilitated the observation of subcellular membrane structures. GFP labeling of marker proteins has provided details of the real-time intracellular dynamics within live cells. In the model plant Arabidopsis, the large number of available T-DNA inactivation mutants or ethyl methane sulfonate (EMS)-induced point mutation stocks provide valuable genetic resources to screen for specific genes involved in endomembrane trafficking process. However, only a limited number of marker lines are available, almost all based on the published literature, with which is possible to describe the continuous and complicated endomembrane system. Also, fusions with fluorescent protein could impair normal subcellular localization due to structural alteration. Moreover, T-DNA insertions in many plant endomembrane system genes are lethal since many genes involved in endocytosis are of vital importance during embryo patterning or meristem differentiation (Xu and Scheres 2005; Dhonukshe et al. 2008a), although a few point mutants are viable and exert clear phenotypes (Jaillais et al. 2007). In many cases, no phenotypes are observed in loss-of-function mutants due to redundancies in gene functions. In the case of proteins which may have several functions in a cell, point mutants may exhibit phenotypes only in a particular gene function, which further limits the characterization of gene properties.

Compared to the classic technologies mentioned above, chemical biology which has emerged in the past 5–10 years serves as a complementary approach to illuminate biological functions through the integration of organic chemistry, biology, and bioinformatics, and displays high capacity to overcome important limitations inherent to mutation-based genetic screening (Sharma and Swarup 2003; Aikawa et al. 2006; Robert et al. 2009). Several terms of chemical biology have been defined according to application scope. While chemical genetics makes use of small molecules to identify protein targets through the induction and modulation of phenotypes, the term chemical genomics explores a wider field with the combination of many modern genomics tools, such as high-throughput screening approaches, sequence and expression databases, and next-generation sequencing for the identification of cognate targets (Hicks and Raikhel 2012). Chemical genomics displays advantages over mutagenesis screening through the easy handling of compounds and rapid screening methods. Dosage dependence and reversibility provide the means to circumvent problems of lethality and redundancy (Drakakaki et al. 2009; Hicks and Raikhel 2009; Norambuena et al. 2009). Chemical genomics also offers the ideal ability to study dynamic endomembrane trafficking processes in real time which often is within minutes (Dhonukshe et al. 2007a). This can be achieved by the observing interference of GFP-tagged protein trafficking through endomembrane compartments via confocal microscopy (Drakakaki et al. 2011; Hicks and Raikhel 2012). High-throughput chemical primary screening can also expand the capacity of chemical genomics by developing automated microscopy. In this manner, endomembrane phenotypes of large EMS populations could be scored instead of using time-consuming manual confocal microscopy. These developments may further facilitate the systematic understanding of the endomembrane trafficking machinery (Hicks and Raikhel 2012).

Even before the advent of chemical screening approaches in plants, a variety of inhibitors including BFA and wortmannin were used intensively to define endosomal compartments and illustrated corresponding biological functions. The fungal toxin BFA inhibits ADP ribosylation factor (ARF) GTPases by interacting with their associated guanine nucleotide exchange factors (GEFs) (Dambournet et al. 2011). In Arabidopsis, the cognate in vivo target of BFA includes the BFA-sensitive ARF-GEF, GNOM, which is a key component involved in cargo recycling (Geldner et al. 2003). BFA has been variously described as an inhibitor of secretion (Nebenfuhr 2002) and endocytosis (Baluška et al. 2002; Geldner 2004) and has been used to dissect trafficking pathways between different endomembrane compartments (Geldner et al. 2003; Grebe et al. 2003; Kleine-Vehn et al. 2008). Another inhibitor Tyrphostin A23, but not Tyrphostin A51, interferes with the recognition of the YxxΦ internalization motif in the cytosolic domain of plasma membrane receptors by the μ-adaptin of the AP-2 adaptor complex (Titapiwatanakun et al. 2009; Dambournet et al. 2011) and is widely used as a competitive inhibitor of receptor Tyr kinases in mammalian cells due to its structural similarity to Tyr (Aniento and Robinson 2005). In plants, tyrphostin A23 has been reported to inhibit endocytosis in protoplasts (Ortiz-Zapater et al. 2006), prevent the accumulation of PIN2 in BFA compartments in Arabidopsis roots (Dhonukshe et al. 2007a), and interfere with cytokinesis and cell plate formation

(Reichardt et al. 2007), indicating that both endocytosis and Golgi-based secretion are targets of the inhibitor. Other inhibitors, such as the phosphatidylinositol-3/4 kinase (PI-3/4 kinase) inhibitor wortmannin, interfere with protein trafficking to the plant vacuole (Dhonukshe et al. 2008b; Takáč et al. 2012) and with endocytosis (Irani and Russinova 2009). Concanamycin A (Conc A) and Bafilomycin A (BafA) bind to V-ATPase subunits c and a (Huss et al. 2002; Wang et al. 2005) leading to massive alteration of Golgi morphology, aggregation of vesicles and a block of transport from the TGN/EE to the vacuole (Strompen et al. 2005; Dettmer et al. 2006). Other compounds affecting endomembrane trafficking are listed in Table 1. Thus, small molecules play important roles in understanding endomembrane compartmentalization and machinery of endomembrane trafficking.

However, most of the inhibitors mentioned above interfere with multiple membrane trafficking processes that have a broad range of functions. In the case of BFA, which is extensively used as endocytosis inhibitor, the compound targets a wide range of endomembrane trafficking processes including ER-to-Golgi, endosome-to-MVB/PVC, and recycling back to the plasma membrane. BFA compartments are actually composed of remnant Golgi stacks surrounding TGN vesicles, and partially of MVB fragments (Satiat-Jeunemaitre et al. 1996; Merigout et al. 2002). To better elucidate the highly specific trafficking routes supported by different endomembrane compartments, including responses to either environmental stimuli or developmental signals, it is important to develop chemicals showing more selectivity for specific targets or particular functions. A step forward is to generate extensive small molecule collections that cover a wide range of protein targets within networks of interests. This would present substantial and diverse phenotypes for systematic organization and provide plentiful information for further studies. Here, we will introduce several successful examples of chemical screens which target the endomembrane system resulting in new inhibitors of endocytosis and endomembrane trafficking.

3 Chemical Genomic Screens to Dissect Vacuole Transport and Biogenesis

The earliest screen for plant bioactive molecules affecting endomembrane trafficking was focused on vacuole transport and biogenesis. Taking advantage of the evolutionary conservation between plants and the budding yeast *Saccharomyces cerevisiae*, a library of 4,800 diverse compounds was screened for those that induced aberrant secretion of carboxypeptidase Y (CPY) in yeast through a high-throughput 96-well assay. This screen identified 14 compounds, termed sorting inhibitors (sortins), which affected the trafficking of proteins to the vacuole in yeast and would phenocopy the vacuole protein sorting (*vps*) mutant phenotypes (Bowers and Stevens 2005) (Table 1). Application of two of these compounds (sortin 1 and sortin 2) in Arabidopsis seedlings resulted in similar effects on CPY secretion and

Table 1 Summary of compounds affecting endomembrane trafficking

Compounds	Organisms	Target	Mode of action	Source
BFA	Plant/animal	ARF-GEFs	Block exocytosis/endomembrane recycling	Dinter and Berger (1998), Robinson et al. (2008)
Wortmannin	Plant/animal	PI3Ks and PI4Ks	Block vacuolar trafficking, MVBs and endocytosis	Robinson et al. (2008), Takáč et al. (2012)
Tyrphostin A23	Plant/animal	Cargo recognition (YXXΦ) by μ subunit of AP complex	Blocks cargo recruitment into CCVs from PM, TGN	Ortiz-Zaoater et al. (2006), Robinson et al. (2008)
Concanamycin A	Plant/animal	V ATPase	Block trafficking at TGN, endosome acidification	Robinson et al. (2008)
Dynasore	Animal	GTPase activity of dynamin	Blocks endocytosis	Macia et al. (2006)
Sortin 1	Yeast/plant	Unknown	Affects vacuole biogenesis	Zouhar et al. (2004)
Sortin 2	Yeast/plant	Unknown	Interference with ESCRT complex components	Zouhar et al. (2004), Norambuena et al. (2008)
Sortin 3	Yeast	Unknown	Possibly target late endosomal compartments	Chanda et al. (2009)
Endosidin 1	Plant	Unknown (stablizes actin)	Blocks trafficking at the TGN	Robert et al. (2008), Toth et al. (2012)
Endosidin 3	Plant	Unknown	Blocks protein trafficking from the PM by affecting TGN/EE compartment	Drakakaki et al. (2011)
Endosidin 5	Plant/animal	Unknown	Blocks recycling	Drakakaki et al. (2011)
Endosidin 7	Plant	Unknown	Disturbs cell plate formation, cell wall maturation and expansion through the perturbation of secretory pathways	Drakakaki et al. (2011)
Gravicin	Plant	PGP19 and other unknown target	Inhibition of auxin transport and trafficking to the vacuole	Surpin et al. (2005), Rojas-Pierce et al. (2007)
LY294002	Plant/animal	PI3Ks	Blocks MVBs, endocytosis	Lee et al. (2008)

(continued)

Table 1 (continued)

Compounds	Organisms	Target	Mode of action	Source
Bafilomycin A	Plant/animal	V ATPase	Blocks trafficking at the TGN, causes endosome acidification	Robinson et al. (2008)
Sulfonamide 16D10	Animal	V ATPase	Blocks endocytosis and exocytosis, causes endosome acidification	Nieland et al. (2004)
Monensin	Plant/animal	Ionophore	Blocks exocytosis, disrupts trafficking at Golgi	Diner et al. (1998)
Cantharidin	Plant/animal	PP2A-specific inhibitor	Blocks FLS2 endocytic trafficking	Serrano et al. (2007)
Triclosan	Plant	Enoyl-acyl carrier protein (acp) reductases (ENRs)	Blocks FLS2 endocytic trafficking, flg22-induced oxidative burst; inhibitor of fatty-acid synthesis	Serrano et al. (2007)
Fluazinam	Plant	Unknown	Blocks FLS2 endocytic trafficking, flg22-induced oxidative burst	Serrano et al. (2007)
Exo1and Exo2	Animal	Unknown	Block exocytosis, ER to Golgi transport	Nieland et al. (2004)
Cobtorin	Plant	Unknown	Affects cortical microtubule alignment	Yoneda et al. (2007)
Triclosan, Fluazinam and Cantharidin	Plant	PP2A (cantharidin)	Affect flg22-mediated FLS2 endocytosis	Serrano et al. (2007)
Morlin	Plant	Unknown	Affects cytoskeletal organization/ interaction with cellulose synthase	DeBolt et al. (2007)

vacuole biogenesis, and additional defects of root development. This suggests that it is possible to take advantage of simpler single cell eukaryotes such as yeast to dissect endomembrane trafficking pathways in multi-cellular organisms. However, the fact that only two of the 14 sortins were active in a plant system may reflect the difference in chemical uptake, drug targets, and the evolutionary differentiation between plants and yeast. Further studies utilizing electron microscopy of sortin 1 effects showed that unlike other drugs such as BFA, this compound only showed specific interference with vacuole morphology but not broad disruption of other endomembrane organelles such as ER and Golgi (Zouhar et al. 2004). This indeed sets a good example of a chemical screen to identify specific inhibitors. More recent studies using GFP fusion markers support the preliminary screening conclusion that sortin 1 primarily affects vacuole biogenesis (Zouhar et al. 2004).

The follow-up genetic screen for hypersensitive lines in EMS-mutagenized GFP: δ-TIP Arabidopsis collections identified mutants defective in both vacuole biogenesis and accumulation of anthocyanin (Rosado et al. 2011). The defects of anthocyanin accumulation in vacuoles can be mimicked by the known glutathione inhibitor buthioninesulfoximine, and this demonstrates that sortin 1 or its metabolite may serve as a xenobiotic leading to the depletion of glutathione pools by oxidative stress. Based on the previous evidence that anthocyanins are transported to the vacuole as glutathione conjugates via ATP-binding cassette (ABC)-type transporters (Zhao and Dixon 2009) and defective anthocyanin accumulation results in tonoplast vesiculation caused by autophagy (Rosado et al. 2011), this suggests a working model that anthocyanin transport to the vacuole is essential for normal vacuole biogenesis through the combination of oxidative stress and autophagic processes. In this case, chemical genomics provides a tool to discover new pathways linking vacuole trafficking and vacuole biogenesis to secondary metabolism and oxidative response (Rosado et al. 2011; Hicks and Raikhel 2012).

The genetic characterization of sortin 2 was carried out by screening for sortin 2 hypersensitivity in a yeast haploid deletion library. The resulting hypersensitive mutants are mainly defective in the endomembrane system, such as some of the ESCRT complex components. Further bioinformatics analysis and subcellular localization showed higher expression of genes in the most hypersensitive mutants that are annotated as TGN and endosome related. This reflects the essential functions of these compartments for vacuole targeting or secretion (Norambuena et al. 2008). Thus, this approach verified the role of above compartments in the regulation of vesicular trafficking toward vacuole.

4 Chemical Investigation of Development and Defense Responses

Several studies have clearly demonstrated the crucial role of endosomes in several plant processes including cell fate specification, abscisic acid and auxin signalling, tropic responses, and pathogen defense (Carter et al. 2004; Surpin and Raikhel 2004).

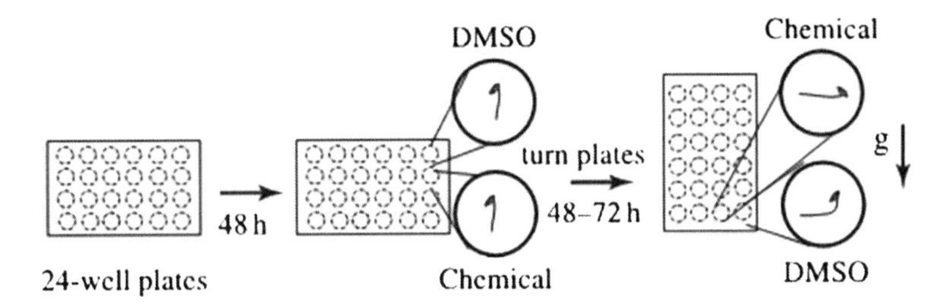

Fig. 2 Screen for chemicals that affect gravitropism. The chemical library was screened in a 24-well format, and seedlings were scored for gravitropic response of both roots and hypocotyls after reorientation. Chemicals dissolved in 20 % DMSO were added to wells. Control wells contained an equivalent concentration of the solvent. The gravity vector (g) is indicated by an *arrow* (*on the right*)

Classical mutant screens have shown that the endomembrane system is intimately involved in auxin-dependent responses such as gravitropism. Mutants in genes encoding SNARE (*S*oluble *N*-ethylmaleimide-sensitive *a*daptor protein *re*ceptor) proteins are both agravitropic and defective in vacuolar morphology (Morita et al. 2002; Yano et al. 2003). Furthermore, polar auxin transport which is crucial for tissue specification and organ formation throughout plant development is highly dependent on endomembrane systems. Recent studies have revealed the constitutive endocytosis and recycling of PIN1 by a clathrin-dependent mechanism (Dhonukshe et al. 2007b; Robert et al. 2010; Kitakura et al. 2011). Disruption of the dynamic cycling between the plasma membrane and endosomal compartments via either pharmacological treatment or interference with gene function results in loss of proper PIN polarity and disruption of normal developmental processes (Jaillais et al. 2007; Kleine-Vehn et al. 2010).

A screen based upon the gravitropic response of *Arabidopsis thaliana* roots and hypocotyls was performed to identify chemical effectors. A chemical library (ChemBridgeDIVERSet) was screened in a 24-well format, and seedlings were scored for both root and hypocotyl responses after reorientation (Fig. 2). The secondary screen was performed under the same conditions as the primary screen, but at variable compound concentrations. The impact on the morphology and targeting to endomembrane compartments was tested using GFP fusion markers, so the effect of these compounds could be used to examine the link between gravitropism and endomembrane system morphology (Norambuena et al. 2009; Robert et al. 2009). The semi-high-throughput primary screen with 10, 000 diverse compounds led to the identification of four compounds affecting vacuolar targeting and morphology and shoot gravitropism (Surpin et al. 2005). One of the screens resulted in the identification of gravicin as a strong inhibitor of gravitropism via disruption of auxin signalling as well as inhibition of δ-TIP delivery to the tonoplast (Table 1). Subsequent screens for resistant and hypersensitive mutants to gravicin resulted in the identification of the multi-drug transporter PGP19 as the cognate target for the inhibition of gravitropism but, interestingly, not for tonoplast mistargeting (Rojas-Pierce et al. 2007).

PGP 19 is an ATP-binding cassette protein that functions as an auxin transporter and participates in the regulation of gravitropic response through interaction with PIN1 (Noh et al. 2001; Titapiwatanakun et al. 2009). However, a second unknown target is responsible for the mistargeting of δ-TIP from the tonoplast to an ER-like compartment. The identification of PGP19 as the gravicin target showed the clear advantage of chemical genomics to dissect links between different pathways. However, the observation that gravicin displays distinct gravitropic and tonoplast targeting defects indicates the potential for multiple targets when dealing with small molecules, possibly due to in vivo metabolism of these compounds resulting in multiple active products and leading to the challenge of defining these active forms (Rojas-Pierce et al. 2007).

The induction of pathogen immunity in plants involves several pathways, including pathogen-associated molecular patterns (PAMP) pathways and the resistance (R-mediated) pathway (Chisholm et al. 2006; Jones and Dangl 2006; Bernoux et al. 2011). In Arabidopsis, the LRR-RLK receptor FLAGELLIN-SENSITIVE 2 (FLS2) is an established receptor for bacterial flagellin (flg22). Using GFP-tagged FLS2, it has been shown that plasma membrane localized FLS2 undergoes endocytosis upon ligand induction. A potentially phosphorylated threonine residue in the juxta membrane region of the receptor is required for both signalling and endocytosis. After internalization, GFP-FLS2 accumulates in intracellular vesicles and is later targeted for degradation, which subsequently attenuates the signalling process (Robatzek et al. 2006). Recently, a chemical screen using a library composed of 120 small molecules with known biological activities was performed to identify candidate compounds which could modify elicitor-responsive gene expression. Seven-day-old Arabidopsis seedlings were used in the screen by growing them in submerged culture in 96-well microtiter plates. In this way it was easy to monitor the transcriptional activation of GUS fused to early PAMP-responsive genes (Serrano et al. 2007). Four compounds were identified that inhibit GUS reporter activity. Among them, three compounds clearly affected flg22-mediated FLS2 endocytosis (Table 1), and two of the three chemicals, triclosan and fluazinam, also strongly inhibited the production of flg22-stimulated reactive oxygen species (Serrano et al. 2007). Thus, the identified chemicals may decrease elicitor-responsive gene induction by interfering with endocytic processes of PAMP receptors at the cell periphery. The chemical cantharidin was identified in the screen, and it induced sustained rather than transient accumulation of reactive oxygen species upon flg22 treatment and inhibition of FLS2 endocytosis (Serrano et al. 2007). These compounds can be used for further investigation on the role of FLS2 endocytosis in downstream signalling and elicitor-mediated immune response.

5 Identifying New Inhibitors of Endosomes Involved in Polar Cell Growth

Polarized tip growth of pollen tubes and root hairs involves targeted transport and vesicle fusion which is necessary to form and maintain a tubular structure. This involves processes including calcium gradient-dependent cytoskeleton organization, vesicle targeting, and recycling at the apex (Drakakaki et al. 2009; Robert et al. 2009). Pollen is an ideal system for chemical screens based on the fact that endomembrane components in large tobacco pollen tubes are relatively easy to view by confocal microscopy, and the organization of endocytic and exocytic vesicles are clearly visible within the growing tip by membrane-staining styryl dyes such as FM4-64 and FM1-43 (Zonia and Munnik 2008; Robert et al. 2009). Chemical genomics is especially valuable for pollen biology because chemical phenotypes are dose dependent, reversible, and can be applied in time course experiments, whereas genetic approaches using mutants defective in endomembrane trafficking in Arabidopsis pollen typically result in gametophyte lethality. A high-throughput pollen chemical screen was recently developed using a 384-well microtiter plate format (Robert et al. 2008). More than 2,000 compounds from a collection of known bioactive molecules were scored for inhibition of tobacco pollen germination in the primary screen. The identified germination inhibitors in the first screen represented less than 1 % of the componds, and could possibly affect exocytosis, endocytosis, cytoskeleton, calcium signalling, or protein targeting, among other factors. These chemicals were subsequently scored in a secondary screen for the perturbation of a plasma membrane-associated tip-focused marker GFP-ROP1-interacting partner 1 (RIP1 or ICR1 (Lavy et al. 2007)) for polarity loss phenotypes such as mislocalization and isodiametric growth. From the initial screen of 2016 chemicals, four resulted in both pollen germination inhibition and mislocalization of GFP-RIP1.

One of these bioactive compounds termed endosidin1 (ES1) was further investigated in detail in Arabidopsis roots with available markers which are known to recycle between the plasma membrane and endomembrane compartments (Robert et al. 2008). Two-hour treatment with ES1 resulted in specific effects on the auxin transporters PIN2-GFP (Xu and Scheres 2005) and AUX1-YFP (Swarup and Bennett 2003), and the brassinosteroid (BR) receptor BRI1-GFP (Geldner et al. 2007). In each case, intracellular aggregates termed "endosidin bodies" were formed. However, other plasma membrane markers such as PIN1-GFP and PIN7-GFP were unaffected by ES1 treatment, indicating target specificity and selectivity for endocytic pathways of diverse plasma membrane proteins (Robert et al. 2008, 2009). Differential sensitivity of plasma membrane proteins to ES1 may reflect different endocytic/recycling routes during cargo transport which is consistent with a previous study using BFA (Kleine-Vehn et al. 2008) and suggesting that AUX1, PIN2 and PIN1, PIN7 are trafficking via different pathways (Robert et al. 2008). ES1 antagonizes BR signalling as evidenced by the inhibitory effect on a BR-specific transcriptional factor when examined by quantitative PCR. Seedlings grown in the dark in the presence of ES1 display a light-grown phenotype, similar to the BR receptor mutant *bri1-1* (Li and Chory 1997). However, in the presence of BFA, BR signalling was increased due to the increased pool of BRI1 in endosomes and other compartments, such as Golgi. Thus, ES1 appears to display action that is more specific than BFA for endomembrane compartments.

Further analysis via the examination of different endomembrane markers revealed that ES1 specifically incorporated TGN/EE compartments into endomembrane bodies defined by markers for the syntaxin SYP61 and the ATPase subunit VHA-a1, but had no effect on other markers for endosomal compartments. This indicates that the major intracellular action of ES1 was achieved through the impact on early endosomes. Although PIN2 formed agglomerates during treatments, ES1 had no effect on PIN2 polarity in the epidermis and cortex, suggesting a different mechanism of asymmetric localization between root cells organized in cell files and pollen tube growth polarity. Consistent with this phenomenon, ES1 did not induce an auxin-specific reporter nor inhibit gravitropism, which depends on PIN asymmetry. This demonstrates that endocytosis and plasma membrane polar localization mechanisms may be uncoupled by ES1. Given that PIN1, PIN2, and BRI1 are sorted through a SNX1 endosomal compartment after internalization and ES1 had no effect on GNOM endosomal compartment essential for PIN1 recycling, this revealed the specification of early endocytic compartments for different plasma membrane proteins prior to trafficking through the shared common steps in late endosomes (Robert et al. 2008; Hicks and Raikhel 2012). Thus, the identification of SYP61/VHA-a1 as an EE compartment highlights the power of bioactive chemicals to dissect the dynamic endomembrane trafficking system and allow us to investigate endocytosis, recycling, and degradation pathways in more detail. It further provides a novel inhibitor of endocytosis through the SYP61/VHA1 defined endosomal compartment.

Although the target of ES1 has not been reported, recent evidence indicates that ES1 halts the movement of FLS2-GFP endosomes after flg22 induction. This coincides with the increased bundling of actin filaments (M. Beck and S. Robatzek, unpublished data), suggesting that ES1 may target the cytoskeleton (Hicks and Raikhel 2012). A recent chemical screen for circadian clock effectors in *Arabidopsis thialiana* identified ES1 as the stimulator of shortened rhythm periodicity. Although there is no direct evidence to link ES1 action with rhythm regulation, Toth and coworkers show that ES1 primarily affects actin filament dynamics. This may later result in reduced severing and depolymerization of actin filaments in both Arabidopsis seedlings and mammalian cells. This stabilization of the actin cytoskeleton is the likely cause of changes in vesicular trafficking (Toth et al. 2012). Interestingly, ES1 does not act in the same manner as known actin drugs binding G-actin (e.g. latrunculin B), suggesting that it may rather target an actin-binding protein (ABP). Such an ABP could link cytoskeleton to specific vesicle transport. ES1 presents a clear pharmacological advantage linking endomembrane trafficking and cytoskeleton dynamics with developmental regulation and response to environmental stimulus. In this respect, a recent proteomic screen using BFA already identified the ABP profilin 2 as a crucial cytoskeletal protein regulating vesicular trafficking in Arabidopsis (Takáč et al. 2011).

6 Systematic Approach to Understand the Endomembrane Network

To fully understand the dynamic endomembrane trafficking network and obtain novel effectors of endocytosis, chemical screens need to be combined with real-time microscopy which is essential to observe the effects of compounds on markers rapidly and directly. A step forward is to generate extensive small molecule collections that cover a wide range of protein targets, so the scaled-up chemical genomics would provide systematic information through the evaluation of subcellular phenotypes based on multiple endomembrane-specific markers (Hicks and Raikhel 2010, 2012).

In this regard, Drakakaki et al. (2011) developed a modified high-throughput confocal-based screen focused on the rapid cycling of plasma membrane markers to discover molecules that target endomembrane trafficking in vivo in a complex eukaryote. Compared with the previous semi-high-throughput screen performed by the same group, the modified screening expanded the scale of chemicals to 46,418 compounds in the primary screen. Based on the germination and growth of tobacco pollen, 360 (0.77 %) inhibitors of pollen germination were identified in the first screen by using a 384-well format and automated microscopy (Drakakaki et al. 2011). By using CLSM, the molecules were then used to challenge model cargoes represented by polar plasma membrane markers PIN2::PIN1:GFP, PIN2::-PIN2:GFP, and non-polar receptor BRI1::BRI1:GFP expressed in Arabidopsis root tips. The secondary screen resulted in about 7,500 CLSM images representing the marker behaviors of 18,570 seedlings. About 123 small molecules (34 % out of 360) were identified in the secondary screen and these were further analyzed against plasma membrane, endosome/TGN, and other endomembrane markers. To evaluate and characterize the intracellular effects of these molecules, a scoring system of chemically induced subcellular phenotypes was developed by using hierarchical clustering that treats data points with single and multiple phenotypes. This clustering grouped structurally diverse compounds into 18 clusters, of which 15 classes represented distinct subcellular phenotypes induced by chemical treatment. This included endomembrane bodies of various qualitative size categories, morphology, aberrant localization of plasma membrane or endosomal markers, changes in the polarity of PIN proteins, altered vacuole morphology, and cell plate defects, among others. A color-coded map incorporating multiple subcellular markers treated at different time points was also developed in a non-biased manner by computational methods (Fig. 3). The resulting clustering of PIN2::-PIN1:GFP and PIN2::PIN2:GFP in different groups at 24 h of treatment supported previous hypotheses that apical, basal, and basolateral targeting of plasma membrane proteins are maintained through distinct endomembrane pathways.

Further characterization of the molecules identified additional endosidins (Table 1), other than ES1. Endosidins are defined as bioactive molecules affecting trafficking or the morphology of endomembrane compartments. Among them, ES3 is a potential inhibitor of protein trafficking from the plasma membrane by

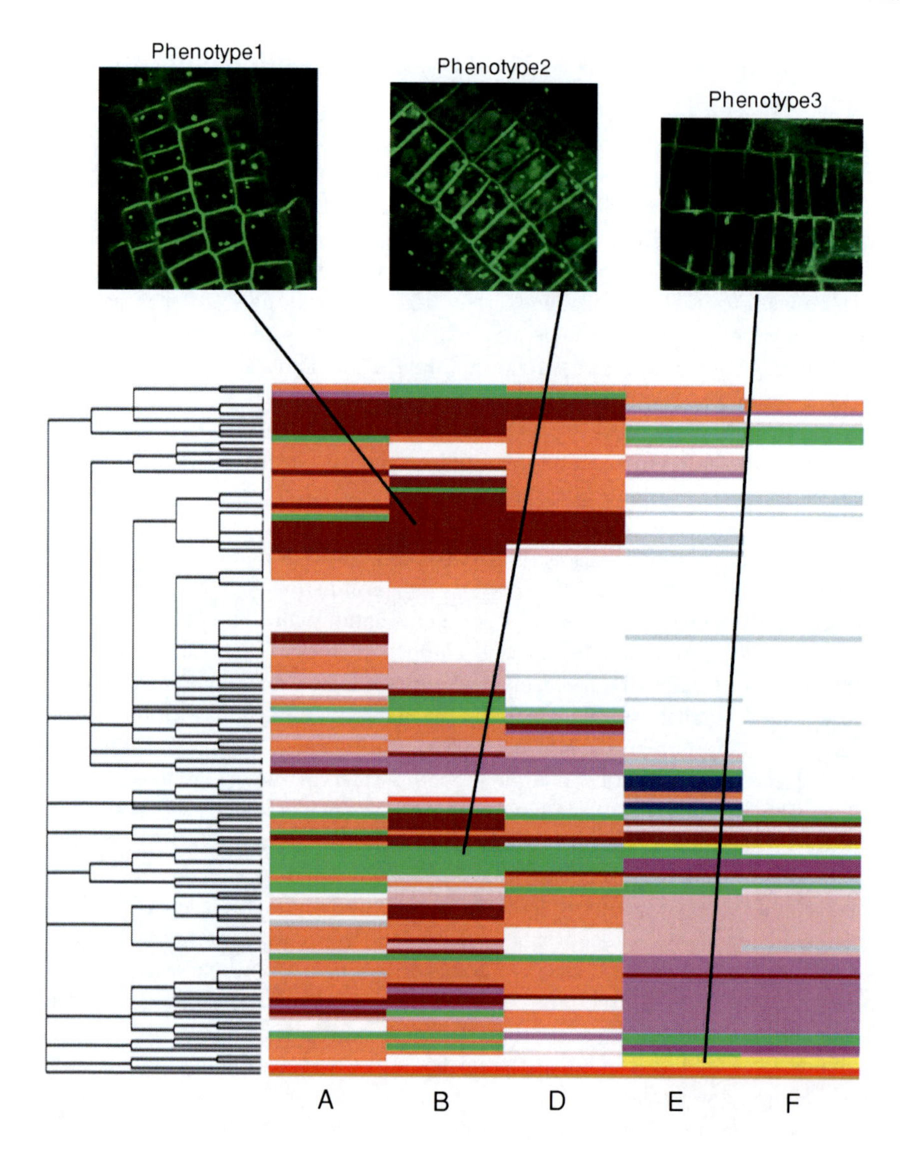

Fig. 3 An illustration of phenotypes and output from a high-content intracellular small molecule screen. The *colored map* presents the results from screening different molecules (*dendrogram at left*) on a collection of Arabidopsis lines expressing different fluorescent protein markers in roots (A–F). *Color coding* represents distinct phenotypes observed upon chemical treatment. In this display, a single color-coded phenotype can be produced within a single marker line or across multiple marker lines, which leads to clustering of bioactive compounds based on phenotype. *Dashed lines* point to clusters represented by phenotypes 1, 2, or 3. Within each cluster, more in-depth structural analysis defines molecules of similar or dissimilar structure. Thus, this approach is also potentially predictive of the activity of uncharacterized molecules. Marker lines in this example can also be screened after different time of treatment or following other treatments, such as stress or pathogen challenge. Figure modified from Hicks and Raikhel (2012) with permission

affecting TGN/EE compartments without effecting Golgi or late endosome/ prevacuolar compartments. One interesting thing is that ES3 can also distinguish between two antagonist ROP-signalling pathways necessary to establish polar growth via the auxin-responsive remodeling of cytoskeleton, possibly through the affection of lipid modification. ES5 induces a strong accumulation of plasma membrane cargoes in the vacuole and also affects gravitropism in a dose-dependent manner. The further investigation of ES5 function in Hela cells demonstrated that ES5 targeted recycling in both plant and animal cells, providing insights into recycling mechanisms across kingdoms. ES7 displays a capacity to disturb cell plate formation, cell wall maturation, and expansion through the perturbation of secretory pathways. This demonstrated the ability of small molecules to dissect specificity of endomembrane system in late cytokinesis (Drakakaki et al. 2011).

Taken together, the high-content intracellular screening paralleled with evaluation and elegant clustering of complex intracellular phenotypes visualized by one or multiple endomembrane-specific molecular markers permits a comprehensive view of the endomembrane network. Also, the database of phenotypic clusters presents a good resource for future investigations. Furthermore, all three ES compounds mentioned above linked intracellular trafficking to certain developmental or physiological phenotypes, which would be a challenge for forward genetic screens. Phenotypic data combined with structural databases can be used for the design of new molecules. The common effect of certain chemicals between plants and other organisms also provides the potential to use plant-based systems for drug discovery. Critically, through the assembly of clustered phenotypic data from intracellular dynamics to whole seedlings, it is possible to investigate the network of endomembrane trafficking, developmental regulation and response to environmental stimulus (Drakakaki et al. 2011; Hicks and Raikhel 2012).

7 Finding New Effectors and Pathways

Plant researchers in traditional genetics tend to prefer the one-to-one principle that chemicals should display strong activity against a single cognate target. Although this will be helpful in subsequent genetic screens to identify resistance genes, it does not necessarily block the value of compounds with well-characterized cellular effects. In the case of gravicin, although the identified resistant gene PGP19 is responsible for one of the compound effects, inhibition of gravitropism, whereas the other target for vacuole trafficking defects is unknown (Rojas-Pierce et al. 2007), it still provides a good platform for the investigation of intracellular trafficking and physiological response. Another example is the well-characterized fungal toxin BFA (Satiat-Jeunemaitre et al. 1996). The best known target of this chemical in plants is the SEC7 family member ARF/GEF (ADP-ribosylation factor/GDP/GTP exchange factor) protein GNOM (Geldner et al. 2003). Treatment with BFA results in the agglomeration of TGN/EE and part of the Golgi stacks by the disruption of Golgi-based secretion and indirect interference with endocytosis and exocytosis (Nebenfuhr et al. 2002; Ritzenthaler et al. 2002). However, BFA is still one of the most widely used chemical inhibitors and has contributed greatly to the research literature, for example in elucidating differential sensitivities of endosomal vesicles and distinguishing between different endomembrane trafficking routes. Other cases such as ES1, ES3, ES5, and ES7, as mentioned in the above sections, have already drawn wide interest for their specific effects on particular intracellular trafficking pathways even before characterization of the cognate targets. By their direct application, these newly identified compounds provide excellent tools by being rapid, reversible, and conditional for the study of highly dynamic processes in endomembrane trafficking (Drakakaki et al. 2011).

The other challenge of chemicals in plant cell biology is target identification since targets are essential to fully understand functionality. Forward genetic screens for resistant or hypersensitive mutants are the most widely used methods for target identification after primary or secondary chemical screening. Structure–activity relationship (SAR) studies are often helpful in selecting related molecules which might be a better choice in subsequent genetic screens (Peddibhotla et al. 2007; Rojas-Pierce et al. 2007). Identification of core structures of chemicals by SAR studies also facilitates the design of biotin-tagged derivatives in subsequent affinity purification (Khersonsky et al. 2003; Wang et al. 2004). Affinity chromatography is a primary approach in the animal field and has gained some success in plant biology as well, such as the identification of the terfestatin A target (Hayashi et al. 2008). Other strategies have been developed, although most of them are only applied in mammalian systems, such as the improved affinity chromatography-tagged library, yeast-three-hybrid (Y3H), phage display, protein microarray, and transcriptome analysis (Toth and van der Hoorn 2010). Although EMS-based mutagenesis is still the most prevalent strategy in plant chemical genomics, these established methods will eventually enable plant scientists to develop more advanced technologies which are suitable for application in plants. The combination of chemical screening, proteomics analysis, next-generation sequencing, and mapping will also facilitate target identification and expand the range of chemical screening to wider application scope (Hicks and Raikhel 2009; Schneeberger et al. 2009).

In the future, to fully capitalize on the potential of chemical biology to uncover new pathways and networks in endomembrane trafficking, the development of more advanced techniques which serve automated plate preparation, seed plating, and imaging to effectively screen large, diverse chemical library will be required. Hardware and software for image-based intracellular phenotypes will simultaneously collect large sets of image and video data. It will be necessary to develop automatic image analysis tools to correlate intracellular bodies, vesicle movement, and signal intensity with different chemical treatments. An important goal is to shift from a qualitative description to quantification of vesicles and organelles upon automatic real-time tracking rather than tedious manual measurements of images (Agee and Carter 2009). An example of this approach is the recent work in zebrafish, in which neuroactive small molecules were screened by automatic record of 14 behavioral responses (Kokel et al. 2010). However, most of the commercial tools are designed for mammalian drug screening in which the cells are relatively two dimensional (Kvilekval et al. 2010), whereas plant tissues with multiple cell layers and cell types present a much greater challenge for instrumentation and computational analysis. However, recent work from the Robatzek laboratory sets a good example of automated image capture and quantification in leaf pavement cells, inspiring scientists to devote more effort on designing such tools in plant field (Salomon et al. 2010). The Raikhel group at UC Riverside is developing image processing methods to facilitate the automatic measurement and quantification of vesicle movement during pollen germination (Ung et al. 2012).

With a large array of data collected from screening, it is necessary to design websites to assist with the structure-based hierarchical clustering of molecules. The well-established web-based systems are ChemMine (Girke et al. 2005; Backman et al. 2011) (http://chemmine.ucr.edu/) and ChemMineR (Cao et al. 2008) (http://manuals.bioinformatics.ucr.edu/home/chemminer; http://cobra20.fhcrc.org/packages/release/bioc/html/ChennineR.html). The ChemMine portal allows users to query compound information by chemical property, structure, and activity and also allows the release of efficiently managed phenotype database from different screening projects. The most important benefit of the website is that it is an open database that allows scientists to upload their collected data and share their results with the community. As with the example of ChemMine, housing and sharing common databases is of vital importance in chemical genomics. This will permit non-expert laboratories to bypass some of the most laborious initial screening necessary for the discovery of chemicals with bioactive functions. ChemMine is housed at UC Riverside's Center for Plant Cell Biology (http://cepceb.ucr.edu). Some other sites located in Europe include the Department of Plant Systems Biology at the University of Ghent (http://www.psb.ugent.be/roots-projects/157-chemical-genetics). This worldwide contribution and arrangement will facilitate the better utilization of chemical biology by the plant community and provide broad dissemination of information and tools such as endomembrane trafficking inhibitors.

Overall, with the development of sophisticated microscope technology, automatic software analysis, systematic data assembly, and community-based databases, chemical genomics will stand in a more powerful position in the research of endomembrane trafficking in plant cell biology.

References

Agee A, Carter D (2009) Whole-organism screening: plants. Methods Mol Biol 486:77–95

Aikawa Y, Xia X, Martin TF (2006) SNAP25, but not syntaxin 1A, recycles via an ARF6-regulated pathway in neuroendocrine cells. Mol Biol Cell 17:711–722

Aniento F, Robinson DG (2005) Testing for endocytosis in plants. Protoplasma 226:3–11

Backman TW, Cao Y, Girke T (2011) ChemMine tools: an online service for analyzing and clustering small molecules. Nucleic Acids Res 39:486–491

Baluška F, Hlavačka A, Šamaj J, Palme K, Robinson DG, Matoh T, McCurdy DW, Menzel D, Volkmann D (2002) F-actin-dependent endocytosis of cell wall pectins in meristematic root cells. Insights from brefeldin A-induced compartments. Plant Physiol 130:422–431

Baluška F, Šamaj J, Hlavačka A, Kendrick-Jones J, Volkmann D (2004) Actin-dependent fluid-phase endocytosis in inner cortex cells of maize root apices. J Exp Bot 55:463–473

Bassham DC, Brandizzi F, Otegui MS, Sanderfoot AA (2008) The secretory system of Arabidopsis. Arabidopsis Book/Am Soc Plant Biol 6:e0116

Benkova E, Michniewicz M, Sauer M, Teichmann T, Seifertova D, Jurgens G, Friml J (2003) Local, efflux-dependent auxin gradients as a common module for plant organ formation. Cell 115:591–602

Bernoux M, Ellis JG, Dodds PN (2011) New insights in plant immunity signaling activation. Curr Opin Plant Biol 14:512–518

Beutler JA, Kang MI, Robert F, Clement JA, Pelletier J, Colburn NH, McKee TC, Goncharova E, McMahon JB, Henrich CJ (2009) Quassinoid inhibition of AP-1 function does not correlate with cytotoxicity or protein synthesis inhibition. J Nat Prod 72:503–506

Bowers K, Stevens TH (2005) Protein transport from the late Golgi to the vacuole in the yeast *Saccharomyces cerevisiae*. Biochim Biophys Acta 1744:438–454

Cao Y, Charisi A, Cheng LC, Jiang T, Girke T (2008) ChemmineR: a compound mining framework for R. Bioinformatics 24:1733–1734

Carter CJ, Bednarek SY, Raikhel NV (2004) Membrane trafficking in plants: new discoveries and approaches. Curr Opin Plant Biol 7:701–707

Chalfie M, Tu Y, Euskirchen G, Ward WW, Prasher DC (1994) Green fluorescent protein as a marker for gene expression. Science 263:802–805

Chanda A, Roze LV, Kang S, Artymovich KA, Hicks GR, Raikhel NV, Calvo AM, Linz JE (2009) A key role for vesicles in fungal secondary metabolism. Proc Natl Acad Sci U S A 106:19533–19538

Chinchilla D, Zipfel C, Robatzek S, Kemmerling B, Nurnberger T, Jones JD, Felix G, Boller T (2007) A flagellin-induced complex of the receptor FLS2 and BAK1 initiates plant defence. Nature 448:497–500

Chisholm ST, Coaker G, Day B, Staskawicz BJ (2006) Host-microbe interactions: shaping the evolution of the plant immune response. Cell 124:803–814

Christie JM (2007) Phototropin blue-light receptors. Annu Rev Plant Biol 58:21–45

Cram WJ (1980) Pinocytosis in Plant. New Phytol 84:1–17

Dambournet D, Machicoane M, Chesneau L, Sachse M, Rocancourt M, El Marjou A, Formstecher E, Salomon R, Goud B, Echard A (2011) Rab35 GTPase and OCRL phosphatase remodel lipids and F-actin for successful cytokinesis. Nat Cell Biol 13:981–988

DeBolt S, Gutierrez R, Ehrhardt DW, Melo CV, Ross L, Cutler SR, Somerville C, Bonetta D (2007) Morlin, an inhibitor of cortical microtubule dynamics and cellulose synthase movement. Proc Natl Acad Sci U S A 104:5854–5859

Dettmer J, Hong-Hermesdorf A, Stierhof YD, Schumacher K (2006) Vacuolar H+-ATPase activity is required for endocytic and secretory trafficking in Arabidopsis. Plant Cell 18:715–730

Dhonukshe P, Šamaj J, Baluška F, Friml J (2007a) A unifying new model of cytokinesis for the dividing plant and animal cells. BioEssays 29:371–381

Dhonukshe P, Baluška F, Schlicht M, Hlavačka A, Šamaj J, Friml J, Gadella TW Jr (2006) Endocytosis of cell surface material mediates cell plate formation during plant cytokinesis. Dev Cell 10:137–150

Dhonukshe P, Aniento F, Hwang I, Robinson DG, Mravec J, Stierhof YD, Friml J (2007b) Clathrin-mediated constitutive endocytosis of PIN auxin efflux carriers in Arabidopsis. Current Biol: CB 17:520–527

Dhonukshe P, Tanaka H, Goh T, Ebine K, Mahonen AP, Prasad K, Blilou I, Geldner N, Xu J, Uemura T, Chory J, Ueda T, Nakano A, Scheres B, Friml J (2008a) Generation of cell polarity in plants links endocytosis, auxin distribution and cell fate decisions. Nature 456:962–966

Dhonukshe P, Grigoriev I, Fischer R, Tominaga M, Robinson DG, Hasek J, Paciorek T, Petrášek J, Seifertova D, Tejos R, Meisel LA, Zazimalova E, Gadella TW Jr, Stierhof YD, Ueda T, Oiwa K, Akhmanova A, Brock R, Spang A, Friml J (2008b) Auxin transport inhibitors impair vesicle motility and actin cytoskeleton dynamics in diverse eukaryotes. Proc Natl Acad Sci U S A 105:4489–4494

Dinter A, Berger EG (1998) Golgi-disturbing agents. Histochem Cell Biol 109:571–590

Drakakaki G, Robert S, Raikhel NV, Hicks GR (2009) Chemical dissection of endosomal pathways. Plant Signal Behav 4:57–62

Drakakaki G, Robert S, Szatmari AM, Brown MQ, Nagawa S, Van Damme D, Leonard M, Yang Z, Girke T, Schmid SL, Russinova E, Friml J, Raikhel NV, Hicks GR (2011) Clusters of bioactive compounds target dynamic endomembrane networks in vivo. Proc Natl Acad Sci USA 108:17850–17855

Geldner N (2004) The plant endosomal system—its structure and role in signal transduction and plant development. Planta 219:547–560

Geldner N, Friml J, Stierhof YD, Jurgens G, Palme K (2001) Auxin transport inhibitors block PIN1 cycling and vesicle trafficking. Nature 413:425–428

Geldner N, Hyman DL, Wang X, Schumacher K, Chory J (2007) Endosomal signaling of plant steroid receptor kinase BRI1. Genes Dev 21:1598–1602

Geldner N, Anders N, Wolters H, Keicher J, Kornberger W, Muller P, Delbarre A, Ueda T, Nakano A, Jurgens G (2003) The Arabidopsis GNOM ARF-GEF mediates endosomal recycling, auxin transport, and auxin-dependent plant growth. Cell 112:219–230

Girke T, Cheng LC, Raikhel N (2005) ChemMine. A compound mining database for chemical genomics. Plant Physiol 138:573–577

Gohre V, Spallek T, Haweker H, Mersmann S, Mentzel T, Boller T, de Torres M, Mansfield JW, Robatzek S (2008) Plant pattern-recognition receptor FLS2 is directed for degradation by the bacterial ubiquitin ligase AvrPtoB. Current Biol: CB 18:1824–1832

Gradmann D, Robison DG (1989) Does turgor prevent endocytosis in plant-cells. Plant Cell Environ 12:151–154

Grebe M, Xu J, Mobius W, Ueda T, Nakano A, Geuze HJ, Rook MB, Scheres B (2003) Arabidopsis sterol endocytosis involves actin-mediated trafficking via ARA6-positive early endosomes. Current Biol: CB 13:1378–1387

Hayashi K, Yamazoe A, Ishibashi Y, Kusaka N, Oono Y, Nozaki H (2008) Active core structure of terfestatin A, a new specific inhibitor of auxin signaling. Bioorg Med Chem 16:5331–5344

Hicks GR, Raikhel NV (2009) Opportunities and challenges in plant chemical biology. Nat Chem Biol 5:268–272

Hicks GR, Raikhel NV (2010) Advances in dissecting endomembrane trafficking with small molecules. Curr Opin Plant Biol 13:706–713

Hicks GR, Raikhel NV (2012) Small molecules present large opportunities in plant biology. Ann Rev Plant Biol 63:261–282

Hillmer S, Depta H, Robinson DG (1986) Conformation of endocytosis in higher-plant protoplasts using lectin-gold conjugates. Eur J Cell Biol 41:142–149

Hurst AC, Meckel T, Tayefeh S, Thiel G, Homann U (2004) Trafficking of the plant potassium inward rectifier KAT1 in guard cell protoplasts of *Vicia faba*. Plant J Cell Mol Biol 37:391–397

Huss M, Ingenhorst G, Konig S, Gassel M, Drose S, Zeeck A, Altendorf K, Wieczorek H (2002) Concanamycin A, the specific inhibitor of V-ATPases, binds to the V(o) subunit c. J Biol Chem 277:40544–40548

Irani NG, Russinova E (2009) Receptor endocytosis and signaling in plants. Curr Opin Plant Biol 12:653–659

Jaillais Y, Santambrogio M, Rozier F, Fobis-Loisy I, Miege C, Gaude T (2007) The retromer protein VPS29 links cell polarity and organ initiation in plants. Cell 130:1057–1070

Jones JD, Dangl JL (2006) The plant immune system. Nature 444:323–329

Jürgens G (2004) Membrane trafficking in plants. Annu Rev Cell Dev Biol 20:481–504

Khersonsky SM, Jung DW, Kang TW, Walsh DP, Moon HS, Jo H, Jacobson EM, Shetty V, Neubert TA, Chang YT (2003) Facilitated forward chemical genetics using a tagged triazine library and zebrafish embryo screening. J Am Chem Soc 125:11804–11805

Kitakura S, Vanneste S, Robert S, Lofke C, Teichmann T, Tanaka H, Friml J (2011) Clathrin mediates endocytosis and polar distribution of PIN auxin transporters in Arabidopsis. Plant Cell 23:1920–1931

Kleine-Vehn J, Ding Z, Jones AR, Tasaka M, Morita MT, Friml J (2010) Gravity-induced PIN transcytosis for polarization of auxin fluxes in gravity-sensing root cells. Proc Natl Acad Sci U S A 107:22344–22349

Kleine-Vehn J, Dhonukshe P, Sauer M, Brewer PB, Wisniewska J, Paciorek T, Benkova E, Friml J (2008) ARF GEF-dependent transcytosis and polar delivery of PIN auxin carriers in Arabidopsis. Current Biol: CB 18:526–531

Kokel D, Bryan J, Laggner C, White R, Cheung CY, Mateus R, Healey D, Kim S, Werdich AA, Haggarty SJ, Macrae CA, Shoichet B, Peterson RT (2010) Rapid behavior-based identification of neuroactive small molecules in the zebrafish. Nat Chem Biol 6:231–237

Kvilekval K, Fedorov D, Obara B, Singh A, Manjunath BS (2010) Bisque: a platform for bioimage analysis and management. Bioinformatics 26:544–552

Lavy M, Bloch D, Hazak O, Gutman I, Poraty L, Sorek N, Sternberg H, Yalovsky S (2007) A novel ROP/RAC effector links cell polarity, root-meristem maintenance, and vesicle trafficking. Current Biol: CB 17:947–952

Lee Y, Bak G, Choi Y, Chuang WI, Cho HT, Lee Y (2008) Roles of phosphatidylinositol 3-kinase in root hair growth. Plant Physiol 147:624–635

Li J, Chory J (1997) A putative leucine-rich repeat receptor kinase involved in brassinosteroid signal transduction. Cell 90:929–938

Low PS, Chandra S (1994) Endocytosis in Plant. Annu Rev Plant Physiol Plant Mol Biol 45:609–631

Macia E, Ehrlich M, Massol R, Boucrot E, Brunner C, Kirchhausen T (2006) Dynasore, a cell-permeable inhibitor of dynamin. Dev Cell 10:839–850

Marles MA, Ray H, Gruber MY (2003) New perspectives on proanthocyanidin biochemistry and molecular regulation. Phytochemistry 64:367–383

Merigout P, Kepes F, Perret AM, Satiat-Jeunemaitre B, Moreau P (2002) Effects of brefeldin A and nordihydroguaiaretic acid on endomembrane dynamics and lipid synthesis in plant cells. FEBS Lett 518:88–92

Morita MT, Kato T, Nagafusa K, Saito C, Ueda T, Nakano A, Tasaka M (2002) Involvement of the vacuoles of the endodermis in the early process of shoot gravitropism in Arabidopsis. Plant Cell 14:47–56

Murphy AS, Bandyopadhyay A, Holstein SE, Peer WA (2005) Endocytotic cycling of PM proteins. Annu Rev Plant Biol 56:221–251

Nebenfuhr A (2002) Vesicle traffic in the endomembrane system: a tale of COPs, Rabs and SNAREs. Curr Opin Plant Biol 5:507–512

Nebenfuhr A, Ritzenthaler C, Robinson DG (2002) Brefeldin A: deciphering an enigmatic inhibitor of secretion. Plant Physiol 130:1102–1108

Nieland TJ, Feng Y, Brown JX, Chuang TD, Buckett PD, Wang J, Xie XS, McGraw TE, Kirchhausen T, Wessling-Resnick M (2004) Chemical genetic screening identifies sulfonamides that raise organellar pH and interfere with membrane traffic. Traffic 5:478–492

Noh B, Murphy AS, Spalding EP (2001) Multidrug resistance-like genes of Arabidopsis required for auxin transport and auxin-mediated development. Plant Cell 13:2441–2454

Norambuena L, Hicks GR, Raikhel NV (2009) The use of chemical genomics to investigate pathways intersecting auxin-dependent responses and endomembrane trafficking in *Arabidopsis thaliana*. Methods Mol Biol 495:133–143

Norambuena L, Zouhar J, Hicks GR, Raikhel NV (2008) Identification of cellular pathways affected by Sortin2, a synthetic compound that affects protein targeting to the vacuole in *Saccharomyces cerevisiae*. BMC Chem Biol 8:1

Ortiz-Zapater E, Soriano-Ortega E, Marcote MJ, Ortiz-Masia D, Aniento F (2006) Trafficking of the human transferrin receptor in plant cells: effects of tyrphostin A23 and brefeldin A. Plant J Cell Mol Biol 48:757–770

Peddibhotla S, Dang Y, Liu JO, Romo D (2007) Simultaneous arming and structure/activity studies of natural products employing O–H insertions: an expedient and versatile strategy for natural products-based chemical genetics. J Am Chem Soc 129:12222–12231

Reichardt I, Stierhof YD, Mayer U, Richter S, Schwarz H, Schumacher K, Jurgens G (2007) Plant cytokinesis requires de novo secretory trafficking but not endocytosis. Current Biol: CB 17:2047–2053

Ritzenthaler C, Nebenfuhr A, Movafeghi A, Stussi-Garaud C, Behnia L, Pimpl P, Staehelin LA, Robinson DG (2002) Reevaluation of the effects of brefeldin A on plant cells using tobacco bright yellow 2 cells expressing Golgi-targeted green fluorescent protein and COPI antisera. Plant Cell 14:237–261

Robatzek S, Chinchilla D, Boller T (2006) Ligand-induced endocytosis of the pattern recognition receptor FLS2 in Arabidopsis. Genes Dev 20:537–542

Robert S, Raikhel NV, Hicks GR (2009) Powerful partners: Arabidopsis and chemical genomics. Arabidopsis Book/Am Soc Plant Biol 7:e0109

Robert S, Chary SN, Drakakaki G, Li S, Yang Z, Raikhel NV, Hicks GR (2008) Endosidin1 defines a compartment involved in endocytosis of the brassinosteroid receptor BRI1 and the auxin transporters PIN2 and AUX1. Proc Natl Acad Sci U S A 105:8464–8469

Robert S, Kleine-Vehn J, Barbez E, Sauer M, Paciorek T, Baster P, Vanneste S, Zhang J, Simon S, Covanova M, Hayashi K, Dhonukshe P, Yang Z, Bednarek SY, Jones AM, Luschnig C, Aniento F, Zazimalova E, Friml J (2010) ABP1 mediates auxin inhibition of clathrin-dependent endocytosis in Arabidopsis. Cell 143:111–121

Robinson DG, Jiang L, Schumacher K (2008) The endosomal system of plants: charting new and familiar territories. Plant Physiol 147:1482–1492

Rojas-Pierce M, Titapiwatanakun B, Sohn EJ, Fang F, Larive CK, Blakeslee J, Cheng Y, Cutler SR, Peer WA, Murphy AS, Raikhel NV (2007) Arabidopsis P-glycoprotein19 participates in the inhibition of gravitropism by gravacin. Chem Biol 14:1366–1376

Rosado A, Hicks GR, Norambuena L, Rogachev I, Meir S, Pourcel L, Zouhar J, Brown MQ, Boirsdore MP, Puckrin RS, Cutler SR, Rojo E, Aharoni A, Raikhel NV (2011) Sortin1-hypersensitive mutants link vacuolar-trafficking defects and flavonoid metabolism in Arabidopsis vegetative tissues. Chem Biol 18:187–197

Šamaj J, Baluška F, Voigt B, Schlicht M, Volkmann D, Menzel D (2004) Endocytosis, actin cytoskeleton, and signaling. Plant Physiol 135:1150–1161

Šamaj J, Read ND, Volkmann D, Menzel D, Baluška F (2005) The endocytic network in plants. Trends Cell Biol 15:425–433

Salomon S, Grunewald D, Stuber K, Schaaf S, MacLean D, Schulze-Lefert P, Robatzek S (2010) High-throughput confocal imaging of intact live tissue enables quantification of membrane trafficking in Arabidopsis. Plant Physiol 154:1096–1104

Satiat-Jeunemaitre B, Cole L, Bourett T, Howard R, Hawes C (1996) Brefeldin A effects in plant and fungal cells: something new about vesicle trafficking? J Microsc 181:162–177

Sauer M, Kleine-Vehn J (2011) Auxin binding protein1: the outsider. Plant Cell 23:2033–2043

Schneeberger K, Ossowski S, Lanz C, Juul T, Petersen AH, Nielsen KL, Jorgensen JE, Weigel D, Andersen SU (2009) SHOREmap: simultaneous mapping and mutation identification by deep sequencing. Nat Methods 6:550–551

Serrano M, Robatzek S, Torres M, Kombrink E, Somssich IE, Robinson M, Schulze-Lefert P (2007) Chemical interference of pathogen-associated molecular pattern-triggered immune responses in Arabidopsis reveals a potential role for fatty-acid synthase type II complex-derived lipid signals. J Biol Chem 282:6803–6811

Sharfman M, Bar M, Ehrlich M, Schuster S, Melech-Bonfil S, Ezer R, Sessa G, Avni A (2011) Endosomal signaling of the tomato leucine-rich repeat receptor-like protein LeEix2. Plant J Cell Mol Biol 68:413–423

Sharma M, Swarup R (2003) The way ahead—the new technology in an old society. Adv Biochem Eng Biotechnol 84:1–48
Strompen G, Dettmer J, Stierhof YD, Schumacher K, Jurgens G, Mayer U (2005) Arabidopsis vacuolar H-ATPase subunit E isoform 1 is required for Golgi organization and vacuole function in embryogenesis. Plant J Cell Mol Biol 41:125–132
Surpin M, Raikhel N (2004) Traffic jams affect plant development and signal transduction. Nat Rev Mol Cell Biol 5:100–109
Surpin M, Rojas-Pierce M, Carter C, Hicks GR, Vasquez J, Raikhel NV (2005) The power of chemical genomics to study the link between endomembrane system components and the gravitropic response. Proc Natl Acad Sci U S A 102:4902–4907
Sutter JU, Sieben C, Hartel A, Eisenach C, Thiel G, Blatt MR (2007) Abscisic acid triggers the endocytosis of the arabidopsis KAT1K+ channel and its recycling to the plasma membrane. Current Biol: CB 17:1396–1402
Swarup R, Bennett M (2003) Auxin transport: the fountain of life in plants? Dev Cell 5:824–826
Takáč T, Pechan T, Richter H, Müller J, Eck C, Böhm N, Obert B, Ren H, Niehaus K, Šamaj J (2011) Proteomics on brefeldin A-treated Arabidopsis roots reveals profilin 2 as a new protein involved in the cross-talk between vesicular trafficking and the actin cytoskeleton. J Proteome Res 10:488–501
Takáč T, Pechan T, Šamajová O, Ovečka M, Richter H, Eck C, Niehaus K, Šamaj J (2012) Wortmannin treatment induces changes in Arabidopsis root proteome and post-Golgi compartments. J Proteome Res 11:3127–3142
Takano J, Miwa K, Fujiwara T (2008) Boron transport mechanisms: collaboration of channels and transporters. Trends Plant Sci 13:451–457
Titapiwatanakun B, Blakeslee JJ, Bandyopadhyay A, Yang H, Mravec J, Sauer M, Cheng Y, Adamec J, Nagashima A, Geisler M, Sakai T, Friml J, Peer WA, Murphy AS (2009) ABCB19/PGP19 stabilises PIN1 in membrane microdomains in Arabidopsis. Plant J Cell Mol Biol 57:27–44
Tonkin CJ, Struck NS, Mullin KA, Stimmler LM, McFadden GI (2006) Evidence for Golgi-independent transport from the early secretory pathway to the plastid in malaria parasites. Mol Microbiol 61:614–630
Toth R, van der Hoorn RA (2010) Emerging principles in plant chemical genetics. Trends Plant Sci 15:81–88
Toth R, Gerding-Reimers C, Deeks MJ, Menninger S, Gallegos RM, Tonaco IA, Hubel K, Hussey PJ, Waldmann H, Coupland G (2012) Prieurianin/endosidin1 is an actin stabilizing small molecule identified from a chemical genetic screen for circadian clock effectors in *Arabidopsis thaliana*. Plant J (in press) doi: 10.1111/j.1365-313X.2012.04991.x
Ung N, Hicks GR, Raikhel NV (2012) An approach to quntify endomembrane dynamics in pollen utilizing bioactive chemicals. Mol Plant (in press)
Voigt B, Timmers AC, Šamaj J, Hlavačka A, Ueda T, Preuss M, Nielsen E, Mathur J, Emans N, Stenmark H, Nakano A, Baluška F, Menzel D (2005) Actin-based motility of endosomes is linked to the polar tip growth of root hairs. Eur J Cell Biol 84:609–621
Wang S, Sim TB, Kim YS, Chang YT (2004) Tools for target identification and validation. Curr Opin Chem Biol 8:371–377
Wang Y, Inoue T, Forgac M (2005) Subunit a of the yeast V-ATPase participates in binding of bafilomycin. J Biol Chem 280:40481–40488
Xu J, Scheres B (2005) Dissection of Arabidopsis ADP-RIBOSYLATION FACTOR 1 function in epidermal cell polarity. Plant Cell 17:525–536
Yano D, Sato M, Saito C, Sato MH, Morita MT, Tasaka M (2003) A SNARE complex containing SGR3/AtVAM3 and ZIG/VTI11 in gravity-sensing cells is important for Arabidopsis shoot gravitropism. Proc Natl Acad Sci U S A 100:8589–8594
Yoneda A, Higaki T, Kutsuna N, Kondo Y, Osada H, Hasezawa S, Matsui M (2007) Chemical genetic screening identifies a novel inhibitor of parallel alignment of cortical microtubules and cellulose microfibrils. Plant Cell Physiol 48:1393–1403

Zhao J, Dixon RA (2009) MATE transporters facilitate vacuolar uptake of epicatechin 3'-O-glucoside for proanthocyanidin biosynthesis in *Medicago truncatula* and Arabidopsis. Plant Cell 21:2323–2340

Zonia L, Munnik T (2008) Vesicle trafficking dynamics and visualization of zones of exocytosis and endocytosis in tobacco pollen tubes. J Exp Bot 59:861–873

Zouhar J, Hicks GR, Raikhel NV (2004) Sorting inhibitors (sortins): chemical compounds to study vacuolar sorting in Arabidopsis. Proc Natl Acad Sci U S A 101:9497–9501

Cell Polarity and Endocytosis

Ricardo Tejos and Jiří Friml

Abstract Multicellular organisms have to generate asymmetries in cells and tissues to create different organs. Moreover, several responses to environmental factors are directional and hence require an equal directional response. In plants, such a challenge is accomplished by a multitude of polarly localized proteins that are involved in embryonic and post-embryonic development and dynamic polar responses to the environment. The phytohormone auxin and its polar cell to cell transport play a key role in several of those events providing a mean to coordinate cell and tissue polarities through regulating the polar localization of plasma membrane localized PIN auxin transporters. In this chapter, we discuss the crosstalk between cell trafficking and polarity as a way to integrate external as well as internal signals into asymmetry generation and directional responses in the context of PIN localizations and auxin-dependent processes.

Keywords Cell polarity · Polarly localized proteins · Auxin · PIN proteins

1 Introduction

A cell or tissue is considered to be polarized if any of its characteristics is orientated or more pronounced along one preferred direction. In plants, cell polarity is essential for many different developmental processes including asymmetric cell division,

R. Tejos · J. Friml
Department of Plant Systems Biology, VIB, 9052, Ghent, Belgium

R. Tejos · J. Friml (✉)
Department of Plant Biotechnology and Bioinformatics,
Ghent University, 9052 Ghent, Belgium
e-mail: jifri@psb.vib-ugent.be

J. Šamaj (ed.), *Endocytosis in Plants*,
DOI: 10.1007/978-3-642-32463-5_3, © Springer-Verlag Berlin Heidelberg 2012

patterning in early embryogenesis, post-embryonic generation of new organs such as flowers and lateral roots, epidermal cell interdigitation, vascular tissue formation and re-generation, and others. In addition, nutrient uptake, gravi- and phototropism as well as plant's defense to pathogen attack involve a polarized response.

One manifestation of cell polarity is the asymmetric distribution of proteins in the plasma membrane (PM). Several examples of polarly localized proteins in plants are documented to date being involved in a variety of processes which often require a directional response and/or are essential for the establishment of tissue polarity and patterning (Table 1). We will use the embryo-derived terminology to describe the different polar domains in a cell. During embryogenesis, two axons of polarity are formed: an apical/basal axis following the shoot and root meristems and a radial axis of concentric cell layers. In a homologous way, we will call apical to the upper and basal to the bottom side in a cell, and outer lateral to the cell side facing the environment and inner lateral to the side oriented to the inner cell layers (Dettmer and Friml 2011)

2 Polarly Localized Proteins and their Roles

Multicellular organisms have to create differences between cells to generate patterns and asymmetry. Asymmetric somatic divisions play a key role in shaping tissues and organs and this is especially important in plants where rigid cell walls do not allow cell migration. A good example of tissue patterning formation through asymmetric division occurs in leaf epidermis. Leaf epidermis contains a cell lineage that generates the stomata, a structure that participates in gas exchange and transpiration consisting of two bean-like cells called guard cells (GC) that surround a pore (Dong and Bergmann 2010). The protein breaking of asymetry in the stomatal lineage (BASL) dynamically switches between nuclear and polar PM localization regulating the asymmetric divisions necessary for stomata formation (Dong et al. 2009). The delivery and maintenance of BASL in a small region of the cell periphery is crucial for the asymmetric division and cell fate determination. Stomata lineage cells all express BASL in nucleus and they undergo symmetric division eventually generating the GC and stomatal pore. If BASL is only at the periphery, cells differentiate into pavement cells (PC). When BASL is nuclear and PM localized cell divides asymmetrically again. How BASL exerts its function is unknown, but it is hypothesized that BASL helps to establish a local cell outgrowth upstream or independently from the known small GTPase ROP2 which is involved in the lobe outgrowth during PC formation (Fu et al. 2005). However, due to a rather specific expression pattern to stomatal cell lineage we cannot extrapolate BASL activity into a wider developmental context, although when ectopically expressed in roots, hypocotyls and other leaf epidermal cells BASL is able to retain its polar localization (Dong et al. 2009) suggesting that polar determinants important for its localization are common to other cell types.

A more widely used mechanism for asymmetry generation in plants involves the hormone auxin. Auxin signalling is part of a multitude of processes including

Table 1 Polarly localized proteins in plants

Protein family	Function	Polar localization
PIN	Auxin efflux carriers	**PIN1** basal in vasculature in embryos, leaves, and root; apical in epidermis of shoot apices, embryos, and gynoecium **PIN2** apical in lateral root cap and epidermis and basal in young cortex cells **PIN3** inner lateral in shoot endodermis cells, columella: nonpolar, but polarly localized after gravistimulation, lateral in pericycle cells during lateral root formation **PIN4** basal localization in embryo and in root meristem **PIN7** apical localization until 32 cell stage then switch to basal localization in the suspensor
AUX/LAX	Auxin influx carrier	**AUX1** preferentially apical in protophloem in roots, axial in epidermis; polarly localized in shoot apical meristem
ATP binding cassette (ABC) transporters	Pleiotropic plasma membrane transporters. Some of them shown to specifically transport auxin.	**PGP1/ABCB1** above the distal elongation zone in roots is preferentially basally localized in endodermis and cortex **PGP4/ABCB4** basal or apical side in epidermal root cells **PGP19/ABCB19** polar in procambial cells in roots **PIS1/ABCG37** outer lateral in root epidermis **DSO/ABCG11** outer lateral in leaf epidermis **CER5/ABCG12** requires DSO/ABCG11 for its plasma membrane localization **PEN3/PDR8/ABCG36** outer lateral in root epidermis
AGC kinases	Protein phosphorylation, regulate auxin flow in various developmental contexts	**PINOID** enriched at apical and basal cell sides **PHOT1** enriched at apical and basal cell sides
NHP3-like	Functionally interact with PINOID to regulate PIN phosphorylation	**MAB4/ENP, MEL1-4** polarly localized following PIN localization
PIP5K	Phosphatidylinositol phosphate kinases	**PIP5K3** apical in growing root hairs **PIP5K4/5** subapical region of pollen tubes **PIP5K6** subapical region in growing pollen tubes, and switches to apical when growth stops

(continued)

Table 1 (continued)

Protein family	Function	Polar localization
Rho-GTPases	Molecular switches for multiple cellular mechanisms	Various location, tip of pollen tubes and root hairs, lobes in leaf epidermis cells
BOR	Boron efflux	**BOR1** inner lateral in roots **BOR4** outer lateral in root epidermis
CASP	Casparian strip formation	**CASP1** localized along the Casparian strip in the root endodermis
NIP5;1	Boron influx	**NIP5;1**root epidermis and lateral root cap; outer lateral
LSI	Silicon efflux	**LSI1** and **LSI2** outer lateral exodermis and endodermis in rice root
NRT1;1	Nitrate/auxin influx	Epidermis of root primordia possibly axial (anticlinal)
PEN1	SNARE domain-containing syntaxin, pathogen defense	Polar PM domain facing pathogen attack site
PEN2	Glycosyl hydrolase, pathogen defense	Polar PM domain facing pathogen attack site
BASL	Regulator for asymmetric cell division	Nucleus and polar domains at plasma membrane in stomata cell lineage cells;

See the text for references and extended description

de novo generation of cell and tissue asymmetry, re-establishment of axial polarity (e.g., during regeneration of vascular tissues), or directional responses to developmental or environmental cues (Grunewald and Friml 2010). In most of these events, auxin is distributed differentially within tissues, eliciting cell-specific responses depending on hormone levels and developmental context (Calderón-Villalobos et al. 2012). This differential auxin distribution results mainly from a directional (polar) symplastic transport which depends on PM carriers. These include several ATP binding cassette transporters of the subfamily B (ABCB, Geisler and Murphy 2006), influx carriers constituting the AUXIN RESISTANT1/LIKE AUX1 (AUX/LAX) family, and on the efflux carriers of the PIN-FORMED (PIN) family (Grunewald and Friml 2010). Though some of the ABCB and AUX/LAX transporters are polarly localized in certain tissues (Swarup et al. 2001, 2008; Geisler et al. 2005), the direction of auxin transport seems to be generated by the polarly localized PIN proteins (Wiśniewska et al. 2006). While some of these PIN carriers reside in the endoplasmic reticulum where they contribute to the auxin cellular homeostasis (Mravec et al. 2009; Dal Bosco et al. 2012; Ding et al. 2012), the PM localized PINs can be apically, basally, or laterally distributed depending on the developmental and tissue context (Kleine-Vehn and Friml 2008). Already, early in embryogenesis, after division of the zygote, auxin preferentially accumulates in the apical cell which will form the future proembryo. Later on, auxin maximum switches from apical to basal part of the embryo, where future root pole will be established. Dynamic PIN relocalization during those early ontogenetic stages redirects auxin flow generating distinct maxima at different sites of the developing embryo that result in specification of the main apical-basal plant axis and the embryonic leaves, the cotyledons (Friml et al. 2003). After germination, polar auxin transport and differential auxin distribution coordinate organogenesis (Benková et al. 2003; Reinhardt et al. 2003; Heisler et al. 2005), vascular tissue formation (Scarpella et al. 2006), and growth responses to light and gravity (Friml et al. 2002; Harrison and Masson 2008; Kleine-Vehn et al. 2010; Ding et al. 2011; Rakusová et al. 2011) that involve PIN polarity switches in response to intrinsic or extrinsic cues. On top of these regulations, the feedback loop between auxin signalling and directional auxin transport can coordinate individual cell polarities within tissues providing a versatile response to the ever changing environmental conditions.

Once the tissue asymmetry is formed, cells differentiate into highly specialized tissues with particular functions and characteristics. For instance, polarized epithelia constitute an interface between the environment and the organism, being the first barrier protecting against invasion of pathogens. A passive mechanism for pathogen defense is a waxy layer that covers the entire surface of the plant exposed to air, the cuticle. Two ABC transporters, DSO/ABCG11 and CER5/ABCG12, polarly localize exclusively to the outer PM facing the environment and are crucial in exporting lipids necessary for the cuticle synthesis at the epidermal cells surface (Panikashvili et al. 2007; McFarlane et al. 2010). On the other hand, a more active mechanism of plant immunity involves the focal secretion at the cell periphery to the actual infection sites. The β-glycosyl hydrolase PENETRATION2 (PEN2), together with the transporter PEN3/PDR8/ABCG36 is necessary for the cytoplasmic synthesis and transport of antimicrobial toxins, while the PM-resident syntaxin PEN1 mediates vesicle fusion processes (Lipka et al. 2005; Stein et al. 2007; Kwon et al. 2008). Each of these proteins relocalizes in response to pathogen attack and becomes concentrated at the invasion point which is enriched in structural sterols in the form of PM micro domains (Meyer et al. 2009).

On the other hand, the root epidermis works as an uptake surface for nutrients and a layer where detoxification of nocive substances occurs. Here, several proteins have been reported to reside at the cell's outer lateral side, facing the soil, as well as inner lateral side oriented toward the internal cell layers and vascular tissues. It is at this layer where uptake of the essential element boron is regulated by the outer lateral localized importers BOR4 and NIP5;1, and the inner localized exporter BOR1 (Takano et al. 2008). A second barrier controlling the diffusion of substances from soil to root inner tissues is the endodermis, a cell layer common to all higher plants. Plant endodermis develops early during its differentiation a belt of specialized cell wall material, the so-called Casparian strip. The CASP1 protein was reported to be polarly localized to the Casparian strip together with other members of the CASP family defining a physical border at the outer and inner lateral PM domains, where the boron transporters NIP5;1 and BOR1 are polarly localized in the same way as they are in epidermis (Roppolo et al. 2011). In rice and other higher plants, the exodermis is the outer cell layer in the root and constitutes the first diffusion barrier and, as well as the endodermis, it contains a Casparian band (Enstone et al. 2003). Here, similarly as in endodermis, the silicon transporters LSI1 and LSI2 and the boron translocators OsNIP3:1 and OsBOR1 are laterally localized facilitating a controlled uptake of these two important nutrients (Ma et al. 2006 and 2007). Laterally localized proteins in root outer layers are also involved in export of toxic compounds, such as the transporter PIS1/PDR9/ABCG37 and the pathogen defense-related protein PEN3/PDR8/ABCG36 (Strader and Bartel 2009; Růzicka et al. 2010).

Other specialized cell types are root hairs and pollen tubes. Both exhibit a highly polarized tip growth necessary for their functions (Šamaj et al. 2006; Lee and Yang 2008). This tip growth depends on several polarly localized proteins including ROP GTPases, phospholipases, phosphatidylinositol kinases, and phosphatases which contribute to cytoskeleton orientation, generation of Ca^{+2} gradient at the tip, and the synthesis of phospholipids at the PM which all together are essential for the fast and highly directional secretion of vesicles toward the growing tip which is tightly coupled with compensatory endocytosis (for details see chapter by Ovečka et al. in this volume).

3 Subcellular Trafficking in PIN Polarity

Despite the profound importance of cell polarity in the sessile life style of plants, the molecular mechanisms for its generation, maintenance, and re-specification are not well understood. Nonetheless, some important factors and hypothetical

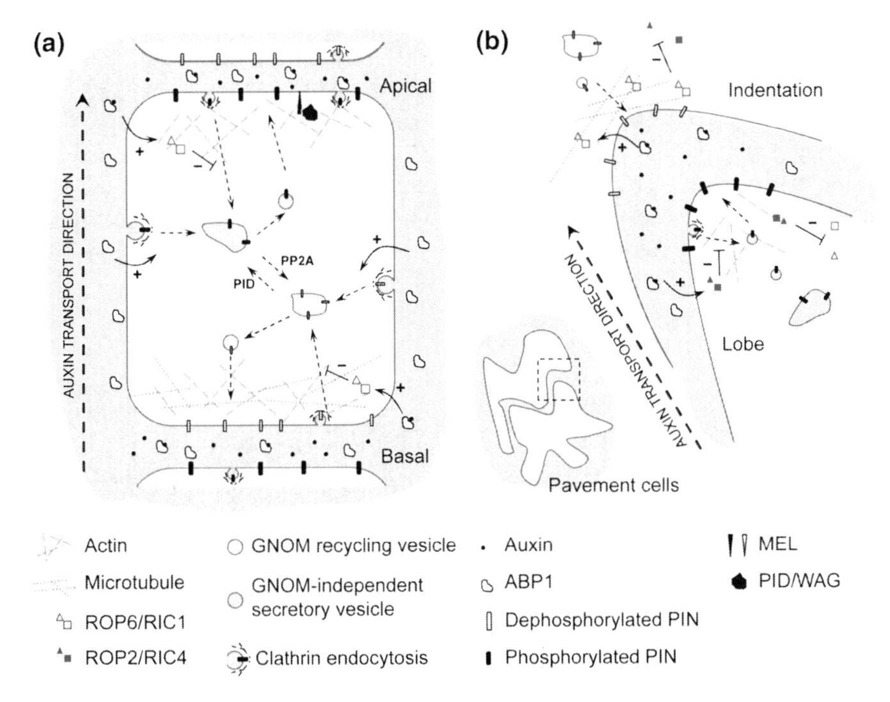

Fig. 1 Trafficking regulated polar delivery of PIN proteins. **a** In roots, the PID/WAG-phosphorylated PINs are recruited into a GNOM-independent recycling pathway, which is thought to depend preferentially on the actin cytoskeleton. The PID/WAG trafficking pathway is enhanced by the action of the MEL proteins possibly at the place of endocytosis. On the other hand, dephosphorylated PINs are trafficking by GNOM pathway which depends on both microtubules and actin cytoskeleton. The putative extracellular ABP1 signalling, which positively regulates PIN endocytosis, is inhibited by auxin thereby enhancing PIN polar localization at the places of auxin efflux. **b** In the pavement leaf cells, the lobe tip PIN localization depends on the PID activity and is regulated by the remodeling of the actin cytoskeleton and the PIN trafficking by the ROP6/RIC1 module, which acts antagonistically to the microtubule remodeling ROP2/RIC1 pathway. Here, the ABP1 auxin perception maintains the ROP/RIC signalling by enhancing their activity

mechanisms had been proposed over the last years which to some extent resemble those described in animals. Interestingly, plants also developed independent specific polarity determinants. For instance, the phytohormone auxin works in many processes as a cue governing the polar localization of its efflux carriers, the PIN proteins, by which auxin modulates its own transport and polarity of cells and tissues. We focus the following discussion on the interplay between trafficking and polarity of PINs as they are the most studied polar cargos in plants (see Fig. 1, and chapter by Nodzyński et al. for a detailed review on PIN trafficking)

The current notion of PIN polarity establishment highlights the importance of constitutive endocytic recycling (Geldner et al. 2001). PIN endocytosis is clathrin-mediated (Dhonukshe et al. 2007; Chen et al. 2011); internalization also depends on the ARF GEFs GNOM and GNL1 (Teh and Moore 2007; Naramoto et al. 2010) and the activity of the Rab GTPase ARA7 (Dhonukshe et al. 2008a). The recycling back to the PM also requires the activity of the GNOM (Geldner et al. 2001) and early endosomal protein BEN1/MIN7 (Tanaka et al. 2009). Such constitutive PIN recycling might provide a mechanism for rapid changes in polarity and concomitant redirectioning of auxin flow at certain developmental events (Dettmer and Friml 2011). This polarity switch does not involve de novo PIN protein synthesis but follows a transcytosis-like mechanism from one cell side to another as described for PIN3 repolarization during gravitropic stimulation (Kleine-Vehn et al. 2010) and for PIN basal-to-apical translocation following inhibition of the basal GNOM-mediated recycling by brefeldin A (BFA) (Kleine-Vehn et al. 2008a and 2008b). Endocytosis coupled with ongoing recycling are clearly required also for the maintenance of polar PIN distribution (Dhonukshe et al. 2008a; Kitakura et al. 2011; Kleine-Vehn et al. 2011); however, it still remains unclear how PIN polarity is originally established. Indirect observations suggest that newly synthesized PINs are initially secreted in a non-polar manner and subsequently polarized by a recycling mechanism (Dhonukshe et al. 2008a), though a default polar secretion cannot be excluded based on available data.

Genetic analysis of the components encoding the clathrin machinery in Arabidopsis clearly established a role of clathrin-mediated endocytosis (CME) in PIN polarity establishment impacting plant development (Kitakura et al. 2011). CME is essential for the previously mentioned PIN constitutive recycling (Dhonukshe et al. 2007) and, together with dynamin-related proteins, for the re-establishment of PIN polarity after cytokinesis (Mravec et al. 2011). Moreover, the feedback loop between auxin signalling and transport seems to involve the putative auxin receptor AUXIN BINDING PROTEIN 1(ABP1)-mediated regulation of CME (Robert et al. 2010). Auxin itself is able to promote its own transport by inhibiting CME endocytosis (Paciorek et al. 2005) and putative extracellular ABP1 signalling has a positive effect on PIN endocytosis. This enhancing effect of ABP1 on PIN endocytosis is inhibited by auxin resulting in elevated levels of those efflux carriers on PM of cells with higher auxin content subsequently leading to intensified hormone efflux rates. Although there is no experimental demonstration of the effect of ABP1 on PIN polarity, a mathematical model that takes in account this hypothetical extracellular auxin signalling can reproduce PIN polarity and auxin gradients seen *in planta* (Wabnik et al. 2010).

The phosphorylation status of PIN proteins is an important cue for their polar localization. The phosphorylation of serine residues in the PIN hydrophilic loop (Huang et al. 2010; Zhang et al. 2010) by the AGC3 kinases PINOID (PID) and its homologs WAG1 and WAG2 directs PINs for apical targeting in root and embryonic protodermal cells (Friml et al. 2004; Dhonukshe et al. 2010). The opposite effect is exerted by protein phosphatase 2A (PP2A) which preferentially leads to basal delivery of dephosphorylated PINs (Michniewicz et al. 2007). The gene ENHANCER OF PINOID (ENP)/MACCHI-BOU4 (MAB4) is suggested to control PIN1 polarity in concert with PID, as the reversal from apical to basal localization seen sporadically in *pid* embryos is much more pronounced in the double *pid enp* mutant (Treml et al. 2005). The gene ENP/MAB4 encodes a protein similar to

NON-PHOTOTROPIC HYPOCOTYL 3 (NPH3), an interacting partner of the light-activated Ser/Thr kinase PHOTOTROPIN1 (PHOT1), which functions as a scaffold protein involved in signal transduction (Motchoulski and Liscum 1999) and has been implicated in modulating PIN2 trafficking in response to blue light (Wan et al. 2012). The Arabidopsis genome contains other four ENP/MAB4-like (MEL) proteins whose asymmetric localization resembles PIN polarity in all tissues and cell types examined. These molecular players are regarded to modulate PIN trafficking subsequently regulating the polarity of those carriers (Furutani et al. 2011).

The process of polar sorting in plants is not known in detail and thus mechanistic understanding how phosphorylation influences PIN targeting is lacking. It has been suggested that PINs are phosphorylated at PM by the AGC3 kinases and from there internalized and recruited into the GNOM-independent apical recycling pathway (Kleine-Vehn et al. 2009; Dhonukshe et al. 2010). Consequently, GNOM recycling pathway preferentially recruits dephosphorylated basal cargoes. Besides influencing apical and basal PIN targeting, PID-mediated phosphorylation also provides entry points for various external (Ding et al. 2011; Rakusová et al. 2011) and internal (Sorefan et al. 2009) signals that influence PIN localization and auxin response.

4 PINOID Phosphorylation and ROP/RIC Interactions with Cytoskeleton: Common Mechanisms for PIN Polar Localization

In animal and plant cells, actin and microtubule cytoskeleton cooperatively interact to regulate cell polarity and vesicle trafficking (Goode et al. 2000; Petrásek and Schwarzerová 2009, see chapter by Baral and Dhonukshe for more information). Actin filaments are considered to provide guidance for vesicle trafficking and polar growth in plants (Voigt et al. 2005) and have been shown to play important roles in auxin transport and PIN recycling (Geldner et al. 2001; Dhonukshe et al. 2008b) and polarity (Baluška et al. 2001; Kleine-Vehn et al. 2008a). Although intact actin cytoskeleton is necessary for endocytic uptake (Baluška et al. 2002; Baluška et al. 2004) and PIN recycling (Geldner et al. 2001) and thus for both apical and basal PIN localization in root cells, it is the apical delivery route that is more sensitive to actin filament disruption by the actin depolymerizing agent latrunculin B. On the other hand, the basal localized PIN1 in stele and PIN2 in young cortex cells are sensitive to microtubule disruption by oryzalin while apical PIN2 in epidermis is largely insensitive to the disruption of the microtubule arrays (Kleine-Vehn et al. 2008a). These data suggest that the delivery and maintenance of polar cargos to apical or basal cell domains are dependent on different arrays of cytoskeletal components.

Similarly, in leaf PC, the coordinated communication between adjacent cells permits local inhibition of growth concomitant to the activation of the outgrowth of the neighbor cell by the regulation of cytoskeletal components. Two mutually exclusive ROP pathways modulate this process (Fu et al. 2005, 2009). While cortical microtubules inhibit the growth at the indentation, meanwhile the F-actin array at the lobe promotes the local elongation. Active ROP2 locally stimulates RIC4 at the lobe tips to allow the formation of fine cortical actin filaments and lobe development, while ROP6 activates RIC1 leading to organized microtubule formation at indentation which constrains the growth. RIC1-dependent microtubule organization not only locally inhibits the outgrowth but in turn also suppresses the activity of ROP2, while lobe-activated ROP2 inhibits RIC1 activity and microtubule organization. Thus, both pathways antagonize each other for cell shape formation using different cytoskeletal components.

Auxin promotes and is required for the PC interdigitation. Auxin can increase the average number of cell lobes, and this response is dependent on ROP2/ROP6 action and the ABP1 auxin perception. Indeed, auxin rapidly activates ROP2-RIC4 and ROP6-RIC1 pathways through ABP1 activity (Xu et al. 2010). The enrichment of PIN1 at the expanding lobe tip is necessary for the generation of a local gradient of auxin and for auxin-promoted PC interdigitation. At the lobe tip, ROP2 attenuates the mechanism of auxin inhibited clathrin-mediated PIN1 endocytosis (Nagawa et al. 2012). ROP2-activated RIC4-dependent cortical F-actin formation at the lobe mediates inhibition of PIN1 endocytosis in a similar way shown for chemical stabilization of actin filaments in root tips (Dhonukshe et al. 2008b) suggesting a conserved function for the actin cytoskeleton in the regulation of PIN internalization. This is further supported by observations that ROP/RIC signalling is required for clathrin-mediated PIN internalization in roots (Chen et al. 2012; Lin et al. 2012). PIN1 polar localization in PC also depends on phosphorylation status. The degree of PC interdigitation was greatly reduced when the PP2A phosphatase FYPP1 was knocked out or when PID was overexpressed. PIN1 localization in those genetic backgrounds changed from lobe to indentation (Li et al. 2011), pointing out the importance of PIN phosphorylation for the polar localization in this developmental context as it is for the apical-basal axis in root cells.

The regulation of PIN polarity in root cells and PC shares some common components (see Fig. 1). In roots, phosphorylated PIN2 is recruited to apical and PIN1 to basal surfaces in stele, in both cases pointing toward the direction of auxin flow; while in PC the phosphorylated PIN1 is located at the lobe tip also here being polarized with auxin flow. In both cases, the genetic interference of PIN phosphorylation recruits dephosphorylated PINs to the 'opposite pole' being in this case the cell side facing the auxin efflux from the neighbor cell. Actin filaments and microtubules have a different contribution in opposite cell sites. While actin filaments contribute to polar localization at cell site where phosphorylated PINs are targeted, the microtubule array contributes to the localization at the 'opposite pole'. ABP1-auxin effect on clathrin-mediated PIN endocytosis through modulation of ROP/RIC activity is also a common module for trafficking regulation in both cell types. To which extent the ABP1/ROP-RIC and the PID/GNOM pathways overlap to regulate PIN trafficking and polarity is still unclear, but it may account for a general mechanism for polarity establishment.

5 PIN Polarity Regulation in Response to Environmental Signals

Plants are able to respond to environmental signals such as light and gravity by modulating their growth. These growth changes result from asymmetric auxin distribution facilitated by PIN-dependent auxin transport. One of the typical adaptation responses to light is the so-called shade avoidance response. Wavelength ratio changes from red to far red (R:FR) occur typically in areas with dense vegetation where light availability is limited. The change in R:FR is perceived by the plant and translated into a switch in polar PIN3 localization to the outer lateral side in endodermis resulting in a transient increase in the auxin levels at the outer layers which in turn induce cell elongation and permits shade avoidance (Keuskamp et al. 2010). Another example is phototropism, a directional growth by which plants respond to a light stimulus. When exposed to unidirectional light, auxin accumulates on the shaded side leading to a differential elongation and bending toward the light. Arabidopsis PHOT1 and its homolog PHOT2 are PM localized Ser/Thr kinases that function as blue light receptors (Christie 2007). The auxin transporter ABCB19/PGP19 is phosphorylated by PHOT1 in response to light thus inhibiting its export activity thereby increasing the auxin content in and above the hypocotyl (Christie et al. 2011) priming the phototropic bending. In a subsequent process, a blue light-dependent signalling polarizes the cellular localization of PIN3 in hypocotyl endodermis cells away from light coinciding with the establishment of asymmetric auxin distribution with a maximum at the shaded side (Ding et al. 2011). This PIN3 polarization is dependent on GNOM-mediated trafficking and PID activity that phosphorylates PIN3. Roots are also able to respond to unidirectional light but in this case, they show a negative phototropism. Here, PIN2 mediates a PHOT1/NPH3 transduction pathway to generate asymmetric auxin distribution and root bending away from the light source (Wan et al. 2012). Since PHOT1 does not phosphorylate PIN3 directly (Ding et al. 2011), and probably neither PIN2, somehow downstream PHOT1-dependent signalling cascade is able to modulate GNOM-dependent trafficking or PIN phosphorylation by PID, or both, by a mechanism that is not fully understood but might involve the activity of NPH3-like proteins (Furutani et al. 2011; Wan et al. 2012).

Similar to phototropism, also during gravitropic response, PIN proteins are central players in auxin-mediated gravitropic bending. PIN3 and PIN7 are symmetrically localized in the first two rows of the root columella, the gravity sensing tissue. Following the gravity stimulation, PIN3 and PIN7 become polarized toward the bottom side of the cells. This gravity-induced PIN3 polarization requires the activity of GNOM and might involve endosome-based PIN3 transcytosis from one cell side to another (Kleine-Vehn 2010). A similar situation occurs in the hypocotyl gravitropic response. Here, also PIN3 together with PIN7 polarize to the bottom side of gravity-sensing endodermis cells and mediate the differential auxin accumulation at the lower side of hypocotyl for asymmetric growth and bending (Rakusová et al. 2011). Similar to light, also gravity-mediated PIN relocalization depends on GNOM and PID activity but how the sedimentation of amyloplasts triggers changes in subcellular PIN trafficking is unclear. Overall, these examples show that external signals, such as light and gravity, have potential to regulate auxin-mediated response via the modulation of polar PIN trafficking representing a plant-specific mechanism for adaptive development.

6 A Mechanism for PIN Polar Maintenance

The examples mentioned above demonstrated how flexible the PIN polarity can be. However, any polarly localized protein requires a mechanism to maintain its asymmetric disposition within a fluid membrane. With the only known exception of the Casparian band in the endodermis, in plant cells it is not possible to detect a diffusion barrier separating the PM domains. In addition, the canonical polarity modules and regulators highly conserved among animals (Assémat et al. 2008) are absent in plant genomes, though some functional homologs to those proteins are present and partially characterized in terms of cell trafficking or polarity (e.g., the Ser/Thr kinases). This strongly suggests an existence of a plant-specific mechanism for maintaining different polar PM domains. Recently, a model which involves combination of directional exocytosis to the middle of the polar domain, reduced lateral diffusion at the PM and constitutive endocytosis at margins of polar domain has been proposed to maintain PIN localization (Fig. 2, Kleine-Vehn et al. 2011). PIN2 is apically localized in epidermal root cells with some degree of enrichment at the central part of the apical surface of the cell, the so-called 'super-polar domain'. This focused localization is gradually becoming more pronounced during the cell differentiation and has not been observed in other PM proteins. Apparently, this super-polar localization is generated by preferential polar exocytosis to the center of the polar domain and limited lateral diffusion within this domain. This poor lateral diffusion is presumably related to the enrichment of PIN proteins into immobile "clusters" within the PM. Super resolution microscopy confirmed that these clusters are of 100–200 nm size and pharmacological treatment suggested that they are enriched in structural sterols. It is unclear how those clusters are immobilized but this might be due to connections with the cell wall that has been shown to contribute to maintain the polar domain and/or slowdown the diffusion rate of the PINs and other proteins at the PM (Feraru et al. 2011; Martinière et al. 2012). In addition to super-polar exocytosis and limited lateral diffusion, endocytosis has been shown to be absolutely essential for PIN polar localization (Dhonukshe et al. 2008a; Kitakura et al. 2011), probably by retrieving PINs that escaped from the polar domain back into the super-polar exocytosis trafficking route.

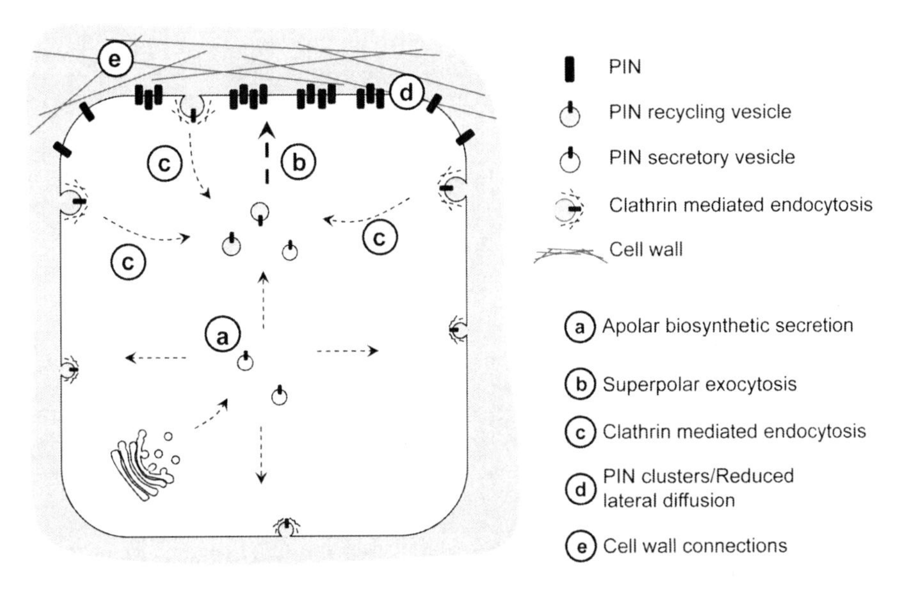

Fig. 2 Mechanism of PIN2 polar localization and maintenance of its polarity. After initial post-synthetic secretion that can be apolar or polar (**a**), the PIN proteins get recruited into an apical recycling pathway by clathrin-mediated endocytosis (CME). The apical domain is generated by the contribution of the super-polar exocytosis (**b**) and the contribution of CME (**c**). The PIN clustering (**d**) presumably reduces the lateral diffusion within the plasma membrane while the connections with the cell wall (**e**) contribute to maintain this polar domain

7 Conclusions and Future Prospects

Plants rely on polarly localized proteins to mediate several directional responses and de novo establishment of cellular and tissue asymmetries. Environmental as well as developmental cues are sensed and integrated into trafficking processes thereby influencing the localization and abundance of polar proteins. It is important to note that despite increasing wealth of information about the mechanisms for generation and maintenance of PIN polarity and the interplay between cell trafficking and polarity, it is currently unclear to which extent this information can be translated to other polar proteins or if the outer and inner lateral domains share some cellular mechanisms and molecular components with the apical-basal polar targeting machinery.

Acknowledgments We thank Tomasz Nodzyński for his valuable comments during the writing of this chapter. Work at Friml's lab is supported by the Odysseus Program of the Research Foundation-Flanders (Grant no. G091608)

References

Assémat E, Bazellières E, Pallesi-Pocachard E, Le Bivic A, Massey-Harroche D (2008) Polarity complex proteins. Biochim Biophys Acta 1778:614–630

Baluška F, Busti E, Dolfini S, Gavazzi G, Volkmann D (2001) Lilliputian mutant of maize lacks cell elongation and shows defects in organization of actin cytoskeleton. Dev Biol 236:478–491

Baluška F, Hlavačka A, Šamaj J, Palme K, Robinson DG, Matoh T, McCurdy DW, Menzel D, Volkmann D (2002) F-actin-dependent endocytosis of cell wall pectins in meristematic root cells. Insights from brefeldin A-induced compartments. Plant Physiol 130:422–431

Baluška F, Šamaj J, Hlavačka A, Kendrick-Jones J, Volkmann D (2004) Actin-dependent fluid-phase endocytosis in inner cortex cells of maize root apices. J Exp Bot 55:463–473

Benková E, Michniewicz M, Sauer M, Teichmann T, Seifertová D, Jürgens G, Friml J (2003) Local, efflux-dependent auxin gradients as a common module for plant organ formation. Cell 115:591–602

Calderón-Villalobos LIA, Lee S, De Oliveira C, Ivetac A, Brandt W, Armitage L, Sheard LB, Tan X, Parry G, Mao H, Zheng N, Napier R, Kepinski S, Estelle M (2012) A combinatorial TIR1/AFB–Aux/IAA co-receptor system for differential sensing of auxin. Nat Chem Biol 8:477–485

Chen X, Irani NG, Friml J (2011) Clathrin-mediated endocytosis: the gateway into plant cells. Curr Opin Plant Biol 14:674–682

Chen X, Naramoto S, Robert S, Tejos R, Löfke C, Lin D, Yang Z, Friml J (2012) ABP1 and ROP6 GTPase signaling regulate clathrin-mediated endocytosis in Arabidopsis roots. Curr Biol 22:1326–1332

Christie JM (2007) Phototropin blue-light receptors. Annu Rev Plant Biol 58:21–45

Christie JM, Yang H, Richter GL, Sullivan S, Thomson CE, Lin J, Titapiwatanakun B, Ennis M, Kaiserli E, Lee OR, Adamec J, Peer WA, Murphy AS (2011) phot1 inhibition of ABCB19 primes lateral auxin fluxes in the shoot apex required for phototropism. PLoS Biol 9:e1001076

Dal Bosco C, Dovzhenko A, Liu X, Woerner N, Rensch T, Eismann M, Eimer S, Hegermann J, Paponov I, Ruperti B, Heberle-Bors E, Touraev A, Cohen J, Palme K (2012) The endoplasmic reticulum localized PIN8 is a pollen specific auxin carrier involved in intracellular auxin homeostasis. Plant J, doi:10.1111/j.1365-313X.2012.05037.x

Dettmer J, Friml J (2011) Cell polarity in plants: when two do the same, it is not the same. Curr Opin Cell Biol 23:686–696

Dhonukshe P, Aniento F, Hwang I, Robinson DG, Mravec J, Stierhof YD, Friml J (2007) Clathrin-mediated constitutive endocytosis of PIN auxin efflux carriers in Arabidopsis. Curr Biol 17:520–527

Dhonukshe P, Tanaka H, Goh T, Ebine K, Mähönen AP, Prasad K, Blilou I, Geldner N, Xu J, Uemura T, Chory J, Ueda T, Nakano A, Scheres B, Friml J (2008a) Generation of cell polarity in plants links endocytosis, auxin distribution and cell fate decisions. Nature 456:962–966

Dhonukshe P, Grigoriev I, Fischer R, Tominaga M, Robinson DG, Hasek J, Paciorek T, Petrásek J, Seifertová D, Tejos R, Meisel LA, Zazímalová E, Gadella TW Jr, Stierhof YD, Ueda T, Oiwa K, Akhmanova A, Brock R, Spang A, Friml J (2008b) Auxin transport inhibitors impair vesicle motility and actin cytoskeleton dynamics in diverse eukaryotes. Proc Natl Acad Sci USA 105:4489–4494

Dhonukshe P, Huang F, Galvan-Ampudia CS, Mähönen AP, Kleine-Vehn J, Xu J, Quint A, Prasad K, Friml J, Scheres B, Offringa R (2010) Plasma membrane-bound AGC3 kinases phosphorylate PIN auxin carriers at TPRXS(N/S) motifs to direct apical PIN recycling. Development 137:3245–3255

Ding Z, Galván-Ampudia CS, Demarsy E, Łangowski Ł, Kleine-Vehn J, Fan Y, Morita MT, Tasaka M, Fankhauser C, Offringa R, Friml J (2011) Light-mediated polarization of the PIN3 auxin transporter for the phototropic response in Arabidopsis. Nat Cell Biol 13:447–452

Ding Z, Wang B, Moreno I, Dupláková N, Simon S, Carraro N, Reemmer J, Pěnčík A, Chen X, Tejos R, Skůpa P, Pollmann S, Mravec J, Petrášek J, Zažímalová E, Honys D, Rolčík J, Murphy A, Orellana A, Geisler M, Friml J (2012) ER-localized auxin transporter PIN8 regulates auxin homeostasis and male gametophyte development in Arabidopsis. Nat Commun 3:941

Dong J, Bergmann DC (2010) Stomatal patterning and development. Curr Top Dev Biol 91:267–297

Dong J, MacAlister CA, Bergmann DC (2009) BASL controlsasymmetric cell division in Arabidopsis. Cell 137:1320–1330

Enstone D, Peterson CA, Ma F (2003) Root endodermis and exodermis: structure, function, and responsesto the environment. J Plant Growth Regul 21:335–351

Feraru E, Feraru MI, Kleine-Vehn J, Martinière A, Mouille G, Vanneste S, Vernhettes S, Runions J, Friml J (2011) PIN polarity maintenance by the cell wall in Arabidopsis. Curr Biol 21:338–343

Friml J, Wiśniewska J, Benková E, Mendgen K, Palme K (2002) Lateral relocation of auxin efflux regulator PIN3 mediates tropism in Arabidopsis. Nature 415:806–809

Friml J, Vieten A, Sauer M, Weijers D, Schwarz H, Hamann T, Offringa R, Jurgens G (2003) Efflux-dependent auxin gradients establish the apical-basal axis of Arabidopsis. Nature 426:147–153

Friml J, Yang X, Michniewicz M, Weijers D, Quint A, Tietz O, Benjamins R, Ouwerkerk PB, Ljung K, Sandberg G, Hooykaas PJ, Palme K, Offringa R (2004) APINOID-dependent binary switch in apical-basal PIN polar targeting directs auxin efflux. Science 306:862–865

Fu Y, Gu Y, Zheng Z, Wasteneys G, Yang Z (2005) Arabidopsis interdigitating cell growth requires two antagonistic pathways with opposing action on cell morphogenesis. Cell 120:687–700

Fu Y, Xu T, Zhu L, Wen M, Yang Z (2009) A ROP GTPase signaling pathway controls cortical microtubule ordering and cell expansion in Arabidopsis. Curr Biol 19:1827–1832

Furutani M, Sakamoto N, Yoshida S, Kajiwara T, Robert HS, Friml J, Tasaka M (2011) Polar-localized NPH3-like proteins regulate polarity and endocytosis of PIN-FORMED auxin efflux carriers. Development 138:2069–2078

Geisler M, Murphy AS (2006) The ABC of auxin transport: the role of p-glycoproteins in plant development. FEBS Lett 580:1094–1102

Geisler M, Blakeslee JJ, Bouchard R, Lee OR, Vincenzetti V, Bandyopadhyay A, Titapiwatanakun B, Peer WA, Bailly A, Richards EL, Ejendal KF, Smith AP, Baroux C, Grossniklaus U, Müller A, Hrycyna CA, Dudler R, Murphy AS, Martinoia E (2005) Cellular efflux of auxin catalyzed by the Arabidopsis MDR/PGP transporter AtPGP1. Plant J 44:179–194

Geldner N, Friml J, Stierhof Y-D, Jürgens G, Palme K (2001) Auxin transport inhibitors block PIN1 cycling and vesicle trafficking. Nature 413:425–428

Goode BL, Drubin DG, Barnes G (2000) Functional cooperation between the microtubule and actin cytoskeletons. Curr Opin Cell Biol 12:63–71

Grunewald W, Friml J (2010) The march of the PINs: developmental plasticity by dynamic polar targeting in plant cells. EMBO J 29:2700–2714

Harrison BR, Masson PH (2008) ARL2, ARG1 and PIN3 define a gravity signal transduction pathway in root statocytes. Plant J 53:380–392

Heisler MG, Ohno C, Das P, Sieber P, Reddy GV, Long JA, Meyerowitz EM (2005) Patterns of auxin transport and gene expression during primordium development revealed by live imaging of the Arabidopsis inflorescence meristem. Curr Biol 15:1899–1911

Huang F, Zago MK, Abas L, van Marion A, Galván-Ampudia CS, Offringa R (2010) Phosphorylation of conserved PIN motifs directs Arabidopsis PIN1 polarity and auxin transport. Plant Cell 22:1129–1142

Keuskamp DH, Pollmann S, Voesenek LA, Peeters AJ, Pierik R (2010) Auxin transport through PIN-FORMED 3 (PIN3) controls shade avoidance and fitness during competition. Proc Natl Acad Sci USA 107:22740–22744

Kitakura S, Vanneste S, Robert S, Lofke C, Teichmann T, Tanaka H, Friml J (2011) Clathrin mediates endocytosis and polar distribution of PIN auxin transporters in Arabidopsis. Plant Cell 23:1920–1931

Kleine-Vehn J, Friml J (2008) Polar targeting and endocytic recycling in auxin-dependent plant development. Annu Rev Cell Dev Biol 24:447–473

Kleine-Vehn J, Łangowski Ł, Wisniewska J, Dhonukshe P, Brewer PB, Friml J (2008a) Cellular and molecular requirements forpolar PIN targeting and transcytosis in plants. Mol Plant 1:1056–1066

Kleine-Vehn J, Dhonukshe P, Sauer M, Brewer PB, Wiśniewska J, Paciorek T, Benková E, Friml J (2008b) ARF GEF-dependent transcytosis and polar delivery of PIN auxin carriers in Arabidopsis. Curr Biol 18:526–531

Kleine-Vehn J, Huang F, Naramoto S, Zhang J, Michniewicz M, Offringa R, Friml J (2009) PIN auxin efflux carrier polarity is regulated by PINOID kinase-mediated recruitment into GNOM independent trafficking in Arabidopsis. Plant Cell 21:3839–3849

Kleine-Vehn J, Ding Z, Jones AR, Tasaka M, Morita MT, Friml J (2010) Gravity-induced PIN transcytosis for polarization of auxin fluxes in gravity-sensing root cells. Proc Natl Acad Sci USA 107:22344–22349

Kleine-Vehn J, Wabnik K, Martinière A, Łangowski Ł, Willig K, Naramoto S, Leitner J, Tanaka H, Jakobs S, Robert S, Luschnig C, Govaerts W, Hell SW, Runions J, Friml J (2011) Recycling, clustering, and endocytosis jointly maintain PIN auxin carrier polarity at the plasma membrane. Mol Syst Biol 25(7):540

Kwon C, Neu C, Pajonk S, Yun HS, Lipka U, Humphry M, Bau S, Straus M, Kwaaitaal M, Rampelt H, El Kasmi F, Jürgens G, Parker J, Panstruga R, Lipka V, Schulze-Lefert P (2008) Co-option of a default secretory pathway for plant immune responses. Nature 451:835–840

Lee YJ, Yang Z (2008) Tip growth: signaling in the apical dome. Curr Opin Plant Biol 11:662–671

Li H, Lin D, Dhonukshe P, Nagawa S, Chen D, Friml J, Scheres B, Guo H, Yang Z (2011) Phosphorylation switch modulates the interdigitated pattern of PIN1 localization and cell expansion in Arabidopsis leaf epidermis. Cell Res 21:970–978

Lin D, Shingo N, Chen J, Chen X, Cao L, Xu T, Li H, Dhonukshe P, Friml J, Scheres B, Fu Y, Yang Z (2012) A ROP GTPase-dependent auxin signaling pathway inhibits PIN2 endocytosis in Arabidopis roots. Curr Biol 22:1319–1325

Lipka V, Dittgen J, Bednarek P, Bhat R, Wiermer M, Stein M, Landtag J, Brandt W, Rosahl S, Scheel D, Llorente F, Molina A, Parker J, Somerville S, Schulze-Lefert P (2005) Pre and post invasion defenses both contribute to non host resistance in Arabidopsis. Science 310:1180–1183

Ma JF, Tamai K, Yamaji N, Mitani N, Konishi S, Katsuhara M, Ishiguro M, Murata Y, Yano M (2006) A silicon transporter in rice. Nature 440:688–691

Ma JF, Yamaji N, Mitani N, Tamai K, Konishi S, Fujiwara T, Katsuhara M, Yano M (2007) An efflux transporter of silicon in rice. Nature 448:209–212

Martinière A, Lavagi I, Nageswaran G, Rolfe DJ, Maneta-Peyret L, Luu DT, Botchway SW, Webb SEW, Mongrand S, Maurel C, Martin-Fernandez ML, Kleine-Vehn J, Friml J, Moreau P, Runions J (2012) The cell wall constrains lateral diffusion of plant plasma-membrane proteins. Proc Natl Acad Sci USA (in press)

McFarlane HE, Shin JJ, Bird DA, Samuels AL (2010) Arabidopsis ABCG transporters, which are required for export of diverse cuticular lipids, dimerize in different combinations. Plant Cell 22:3066–3075

Meyer D, Pajonk S, Micali C, O'Connell R, Schulze-Lefert P (2009) Extracellular transport and integration of plant secretory proteins into pathogen-induced cell wall compartments. Plant J 57:986–999

Michniewicz M, Zago MK, Abas L, Weijers D, Schweighofer A, Meskiene I, Heisler MG, Ohno C, Zhang J, Huang F, Schwab R, Weigel D, Meyerowitz EM, Luschnig C, Offringa R, Friml J (2007) Antagonistic regulation of PIN phosphorylation by PP2A and PINOID directs auxin flux. Cell 130:1044–1056

Motchoulski A, Liscum E (1999) Arabidopsis NPH3: a NPH1 photoreceptor-interacting protein essential for phototropism. Science 286:961–964

Mravec J, Skůpa P, Bailly A, Hoyerová K, Krecek P, Bielach A, Petrásek J, Zhang J, Gaykova V, Stierhof YD, Dobrev PI, Schwarzerová K, Rolcík J, Seifertová D, Luschnig C, Benková E, Zazímalová E, Geisler M, Friml J (2009) Subcellular homeostasis of phytohormoneauxin is mediated by the ER-localized PIN5 transporter. Nature 459:1136–1140

Mravec J, Petrášek J, Li N, Boeren S, Karlova R, Kitakura S, Pařezová M, Naramoto S, Nodzyński T, Dhonukshe P, Bednarek SY, Zažímalová E, de Vries S, Friml J (2011) Cellplate restricted association of DRP1A and PIN proteins isrequired for cell polarity establishment in Arabidopsis. Curr Biol 21:1055–1060

Nagawa S, Xu T, Lin D, Dhonukshe P, Zhang X, Friml J, Scheres B, Fu Y, Yang Z (2012) ROP GTPase-dependent actin microfilaments promote PIN1 polarization by localized inhibition of clathrin-dependent endocytosis. PLoS Biol 10:e1001299

Naramoto S, Kleine-Vehn J, Robert S, Fujimoto M, Dainobu T, Paciorek T, Ueda T, Nakano A, Van Montagu MC, Fukuda H, Friml J (2010) ADP-ribosylation factor machinery mediates endocytosis in plant cells. Proc Natl Acad Sci USA 107:21890–21895

Paciorek T, Zazímalová E, Ruthardt N, Petrásek J, Stierhof YD, Kleine-Vehn J, Morris DA, Emans N, Jürgens G, Geldner N, Friml J (2005) Auxin inhibits endocytosis and promotes its own efflux fromcells. Nature 435:1251–1256

Panikashvili D, Savaldi-Goldstein S, Mandel T, Yifhar T, Franke RB, Höfer R, Schreiber L, Chory J, Aharoni A (2007) The Arabidopsis DESPERADO/AtWBC11 transporter is required for cutin and wax secretion. Plant Physiol 145:1345–1360

Petrášek J, Schwarzerová K (2009) Actin and microtubule cytoskeleton interactions. Curr Opin Plant Biol 12:728–734

Rakusová H, Gallego-Bartolomé J, Vanstraelen M, Robert HS, Alabadí D, Blázquez MA, Benková E, Friml J (2011) Polarization of PIN3-dependent auxin transport for hypocotyl gravitropic response in Arabidopsis thaliana. Plant J 67:817–826

Reinhardt D, Pesce ER, Stieger P, Mandel T, Baltensperger K, Bennett M, Traas J, Friml J, Kuhlemeier C (2003) Regulation of phyllotaxis by polar auxin transport. Nature 426:255–260

Robert S, Kleine-Vehn J, Barbez E, Sauer M, Paciorek T, Baster P, Vanneste S, Zhang J, Simon S, Čovanová M, Hayashi K, Dhonukshe P, Yang Z, Bednarek SY, Jones AM, Luschnig C, Aniento F, Zažímalová E, Friml J (2010) ABP1 mediates auxin inhibition of clathrin-dependent endocytosis in Arabidopsis. Cell 143:111–121

Roppolo D, De Rybel B, Tendon VD, Pfister A, Alassimone J, Vermeer JE, Yamazaki M, Stierhof YD, Beeckman T, Geldner N (2011) A novel protein family mediates Casparian strip formation in the endodermis. Nature 473:380–383

Růzicka K, Strader LC, Bailly A, Yang H, Blakeslee J, ŁangowskiŁ, Nejedlá E, Fujita H, Itoh H, Syono K, Hejátko J, Gray WM, Martinoia E, Geisler M, Bartel B, Murphy AS, Friml J (2010) Arabidopsis PIS1 encodes the ABCG37 transporter of auxinic compounds including the auxin precursor indole-3-butyric acid. Proc Natl Acad Sci USA 107:10749–10753

Šamaj J, Müller J, Beck M, Böhm N, Menzel D (2006) Vesicular trafficking, cytoskeleton and signaling in root hairs and pollen tubes. Trends Plant Sci 11:594–600

Scarpella E, Marcos D, Friml J, Berleth T (2006) Control of leaf vascular patterning by polar auxin transport. Genes Dev 20:1015–1027

Sorefan K, Girin T, Liljegren SJ, Ljung K, Robles P, Galván-Ampudia CS, Offringa R, Friml J, Yanofsky MF, Østergaard L (2009) A regulated auxin minimum is required for seed dispersal in Arabidopsis. Nature 459:583–586

Stein M, Dittgen J, Sánchez-Rodríguez C, Hou BH, Molina A, Schulze-Lefert P, Lipka V, Somerville S (2007) Arabidopsis PEN3/PDR8, an ATP binding cassette transporter, contributes to non-host resistance to inappropriate pathogens that enter by direct penetration. Plant Cell 18:731–746

Strader LC, Bartel B (2009) The Arabidopsis PLEIOTROPIC DRUG RESISTANCE8/ABCG36 ATP binding cassette transporter modulates sensitivity to the auxin precursor indole-3-butyric acid. Plant Cell 21:1992–2007

Swarup R, Friml J, Marchant A, Ljung K, Sandberg G, Palme K, Bennett M (2001) Localization of the auxinpermease AUX1 suggests two functionally distinct hormone transport pathways-operate in the Arabidopsis root apex. Genes Dev 15:2648–2653

Swarup K, Benková E, Swarup R, Casimiro I, Péret B, Yang Y, Parry G, Nielsen E, De Smet I, Vanneste S, Levesque MP, Carrier D, James N, Calvo V, Ljung K, Kramer E, Roberts R, Graham N, Marillonnet S, Patel K, Jones JD, Taylor CG, Schachtman DP, May S, Sandberg G, Benfey P, Friml J, Kerr I, Beeckman T, Laplaze L, Bennett MJ (2008) The auxin influx carrier LAX3 promotes lateral root emergence. Nat Cell Biol 10:946–954

Takano J, Miwa K, Fujiwara T (2008) Boron transport mechanisms: collaboration of channels and transporters. Trends Plant Sci 13:451–457

Tanaka H, Kitakura S, De Rycke R, De Groodt R, Friml J (2009) Fluorescence imaging-based screen identifies ARF GEF component of early endosomal trafficking. Curr Biol 19:391–397

Teh OK, Moore I (2007) An ARF-GEF acting at the Golgi and in selective endocytosis in polarized plant cells. Nature 448:493–496

Treml BS, Winderl S, Radykewicz R, Herz M, Schweizer G, Hutzler P, Glawischnig E, Ruiz RA (2005) The gene ENHANCER OF PINOID controls cotyledon development in the Arabidopsis embryo. Development 132:4063–4074

Voigt B, Timmers AC, Šamaj J, Müller J, Baluška F, Menzel D (2005) GFP-FABD2 fusion construct allows in vivo visualization of the dynamic actin cytoskeleton in all cells of Arabidopsis seedlings. Eur J Cell Biol 84:595–608

Wabnik K, Kleine-Vehn J, Balla J, Sauer M, Naramoto S, Reinöhl V, Merks RMH, Govaerts W, Friml J (2010) Emergence of tissue polarization from synergy of intracellular and extracellularauxin signaling. Mol Syst Biol 6:447

Wan Y, Jasik J, Wang L, Hao H, Volkmann D, Menzel D, Mancuso S, Baluška F, Lin J (2012) The signal transducer NPH3 integrates the phototropin1 photosensor with PIN2-based polar auxin transport in Arabidopsis root phototropism. Plant Cell 24:551–565

Wiśniewska J, Xu J, Seifertová D, Brewer PB, Růzicka K, Blilou I, Rouquié D, Benková E, Scheres B, Friml J (2006) Polar PIN localization directs auxin flow in plants. Science 312:883

Xu T, Wen M, Nagawa S, Fu Y, Chen JG, Wu MJ, Perrot-Rechenmann C, Friml J, Jones AM, Yang Z (2010) Cell surface- and rho GTPase-based auxin signaling controls cellular interdigitationin Arabidopsis. Cell 143:99–110

Zhang J, Nodzynski T, Penčík A, Rolčík J, Friml J (2010) PIN phosphorylation is sufficient to mediate PIN polarity and direct auxin transport. Proc Natl Acad Sci USA 107:918–922

Endocytosis and Vesicular Recycling in Root Hairs and Pollen Tubes

Miroslav Ovečka, Peter Illés, Irene Lichtscheidl, Jan Derksen and Jozef Šamaj

Abstract Tip growth of root hairs and pollen tubes provides an excellent model for the study of plant cell growth since all the critical molecular players and essential signalling pathways operate simultaneously at the tip of the growing cell. Though not all molecular mechanisms are yet fully understood, many factors have been identified that control and integrate the various structural and physiological networks involved and that thus initiate and maintain polar organization and tip growth. Presently, we will focus on membrane trafficking, in particular on the spatio-temporal regulation of endocytosis and selective recycling by vesicle trafficking and its meaning for the regulation of tip growth. The current state of the art allows to in detail comparing tip growth in root hairs and pollen tubes. Data from both systems may be mutually complementary and used for a better understanding of the complex phenomenon of tip growth.

M. Ovečka (✉) · P. Illés · J. Šamaj
Faculty of Science, Centre of the Region Haná for Biotechnological and Agricultural Research, Palacký University, Šlechtitelů 11, 783 71 Olomouc, Czech Republic
e-mail: miroslav.ovecka@upol.cz

I. Lichtscheidl
Core Facility of Cell Imaging and Ultrastructure Research, University of Vienna, Althanstrasse 14, A-1090 Vienna, Austria

J. Derksen
Department of Plant Cell Biology, Radboud University, Toernooiveld 1, 6525 ED, Nijmegen, The Netherlands

J. Šamaj (ed.), *Endocytosis in Plants*,
DOI: 10.1007/978-3-642-32463-5_4,

1 Introduction

Tip growth, a strictly localized mode of cell expansion, is a very attractive subject in experimental biology since all determining factors are constricted to a well-defined cellular domain. In addition, available model systems, i.e., pollen tubes and root hairs, meet all the demands needed for experimental approaches.

Pollen tubes, rapidly growing male gametophytes delivering the sperm cells to the ovule in flowering plants (Heslop-Harrison 1987), became widely used model systems because of their ability to germinate and grow in vitro. Germination of the pollen grain leads to the formation of a long tubular cell with highly polarized cytoplasm where organelles move in distinct patterns (Cai and Cresti 2009). Growth speed of pollen tubes in vitro depends on species and culture conditions and is often much lower than in vivo in stigma and in style. Nevertheless, the ability to autonomously and almost synchronously germinate and grow under controlled culture conditions, the high amounts available, and the ease of in vitro manipulation make pollen tubes the pre-eminent subject in studies involving microscopy.

Root hairs represent an alternative model of tip-growing cells in plants. Their formation is developmentally based on the sudden switch in cell expansion polarity of specialized root epidermal cells, the trichoblasts. Development of root hair starts with the site selection of the future hair outgrowth followed by subsequent bulge formation. The bulge further develops into long tubular root hair growing out in lateral direction of the root. The cytoplasm of the emerging root hair has a distinct polar internal organization that is established already during the first steps of bulge formation (Baluška et al. 2000). The lateral position makes root hairs easily accessible for experimental studies and microscopic observations. The process of root hair formation may occur under different conditions, on fluid or solidified media or even in humid air. The continual development of new root hairs along the root differentiation zone ensures sufficient quantity of easily accessible single tip-growing cells available for both qualitative and quantitative studies.

The aim of this chapter is to summarize the current state of art in the study of tip-growing plant cells, focusing on the integration of recent molecular and cell biological studies about dynamic vesicular trafficking processes. This topic is highlighted by a brief summary of data gained over the last five years elucidating critical roles of the cytoskeleton, molecular determinants of membrane identity, and membrane signalling lipids. From these data, cell biological models describing endocytosis and vesicular recycling as central components of the tip growth driving force, are compared for pollen tubes and root hairs (Fig. 1). Functional similarities and differences of regulatory processes are discussed in order to improve our understanding of general and system-specific regulation of polar cell growth in plants.

2 Structural and Functional Organization of Plant Tip-Growing Cells

Tip growth is characterized by cell elongation at the tip region and accordingly requires polar organization of cytoplasm and cell wall at the structural, physiological, and molecular level especially in tip-growing domain. Polar delivery of membranous and cell wall materials to the growing cell domain is under strict spatial and temporal regulation. This facilitates the formation of an elongated tubular cell through rapid tip growth. Many cellular activities contribute to the tip growth, and among them the most important are dynamic properties of the cytoskeleton together with targeted exo- and endocytosis and a proportional wall extension in the tip (Hepler et al. 2001). The underlying physiological processes concern the transmembrane exchange of ions and other compounds regulating membrane functions. They allow for cell expansion while maintaining turgor pressure, and rule wall deposition hence balancing expansion of the wall in the tip. The control over plastic and especially elastic properties of the wall maintains mechanical integrity during expansion (Winship et al. 2010; Zhang et al. 2010; Derksen et al. 2011). Particularly important appears the steep tip-focused Ca^{2+} gradient in the cytosol that regulates the activity of several key components in the growing tip, including exocytosis, ion flux, and homeostasis, generation of reactive oxygen and organization and activity of the cytoskeleton. Indeed, Ca^{2+}-permeable ion channels at the plasma membrane are one of the main effectors in the regulation of root hair elongation, signalling, and organellar movements (Hepler et al. 2001; Miedema et al. 2008; Cai and Cresti 2009; Zhang et al. 2010). The coexistence of both hyperpolarization-activated Ca^{2+} channels and depolarization-activated Ca^{2+} channels in the apical plasma membrane of root hairs of *Arabidopsis thaliana* supports satisfactory Ca^{2+} influx in a range, which is consistent with the supposed signalling role of Ca^{2+} ions in the tip growth (Miedema et al. 2008).

One of the main aspects of tip growth is a fast and targeted delivery of vesicles depositing membranes and cell wall precursors to the expanding tip. To that purpose, growing root hairs and pollen tubes contain dense populations of motile vesicles in the tip, whereas larger organelles are excluded (Bove et al. 2008). Vesicles disappear from the tip after termination of growth. Since the size of these vesicles is at the border or even below the resolution limit of the light microscope, this zone appears empty, wherefore it is also described as clear zone. In the subapical region, larger organelles, such as ER, Golgi apparatus, and mitochondria are abundant and they are followed at some distance by central vacuole and the nucleus (Galway et al. 1997; Lennon and Lord 2000; Ketelaar et al. 2002). Concerning the nucleus in growing root hairs this distance from the tip might be fixed while in the case of vacuole it is variable due to dynamic vacuolar protrusions invading subapical regions. Membrane transport is controlled and orchestrated by membrane-associated small GTPases belonging to the Rab, Arf, and Rop/Rac families. Together with other regulatory proteins they regulate polar transport, crucial for polarized growth of root hairs and pollen tubes (Lin et al. 1996; Molendijk et al. 2004; Šamaj et al. 2006; Kost 2008).

Organelles and vesicles of different size move along the length of tubular cells via cytoplasmic streaming. Analysis of FM1-43-positive organelles and their movements in the cytoplasm of a lily pollen tube showed that, apart from the extreme tip, the prevalent motions at the periphery occur in a tip-forward direction, whereas in the center of the tube they return in the opposite direction (Bove et al. 2008). This pattern agrees with the older observations of a reverse fountain-like plasma streaming of large organelles. Vesicles show the same motion pattern as the larger organelles, but unlike these, the vesicles may enter the clear zone, while other organelles reverse their motion and return through the central cytoplasm (Lovy-Wheeler et al. 2007; Bove et al. 2008; Emons and Ketelaar 2008).

2.1 Tip Growth Requires Dynamic Cytoskeleton

The cytoskeleton of both pollen tubes and root hairs organizes polarity of the cytoplasm as well as organelle and vesicle transport and it is instrumental in localized cell wall deposition. In emerging root hairs, distinct polar organization is established already during the first steps of bulge formation (Baluška et al. 2000). The cytological changes in pollen grains preparing for germination and tube formation in principle are very similar, including stage-specific rearrangements of the cytoskeleton (Cheung and Wu 2008). Especially, the actin cytoskeleton is involved in all early stages of root hair formation and pollen tube germination undergoing a sequence of rearrangements that initiate and support cellular organization during development. Before root hair emergence, actin dynamics is increased at the site of the bulge and actin from the cortical filaments accumulates in the bulging domain. Subsequently, a dynamic actin meshwork is formed in the emerging tip (Miller et al. 1999; Baluška et al. 2000). In growing root hairs, fine, highly dynamic F-actin is present at the tip. In the subapical region, dense actin filaments occur, including fine actin bundles and shorter actin cables. They are arranged in arrays that represent an intermediary state between the fine, dynamic apical F-actin and the more stable actin microfilaments and microfilament bundles in the more basal parts. The dense subapical actin arrays participate in the reversal of organelle motion and in the release of secretory vesicles from cytoplasmic streaming into the clear zone with potential access to the plasma membrane (Baluška et al. 2000; Cheung et al. 2008). The thick and long bundles of actin filaments in the shank of the tube provide the tracks for bi-directional vesicle and organelle movement parallel to the longitudinal axis.

The actin cytoskeleton has always been considered a tracking element providing structural support for organelles equipped with actin-binding motor proteins (Hepler et al. 2001). Cytoskeletal dynamic properties, however, may be equally important, especially in the tips of root hairs and pollen tubes. Pharmacological and genetic perturbation experiments of actin assembly, distribution, and dynamics all clearly established the pivotal role of actin itself (Gibbon et al. 1999; Miller et al. 1999; Fu et al. 2001; Vidali et al. 2001; Ketelaar et al. 2003; Šamaj et al. 2004; Voigt et al. 2005). Both motor protein activity and actin dynamics appear to be regulated by cytosolic Ca^{2+}. Its steep tip-focused gradient provides a major, multifunctional signalling cue in the regulatory system of tip growth. The inactivation of myosin motor activity by the increased Ca^{2+} concentration (or releasing of cargo) at the tip prevents larger organelles but allows secretory vesicles to enter the clear zone. The high Ca^{2+} concentration also activates actin binding proteins (ABPs) from the villin/gelsolin/profilin superfamily. These ABPs shift the G-/F-actin equilibrium toward the monomeric G-actin form, thus releasing organelles that had been attached and allowing for a more or less free motion of the vesicles. The larger organelles may take part in the downward motion by binding to the central, longitudinal actin filaments (Cai and Cresti 2009). Changes in the cytosolic Ca^{2+} concentration thus is a signal that profoundly determines the state of the actin cytoskeleton, and hence also the cytoarchitecture and motility in tip-growing cells.

The actin-dependent transport of the highly motile endosomes is indispensable for tip growth. The protein domain FYVE (acronym for Fab1, YOTB, Vac1 and EEA1) binds highly specifically to phosphatidylinositol-3-phosphate (PI-3P) in endosomal membranes. Therefore, a GFP-tagged construct of FYVE may be used as a reliable marker for PI-3P and endosomes, particularly for late endosomes/ multivesicular bodies (MVBs), as was shown by expression of a *2xFYVE-GFP* transgene (Voigt et al. 2005).

RabF2a, a member of the RabF clade of small Rab GTPases closely related to Rab5 in mammalian cells, is another useful molecular marker for MVBs/late endosomes (Voigt et al. 2005). Transgenic Arabidopsis lines stably expressing the double *FYVE-GFP* construct showed that endosomes visualized by this construct were present in the entire root hair, but with a higher abundance in the subapical zone. Endosomes containing GFP-RabF2a moved actively also into apical and subapical regions of growing root hairs (Ovečka et al. 2010). Though they do not frequently enter the apical dome, they represent an integral, indispensable part of the actin-dependent vesicular trafficking pathways in root hairs (Voigt et al. 2005) but also in plant cells in general (Dettmer et al. 2006; Lam et al. 2007; Müller et al. 2007).

RabA1d, a member of the RabA subfamily of small Rab GTPases fused with GFP, locates in the *trans*-Golgi-network (TGN)/early endosomal compartment (Ovečka et al. 2010; Takáč et al. 2012). In Arabidopsis, GFP-RabA1d accumulates in the vesicle-rich apical dome of growing root hairs, providing the evidence that RabA1d is a reliable marker for secretory/recycling vesicles in the clear zone. GFP-RabA1d colocalizes with the endocytic tracer dye FM4-64, providing further proof that these two markers identify early endosomal/TGN compartments and delineate endocytic membrane trafficking and vesicle recycling at the growing tip of root hairs (Ovečka et al. 2010). A fine net of dynamic F-actin close to the plasma membrane might contribute to endocytosis by drawing coated vesicles away from the plasma membrane (Šamaj et al. 2004). Accordingly, the movement of endosomal compartments and the spatial distribution of FM4-64-positive vesicles are both compromised by preventing actin polymerization using latrunculin B (Ovečka et al. 2005; Voigt et al. 2005). Recently, interactions of the actin cytoskeleton with vesicle transport, endomembrane trafficking and organelle movement have been documented in both pollen tubes (Wang et al. 2006, 2009; Chen et al. 2007; Zhang et al. 2010; Zheng et al. 2010) and root hairs (Zheng et al. 2009; Wang et al. 2010b).

Direct evidence for the involvement of the actin cytoskeleton in root hair elongation is provided by genetic experiments affecting the expression level of actin isoforms. The root hair phenotypes of the respective mutants revealed that ACT2 and ACT8 are involved in root hair tip growth (Ringli et al. 2002; Kandasamy et al. 2009). Regulation of spatial and temporal arrangements and behavior of the actin cytoskeleton is largely modulated by diverse ABPs, also in non-tip-growing cells. Upstream components of the entire signalling network such as calcium, phospholipids, and small GTPases regulate the activity of the ABPs and provide a direct link between proper actin configuration and signalling events during root hair development. A large number of proteins may affect and specify structure and function of the actin cytoskeleton in root hairs, from bulge formation to tip growth termination (Braun et al. 1999; Pei et al. 2012). These proteins include formins (actin-nucleating proteins), Arp2/3 complex (actin nucleation factor initiating a new actin filament branching), profilins (actin monomer binding proteins), actin depolymerizing factors (ADFs severing actin filaments into fragments), and villins (calcium responsive ABPs) (reviewed in Pei et al. 2012).

Modulation of the steady-state level of ABPs in plants by mutation, down-regulation, or overexpression approaches revealed that mostly total length or the shape of root hairs and pollen tubes is affected. Deregulation of ABP expression altered the effectiveness of tip growth and, not surprisingly, also vesicular trafficking. In both pollen tubes and root hairs, impaired actin nucleation by altered formin levels disrupted tip growth (Cheung and Wu 2004; Deeks et al. 2005). In growing pollen tubes, formin stimulates actin assembly from the subapical plasma membrane which in turn affects motility of organelles in the subapical zone. Subapical actin structure assembled from the subapical plasma membrane prevents larger organelles from reaching the apical zone, while this dynamic F-actin assembly does not restrict the small vesicles to move to the clear zone in the apex (Cheung et al. 2010). Thus, formins anchored to the plasma membrane in the apical dome provide a structural platform for actin-based vesicular trafficking in tip-growing cells. On the proteome level, F-actin disruption by the drug latrunculin B caused complex changes not only in the abundance of actin, ABPs but also in proteins related to signalling, metabolism, cell wall, and vesicular trafficking in gymnosperm pollen tubes (Chen et al. 2006). Altogether, these data support essential functions of the actin cytoskeleton in tip growth.

The function of the microtubular cytoskeleton in the tip growth is less clear and often has been considered negligible, at least in pollen tubes. Other studies demonstrated that its function in root hairs is maintaining tip growth direction (Ketelaar et al. 2003; Sieberer et al. 2005). Imaging of in vivo dynamic microtubule rearrangements in Arabidopsis transgenic plants expressing GFP-MBD (molecular marker for MTs) revealed distinct configurations and dynamic reorganizations of both cortical and endoplasmic microtubules (MTs) during the root hair formation (Van Bruaene et al. 2004), which may relate to the change in the cell polarity. Experiments with MT-destabilizing drugs showed that treated root hairs continue to grow at normal growth rate. Growth orientation, however, is unstable and root hairs show a wavy growth pattern (Bibikova et al. 1999; Ovečka et al. 2000). Visualization of endocytosis in growing pollen tubes by fluorescent phospholipids revealed internalization of labeled plasma membrane and redistribution of recycling vesicles at the growing apex. While depolymerization of actin filaments prevented probe internalization, depolymerization of MTs by oryzalin had no effect (Lisboa et al. 2008). This clearly indicates that MTs do not directly support endocytosis and vesicular trafficking (Traas et al. 1985; Emons et al. 1987; Hepler et al. 2001).

2.2 *Vesicular Trafficking as an Integral Part of the Subcellular Membrane Flow*

In all plant cells, growing cell domains are established and maintained by delivery of new building material in the form of vesicles. In tip-growing cells, current models propose targeted membrane trafficking and generation of dense populations of vesicles in the apical domain as fundamental structural aspects of tip growth (Hepler et al. 2001).

Exocytosis of secretory vesicles in root hairs and pollen tubes delivers lipids, proteins, and cell wall matrix material to enable tip extension. However, quantitative estimations showed differences between the amount of membrane material delivered and that required for plasma membrane extension. Retrieval of the excess membrane material has been proposed to occur with an active endocytic system (Picton and Steer 1983; Derksen et al. 1995; Hepler et al. 2001). Preliminary evidence for endocytosis in the apical region of the tip was based on the presence and localization of clathrin, clathrin-coated pits, and clathrin-coated vesicles in both pollen tubes (Derksen et al. 1995; Blackbourn and Jackson 1996) and root hairs (Robertson and Lyttleton 1982; Emons and Traas 1986; Ridge 1995; Galway et al. 1997). Further proof has been provided by studying site and speed of the internalization of fluorescent markers and nanogold particles (Camacho and Malhó 2003; Ovečka et al. 2005; Moscatelli et al. 2007; Lisboa et al. 2008; Zonia and Munnik 2008). Precise time course microscopy focused on the uptake of endocytic markers offered the necessary details to compute and model the spatial and temporal details of vesicle trafficking pathways together with the rates of the exo- and endocytosis. Both exo- and endocytosis appeared to operate in the apical and subapical regions of both pollen tubes and root hairs (Moscatelli et al. 2007; Bove et al. 2008; Ketelaar et al. 2008; Zonia and Munnik 2008; Kato et al. 2010).

Vesicles from different endocytic routes are redistributed by membrane recycling within the apical dome, including the clear zone (Fig. 1). Thus, the vesicles at the tip are not only involved in transport, but also in active membrane recycling and hence play a crucial role in tip growth.

3 Small GTPases: Master Determinants of Endocytosis and Vesicular Trafficking in Tip-Growing Cells

3.1 Membrane Definition by Rab GTPases

Rab GTPases, members of the Ras-related superfamily of small GTPases, regulate vesicular trafficking. Thus, they represent one of the basic regulatory components spatially and temporally organizing exocytosis, endocytosis, and membrane recycling (Nielsen et al. 2008; Stenmark 2009, see also chapter by Ueda et al. in this volume). A tip-localized distribution of Rab GTPases has been demonstrated for RabA4b in Arabidopsis root hairs (Preuss et al. 2004) and Rab11b in tobacco pollen tubes (de Graaf et al. 2005), which implicates a function in the activity of TGN compartments, particularly within the clear zone. Endosomal Rab GTPases Rha1/RabF2a and Ara6/RabF1 are distributed not only along the whole length of the root hair, but also in the clear zone (Voigt et al. 2005). RabA4b, which defines TGN compartments and TGN-based polarized growth of root hair tips, is involved in recruiting phosphoinositide kinase PI-4Kb1 (Preuss et al. 2006). Similarly, AtRabA4d has been shown to interact with phosphoinositide kinase PI-4Kb1 in pollen tubes (Szumlanski and Nielsen 2009).

Arabidopsis Rab GTPases of subclass RabE are implicated in membrane trafficking from the Golgi. In particular RabE1d, acting in post-Golgi biosynthetic trafficking to the plasma membrane (Zheng et al. 2005), has been thoroughly characterized including a possible interaction with the Arabidopsis phosphatidylinositol-4-phosphate 5-kinase (PIP5K2). This interaction has been confirmed using a yeast two hybrid assay, which suggests that PIP5K2 interacts with the GTP-bound form of RabE1d (Camacho et al. 2009). It was shown that PIP5K2 interacts specifically with all five members of the RabE subclass but not with the other closely related members of the Golgi- or *trans*-Golgi-localized subclasses (RabA2a, RabA2d, RabB1b and RabD2a). In contrast, interaction of the RabE subclass GTPases with Arabidopsis phosphatidylinositol 4-kinase (PI-4Kβ1), an interacting partner of RabA4b localized to TGN (Preuss et al. 2006) could not be detected (Camacho et al. 2009). Since RabA4 and RabE proteins appear to interact at the plasma membrane with enzymes involved in the sequential conversion of PtdIns to PtdIns(4,5)*P*2 (Camacho et al. 2009), spatial and functional interactions between Rab GTPases and PIP5K2 emerge as an important aspect of tip growth regulation. This may be further corroborated by the implication of Arabidopsis PtdIns(4)*P* 5-kinases in the regulation of tip growth in both root hairs (Kusano et al. 2008;

Stenzel et al. 2008) and pollen tubes (Ischebeck et al. 2008; Sousa et al. 2008). PIP5K2 accumulates at the plasma membrane and although YFP-tagged RabE proteins were originally detected in the Golgi apparatus in Arabidopsis root tips (Chow et al. 2008), PIP5K2 can cause redistribution of RabE1d from the Golgi. Indeed, in tobacco leaf epidermis, co-expression with GFP-PIP5K2 caused relocation of YFP-RabE1d from Golgi to the plasma membrane. Activation of RabE1d (GTP bound state) was necessary for its recruitment to the plasma membrane by GFP-PIP5K2. In addition, activity of PIP5K2 kinase toward phosphatidylinositol phosphates was stimulated by Rab binding. Although the direct interaction of PIP5K2 with GTP-bound RabE1d was demonstrated only in vitro, it is likely that RabE proteins targeted to the plasma membrane stimulate the production of PtdIns(4,5)*P*2 by PIP5K2 also in vivo, which may help to balance vesicular trafficking rates at the membrane. Importantly, tips of growing root hairs show plasma membrane localization of YFP-PIP5K2 (Camacho et al. 2009), where it may provide a platform for spatio-temporal interaction with RabE1d. Altogether, Rab GTPases in both root hairs (such as RabA4b and RabE1d) and pollen tubes (such as Rab11b) regulate vesicular trafficking within the clear zone through the molecular definition of the TGN compartments and by targeted, polarized secretion and endocytosis (Šamaj et al. 2006).

RabA1d, a member of the RabA subfamily of small Rab GTPases, has been identified in the TGN/early endosomal compartments. In growing root hairs, GFP-RabA1d fusion protein was concentrated in the vesicle-rich apical dome, suggesting that RabA1d is a reliable marker for secretory/recycling vesicle pools in the clear zone (Ovečka et al. 2010). Subcellular distribution of another molecular marker, the SNARE protein Wave line 13 (VTI12) fused with YFP (Geldner et al. 2009) demonstrated a close association of TGN/early endosomal compartments with the recycling pathway in the tip of growing root hairs. Complexation of plasma membrane sterols, polarly accumulated in the tip of growing root hairs, caused blocking of movement and artificial accumulation of both RabA1d- and VTI12-positive compartments at the cytoplasmic face of the plasma membrane in the tip, which in turn leads to rapid cessation of tip growth (Ovečka et al. 2010). Recently, RabA1d has been shown to be sensitive to wortmannin. The downregulation of RabA1d abundance by wortmannin indicates a direct link between the phosphatidylinositol 3-kinase and phosphatidylinositol 4-kinases-related signalling pathways and the subcellular distribution and functionality of the TGN/early endosome system (Takáč et al. 2012). Taking together, it appears that tip growth depends on distribution, motility, and structural and functional integrity of the TGN, where membrane definition through recruitment of Rab GTPases plays an important specification role.

Dynamin and dynamin-related proteins are part of the superfamily of large GTPases and represent important executors of plasma membrane- and endomembrane-related functions, due to their ability to interact with proteins and lipids through the proline-rich and pleckstrin homology domains. Because of variations in specificity to interaction partners and in particular their subcellular distribution, dynamins, and dynamin-related proteins are involved in endomembrane dynamics including endocytosis and vesicular trafficking (reviewed in Konopka et al. 2006; Bednarek and Backues 2010). In Arabidopsis, there are two dynamin-related proteins called DRP2A and DRP2B which share functional domain similarity with the animal dynamins. Recently, it was shown that DRP2B colocalizes with DRP1A and also with the clathrin light chain at the plasma membrane (Fujimoto et al. 2010). This is a strong indication for an important role of dynamin-related proteins in clathrin-mediated endocytosis. In root hairs, DRP2A and DRP2B were localized to the growing tips using an antibody raised against the pleckstrin homology domain. Importantly, expression of the dominant negative form of DRP2A/B under an inducible promoter resulted in reduced endocytosis of a tracer dye at the tips of root hairs (Taylor 2011). The distribution of GFP-tagged DRP1C in root hairs visualized with a combination of confocal laser scanning microscopy and variable angle epifluorescence microscopy revealed the presence of DRP1C-GFP throughout the circulating cytoplasm and at the apical plasma membrane, but most prominently it was seen along the lateral plasma membrane in the subapical and lateral flank regions (Konopka et al. 2008). Interestingly, the DRP1C-GFP plasma membrane-associated fluorescence intensity fluctuated with the growth rate. When the growth rate of the root hair was low, DRP1C-GFP fluorescence was recognizably associated with the apical plasma membrane, but at high growth rate, membrane-associated DRP1C-GFP fluorescence was more abundant at the flanks. In the same study, authors demonstrated that CLC-GFP was localized to the subapical and flank regions in growing root hairs, thus plasma membrane-associated CLC-GFP exhibits a similar distribution as DRP1C-GFP in growing root hairs. Fluorescence recovery after photobleaching (FRAP) and physiological inhibitor studies support the hypothesis that DRP1C is a component of the clathrin-associated machinery in plants (Konopka et al. 2008). Extensive exchange of DRP1C between apex and flank regions clearly indicates the interdependence of membrane retrieval activity on fast tip growth. Pollen grains of mutants lacking DRP2A, DRP2B, or both displayed a variety of physiological and morphological defects, which indicates an important role of DRP2-dependent clathrin-mediated trafficking in the gametophyte-specific transduction of developmental signals (Backues et al. 2010). Likewise, plasma membrane-localized DRP1C-GFP accumulates near the aperture during pollen grain germination. In growing pollen tubes, DRP1C-GFP is distributed in the cytoplasm and along the plasma membrane at the subapex (Konopka et al. 2008). These observations clearly indicate the indispensable and important roles of dynamins and dynamin-related proteins in endocytosis in tip-growing cells.

ADP-ribosylation factor (ARF) GTPases represent an important group of the small GTPase superfamily in Arabidopsis (Vernoud et al. 2003). The activity of ARFs is modulated by regulatory proteins such as guanine nucleotide exchange factors (GEFs, activators) and GTPase-activating proteins (GAPs, deactivators). Recent studies showed that both GAPs and GEFs are involved in the regulation of membrane dynamics responsible for correct polar development of both root hairs and pollen tubes. The Arabidopsis mutant *agd1* (knockout of one ARF-GAP) develops wavy and branched root hairs. Although it was shown that actin dynamics is altered in these root hairs (Yoo et al. 2008), functional studies showed that negative effects on the actin cytoskeleton are mediated through direct links of AGD1 to membrane trafficking effectors. Phosphatidylinositol-4-phosphate (PI-4P) is preferentially localized along the lateral subapical plasma membrane in wild-type root hairs (Yoo et al. 2012). AGD1 has the ability to bind to phosphoinositides and thus, AGD1 stabilizes domains of PI-4P at the lateral subapical plasma membrane of growing root hairs. The small regulatory protein ROP2 has been localized to the apical plasma membrane in wild-type root hairs, while this targeting of ROP2 was altered in the *agd1* mutant (Yoo et al. 2012). AGD1 thus performs an important role in the stabilization and proper localization of membrane modulators in the vesicular secretion at the tip. Taking together, AGD1 is functioning in common phosphatidylinositol-dependent signalling pathway, regulating the distribution of RAC/ROP components and maintaining directional membrane trafficking during root hair development by remodeling of the actin cytoskeleton (Yoo et al. 2012).

ARF-GEFs are also essential components of the intracellular vesicular trafficking machinery in plants. Some of them, like GNOM, are targets of brefeldin A (BFA). It is known that inhibition of secretion and vesicular recycling by BFA treatment normally terminates the tip growth of root hairs (Šamaj et al. 2002; Ovečka et al. 2005). GNOM, an Arabidopsis GBF1-related ARF GEF sensitive to BFA performs a plant-specific function in the vesicular recycling from endosomes to the plasma membrane (Geldner et al. 2003; Richter et al. 2007). Expression of the BFA-resistant form of GNOM in Arabidopsis prevents BFA-induced inhibition of root hair elongation, which clearly shows that tip growth of root hairs in Arabidopsis needs GNOM-dependent endosomal recycling. Endosomal recycling could be even more important than secretory trafficking, at least in growing root hairs. Recently, an essential role of vesicular recycling for tip growth has also been documented in germinating pollen. BFA treatment prevented pollen grains to germinate and in growing pollen tubes it blocked the tip growth (Richter et al. 2012). Thus, endosomal recycling seems to be of critical importance for the basic mechanisms of tip growth in both root hairs and pollen tubes. GNOM-LIKE2 (GNL2), a pollen-specific ARF GEF, is another marker of polar development because a GFP-GNL2 fusion protein decorates the emerging tip of the germinating pollen grain and the growing tip in the more developed pollen tubes. Experiments with the BFA-resistant GNL2 showed that GNL2 is the critical ARF GEF involved in pollen germination. In conclusion, the ARF GEFs GNL2 in pollen tubes and GNOM in root hairs are both involved in the promotion of endosomal recycling and thus indispensable for tip growth. The GNL2-dependent trafficking pathway is also involved in the deposition and accumulation of pectins in the tip of growing pollen tubes.

3.2 Signalling by Rop GTPases

Regulation of the cytoskeleton, ion fluxes including calcium homeostasis, generation of reactive oxygen species (ROS), and vesicle trafficking in both pollen tubes and root hairs requires ROP/RAC GTPases, master regulators of cell polarity in plants. The ROP/RAC signalling system plays a pivotal role in maintaining cell polarity through co-ordination of positive and negative feedback signalling pathways, performed by two interlinked RIC3- and RIC4-dependent cascades, affecting the actin cytoskeleton and vesicular trafficking during targeted exocytosis (Gu et al. 2005; Lee et al. 2008a). In principle, ROP/RAC GTPases regulate the structure of F-actin in the tip and membrane trafficking necessary for polarized cell expansion. Also, ROS production by tip-localized NADPH oxidase that promotes cytoplasmic Ca^{2+} gradient formation is ROP/RAC dependent. Additionally, some ABPs are controlled by the ROP/RAC system as well (Yalovsky et al. 2008). Distribution of tagged proteins of ROP/RAC GTPases responsible for restriction of the tip growth shows a similar, cap-like pattern. Both ROP2 in root hairs (Yoo et al. 2012) and ROP1 in pollen tubes (Hwang et al. 2010) are localized to the plasma membrane in the apical dome.

ROP/RAC activity supports the distribution of ROP1 in the apical part of pollen tubes and promotes signalling lipids to accumulate in the tip (Ischebeck et al. 2010). In pollen tubes, the ROP/RAC GAPs of the plasma membrane in the flanks inactivate ROP1, while RhoGDI mediates the relocation of the inactive ROP/RAC to the apex, where they are re-activated by ROP/RAC-GEFs. Such completed cycle leads to targeted accumulation of active ROP/RAC in the growing region (Klahre et al. 2006; Kost 2008; Yalovsky et al. 2008; Zou et al. 2011). Notably, overexpression of the constitutively active form of AtROP11/RAC10^{CA} induces root hair swelling and compromises the internalization of the endocytic tracer dye FM4-64 (Bloch et al. 2005). Apparently, altering the expression or activation rates of the ROP/RAC signalling system at the tip might shift exocytosis/endocytosis ratio.

4 Other Molecular Players Involved in Tip Growth

4.1 Membrane Lipids: Phosphoinositides and Sterols

Delivery and exocytic release of new building components at the tip as well as retrieval of plasma membrane and cell wall material by endocytosis must be facilitated by determinants in the apical plasma membrane. Tip expansion demands that plasma membrane-localized molecular determinants continuously relocate in parallel with growth toward the tip. Lipids are indispensable for the determination of structural and signalling properties of the membranes. As shown in growing tobacco pollen tubes, an isoform of phospholipase called NtPLC3 specifically localizes in the flanks of the tip, its absence at the extreme apex was maintained by endocytosis. PLC3 may maintain the tip-focused distribution of phosphatidylinositol 4,5-bisphosphate (PIP2) by limiting its lateral spreading through generation of DAG (diacylglycerol), which is also continuously retrieved from subapical parts by endocytosis and reinserted to the apical domain (Helling et al. 2006). PIP2 is a key phospholipid for the regulation of tip growth in plants. In addition to the signalling role in vesicle docking at the plasma membrane, its local concentration at the apex was proposed to directly remodel the actin cytoskeleton in this domain. It may prevent actin depolymerization by inhibiting ADF activity and it may promote actin polymerization by destabilizing the profilin:actin complexes (Pei et al. 2012).

PI-4P might serve as a substrate for the formation of PIP2 (Kusano et al. 2008; Stenzel et al. 2008). Therefore, regulation of the endogenous levels of both PI-4P and PIP2 is required for proper tip growth and spatial distribution of source lipids is under tight control. The enrichment with PI-4P of the subapical plasma membrane of root hairs is stabilized by AGD1 (Yoo et al. 2012), which provides a direct link between lipid signalling and rapid endocytosis in this region (Konopka et al. 2008). Phospholipid PI-3P is generated by phosphatidylinositol 3-kinase (PI-3K) and modulates tip growth in root hairs. Since PI-3K-specific inhibitors reduced not only tip growth but also ROS production and endocytosis, it may be concluded that PI-3P-mediated vesicle trafficking is important for root hair tip growth (Lee et al. 2008b). PI-3K is also essential for vacuole biogenesis and reorganization during pollen development, for pollen germination and for pollen tube growth (Lee et al. 2008c). Thus, polarized growth requires a specific lipid composition of the apical plasma membrane, which is maintained by endocytosis and vesicular recycling.

Membrane lipids determine the mode of endocytic internalization also by delimiting clathrin-independent, sterol/lipid raft-mediated endocytosis in plant cells. Sterol-mediated trafficking pathways participate in the establishment of cell polarity in diffusely expanding root cells by polar endocytic recycling of the auxin efflux carrier PIN2 (Grebe et al. 2003). Rapid sterol-mediated endocytosis in root cells during cytokinesis revealed involvement of the TGN/early endosome compartments in the redistribution of plasma membrane sterols within the recycling pathway (Boutté et al. 2009). Sterol-enriched lipid rafts are essential for cellular polarity and proper growth pattern in a large variety of tip-growing cells, including fungal, algal, and animal cells (Ovečka and Lichtscheidl 2006). Depletion of structural sterols shows that they are indispensable also for tip growth of Arabidopsis root hairs (Ovečka et al. 2010). Complexation of sterols by filipin inhibits tip growth and causes extensive disintegration of vesicles in the clear zone. Endocytic internalization of sterols and FM4-64 became uncoupled from further redistribution. Vesicles/early endocytic compartments remained blocked at the plasma membrane when sterols were complexed after filipin treatment. This phenotype was restricted to the apex and annular zone encompassing a vesicle-rich region, but not to the plasma membrane at the subapical zone. Domains of the plasma membrane, sensitive to sterol blockage, were located at the apex of root hairs, which may indicate a preferential location of sterol-mediated endocytosis in the apical membrane of growing root hairs. Local accumulation of structural sterols in trichoblasts, in root hair bulges and emerging tips, may also regulate apical plasma membrane properties, vesicular trafficking and the continuation of polarized growth. Taking together, structural sterols are involved in endocytosis and in vesicular recycling in the tip of growing root hairs in Arabidopsis, which is further supported by the specific sterol accumulation in the apical plasma membrane (Ovečka et al. 2010).

Flotillins, markers of cholesterol-rich and detergent-resistant membrane microdomains, are involved in endocytic pathway in mammalian cells (Otto and Nichols 2011). Flotillin-like proteins (FLOTs) were found in detergent-resistant membrane domains also in plant cells (Borner et al. 2005). FLOTs in *Medicago truncatula* were studied in the establishment of plant-bacteria symbiosis. It was demonstrated that two members of this family, FLOT2 and FLOT4, are required for early symbiotic events. Upon inoculation with *Sinorhizobium meliloti* they were strongly upregulated and FLOT4 changes their subcellular localization. FLOT4:GFP is evenly distributed in the plasma membrane of control root hairs, while upon inoculation with *S. meliloti* becomes unequally localized and enriched to the tips of elongating root hairs, implicating possible roles of flotillin-dependent pathways in the complex process of bacteria internalization (Haney and Long 2010).

Differences between clathrin-mediated and sterol-mediated endocytic pathways can be documented by differences in size of the vesicles involved. The size of GFP-Flot1-positive vesicles in Arabidopsis root cells was estimated to be around 100 nm, while a size of 30 nm was estimated for clathrin-coated vesicles. Moreover, Flot1-positive vesicles and clathrin-positive vesicles show clear differences in their dynamic properties (Li et al. 2012). These data suggest that Flot1 in Arabidopsis participates in non-clathrin-mediated endocytosis. An important role of sterol-enriched microdomains in the apical plasma membrane was further indicated by the association of NADPH oxidase, responsible for reactive oxygen production, with detergent-resistant membranes of *Picea meyeri* pollen tube tips. Importantly, association of NADPH oxidase with detergent-resistant membranes increased its enzymatic activity (Liu et al. 2009).

In conclusion, lipids in tip-growing cells constitute an integral part of the complex mechanism to ensure proper polarized distribution of regulatory proteins including their selective endocytosis and recycling. Comprehensive data about involvement of membrane lipids in the regulation of tip growth are provided in recent reviews (Kost 2008; Thole and Nielsen 2008; Zhang and McCormick 2010; Zonia 2010).

5 Root Hairs Versus Pollen Tubes: How Similar are Current Models of Endocytosis and Vesicular Recycling in the Tip?

The most specific feature of all tip-growing cells is their highly dynamic polar organization. In the dome-shaped apex of growing pollen tubes and root hairs numerous secretory vesicles accumulate in the extreme tip (clear zone). The typical V-shaped form of the clear zone is characteristic for pollen tubes and does not occur as such in root hairs (Derksen et al. 1995; Bove et al. 2008; Ketelaar et al. 2008, reviewed by Miller et al. 1997). Vesicle distributions in the tips of growing pollen tubes and root hairs are summarized in Fig. 1. Locations and main directions of the movements between membrane donors, receiving-, and recycling compartments are represented. The pool of the vesicles in the growing tip is permanently supplied by exocytic vesicles delivered to the apex by anterograde cytoplasmic streaming along the cortical region of the tube, by new vesicles derived from the apical plasma membrane by endocytosis, and by vesicles undergoing recycling in the apex. The retrograde cytoplasmic streaming from the apex to subapical region within the central area helps to maintain continuous exchange and cycling of vesicles, which is completed by transporting of retrieved vesicles back to the tip (Fig. 1). Larger organelles carried by anterograde cytoplasmic streaming do not enter the clear zone. After reaching the subapical region, termed also organelle-rich zone, these organelles reverse their direction and move back through central part of the tube (Lovy-Wheeler et al. 2007).

Delivery of membranes and cell wall material to the cell surface is a prerequisite for tip growth. According to observations in pollen tubes and root hairs of different plant species, the amount of membrane material delivered to their apex by exocytosis exceeds several times the amount exactly utilized for membrane/cell surface enlargement (Derksen et al. 1995; Bove et al. 2008; Ketelaar et al. 2008). Excess of plasma membrane and cell wall material is internalized back to the cell via endocytosis and is directed either to recycling pathway or to degradation pathway.

Recent studies revealed the existence of two distinct, precisely defined zones of endocytosis in growing pollen tubes (Moscatelli et al. 2007; Zonia and Munnik 2008). The first, apical endocytic zone is restricted to extreme apex in pollen tubes and is adjacent to the zone of exocytosis (Fig. 1a). The second, lateral endocytic zone is localized in subapical region of pollen tube (Fig. 1a). Use of positively and negatively charged nanogold particles and specific fluorescent markers in combination with ikarugamycin, an inhibitor of clathrin-mediated endocytosis, showed that different internalization events occur in these two zones (Moscatelli et al. 2007). The predominant mechanism responsible for retrieval and delivery of material from the apical endocytic zone is clathrin-mediated endocytosis (Moscatelli et al. 2007). Clathrin-coated vesicles are transported by retrograde cytoplasmic streaming to TGN and continue on their path to MVBs and vacuole for degradation. In the presence of ikarugamycin, still a though limited amount of vesicles from the apical plasma membrane remains in the pollen tube tip, indicating that besides clathrin-mediated also clathrin-independent pathway is

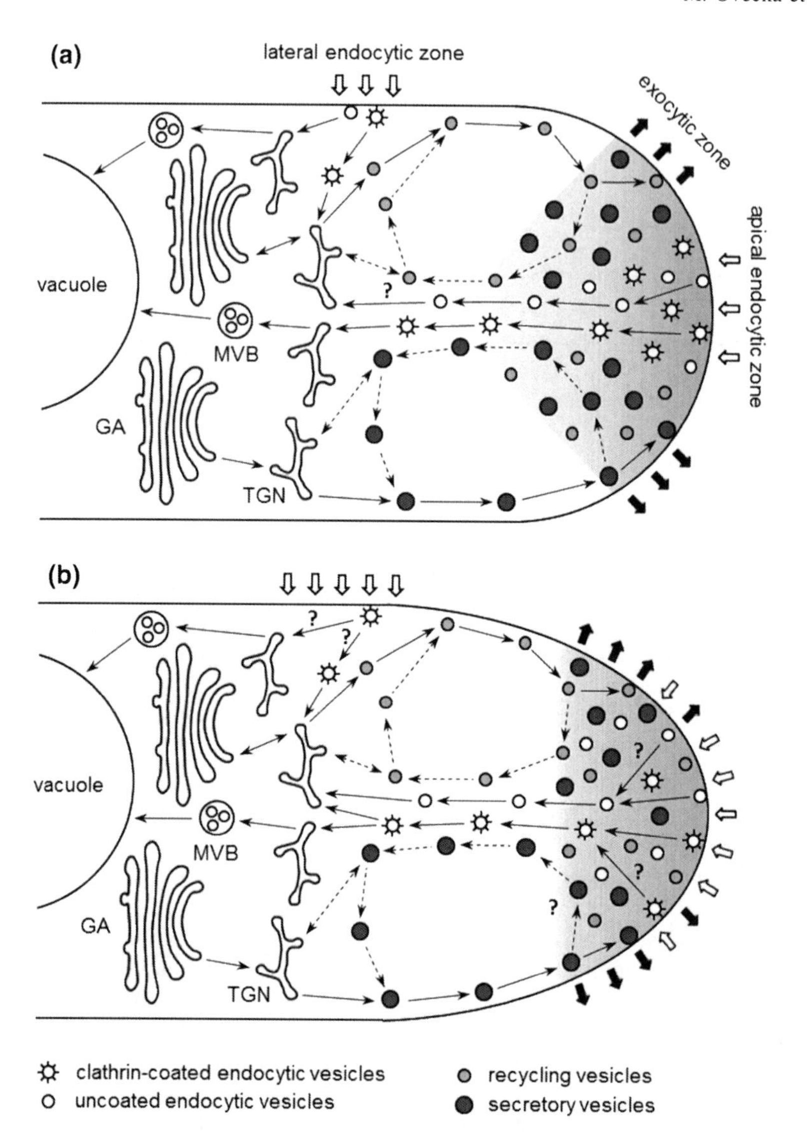

involved in the internalization of plasma membrane material. However, uncoated vesicles remain restricted to the apical region and they are never transported to vacuoles (Moscatelli et al. 2007), suggesting a possible involvement in recycling pathway (Zonia and Munnik 2008). The exocytic zone is the region of accumulation and fusion of Golgi-derived secretory vesicles with plasma membrane. It is localized in the apical part of pollen tube next to the apical endocytic zone. Secretory vesicles delivered from Golgi by anterograde cytoplasmic streaming are released to cytoplasm in an annulus-shaped zone defining the apical growth domain.

◀ **Fig. 1** Comparison of vesicular trafficking maps in growing pollen tubes and root hairs. **a** Model of vesicular trafficking in growing pollen tube. In pollen tubes, three distinct zones of vesicle-plasma membrane interactions are precisely defined: apical endocytic zone, exocytic zone, and lateral endocytic zone. Clathrin-coated vesicles originating in the apical endocytic zone are delivered via the TGN and MVBs to the vacuole (degradation pathway). Uncoated vesicles probably enter the recycling pathway through the TGN. Golgi-derived secretory vesicles are transported to the exocytic zone where they undergo exocytosis. Secretory vesicles that do not get in contact with the plasma membrane return to subapical zone from where they may repeatedly be directed to the exocytic zone (*dashed arrows*). Clathrin-coated vesicles internalized from the lateral endocytic zone are delivered to the TGN and consigned to the recycling pathway toward the exocytic zone. Part of the recycling vesicles that do not fuse with the plasma membrane do return and may re-enter the recycling pathway (*dashed arrows*). Uncoated vesicles derived from the lateral endocytic zone are directed through the TGN and MVBs to the vacuole for degradation (degradation pathway). **b** Model of vesicular trafficking in growing root hairs. A clearly defined discrete apical endocytic zone has not yet been described. Clathrin-coated and uncoated vesicles are probably internalized along the whole apical part of the root hair. While clathrin-coated vesicles retrieved from root hair apex undergo both degradation and recycling pathways, uncoated vesicles are directed for recycling. Although recycling vesicles transported to exocytic zone can repeatedly be recycled like in pollen tubes (*dashed arrows*), a possible rapid recirculation of Golgi-derived secretory vesicles remains to be elucidated (*dashed arrows*). Internalization of clathrin-coated vesicles from the lateral endocytic zone has already been characterized, but the destination of these vesicles is still unclear. *Gray areas* in both pollen tubes and root hairs represent the shape of the vesicle-rich regions. Question marks depict possible pathways that need to be experimentally approved. Objects are not drawn to scale. *GA* Golgi apparatus, *MVB* multivesicular body, *TGN trans*-Golgi network

After the contact between vesicles and plasma membrane, the membranes fuse and the vesicle's content is secreted. Complete fusion of secretory vesicles with the plasma membrane does not always occur. A substantial portion of the vesicles do not completely fuse but deliver and release material by "kiss-and-run" mechanism (Bove et al. 2008). Vesicles that do not contact or fuse with the plasma membrane are quickly redirected and transferred to the subapical region by retrograde cytoplasmic streaming to be passed again to the exocytic zone (Bove et al. 2008). Comparison of the computed rate of vesicle turnover in the apical region with data obtained by FRAP analysis indicated that in pollen tubes more than one passage through the apex may be needed before a significant number of secretory vesicles has fused with the plasma membrane and the contents are delivered (Bove et al. 2008). In the lateral endocytic zone both types, clathrin-mediated and clathrin-independent mechanisms, are responsible for internalization of plasma membrane material (Moscatelli et al. 2007; Moscatelli and Idilli 2009). Lateral plasma membrane-derived clathrin-coated and uncoated vesicles are delivered to the TGN, a conversion point of endocytic and secretory pathways. Membrane material and content of clathrin-coated vesicles is sorted and consigned for further recycling. Recycling vesicles are directed by anterograde cytoplasmic streaming to the zone of exocytosis where they behave similar to that of the secretory vesicles. Part of the recycling vesicles fuse with the plasma membrane and release their content. The remaining vesicles return to the subapical zone and may repeat the cycle and move along the lateral plasma membrane toward the apex of the pollen tube. Recent studies confirmed the transport of endocytic vesicles from the lateral plasma membrane to the TGN (Wang et al. 2010a, 2011; Cai et al. 2011). Application of ikarugamycin inhibits recycling of vesicles, but material internalized in lateral endocytic zone will still be found in TGN, MVBs, and vacuoles (Moscatelli et al. 2007). These results indicate that uncoated vesicles originating from the clathrin-independent endocytic pathway are transported from the lateral plasma membrane to the TGN and then directed through MVBs to the vacuole for degradation. In summary, current models of vesicular trafficking in growing pollen tubes suggest that a major portion of membrane material is internalized by clathrin-mediated endocytosis and eventually recycled. Endocytic vesicles derived from the apical zone are mainly destined for degradation.

In comparison to pollen tubes, complex studies precisely defining sites of endocytosis and detailed descriptions of clathrin-mediated and clathrin-independent endocytic pathways in root hairs are still lacking. However, studies about the spatial organization of the growing zone in root hairs (Shaw et al. 2000) show an annular growth zone between the extreme tip apex and subapical region, which would well fit in with the current scenarios for pollen tubes. The data on endocytosis of material in the tip of root hairs do not allow clearly defining and distinguishing discrete zones of endocytosis similar to those in pollen tubes (Fig. 1b). In fact, massive endocytosis occurs over the entire apical part of root hairs. Studies to the uptake of fluorescent marker dyes FM4-64 and FM1-43 showed rapid internalization of membrane material and transport from the apical region to the subapical part of the root hair and ultimately delivery to the vacuole (Ovečka et al. 2005). However, only part of the material retrieved from apical plasma membrane is directed to the vacuole. A significant portion of the vesicles remains in the apical part of the root hair and becomes delivered through a recycling pathway to the exocytic zone, where vesicles either fuse completely with the plasma membrane or just release their cargo by the "kiss-and-run" mechanism (Ovečka et al. 2005, 2008). Besides secretory vesicles, a substantial part of material utilized for apical growth is delivered to the exocytic zone by recycling vesicles (Richter et al. 2012).

Ultrastructural studies showed the presence of numerous clathrin-coated pits and clathrin-coated vesicles in the apical part of root hairs. This indicates involvement of a clathrin-mediated endocytic pathway in the internalization of plasma membrane material (Emons and Traas 1986; Galway et al. 1997). Clathrin-coated pits and vesicles appeared less abundant in the subapical region (Emons and Traas 1986; Galway et al. 1997), which may suggest the existence of a lateral endocytic zone similar to that in pollen tubes. Later studies confirmed that clathrin-mediated endocytosis occurs in both the apical part and subapical zone of root hairs (Konopka et al. 2008). Intriguingly, both intensity of internalization and localization of endocytic region fluctuate according to growth rate. During slow root hair growth, the region of clathrin-mediated endocytosis is recognized preferentially in the apical zone of the root hair, in the rapid growth phase it is localized in the subapical, lateral zone (Konopka et al. 2008). Unfortunately, precise determination of the proportion of clathrin-coated vesicles involved either in degradation or in recycling pathways is still unclear. However, it seems likely that the lateral endocytic zone plays an important role in root hair tip growth.

In growing root hairs, sterols are found in the extreme tip as well as in the adjacent region that spreads out up to the subapical zone. This distribution supports the idea that clathrin-independent/sterol-mediated endocytosis might operate in the entire apical part of the root hair and thus may represent a substantial portion of endocytic internalization events in all stages of rapid tip growth (Ovečka et al. 2010).

Studies to the effect of fluorescent probe filipin showed that both apical and annular zones of the apical dome are invaded by the artificial sterol complexes. Such behavior is particular only for growing root hairs, whereas non-growing root hairs and diffusely expanding root cells do not show such specific sterol sequestration (Ovečka et al. 2010). These observations are consistent with the model of cell polarization by endosomal/vesicular recycling, which is presently recognized as one of the key components of the tip growth mechanism (Campanoni and Blatt 2007; Žárský et al. 2009; Richter et al. 2012).

6 Conclusions and Future Prospects

Growing root hairs and pollen tubes show characteristic dense populations of vesicles in the apex together with a high rate of membrane trafficking. The traditional view about endocytosis in tip-growing plant cells is based on a compensatory role of endocytosis which would only serve as drainage for excessive material secreted in the tip. Presently, complexity and dynamics of the system are much better understood and our view on the regulation of tip growth has been adapted accordingly. Vesicles in the tip are thought to originate not only from the anterograde, secretory pathway, but also from an active endocytic system which retrieves vesicles from the plasma membrane. In full coordination with the secretion of vesicles that deliver material for expansion, particularly in rapidly tip-growing cells, compensatory endocytosis is needed to retrieve and recycle the material secreted in excess. To balance the retrograde versus anterograde membrane flow the entire system should be tightly controlled. Current results clearly show that endocytosis is involved in key regulatory mechanisms for selective removal and subsequent recycling of important molecular components and determinants for polar growth of both pollen tubes and root hairs. Endocytosis of regulatory molecules from non-growing parts of the apical domain and their repositioning to the growing zone is required to support and maintain cellular polarity and growth. Thus, secretion, endocytosis, and membrane recycling are closely interwoven and tightly controlled events. The high speed of these events is probably best illustrated by the fast internalization and redistribution of FM dyes. Once applied to growing root hairs or pollen tubes, endocytic vesicles carrying these dyes are redistributed in the tip, particularly in the clear zone, within a few minutes. Another purpose of endocytosis and vesicular recycling is to establish and maintain the accumulation of molecular determinants in the membrane. The well-defined membrane domains in the plasma membrane of the growing tip might develop through regulation of the concentrations of these molecular determinants. Cargo uptake by endocytic vesicles for recycling and repositioning to the tip include plasma membrane lipids. Thus, particular types of structural or signalling lipids may become concentrated in the different membrane domains of the tip. Also, lipid redistribution in the rest of the plasma membrane of root hairs and pollen tubes might be regulated in the same way, like for example the specific distribution of structural sterols in growing root hairs (Ovečka et al. 2010) and pollen tubes (Liu et al. 2009). The distinct domains with structural sterols in the apical plasma membrane may be involved in spatial organization of endocytosis and the following vesicular recycling, required to generate and sustain the highly polarized tip-growing zone. Moreover, the plasma membrane at the tip, with a fitting of lipid and protein composition may well act as a platform for recognition and further transduction of external signals.

However, the mechanism of tip growth is still not fully understood and many questions remain, in particular questions related to the integration of endocytosis in tip demand further attention. For example, why is the clear zone invaded by such a high number of endocytic vesicles undergoing recycling? Why is direct fusion of secretory vesicles with the apical plasma membrane a rare event? Could it be that the excess secretory vesicles is necessary because of an inefficient docking and/or fusion mechanism? How long do the endocytic vesicles spend in the clear zone at recycling? How are the endo- and exocytic vesicles in the clear zone recognized at the molecular level? Could it be that some vesicles, designed to fuse with the plasma membrane and to deliver important cargo, are redirected to the degradation route after unsuccessful contact with the plasma membrane? How precisely is the number of vesicles in the clear zone controlled and balanced? Certainly, control mechanisms must be based on the availability of different molecular markers that provide subpopulations of vesicles carrying their own specific molecular signatures. Questions remain as to which markers represent key players, how they achieve their specificity, and how markers shuttle between membrane donor and fusion compartments. The molecular signature of vesicles must be very flexible, complex and suitable for recognition at the plasma membrane. Identification and molecular characterization of new molecular markers would bring more and highly needed answers. Presently, the main message appears to be that endocytosis and vesicular recycling provide spatial and temporal clues for the establishment and maintenance of root hair and pollen tube polarity.

Acknowledgments This work was supported by grant No. ED0007/01/01 to the Centre of the Region Haná for Biotechnological and Agricultural Research, Faculty of Science, Palacký University, Olomouc, Czech Republic, by grant no. P501/11/1764 from the Czech Science Foundation (GAČR) and partly supported by grant No. 2/0200/10 from the Grant Agency VEGA.

References

Backues SK, Korasick DA, Heese A, Bednarek SY (2010) The Arabidopsis dynamin-related Protein2 family is essential for gametophyte development. Plant Cell 22:3218–3231

Baluška F, Salaj J, Mathur J, Braun M, Jasper F, Šamaj J, Chua NH, Barlow PW, Volkmann D (2000) Root hair formation: F-actin-dependent tip growth is initiated by local assembly of profilin-supported F-actin meshworks accumulated within expansin-enriched bulges. Dev Biol 227:618–632

Bednarek SY, Backues SK (2010) Plant dynamin-related protein families DRP1 and DRP2 in plant development. Biochem Soc Trans 38:797–806

Bibikova TN, Blancaflor EB, Gilroy S (1999) Microtubules regulate tip growth and orientation in root hairs of *Arabidopsis thaliana*. Plant J 17:657–665

Blackbourn HD, Jackson AP (1996) Plant clathrin heavy chain: sequence analysis and restricted localisation in growing pollen tubes. J Cell Sci 109:777–787

Bloch D, Lavy M, Efrat Y, Efroni I, Bracha-Drori K, Abu-Abied M, Sadot E, Yalovsky S (2005) Ectopic expression of an activated RAC in Arabidopsis disrupts membrane cycling. Mol Biol Cell 16:1913–1927

Borner GHH, Sherrier DJ, Weimar T, Michaelson LV, Hawkins ND, MacAskill A, Napier JA, Beale MH, Lilley KS, Dupree P (2005) Analysis of detergent-resistant membranes in Arabidopsis. Evidence for plasma membrane lipid rafts. Plant Physiol 137:104–116

Boutté Y, Frescatada-Rosa M, Men S, Chow C-M, Ebine K, Gustavsson A, Johansson L, Ueda T, Moore I, Jürgens G, Grebe M (2009) Endocytosis restricts Arabidopsis KNOLLE syntaxin to the cell division plane during late cytokinesis. EMBO J 29:546–558

Bove J, Vaillancourt B, Kroeger J, Hepler PK, Wiseman PW, Geitmann A (2008) Magnitude and direction of vesicle dynamics in growing pollen tubes using spatiotemporal image correlation spectroscopy and fluorescence recovery after photobleaching. Plant Physiol 147:1646–1658

Braun M, Baluška F, von Witsch M, Menzel D (1999) Redistribution of actin, profilin and phosphatidylinositol-4,5-bisphosphate in growing and maturing root hairs. Planta 209:435–443

Cai G, Cresti M (2009) Organelle motility in the pollen tube: a tale of 20 years. J Exp Bot 60:495–508

Cai G, Faleri C, Del Casino C, Emons AMC, Cresti M (2011) Distribution of callose synthase, cellulose synthase, and sucrose synthase in tobacco pollen tube is controlled in dissimilar ways by actin filaments and microtubules. Plant Physiol 155:1169–1190

Camacho L, Malhó R (2003) Endo/exocytosis in pollen tube apex is differentially regulated by Ca^{2+} and GTPases. J Exp Bot 54:83–92

Camacho L, Smertenko AP, Pérez-Gómez J, Hussey PJ, Moore I (2009) Arabidopsis Rab-E GTPases exhibit a novel interaction with a plasma-membrane phosphatidylinositol-4-phosphate 5-kinase. J Cell Sci 122:4383–4392

Campanoni P, Blatt MR (2007) Membrane trafficking and polar growth in root hairs and pollen tubes. J Exp Bot 58:65–74

Chen Y, Shen S, Guo Y, Lin J, Baluška F, Šamaj J (2006) Differential display proteome analysis of *Picea meyeri* pollen germination and pollen tube growth after actin-depolymerization by Latrunculin B. Plant J 47:174–195

Chen T, Teng N, Wu X, Wang Y, Tang W, Šamaj J, Baluška F, Lin J (2007) Disruption of actin filaments by latrunculin B affects cell wall construction in *Picea meyeri* pollen tube by disturbing vesicle trafficking. Plant Cell Physiol 48:19–30

Cheung AY, Wu HM (2004) Overexpression of an Arabidopsis formin stimulates supernumerary actin cable formation from pollen tube cell membrane. Plant Cell 16:257–269

Cheung AY, Wu HM (2008) Structural and signaling networks for the polar cell growth machinery in pollen tubes. Annu Rev Plant Biol 59:547–572

Cheung AY, Duan Q-H, Costa SS, de Graaf BHJ, Di Stilio VS, Feijo J, Wu H-M (2008) The dynamic pollen tube cytoskeleton: live cell studies using actin-binding and microtubule-binding reporter proteins. Mol Plant 1:686–702

Cheung AY, Niroomand S, Zou Y, Wu HM (2010) A transmembrane formin nucleates subapical actin assembly and controls tip-focused growth in pollen tubes. Proc Natl Acad Sci U S A 107:16390–16395

Chow CM, Neto H, Foucart C, Moore I (2008) Rab-A2 and Rab-A3 GTPases define a trans-Golgi endosomal membrane domain in Arabidopsis that contributes substantially to the cell plate. Plant Cell 20:101–123

de Graaf BH, Cheung AY, Andreyeva T, Levasseur K, Kieliszewski M, Wu HM (2005) Rab11 GTPase-regulated membrane trafficking is crucial for tip-focused pollen tube growth in tobacco. Plant Cell 17:2564–2579

Deeks MJ, Cvrčková F, Machesky LM, Mikitova V, Ketelaar T, Žárský V, Davies B, Hussey PJ (2005) Arabidopsis group Ie formins localize to specific cell membrane domains, interact with actin-binding proteins and cause defects in cell expansion upon aberrant expression. New Phytol 168:529–540

Derksen J, Rutten T, Lichtscheidl IK, deWin AHN, Pierson ES, Rongen G (1995) Quantitative analysis of the distribution of organelles in tobacco pollen tubes: implications for exocytosis and endocytosis. Protoplasma 188:267–276

Derksen J, Janssen G-J, Wolters-Arts M, Lichtscheidl I, Adlassnig W, Ovečka M, Doris F, Steer M (2011) Wall architecture with high porosity is established at the tip and maintained in growing pollen tubes of *Nicotiana tabacum*. Plant J 68:495–506

Dettmer J, Hong-Hermesdorf A, Stierhof YD, Schumacher K (2006) Vacuolar H1-ATPase activity is required for endocytic and secretory trafficking in Arabidopsis. Plant Cell 18:715–730

Emons AMC (1987) The cytoskeleton and secretory vesicles in root hairs of Equisetum andLimnobium and cytoplasmic streaming in root hairs of *Equisetum*. Ann Bot 60:625–632

Emons AMC, Ketelaar T (2008) Intracellular organization: a prerequisite for root hair elongation and cell wall deposition. In: Emons AMC, Ketelaar T (eds) Root hairs: excellent tools for the study of plant molecular cell biology. Springer, Berlin, Heidelberg, pp 27–44. doi:10.1007/7089_2008_4

Emons AMC, Traas JA (1986) Coated pits and coated vesicles on the plasma membrane of plant cells. Eur J Cell Biol 41:57–64

Fu Y, Wu G, Yang Z (2001) Rop GTPase-dependent dynamics of tip-localized F-actin controls tip growth in pollen tubes. J Cell Biol 152:1019–1032

Fujimoto M, Arimura S, Ueda T, Takanashi H, Hayashi Y, Nakano A, Tsutsumi N (2010) Arabidopsis dynamin-related proteins DRP2B and DRP1A participate together in clathrin-coated vesicle formation during endocytosis. Proc Nat Acad Sci U S A 107:6094–6099

Galway ME, Heckman JW, Schiefelbein JW (1997) Growth and ultrastructure of Arabidopsis root hairs: the *rhd3* mutation alters vacuole enlargement and tip growth. Planta 201:209–218

Geldner N, Anders N, Wolters H, Keicher J, Kornberger W, Muller P, Delbarre A, Ueda T, Nakano A, Jürgens G (2003) The Arabidopsis GNOM ARF-GEF mediates endosomal recycling, auxin transport, and auxin-dependent plant growth. Cell 112:219–230

Geldner N, Dénervaud-Tendon V, Hyman DL, Mayer U, Stierhof Y-D, Chory J (2009) Rapid, combinatorial analysis of membrane compartments in intact plants with a multicolor marker set. Plant J 59:169–178

Gibbon BC, Kovar DR, Staiger CJ (1999) Latrunculin B has different effects on pollen germination and tube growth. Plant Cell 11:2349–2363

Grebe M, Xu J, Möbius W, Ueda T, Nakano A, Geuze HJ, Rook MB, Scheres B (2003) Arabidopsis sterol endocytosis involves actin-mediated trafficking via ARA6-positive early endosomes. Curr Biol 13:1378–1387

Gu Y, Fu Y, Dowd P, Li S, Vernoud V, Gilroy S, Yang Z (2005) A Rho family GTPase controls actin dynamics and tip growth via two counteracting downstream pathways in pollen tubes. J Cell Biol 169:127–138

Haney CH, Long SR (2010) Plant flotillins are required for infection by nitrogen-fixing bacteria. Proc Natl Acad Sci U S A 107:478–483

Helling D, Possart A, Cottier S, Klahre U, Kost B (2006) Pollen tube tip growth depends on plasma membrane polarization mediated by tobacco PLC3 activity and endocytic membrane recycling. Plant Cell 18:3519–3534

Hepler PK, Vidali L, Cheung AY (2001) Polarized cell growth in higher plants. Annu Rev Cell and Dev Biol 17:159–187

Heslop-Harrison J (1987) Pollen germination and pollen-tube growth. Int Rev Cytol 107:1–78

Hwang J-U, Wu G, Yan A, Lee Y-J, Grierson CS, Yang Z (2010) Pollen-tube tip growth requires a balance of lateral propagation and global inhibition of Rho-family GTPase activity. J Cell Sci 123:340–350

Ischebeck T, Stenzel I, Heilmann I (2008) Type B phosphatidylinositol-4-phosphate 5-kinases mediate Arabidopsis and *Nicotiana tabacum* pollen tube growth by regulating apical pectin secretion. Plant Cell 20:3312–3330

Ischebeck T, Seiler S, Heilmann I (2010) At the poles across kingdoms: phosphoinositides and polar tip growth. Protoplasma 240:13–31

Kandasamy MK, McKinney EC, Meagher RB (2009) A single vegetative actin isovariant overexpressed under the control of multiple regulatory sequences is sufficient for normal Arabidopsis development. Plant Cell 21:701–718

Kato N, He H, Steger AP (2010) A systems model of vesicle trafficking in Arabidopsis pollen tubes. Plant Physiol 152:590–601

Ketelaar T, Faivre-Moskalenko C, Esseling JJ, de Ruijter NC, Grierson CS, Dogterom M, Emons AMC (2002) Positioning of nuclei in Arabidopsis root hairs: an actin-regulated process of tip growth. Plant Cell 14:2941–2955

Ketelaar T, de Ruijter NC, Emons AMC (2003) Unstable F-actin specifies the area and microtubule direction of cell expansion in Arabidopsis root hairs. Plant Cell 15:285–292

Ketelaar T, Galway ME, Mulder BM, Emons AMC (2008) Rates of exocytosis and endocytosis in Arabidopsis root hairs and pollen tubes. J Microsc 231:265–273

Klahre U, Becker C, Schmitt AC, Kost B (2006) Nt-RhoGDI2 regulates Rac/Rop signaling and polar cell growth in tobacco pollen tubes. Plant J 46:1018–1031

Konopka CA, Schleede JB, Skop AR, Bednarek SY (2006) Dynamin and cytokinesis. Traffic 7:239–247

Konopka CA, Backues SK, Bednarek SY (2008) Dynamics of Arabidopsis dynamin-related protein 1C and a clathrin light chain at the plasma membrane. Plant Cell 20:1363–1380

Kost B (2008) Spatial control of Rho (Rac-Rop) signaling in tip-growing plant cells. Trends Cell Biol 18:119–127

Kusano H, Testerink C, Vermeer JEM, Tsuge T, Shimada H, Oka A, Munnik T, Aoyama T (2008) The Arabidopsis phosphatidylinositol phosphate 5-kinase PIP5K3 is a key regulator of root hair tip growth. Plant Cell 20:367–380

Lam SK, Siu CL, Hillmer S, Jang S, An G, Robinson DG, Jiang L (2007) Rice SCAMP1 defines clathrin-coated, *trans*-Golgi-located tubularvesicular structures as an early endosome in tobacco BY-2 cells. Plant Cell 19:296–319

Lee YJ, Szumlanski A, Nielsen E, Yang Z (2008a) Rho-GTPase-dependent filamentous actin dynamics coordinate vesicle targeting and exocytosis during tip growth. J Cell Biol 181:1155–1168

Lee Y, Bak G, Choi Y, Chuang W-I, Cho H-T, Lee Y (2008b) Roles of phosphatidylinositol 3-kinase in root hair growth. Plant Physiol 147:624–635

Lee Y, Kim E-S, Choi Y, Hwang I, Staiger CJ, Chung Y-Y, Lee Y (2008c) The Arabidopsis phosphatidylinositol 3-kinase is important for pollen development. Plant Physiol 147:1886–1897

Lennon KA, Lord EM (2000) In vivo pollen tube cell of *Arabidopsis thaliana*. I. Tube cell cytoplasm and wall. Protoplasma 214:45–56

Li R, Liu P, Wan Y, Chen T, Wang Q, Mettbach U, Baluška F, Šamaj J, Fang X, Lucas WJ, Lin J (2012) A membrane microdomain-associated protein, Arabidopsis Flot1, is involved in a clathrin-independent endocytic pathway and is required for seedling development. Plant Cell 24:2105–2122

Lin Y, Wang J, Zhu J, Yang Z (1996) Localization of a rho GTPase implies a role in tip growth and movement of the generative cell in pollen tubes. Plant Cell 8:293–303

Lisboa S, Scherer GEF, Quader H (2008) Localized endocytosis in tobacco pollen tubes: visualisation and dynamics of membrane retrieval by a fluorescent phospholipid. Plant Cell Rep 27:21–28

Liu P, Li R-L, Zhang L, Wang Q-L, Niehaus K, Baluška F, Šamaj J, Lin J-X (2009) Lipid microdomain polarization is required for NADPH oxidase-dependent ROS signaling in *Picea meyeri* pollen tube tip growth. Plant J 60:303–313

Lovy-Wheeler A, Cárdenas L, Kunkel JG, Hepler PK (2007) Differential organelle movement on the actin cytoskeleton in lily pollen tubes. Cell Motil Cytoskeleton 64:217–232

Miedema H, Demidchik V, Véry A-A, Bothwell JHF, Brownlee C, Davies JM (2008) Two voltage-dependent calcium channels co-exist in the apical plasma membrane of *Arabidopsis thaliana* root hairs. New Phytol 179:378–385

Miller DD, de Ruijter NCA, Emons AMC (1997) From signal to form: aspects of the cytoskeleton plasma membrane–cell wall continuum in root hair tips. J Exp Bot 48:1881–1896

Miller DD, De Ruijter NCA, Bisseling T (1999) The role of actin in root hair morphogenesis: studies with lipochito oligosaccharide as a growth stimulator and cytochalasin as an actin perturbing drug. Plant J 17:141–154

Molendijk AJ, Ruperti B, Palme K (2004) Small GTPases in vesicle trafficking. Curr Opin Plant Biol 7:694–700

Moscatelli A, Idilli AI (2009) Pollen tube growth: a delicate equilibrium between secretory and endocytic pathways. J Int Plant Biol 51:727–739

Moscatelli A, Ciampolini F, Rodighiero S, Onelli E, Cresti M, Santo N, Idilli A (2007) Distinct endocytic pathways identified in tobacco pollen tubes using charged nanogold. J Cell Sci 120:3804–3819

Müller J, Mettbach U, Menzel D, Šamaj J (2007) Molecular dissection of endosomal compartments in plants. Plant Physiol 145:293–304

Nielsen E, Cheung AY, Ueda T (2008) The regulatory RAB and ARF GTPases for vesicular trafficking. Plant Physiol 147:1516–1526

Otto GP, Nichols BJ (2011) The roles of flotillin microdomains—endocytosis and beyond. J Cell Sci 124:3933–3940

Ovečka M, Lichtscheidl IK (2006) Sterol endocytosis and trafficking in plant cells. In: Šamaj J, Baluška F, Menzel D (eds) Plant endocytosis. Springer, Berlin, Heidelberg, pp 117–137

Ovečka M, Nadubinská M, Volkmann D, Baluška F (2000) Actomyosin and exocytosis inhibitors alter root hair morphology in *Poa annua*. Biologia 55:105–114

Ovečka M, Lang I, Baluška F, Ismail A, Illeš P, Lichtscheidl IK (2005) Endocytosis and vesicle trafficking during tip growth of root hairs. Protoplasma 226:39–54

Ovečka M, Baluška F, Lichtscheidl IK (2008) Non-invasive microscopy of tip growing root hairs as a tool for study of dynamic, cytoskeleton-based processes. Cell Biol Int 32:549–553

Ovečka M, Berson T, Beck M, Derksen J, Šamaj J, Baluška F, Lichtscheidl IK (2010) Structural sterols are involved in both the initiation and tip growth of root hairs in *Arabidopsis thaliana*. Plant Cell 22:2999–3019

Pei W, Du F, Zhang Y, He T, Ren H (2012) Control of the actin cytoskeleton in root hair development. Plant Sci 187:10–18

Picton JM, Steer MW (1983) Membrane recycling and the control of secretory activity in pollen tubes. J Cell Sci 63:303–310

Preuss ML, Serna J, Falbel TG, Bednarek SY, Nielsen E (2004) The Arabidopsis Rab GTPase RabA4b localizes to the tips of growing root hair cells. Plant Cell 16:1589–1603

Preuss ML, Schmitz AJ, Thole JM, Bonner HK, Otegui MS, Nielsen E (2006) A role for the RabA4b effector protein PI-4 Kβ1 in polarized expansion of root hair cells in *Arabidopsis thaliana*. J Cell Biol 172:991–998

Richter S, Geldner N, Schrader J, Wolters H, Stierhof Y-D, Rios G, Koncz C, Robinson DG, Jürgens G (2007) Functional diversification of closely related ARF-GEFs in protein secretion and recycling. Nature 448:488–492

Richter S, Müller LM, Stierhof Y-D, Mayer U, Takada N, Kost B, Vieten A, Geldner N, Koncz C, Jürgens G (2012) Polarized cell growth in Arabidopsis requires endosomal recycling mediated by GBF1-related ARF exchange factors. Nat Cell Biol 14:80–87

Ridge RW (1995) Micro-vesicles, pyriform vesicles and macrovesicles associated with the plasma membrane in the root hairs of *Vicia hirsuta* after freeze-substitution. J Plant Res 108:363–368

Ringli C, Baumberger N, Diet A, Frey B, Keller B (2002) ACTIN2 is essential for bulge site selection and tip growth during root hair development of Arabidopsis. Plant Physiol 129:1464–1472

Robertson JG, Lyttleton P (1982) Coated and smooth vesicles in the biogenesis of cell walls, plasma membranes, infection threads and peribacterioid membranes in root hairs and nodules of white clover. J Cell Sci 58:63–78

Šamaj J, Ovečka M, Hlavačka A, Lecourieux F, Meskiene I, Lichtscheidl I, Lenart P, Salaj J, Volkmann D, Bögre L, Baluška F, Hirt H (2002) Involvement of the mitogen-activated protein kinase SIMK in regulation of root hair tip-growth. EMBO J 21:3296–3306

Šamaj J, Baluška F, Voigt B, Schlicht M, Volkmann D, Menzel D (2004) Endocytosis, actin cytoskeleton and signaling. Plant Physiol 135:1150–1161

Šamaj J, Müller J, Beck M, Böhm N, Menzel D (2006) Vesicular trafficking, cytoskeleton and signalling in root hairs and pollen tubes. Trends Plant Sci 11:594–600

Shaw SL, Dumais J, Long SR (2000) Cell surface expansion in polarly growing root hairs of *Medicago truncatula*. Plant Physiol 124:959–969

Sieberer BJ, Ketelaar T, Esseling JJ, Emons AMC (2005) Microtubules guide root hair tip growth. New Phytol 167:711–719

Sousa E, Kost B, Malhó R (2008) Arabidopsis phosphatidylinositol-4-monophosphate 5-kinase 4 regulates pollen tube growth and polarity by modulating membrane recycling. Plant Cell 20:3050–3064

Stenmark H (2009) Rab GTPases as co-ordinators of vesicle traffic. Nat Rev Mol Cell Biol 10:513–525

Stenzel I, Ischebeck T, Konig S, Holubowska A, Sporysz M, Hause B, Heilmann I (2008) The type B phosphatidylinositol-4-phosphate 5-kinase 3 is essential for root hair formation in *Arabidopsis thaliana*. Plant Cell 20:124–141

Szumlanski AL, Nielsen E (2009) The Rab GTPase RabA4d regulates pollen tube tip growth in *Arabidopsis thaliana*. Plant Cell 21:526–544

Takáč T, Pechan T, Šamajová O, Ovečka M, Richter H, Eck C, Niehaus K, Šamaj J (2012) Wortmannin treatment induces changes in Arabidopsis root proteome and post-Golgi compartments. J Proteome Res 11:3127–3142

Taylor NG (2011) A role for Arabidopsis dynamin related proteins DRP2A/B in endocytosis; DRP2 function is essential for plant growth. Plant Mol Biol 76:117–129

Thole JM, Nielsen E (2008) Phosphoinositides in plants: novel functions in membrane trafficking. Curr Op Plant Biol 11:620–631

Traas JA, Braat P, Emons AMC, Meekes H, Derksen J (1985) Microtubules in root hairs. J Cell Sci 76:303–320

Van Bruaene N, Joss G, Van Oostveldt P (2004) Reorganization and in vivo dynamics of microtubules during Arabidopsis root hair development. Plant Physiol 136:3905–3919

Vernoud V, Horton AC, Yang Z, Nielsen E (2003) Analysis of the small GTPase gene superfamily of Arabidopsis. Plant Physiol 131:1191–1208

Vidali L, McKenna ST, Hepler PK (2001) Actin polymerization is essential for pollen tube growth. Mol Biol Cell 12:2534–2545

Voigt B, Timmers ACJ, Šamaj J, Hlavačka A, Ueda T, Preuss M, Nielsen E, Mathur J, Emans N, Stenmark H, Nakano A, Baluška F, Menzel D (2005) Actin-propelled motility of endosomes is tightly linked to polar tip-growth of root hairs. Eur J Cell Biol 84:609–621

Wang X, Teng Y, Wang Q, Li X, Zheng M, Šamaj J, Baluška F, Lin J (2006) Dynamic imaging of vesicles in living pollen tubes of *Picea meyeri* using evanescent wave microscopy. Plant Physiol 141:1591–1603

Wang Y, Chen T, Hao H, Liu P, Zheng M, Baluška F, Šamaj J, Lin J (2009) Nitric oxide is involved in cell wall construction in *Pinus bungeana* pollen tubes by modulating extracellular Ca^{2+} influx and actin filaments organization. New Phytol 182:851–862

Wang H, Tse YC, Law AHY, Sun SSM, Sun Y-B, Xu Z-F, Hillmer S, Robinson DG, Jiang L (2010a) Vacuolar sorting receptors (VSRs) and secretory carrier membrane proteins (SCAMPs) are essential for pollen tube growth. Plant J 61:826–838

Wang Y, Zhu Y, Ling Y, Zhang H, Liu P, Baluška F, Šamaj J, Lin J, Wang Q (2010b) Disruption of actin filaments induces mitochondrial Ca^{2+} release to the cytoplasm and $[Ca^{2+}]_c$ changes in Arabidopsis root hairs. BMC Plant Biol 10:53

Wang H, Zhuang X-H, Hillmer S, Robinson DG, Jiang L-W (2011) Vacuolar sorting receptor (VSR) proteins reach the plasma membrane in germinating pollen tubes. Mol Plant 4:845–853

Winship LJ, Obermeyer G, Geitmann A, Hepler PK (2010) Under pressure, cell walls set the pace. Trends Plant Sci 15:363–369

Yalovsky S, Bloch D, Sorek N, Kost B (2008) Regulation of membrane trafficking, cytoskeleton dynamics, and cell polarity by ROP/RAC GTPases. Plant Physiol 147:1527–1543

Yoo C-M, Wen J, Motes CM, Sparks JA, Blancaflor EB (2008) A class I ADP-ribosylation factor GTPase-activating protein is critical for maintaining directional root hair growth in Arabidopsis. Plant Physiol 147:1659–1674

Yoo C-M, Quan L, Cannon AE, Wen J, Blancaflor EB (2012) AGD1, a class 1 ARF-GAP, acts in common signaling pathways with phosphoinositide metabolism and the actin cytoskeleton in controlling Arabidopsis root hair polarity. Plant J 69:1064–1076

Žárský V, Cvrčková F, Potocký M, Hála M (2009) Exocytosis and cell polarity in plants-exocyst and recycling domains. New Phytol 183:255–272

Zhang Y, McCormick S (2010) The regulation of vesicle trafficking by small GTPases and phospholipids during pollen tube growth. Sex Plant Reprod 23:87–93

Zhang Y, He J, Lee D, McCormick S (2010) Interdependence of endomembrane trafficking and actin dynamics during polarized growth of Arabidopsis pollen tubes. Plant Physiol 152:2200–2210

Zheng H, Camacho L, Wee E, Batoko H, Legen J, Leaver CJ, Malhó R, Hussey PJ, Moore I (2005) A Rab-E GTPase mutant acts downstream of the Rab-D subclass in biosynthetic membrane traffic to the plasma membrane in tobacco leaf epidermis. Plant Cell 17:2020–2036

Zheng M, Beck M, Müller J, Wang X, Wang Q, Baluška F, Logan DC, Šamaj J, Lin J (2009) Cytoskeleton-dependent mitochondrial movements in Arabidopsis root hairs. PLoS One 4:e5961, 1–14

Zheng M, Wang Q, Teng Y, Wang X, Wang F, Chen T, Šamaj J, Lin J, Logan DC (2010) The speed of mitochondrial movement is regulated by the cytoskeleton and myosin in *Picea wilsonii* pollen tubes. Planta 231:779–791

Zonia L (2010) Spatial and temporal integration of signalling networks regulating pollen tube growth. J Exp Bot 61:1939–1957

Zonia L, Munnik T (2008) Vesicle trafficking dynamics and visualization of zones of exocytosis and endocytosis in tobacco pollen tubes. J Exp Bot 59:861–873

Zou Y, Aggarwal M, Zheng W-G, Wu H-M, Cheung AY (2011) Receptor-like kinases as surface regulators for RAC/ROP-mediated pollen tube growth and interaction with the pistil. AoB Plants: plr017

Fluid-Phase Endocytosis in Plant Cells

Ed Etxeberria, Javier Pozueta-Romero and Edurne Baroja Fernández

Abstract The uptake of nutrients by plant cells has been traditionally believed to be mediated by membrane-bound carriers. However, the last decade has seen an increase in evidence pointing to the parallel uptake by fluid-phase endocytosis (FPE). Recent advances in plant endocytosis reveal that this is true for heterotrophic cells, whether storage parenchyma, cell suspensions, or nutrient absorbing cells of carnivorous plants. Uptake of extracellular matrix components, endocytic markers, and sugar analogs in a wide variety of heterotrophic cells has confirmed the uptake of extracellular fluids and their transport to the vacuole. Furthermore, there is evidence to indicate the passage through an intracellular compartment where solutes are distributed. The precise nature of FPE has not been revealed; however, evidence using specific inhibitors, CdSe/ZnS quantum dots in combination with other FPE markers and inhibitors such as ikargalukin, points to the clathrin-independent nature of FPE and its possible association with flotillin. That FPE operates in conjunction with membrane-bound transporters in the uptake of solutes is supported by experiments analyzing uptake kinetics of the fluorescent endocytic marker Alexa-488 in the presence of sucrose and membrane-bound transporters and endocytic inhibitors. The mechanisms of membrane remodeling to accommodate the addition of membrane and aqueous volume to the vacuole during FPE remain unresolved.

E. Etxeberria (✉)
Department of Horticultural Sciences, Citrus Research and Education Center, University of Florida/IFAS, 700 Experiment Station Road, Lake Alfred, FL 33850, USA
e-mail: eetxeber@ufl.edu

J. Pozueta-Romero · E. B. Fernández
Instituto de Agrobiotecnologia, Universidad Publica de Navarra/Consejo de Investigaciones Cientificas, Gobierno de Navarra, Mutiloako etorbidea zenbaki gabe, 31192, Mutiloabeti, Nafarroa, Spain

J. Šamaj (ed.), *Endocytosis in Plants*,
DOI: 10.1007/978-3-642-32463-5_5,

1 Introduction

Endocytosis, the uptake of membrane proteins and lipids, extracellular fluids, ligands, and soluble molecules, achieved through invagination and vesiculation of the plasma membrane (Šamaj et al. 2004) is a hallmark of all eukaryotic cells (Nichols and Lippincott-Schwartz 2001). In plant cells, however, despite mounting evidence, years of debate delayed the inevitable recognition that endocytosis is also an inherent fundamental mechanism. Several types of endocytic pathways have been identified in animal systems, some of which are just beginning to be uncovered in plant cells. By definition and operational design, all types of endocytosis (whether clathrin-mediated, caveolin-mediated, lipid raft-dependent, flotillin-dependent, and micro- or macropinocytosis; Moscatelli et al. 2007; Lin and Guttman 2010) carry a fluid-phase component. The fluid component comprises the extracellular solution trapped during vesicle formation (Holstein 2002; Kruth et al. 2005). Although the volume engulfed and transported by endocytic vesicles varies according to the type of endocytic pathway (which determines vesicle size), it is difficult to envision the continuous incorporation of extracellular fluids not being a significant contributor to the process of nutrient uptake in sink cells and the restructuring of the external framework during cell growth. In this chapter, we use the term fluid-phase endocytosis (FPE) to describe the transport of extracellular fluids mainly by distinct clathrin- and caveolin-independent pathways, which share characteristics with micropinoytosis. However, fluids entering by clathrin-mediated or any other defined endocytic vesicle system are also considered as part of the overall fluid-phase uptake process.

The existence of some form of FPE in plants cells was first suggested by the observations of Jensen and McLaren (1960) that onion and barley roots accumulated externally supplied ^{14}C and ^{3}H-labeled ribonuclease, hemoglobin, and lysozyme in intracellular compartments. Evidence of FPE was also inferred from the uptake of impermeant molecules such as heavy metals (Hubner et al. 1985; Lazzaro and Thompson 1992), biotinylated molecules (Horn et al. 1990), gold conjugated proteins and lectin (Hilmer et al. 1986; Villanueva et al. 1993), and cationic ferritin (Joachim and Robinson 1984; Tanchak et al. 1984) into a variety of intact plant cells, protoplasts, and cells in culture. In all instances, the marker molecules were internalized and observed in several internal structures including clathrin-coated vesicles (Hubner et al. 1985), larger membrane-bound organelles (Šamaj et al. 2005; Etxeberria et al. 2006; Onelli et al. 2008), multivesicular bodies (Samuels and Bisalputra 1990) and in some instances, the central vacuole (Oparka and Prior 1988; Cholewa and Peterson 2001; Baluska et al. 2004; Etxeberria et al. 2005a). Although the final destination of endocyted substances was not always consistent likely due to time of incubation as well as diversity of tissues, marker molecules, and methods utilized, it became clear from these observations that internalization of external substances had occurred in an endocytic manner.

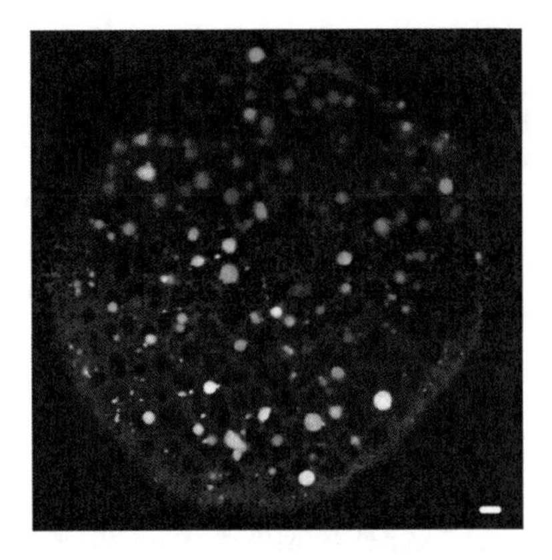

Fig. 1 Nutrient absorbing gland of the carnivorous plant *Nepenthes coccinea* after incubation with FITC-labeled BSA. Fluorescent vesicles and vacuoles indicate the endocytic uptake of the labeled marker into the cells

Contemporary detailed anatomical studies of red beet storage cells during sucrose accumulation revealed the formation of plasma membrane vesicles which were frequently observed penetrating the storage vacuole (Paramonova 1974).

2 Evidence for a Distinct FPE System

From studies elsewhere, it has become evident that endocytosis plays a pivotal role in diverse cellular processes such as membrane recycling (Holstein 2002), protein transport (Lam et al. 2005; Geldner and Jergens 2006), receptor-mediated mechanisms (Russinova et al. 2004; Gross et al. 2005) and cell signalling (Li and Xue 2007). The complexity and variability of these functions are fulfilled by different endocytic systems as demonstrated extensively for animal (Hasumi et al. 1992; Luo et al. 2001) and more recently for plant cells (Moscatti et al. 2007; Onelli et al. 2008). Whereas the majority of these processes are supported by an endocytic system primarily composed of vesicles formed at the plasma membrane with the assistance of a clathrin lattice, a series of clathrin-independent mechanisms have been described that serve overlapping functions (Mayor and Pagano 2007; Sandrig et al. 2008). Aside from the work cited in the previous section, the work of Baluska et al. (2002), Emans et al. (2002), Šamaj et al. (2005), Etxeberria et al. (2005a, b, 2006, 2007a, b), Pozueta et al. (2008) and Adlassnig et al. (2012) offered strong support for a distinctive FPE system in plants. In most instances, uptake of extracellular fluids was inhibited by various endocytic inhibitors and actin depolymerizing agents. Figure 1 depicts the endocytic uptake of FITC-labeled BSA by the nutrient absorbing cells of the carnivorous plant *Nepenthes coccinea*. When the pitcher trap was filled with a nutrient solution containing FITC-BSA, nutrient absorbing cells took up the labeled marker via endocytosis. Fluorescence can be seen in small cytosolic vesicles as well as the central vacuole.

Firm demonstration of a distinct FPE system in plant cells came from the work of Etxeberria et al. (2009) who established the functioning of two parallel endocytic pathways. This conclusion was derived from experiments using a membrane-impermeable form of the Na-dependent fluorescent marker Coro-Na in combination with the fluorescent membrane marker FM 4-64. When protoplasts from sweet lime juice cells were incubated in Na-free solution, FM 4-64, Coro-Na, and 200 mM sucrose, two distinct types of labeled vesicles were evident (Fig. 2). A set of vesicles ($\sim$1 μm in diameter) was intensely labeled with Coro-Na and to a lesser extent with FM 4-64, whereas the second type of 1–7 μm structures appeared exclusively labeled with FM 4-64. These data demonstrate the parallel functioning of two endocytic pathways in these plant cells. In one system, a set of small endocytic vesicles merge with Na-containing endosomal organelles (green fluorescence), whereas a separate set of vesicles apparently fuse to form larger structures independent of the endosome given the absence of Na (FM 4-64 fluorescence only). Although it is likely that both vesicle systems eventually contribute to solutes reaching the vacuole, given their size (1–7 μm) and concomitant volume, the authors concluded that these latter vesicles represent the primary FPE vesicle pathway.

Additional evidence in support of a distinct FPE system independent of clathrin came from observations of endocytic vesicles formation at the plasma membrane of citrus juice cells incubated in Alexa 488 and sucrose (Etxeberria et al. 2007b). Figure 3 shows vesicle formation protruding from the plasma membrane. These vesicles are larger than clathrin-coated vesicles in plant cells of around 70–120 nm (Thiel et al. 1998; Dhonukshe et al. 2007; Gall et al. 2010) and are more in line with micropinocytosis (Kruth et al. 2005; Cao et al. 2007). Although it is arguable that these vesicles could correspond to osmotic excursions (Diekmann et al. 1993), the large vesicles observed in the cytosol and eventual deposition of Alexa 488 in the vacuole (Etxeberria et al. 2005a) argues against this possibility.

Live observations of endocytic vesicles merging with the central vacuole, an event whose brevity makes it difficult to document, adds further evidence for a distinct FPE system. Figure 4 shows this event taking place in turnip protoplasts incubated in Alexa-488 and 200 mM sucrose. Turnip protoplasts took up extracellular fluids in an endocytic manner as depicted by fluorescent cytosolic structures of several sizes. At that time, the vacuole also showed intense fluorescence together with the Alexa 488-containing extracellular fluids. During the observation process, one of the two largest vesicles was captured merging with the vacuole (Fig. 4b), a process that took approximately 2 s. (personal observation). Figure 4c shows the cell after the merging of the FPE vesicle with the vacuole was finalized. The fluorescent vesicle visible at the same location as the one recently merged is a different structure that shifted in during the process, and not a remnant of the merged vesicle. Cholewa and Peterson (2001) also witnessed labeled vesicles merging with the vacuole, although capture images were not presented likely due to the spontaneous and rapid nature of this event.

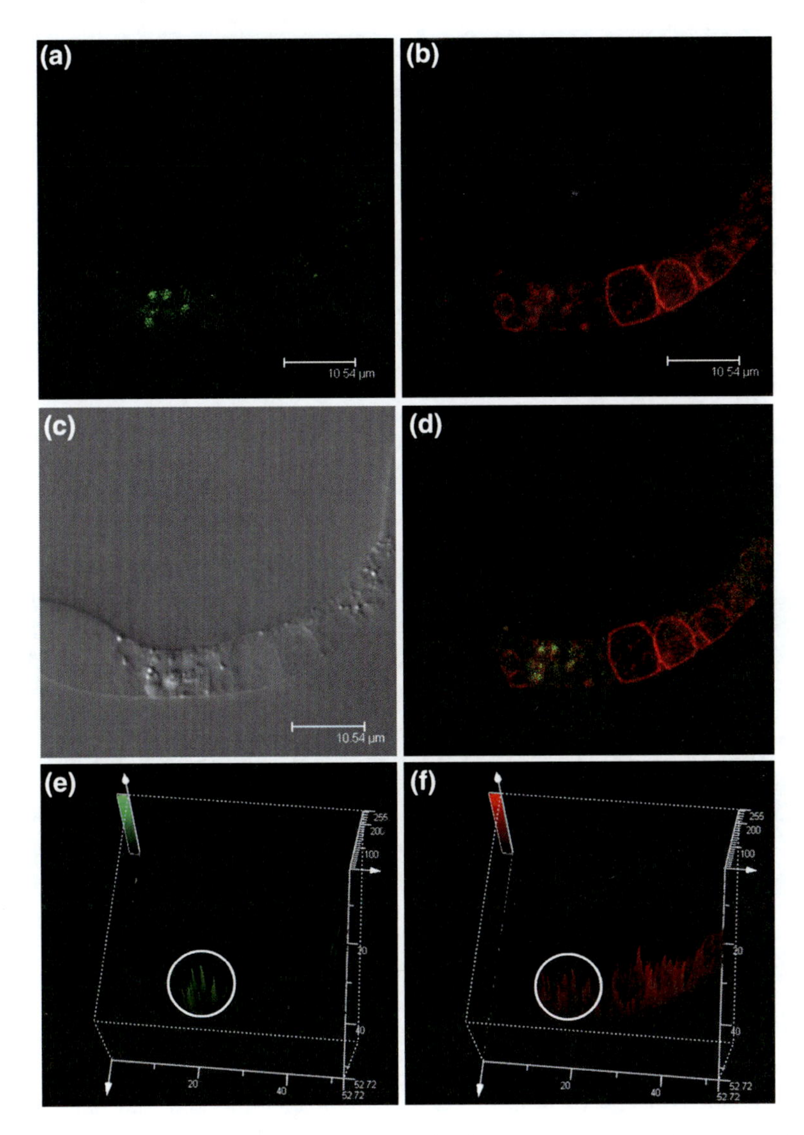

Fig. 2 Composite fluorescent confocal micrograph of a citrus juice protoplast prepared after 48 h incubation of whole tissue in 250 mM sucrose, 40 mM Na-dependent membrane-impermeable fluorescent marker Coro-Na, and 20 mM FM 4-64. Cells show *green fluorescence* of Coro-Na (**a**), *red fluorescence* of FM 4-64 (**b**), and partial overlap of both in (**d**, *yellow color*). Differential interference contrast (DIC) image of cells is shown in (**c**), The persistent lack of *green* overlapping within larger red fluorescent vesicles is best observed in the topographic analysis of (**e**) and (**f**)

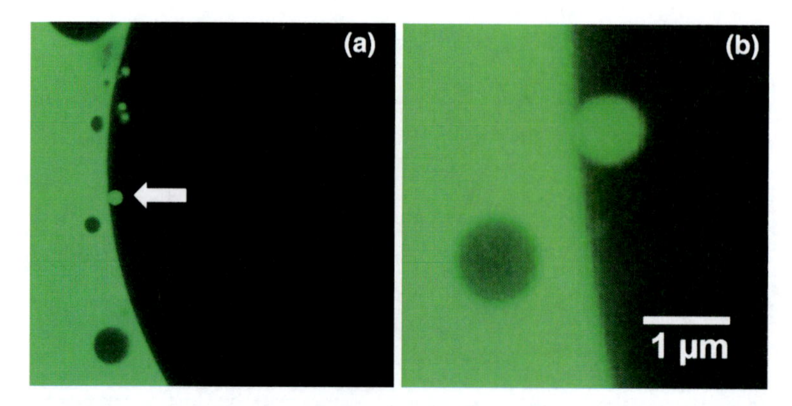

Fig. 3 Endocytic vesicle formation in citrus juice cells. **a** Single Z-plane image showing the formation of endocytic vesicles at the cytosolic side of the plasma membrane in citrus juice cells. Fluorescence (*lighter gray tone*) comes from Alexa 488 in the sucrose-containing incubation media. **b** A close up of vesicle indicated by an arrow in A showing the continuity of fluorescence between the medium and the vesicle interior

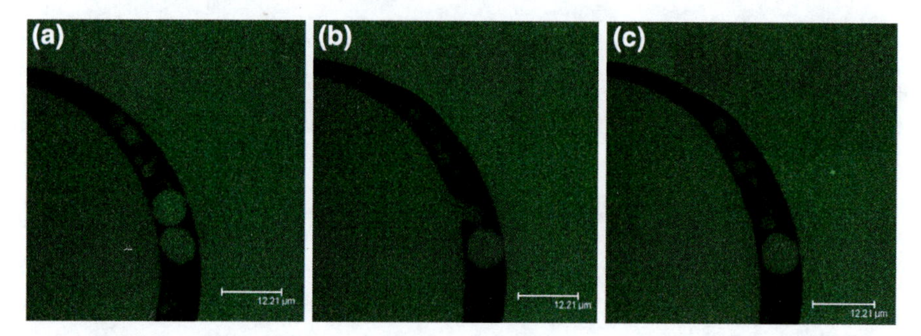

Fig. 4 Merging of endocytic vesicle with the vacuole. Sequence of events in the merging of an endocytic vesicle with the vacuole in a turnip storage parenchyma protoplast

3 FPE is a Clathrin-Independent System

Using ikargalukin (IKA), a specific inhibitor of clathrin-mediated endocytosis, Onelli et al. (2008) and Bandmann and Homann (2012) established that FPE is a clathrin-independent process. In tobacco protoplasts, for example, the differential deposition of positive and negative nanoparticles to different intracellular compartments and differential response to IKA demonstrated two separate endocytic pathways, a clathrin-dependent and a clathrin-independent system (Onelli et al. 2008). A similar conclusion was reached by Bandmann and Homann (2012) using tobacco BY-2 protoplasts. Incubation of protoplasts in the fluorescent glucose-derivative 2-NBDG [2(N-(7-Nitrobenz-2-oxa-1,3-diazol-4-yl)amino)-2-deoxyglucose] resulted in the accumulation of the sugar into endosomal compartments.

Endocytic uptake of 2-NBDG was not impeded by IKA, corroborating the clathrin-independent nature of FPE.

There is now compelling evidence to support the involvement of flotillin1with lipid-raft-dependent FPE in plant cells (Li et al. 2012). In Arabidopsis, the spatial separation of fluorescent signals associated with GFP-AtFlot1 (flotillin 1) and CLC (clathrin light chain)-mOrange structures within transgenic root cells provided strong support for the notion that AtFlot1 is associated with clathrin-independent endocytic pathway in plants. The use of VA-TIRFM imaging (variable-angle total internal reflection fluorescence microscopy) further cemented the independent behavior of AtFlot1-positive structures from CLC-positive puncta. These data, in addition to the demonstration that tyrA23 treatment did not change the dynamics of the GFP-AtFlot1 puncta, further confirmed that AtFlot1 operates independent of clathrin-mediated endocytosis.

Capacitance measurements of endocytic events during FPE in tobacco BY-2 protoplasts estimated the size of FPE vesicles to be an average of 133 nm (Bandman and Homann 2012). These estimates are similar to those obtained by Gall et al. (2010) in turgid guard cells. In their work, Li et al. (2012) also estimated the endocytic vesicles at around 100 nm. These measurements are in disagreement with images captured during vesicle formation at the plasma membrane (Etxeberria et al. 2007b). It is possible that different types of clathrin-independent FPE vesicles may exist including a population of rapidly forming smaller vesicles with an average size 120–130 nm and a slower forming set of vesicles of about 1 μm in diameter. The much larger structures containing endocytic markers commonly observed in most studies are likely merged vesicles or intermediary endosomal organelles (early endosome, TGN) which temporarily accumulate internalized solutes.

Unlike the classical formation of endocytic vesicles at the plasma membrane (events documented for FPE and other endocytic systems), a distinct FPE system has been reported (Kurkova et al. 1994; Neumann and Figuereido 2002; Balnokin et al. 2007). This event is characterized by the protrusion of large plasma membrane invaginations (sometimes over 1 μm) directly into the vacuole forming a vesicle surrounded by two membranes once internalized into the vacuolar lumen. These structures, presumably containing extracellular fluids, are shown to be responsible for the uptake of silicon and zinc in leaf parenchyma cells of *Nicotiana tabacum*, *Arabidopsis thaliana* and other species (Neumann and Figuereido 2002), and salt accumulation in *Suadea altissima* root cells (Balnokin et al. 2007). Aside from extracellular fluids, these vesicles are consistently filled with membranous structures giving the appearance of a multi vesicular body (Kurkova and Balnokin 1994; Neumann and Figuereido 2002; Balnokin et al. 2007).

4 Involvement of FPE in Nutrient and Photoassimilate Uptake

Baluška et al. (2002) revealed for the first time that FPE is intimately involved in physiological processes other than membrane recycling. Working with meristematic maize root cells, Baluška and Co-workers (2002) observed JIM5 reactive pectins accumulating in intracellular compartments and within cell wall plates together with calcium cross-linked RGII pectins. Cell wall components, arabinogalactans, and carbohydrates have also been observed in multivesicular bodies of stylar transmitting tissue of *Datura* (Hudák et al. 1993) and bean root cells (Northcote et al. 1989). In conjunction, these reports demonstrated that external solutes could be trapped by FPE as part of the recycling of membrane components and formation or restructuring of new cell walls.

A series of subsequent reports by Etxeberria and colleagues (Etxeberria et al. 2005a, b, 2006, 2007a, b), and more recently by Bandmann and Homann (2012) and Adlassnig et al. (2012), demonstrated that FPE also mediates nutrient uptake and eventual transport to the vacuole and other compartments in heterotrophic plant cells. According to Paramanova (1974), such a system would allow for faster and more efficient exchange of solutes between the apoplast and the vacuole, and would protect solutes from enzymatic attack or degradation in the cytosol. Storage cells, especially those in specialized long-term storage organs, accumulate photoassimilates during prolonged periods to be subsequently mobilized to support dormancy and resumption of growth. The initial demonstration that under certain circumstances sucrose uptake and accumulation occurs in part by FPE came from experiments using sycamore cultured cells (Etxeberria et al. 2005a). After a starvation period to enhance sucrose uptake, cells incubated in complete growth media containing sucrose took extracellular fluids at rapid rates. However, in the presence of endocytic inhibitors (wortmannin A or LY 294002), uptake was drastically curtailed. Using the endocytic probe Lucifer yellow-CH in conjunction with sucrose measurements, the authors established that both substances are taken up in parallel as demonstrated by their similar uptake kinetics and observations using epifluorescence microscopy (Etxeberria et al. 2005a). Subsequent studies conducted by the same authors provided strong evidence that a sizable pool of sucrose taken up by endocytosis is utilized for starch biosynthesis in heterotrophic cells (Baroja-Fernández et al. 2006). Further confocal microscopy analysis of *Citrus* juice cells using two membrane-impermeable dyes of different size and charge (Alexa 488 and 3,000 mw dextran-Texas Red; d-TR) (Fig. 5) revealed the co-localization of both markers in the vacuole and in smaller cytosolic vesicle-like structures (Etxeberria et al. 2005a). The significance of these data is that in both cases, uptake of appreciable levels of externally supplied soluble fluorescent endocytic markers only took place in the presence of sucrose in the incubation media.

The lack of significant uptake of externally supplied fluorescent endocytic soluble markers in the absence of sucrose was of particular interest in that it

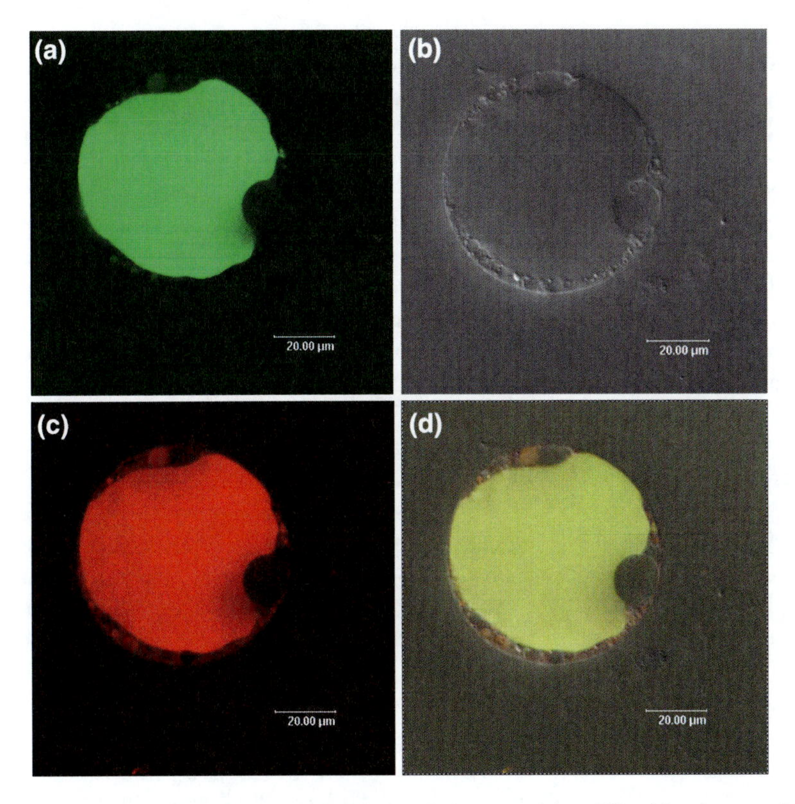

Fig. 5 Citrus juice cells after incubation in a solution containing 250 mM sucrose and two membrane-impermeable dyes (Alexa 488 at 100 μM and dTR at 1 mg ml^{-1}) for 12 h. These fluorescent soluble probes vary considerably in molecular size and charge. Protoplasts were prepared following incubation, and the juice cells observed under fluorescent Nomarski microscopy with appropriate filters (Leica TCS SL). **a**, **c** and **d** showing *green fluorescence* of Alexa 488, *red fluorescence* of dTR and *yellow* merged image of both, respectively

implied either that external sucrose accelerated the existing endocytic machinery or that its presence activated a different endocytic system. Detailed experiments carried out with celery parenchyma cells incubated in the membrane-impermeable fluorescent dextran (Texas Red-labeled mw. 3,000 dextran; d-TR) and mannitol confirmed the increase in endocytic uptake in the presence of an external carbohydrate source (Etxeberria et al. 2007a). When supplied alone in the incubation medium in the absence of mannitol, celery stalk parenchyma accumulated low levels of d-TR in the vacuole. Under these conditions, the uptake of this membrane-impermeable substance was presumed to be mediated by a basal endocytosis system responsible for the turnover of membrane components (Robinson et al. 1998). Uptake of d-TR was almost completely abolished by incubation in the endocytic inhibitor LY294002. Addition of mannitol into the incubation medium (without endocytic inhibitors) tripled basal d-TR uptake levels, corroborating the involvement in carbohydrate uptake. When mannitol and LY 294002 were added together, mannitol uptake was reduced below basal levels. FPE was also induced, albeit not so strongly, by glucose and sucrose in the same cells. Mannitol-induced FPE was substantially reduced in the presence of glucose and sucrose, suggesting that celery stalk parenchyma cells possess a single sensor recognizing different sugars and with different degrees of sugar-sensing capacity. The induction of FPE by external carbohydrate and transport to the vacuole was confirmed by Bandmann and Homann (2012) working with tobacco BY-2 cells and implied by Onelli et al. (2008) where 580 mM sucrose was used to enhance endocytosis in tobacco protoplasts. FPE vesicle formation determined with capacitance was significantly increased when glucose or 2-NBDG was added to the incubation media (Bandmann and Homann, 2012).

By using cytoplasts (cells without vacuoles) incubated with the membrane-impermeable Texas Red-fluorescent dextran (MW 3,000; d-TR) and the green fluorescent glucose analog 2-NBDG, Etxeberria et al. (2005b) demonstrated that participation of FPE in the overall process of extracellular solute uptake does not exclude the involvement of membrane-bound transporters. During incubation, cytoplasts took up both markers differentially. The fluorescent cytosol was indicative of 2-NBGD transport across the plasma membrane, while the overlapping fluorescence of 2-NBDG and d-TR in internal structures demonstrated their co-localization in endocytic vesicles (Etxeberria et al. 2005b). The above results were later confirmed in tobacco BY-2 cells (Bandmann and Homann 2012) using 2-NBDG and FM 4-64. Taken together, the data from sycamore cytoplasts (Etxeberria et al. 2005b) and tobacco BY-2 protoplasts (Bandmann and Homann 2012) demonstrate that glucose uptake into heterotrophic cells involves separate but synchronous transport processes. Each process determines the final destination of the hexose molecule i.e., cytosol or vacuole. Sugars to be stored in the vacuole are transported in bulk by an endocytic transport system, whereas those needed to support immediate metabolic demands are transported to the cytosol by plasma membrane-bound carriers.

The role of FPE in sucrose accumulation was later defined by experiments using phloridzin and latrunculin B (inhibitors of sucrose/H^+ symport and actin polymerization, respectively) and the endocytic marker Alexa 488 (Pozueta et al. 2008). In turnip (*Brassica campestris* L.) storage parenchyma cells incubated in low external sucrose concentration, phloridzin (but not latrunculin B) greatly reduced sucrose accumulation. By contrast, at high external sucrose concentration, both phloridzin and latrunculin B significantly inhibited sucrose accumulation. These results indicated that: (1) within the classic hyperbolic phase of the sucrose accumulation curve in plant tissues (Ayre 2011), most of sucrose enters the cell via plasma membrane-bound carrier(s); (2) within the linear phase, both plasma membrane-bound carriers and FPE participate in sucrose accumulation. As the external sucrose concentration increases, the involvement of fluid-phase endocytosis increases in parallel. In celery tissue slices incubated in mannitol at concentrations within the linear uptake phase, the uptake of d-TR by parenchyma cells was also reduced by two endocytic inhibitors with different modes of action (latrunculin B and LY 294002) (Etxeberria et al. 2007a).

5 Intracellular Routing of FPE

Triggered by the presence of apoplastic sugars, a promary property of FPE is the indiscriminate trapping of external solutes. Solutes trapped by FPE eventually reach the vacuole (Lazzaro and Thompson 1992; Cholewa and Peterson 2001; Šamaj et al. 2002; Baluška et al. 2004; Etxeberria et al. 2005a, 2006); however, some systematic cargo distribution takes place within the endosome. Once internalized in vesicles, solutes do not appear to be transported directly into the vacuole. Vesicles either merge and/or converge at a sorting organelle. The experiments of Baluška et al (2002) and Emans et al. (2002), both using vesicle transport inhibitor brefeldin A (BFA), indicated the passage of solutes through an intermediate compartment. In both cases, BFA arrested delivery in an unidentified compartment, likely the *trans*-Golgi network (TGN) as later described by Bandmann and Homann (2012). When tobacco BY-2 cells were incubated in 2-NBDG and RFP-tagged SYP61 (a TGN marker), external fluids containing 2-NBDG rapidly accumulated in the TGN under the presence of BFA (Bandmann and Homann 2012). The involvement of actin cytoskeleton in this process was inferred from the effect of the actin depolymerizing agent latrunculin B. In its presence, endocytic uptake of cell wall pectins was arrested (Baluška et al. 2002).
That segregation of solutes takes place at an intermediate intracellular structure was confirmed by the results obtained using a system of two fluorescent endocytic markers consisting of CdSe/ZnS quantum dots and soluble fluorescent d-TR (Etxeberria et al. 2006). The data did not allow for the identification of the specific sorting organelle, but offered some insights into the routing process. Although both solutes entered cells in bulk within individual vesicles and co-localized in small cytosolic structures (Fig. 6a), a sorting mechanism occurred at relatively early stage. This conclusion was based on the substantial overlapping fluorescence from CdSe/ZnS quantum dots and d-TR visible in small cytosolic compartments at the beginning of the endocytic process (yellow color in Fig. 6a), but their segregation at a later stage (Fig. 6b). The soluble fluorescent d-TR was eventually delivered to the central vacuole, whereas CdSe/ZnS quantum dots were partitioned to undefined cytosolic structures after 4 h (Etxeberria et al. 2006). Even if incubated alone for 18 h, CdSe/ZnS quantum dots fluorescence was never visualized in the vacuole, indicating an internal recognition process capable of identifying and segregating entrapped substances. Since CdSe/ZnS quantum dots and d-TR were supplied together in solution, separation of fluorescence indicated a physical segregation of fluorescent components.

The segregation of CdSe/ZnS quantum dots away from the vacuole is significant since sycamore protoplasts also took up artificial polystyrene nanospheres

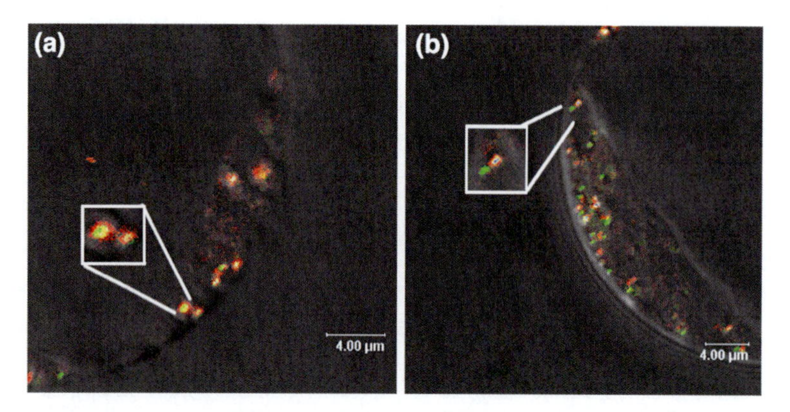

Fig. 6 Laser scanning confocal images showing **a** sycamore cultured cell after 2 h incubation in a medium containing 10^{14} CdSe/ZnS quantum dots per ml and d-TR. **b** A similar cell after 4 h incubation in the same solution

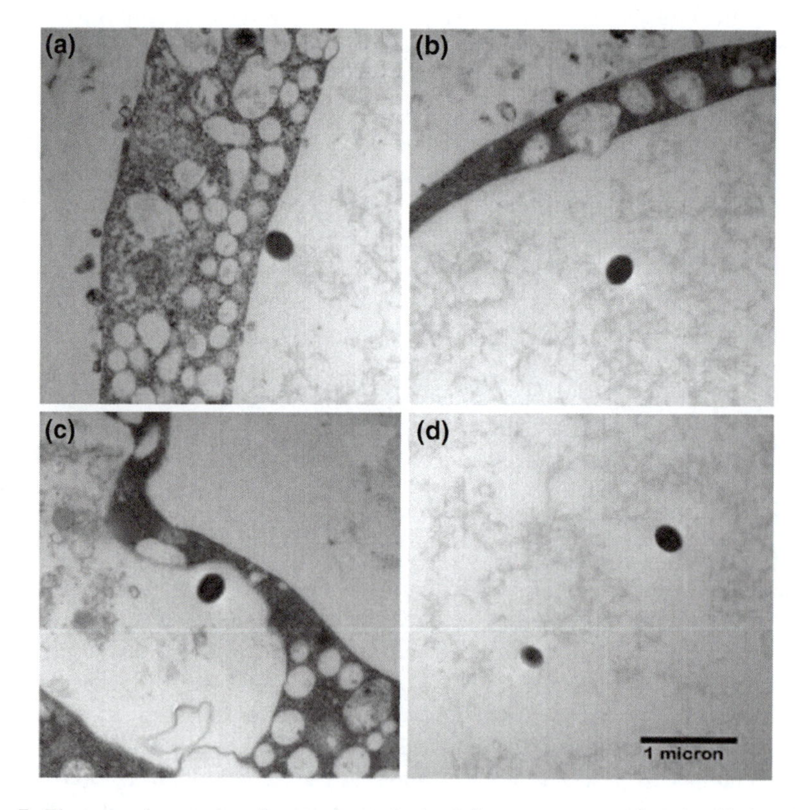

Fig. 7 Electron micrographs of protoplasts obtained from sycamore cultured cells incubated with 40 nm polystyrene nanospheres. **a** and **b** are samples of polystyrene nanospheres inside the vacuole, whereas (**c**) shows a polystyrene nanosphere inside prevacuolar compartment. **d**. Control polystyrene nanospheres immobilized in agar

(40 nm) which were delivered to the central vacuole (Fig. 7). Based on these results, two main conclusions can be reached. First, an intermediate sorting location is capable of recognizing soluble solutes taken up by endocytosis. Second, and most important, the sorting organelle is capable of segregating dissolved and suspended substances trapped in the endocytic solution.

Based on existing evidence, it appears that most solutes are ultimately delivered to the vacuole. In a great number of studies on endocytosis in plant cells, the vacuole has been the final deposit compartment for the endocytic marker. While the presence of endocytic markers have been observed in different intracellular structures, the vacuole appears as the final destination.

6 Conclusions and Future Prospects

Mounting evidence in the last decade has left no doubt that FPE is a primary route for the exchange of solutes between the apoplast and the cell interior, especially the vacuole. It has also become evident that the several endocytic pathways must be intimately interrelated with all other membrane transport systems. One integral aspect of the endocytic process that awaits elucidation is how cells compensate for the increasing uptake of fluids and added membrane, especially to the vacuole. At the moment there are no indications how the homeostasis is maintained in storage cells undergoing FPE. The use of fluorescent probes, transgenic plants, and new imaging techniques will likely give us a clearer picture of a membrane and volume retrieval system.

References

Adlassnig W, Koller-Peroutka M, Bauer S, Koshkin E, Lendl T, Lichtscheild IK (2012) Endocytoic uptake of nutrients in carnivorous plants. Plant J 71:303–313

Ayre B (2011) Membrane-transport systems for sucrose in relation to whole plant carbon partitioning. Mol Plant 4:377–394

Balnokin YV, Kurkova EB, Khalilova LA, Myasoedov NA, Yusofov AG (2007) Pinocytosis in rot cells of salt-accumulating halophyte *Saudea altissima* and its possible involvement in chloride uptake. Russ J Plant Physiol 54:797–805

Baluška F, Hlavačka A, Šamaj J, Palme K, Robinson DR, Matoh DW, McCurdy D, Menzel D, Volkmann D (2002) F-actin-dependent endocytosis of cell wall pectins in meristematic root cells. Insights from brefeldin A-induced compartments. Plant Physiol 130:422–431

Baluška F, Šamaj J, Hlavačka A, Kendrick-Jones J, Volkmann D (2004) Actin dependent fluid-phase endocytosis in inner cortex cells of maize root apices. J Exp Bot 55:463–473

Baroja-Fernández E, Etxeberria E, Muñoz FJ, Morán-Zorzano MT, Alonso-Casajús N, González P, Pozueta-Romero J (2006) An important pool of sucrose linked to starch biosynthesis is taken up by endocytosis in heterotrophic cells. Plant Cell Physiol 47:447–456

Bandmann V, Homann U (2012) Clathrin-independent endocytosis contributes to uptake of glucose into BY-2 protoplast. Plant J 70:578–584

Cao H, Chen J, Awoniyi M, Henley JR, McNiven MA (2007) Dynamin 2 mediates fluid-phase micropinocytosis in epithelial cells. J Cell Sci 120:4167–4177
Cholewa E, Peterson CA (2001) Detecting exodermal casparian bands in vivo and fluid-phase endocytosis in onion (*Allium cepa* L.) roots. Can J Bot 79:30–37
Diekmann W, Hedrich R, Raschke K, Robinson DG (1993) Osmocytosis and vacuolar fragmentation in guard cell protoplasts: their relevance to osmotically-induced volume changes in guard cells. J Exp Bot 267:1569–1577
Dhonukshe P, Aniento F, Hwang I, Robinson DG, Mravec J, Stierhof YD, Friml J (2007) Clathrin-mediated constitutive endocytosis of PIN auxin efflux carriers in *Arabidopsis*. Curr Biol 17:520–527
Emans N, Zimmermann S, Fischer R (2002) Uptake of a fluorescent marker in plant cells sensitive to brefeldin A and wortmannin. Plant Cell 14:71–86
Etxeberria E, Baroja-Fernández E, Muñoz FJ, Pozueta-Romero J (2005a) Sucrose inducible endocytosis as a mechanism for nutrient uptake in heterotrophic plant cells. Plant Cell Physiol 46:474–481
Etxeberria E, Gonzalez PC, Tomlinson P, Pozueta J (2005b) Existence of two parallel mechanisms for glucose uptake in heterotrophic plant cells. J Exp Bot 56:1905–1912
Etxeberria E, Gonzalez P, Baroja-Fernández E, Pozueta-Romero J (2006) Fluid phase uptake of artificial nano-spheres and fluorescent quantum-dots by sycamore cultured cells. Plant Signaling Behavior 1:196–200
Etxeberria E, Gonzalez P, Pozueta J (2007a) Mannitol enhanced fluid-phase endocytosis in storage parenchyma cells of celery (*Apium graveolens*) petioles. Am J Bot 96:1041–1045
Etxeberria E, Gonzalez P, Pozueta J (2007b) Fluid phase endocytosis in Citrus juice cells is independent from vacuolar pH and inhibited by chlorpromazine, a PI-3 kinase and clathrin-mediated endocytosis inhibitor. J Hortic Sci Biotechnol 82:900–907
Etxeberria E, Gonzalez P, Pozueta J (2009) Evidence for two endocytic pathways in plant cells. Plant Sci 177:341–348
Gall L, Stan RC, Kress A, Hertel B, Thiel G, Meckel T (2010) Fluorescent detection of GFP allows for the in vivo estimation of endocytic vesicle sizes in plant cells with sub-diffraction accuracy. Traffic 11:548–559
Geldner N, Jurgens G (2006) Endocytosis in signaling and development. Curr Opin Plant Biol 9:589–594
Gross A, Knapp D, Neihaus K (2005) Endocytosis of xanthomonas campestris pathovar campestris lipopolysaccharides in non-host plant cells of *Nicotiana tabacum*. New Phytol 165:215–226
Hasumi K, Shinohara S, Nagamura S, Endo A (1992) Inhibition of the uptake of oxidized low-density lipoprotein in macrophage J774 by the antibiotic ikarugamycin. Eur J Biochem 205:841–846
Hilmer S, Depta H, Robinson DG (1986) Confirmation of endocytosis in higher plants protoplasts using lectin-gold conjugates. Eur J Cell Biol 41:142–149
Holstein SE (2002) Clathrin and plant endocytosis. Traffic 3:614–620
Horn MA, Heinstein PF, Low PS (1990) Biotin-mediated delivery of exogenous macromolecules into soybean cells. Plant Physiol 93:1492–1496
Hubner R, Depta H, Robinson DG (1985) Endocytosis in maize root cap cells. Evidence obtained using heavy metal salt solutions. Protoplasma 29:214–222
Hudák J, Wales B, Vennigerholz F (1993) The transmitting tissue in *Bugmansia suaveolens* L.: ultrastructure of the stylar transmitting tissue. Ann Bot 71:177–186
Jensen WA, McLaren AD (1960) Uptake of proteins by plant cells-the possible occurrence of pinocytosis in plants. Exp Cell Res 19:414–417
Joachim S, Robinson DG (1984) Endocytosis of cationic ferritin by bean leaf protoplasts. Eur J Cell Biol 34:212–216
Kruth HS, Jones NL, Huang W, Zhao B, Ishii I, Chang J (2005) Macropinocytosis is the endocytic pathway that mediates macrophage foam cell formation with native low density lipoprotein. J Biol Chem 280:2352–2360

Kurkova EB, Balnokin YV (1994) Pinicytosis and its possible role in ion transport in salt accumulating organs of halophytes. Russ J Plant Phys 41:507–511
Lam SK, Tse YC, Jiang L, Oliviusson P, Heinzerling O, Robinson DG (2005) Plant prevacuolar compartment and endocytosis. Plant Cell Monogr 1:37–61
Lazzaro MD, Thompson WW (1992) Endocytosis of lanthanum nitrate in organic acid-secreting trichomes of chickpea (*Cicer arietinum*). Amer J Bot 79:1113–1118
Li G, Xue H-W (2007) Arabidopsis PLD-2 regulates vesicle trafficking and is required for auxin response. Plant Cell 19:281–295
Li R, Liu P, Wan Y, Chen T, Wang Q, Mettbach U, Baluška F, Šamaj J, Fang X, Lucas WL, Lin J (2012) Membrane microdomain-associated protein, AtFlot1, is involved in a clathrin-independent endocytic pathway and is required for seedling development in Arabidopsis. Plant Cell 24:2105–2122
Lin AE-J, Guttman A (2010) Hijacking the endocytic machinery by microbial pathogens. Protoplasma 24:75–90
Luo T, Fredericksen BL, Hasumi K, Endo A, Garcia JV (2001) Human immunodeficiency virus type 1 Nef-induced CD4 cell surface down-regulation is inhibited by ikarugamycin. J Virol 75:2488–2492
Mayor S, Pagano RE (2007) Pathways of clathrin-independent endocytosis. Nat Rev 8:603–612
Moscatelli A, Ciampolini F, Rodighiero S, Onelli E, Cresti M, Santo N, Idilli A (2007) Distinct endocytic pathways identified in tobacco pollen tubes using charged nanogold. J Cell Sci 120:3804–3819
Nichols BJ, Lippincott-Schwatz J (2001) Endocytosis without clathrin coats. Trends Cell Biol 11:406–412
Neumann D, De Figuereido C (2002) A novel mechanism of silicon uptake. Protoplasma 220:59–67
Northcote DH, Davey R, Lay J (1989) Use of antisera to localize callose, xylan and arabinogalactan in cell-plate, primary and secondary walls of plant cells. Planta 178:353–366
Onelli E, Prescianotto-Baschong C, Caccianiga M, Moscatelli A (2008) Clathrin-dependent and independent endocytosis pathways in tobacco protoplasts revealed by labeling with charged nanogold. J Exp Bot 59:3051–3068
Oparka KJ, Prior DAM (1988) Movement of lucifer yellow CH in potato tuber storage tissue: a comparison of symplastic and apoplastic transport. Planta 176:533–540
Paramonova NV (1974) Structural bases of interrelationships between the symplast and apoplast in the root of *Beta vulgaris* during the period of assimilate influx from the leaves. Fiziol Rast 21:578–588
Pozueta D, Gonzalez P, Pozueta J, Etxeberria E (2008) The hyperbolic and linear phases of the sucrose accumulation curve in turnip (*Brassica campestris*) storage cells denote carrier-mediated and fluid-phase endocytic transport, respectively. J Amer Soc Hort Sci 133:612–618
Robinson DC, Hinz G, Holstein SHE (1998) The molecular characterization of transport vesicles. Plant Mol Biol 38:49–76
Russinova E, Borst JW, Kwaaitaal M, Caño-Delgado A, Yin Y, Chory J, de Vries SC (2004) Heterodimerization and endocytosis of Arabidopsis brassinosteroid receptors BRI1 and AtSERK3 (BAK1). Plant Cell 16:3216–3229
Šamaj J, Šamajová O, Peters M, Baluška F, Lichtscheidl IK, Knox JP, Volkman D (2002) Immunolocalization of LM2 arabinogalactan-protein epitope associated with endomembranes of plant cells. Protoplasma 212:186–196
Šamaj J, Baluška F, Voigt B, Schlicht M, Volkmann D, Menzel D (2004) Endocytosis, actin cytoskeleton, and signaling. Plant Physiol 135:1150–1161
Šamaj J, Read ND, Volkmann D, Menzel D, Baluška F (2005) The endocytoic network in plants. Trends Cell Biol 5:425–433
Samuels AL, Bisalputra T (1990) Endocytosis in elongating root cells of *Lobelia erinus*. J Cell Sci 97:157–165
Sandrig K, Torgersen ML, Raa HA, van Deurs B (2008) Clathrin-independent endocytosis: from non-existing to an extreme degree of complexity. Histochem Cell Biol 129:267–276

Tanchak MA, Griffing LR, Mersey BG, Fowke LC (1984) Endocytosis of cationized ferritin by coated vesicles of soybean protoplasts. Planta 162:481–486

Thiel G, Kreft M, Zorec R (1998) Unitary exocytotic and endocytotic events in *Zea mays* L. coleoptile protoplasts. Plant Journal 13:117–120

Villanueva MA, Taylor J, Sui X, Griffing LR (1993) Endocytosis in plant protoplasts. Visualization and quantification of fluid phase endocytosis using silver-enhanced bovine serum albumin-gold. J Exp Bot 44:275–281

Physical Control Over Endocytosis

František Baluška and Ying-Lang Wan

Abstract Plant endocytosis emerges as an active field of contemporary plant sciences. In this chapter, control of plant endocytosis via physical forces is discussed from the perspective of structural homeostasis of the plasma membrane (PM) regulated via vesicular trafficking. Plant cells are very active in endocytosis despite high turgor pressure. Similar to neurons, plant cells are also using endocytosis and endosomes for sensing and processing of sensory information and perhaps also for rapid cell-to-cell communication of this information. As the freshly internalized endosomes are surrounded by their limiting membrane derived from the PM, these endosomes effectively amplify structural and electrical boundaries between the cellular interior and exterior. This feature is critical for the primary processing of sensory information at the PM and its further transduction into signal transduction networks permeating the eukaryotic cell. The higher the number of endosomes a cell generates and recycles, the more it is informed about its environment. Endocytosis is sensitive to diverse physical factors including mechanical, thermal, and electro-magnetic aspects of the PM. Last but not the least, blue light emerges as a physical ligand-like factor for the unique light-induced plant endocytosis.

F. Baluška (✉)
IZMB, University of Bonn, Kirschallee 1, 53115 Bonn, Germany
e-mail: baluska@uni-bonn.de

Y.-L. Wan
College of Biological Sciences and Biotechnology,
Beijing Forestry University, 100083 Beijing, People's Republic of China

J. Šamaj (ed.), *Endocytosis in Plants*,
DOI: 10.1007/978-3-642-32463-5_6,

1 Physical Forces Shape Living Systems: The Case of the Plasma Membrane

Cellular basis of life is based on the plasma membrane (PM) which has two major functions. The first one is protection of the cytoplasm (or protoplasm) from the external environment, being hostile due to numerous fluctuating physical factors such as temperature, ionic imbalances, mechanical insults, and radiation. The second one is sensing of physical and other environmental parameters. Life processes are very sensitive, optimized for a rather narrow range of optimal values of diverse physical parameters. The PM not only provides protection against these physical insults, but it also generates further protective system by synthesis of extracellular polymers such as cellulose and callose. These polymers are part of extracellular matrices (known as cell walls in plant, fungal, and bacterial cells), which are complex assemblies of carbohydrates, proteins, and lipids. The PM acts as a primary sensory system by perceiving diverse physical (abiotic) as well as biotic signals, which are first processed and transformed, and then translocated further deeper into the cellular interior. The most rapid changes are related to ionic and electric aspects of the PM that target downstream cellular signalling processes based on proteins and lipids. Endocytosis is process of internalization of small extracellular space into internal vesicles. This process is associated with the inversion of membrane topology when the outside portion of the PM becomes the inside portion of the vesicle, facing the endosomal interior (Fig. 1). Early endocytic vesicle represents, in fact, a small 'island' of extracellular space, internalized into the cytoplasm. In other words, the 'outside is inside'. This inverted topology proves to be important for effective coupling of sensory events at the outer portion of the PM with early processing/ transduction of this sensory information by early endosomes (Figs. 1 and 2).

2 Thermodynamic Aspects of the PM Integrity, Electricity, and Endocytosis

All life phenomena are based on dynamic equilibrium. PM provides not only a structural but also a chemical and electrical barrier. It allows cells to control their chemical composition, pH and water balances, internal pressures, as well as effective handling of energy. This latter aspect is critical for life phenomena and the living status of organisms (Moore 2012). We will discuss briefly these non-biological aspects at the end of our chapter as they are very relevant for the process of endocytosis and its intimate control via variety of physical factors. Especially, the mechanical status of the PM is crucial for its function as semi-permeable barrier allowing osmotic phenomena and, in fact, for physiology of the whole cell. Vesicle fusion (exocytosis) and vesicle internalization (endocytosis) are two fundamental processes which are shaping that structural and physiological homeostasis of the PM (Fig. 1).

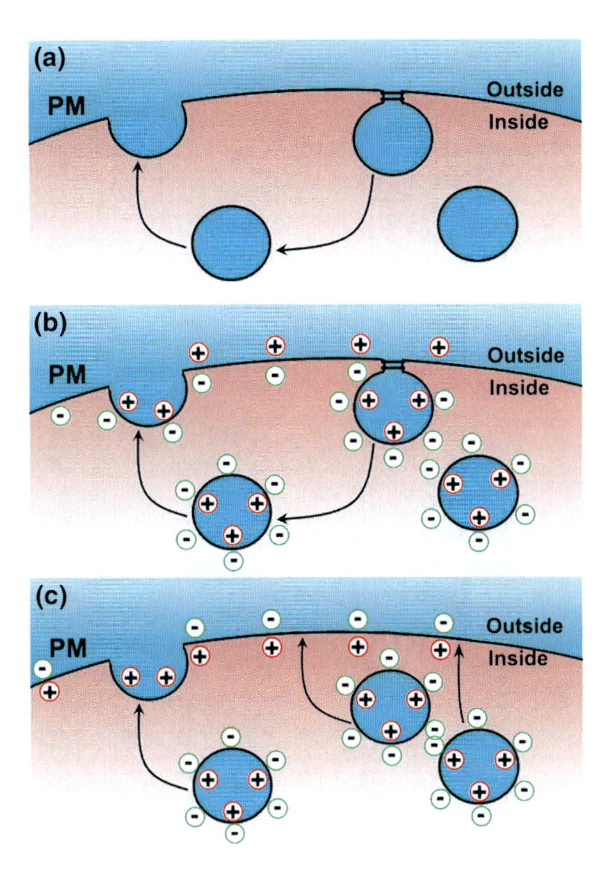

Fig. 1 PM and endosomal membranes act as the sensory interface between the cellular inside and outside. **a** PM represent structural and functional border between the cytoplasm (*red*) and extracellular space (*blue*). Endocytosis not only internalizes extracellular molecules but also portions of extracellular space. Endocytic recycling serves not only to safeguard the structural homeostasis of the PM, and to generate/maintain polarity, but also to modify the extracellular space (allowing modification of cellular niche and communication with adjacent cells). High tension stress of the PM supports exocytosis while low tension stress supports endocytosis. Both the PM and endosomes are equipped with sensors and receptors allowing effective monitoring of extracellular space (*outside*). Cells with high endocytic activity and numerous recycling vesicles have amplified their sensory interface. Such cells (e.g., neurons, plant cells) are well informed about their abiotic and biotic environment and are also superior in cell-to-cell communication. **b** The outside leaflet of the PM is electropositive (which is actively maintained via transporter activities). On the other hand, the inside leaflet is electronegative. This electric polarity of the PM is preserved during endocytosis, so that the outer lipid layer of vesicles is electronegative and, thus, repelled from the PM. **c** During action potential, there is sudden reversal of membrane voltage polarity, when the membrane voltage at the cytoplasmic face of the PM gets electropositive for less than millisecond (Bezanilla 2006; Andersen et al. 2009), allowing rapid fusion (exocytosis) of vesicles in close connection with the PM (Trkanjec 1996; Trkanjec and Demarin 2001)

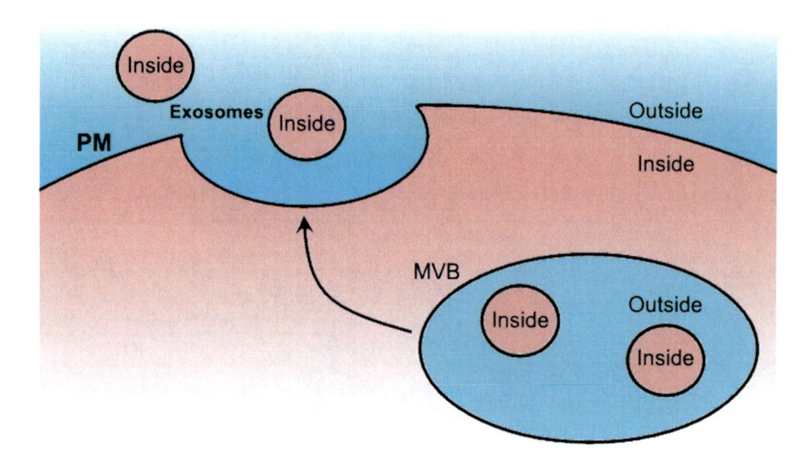

Fig. 2 Multivesicular body (MVB) is special type of late endosomes which accomplishes formation of internal vesicles via their ESCRT complex-driven invagination of the limiting MVB membrane. These internal vesicles represent small cytoplasmic compartments (*inside*) within the endosomal interior corresponding to extracellular space (*outside*). MVBs usually fuse with lysosomes (animal cells) or vacuoles (plant cells), allowing lytic degradation of their contents. But sometimes they fuse with the PM (unconventional exocytosis), releasing their internal vesicles as exosomes. These serve as vehicles for cell–cell communication and also for manipulation of other cells (immune and tumor cells in animals), or defense against pathogens (in plants). For more information, see following articles (An et al. 2006, 2007; Smalheiser 2007; Valadi et al. 2007; Iero et al. 2008; Simons and Raposo 2009; Théry et al. 2009; Wei et al. 2009; Guescini et al. 2010; Babst 2011; Lachenal et al. 2011; Lee et al. 2011; Taylor and Gercel-Taylor 2011; Koles et al. 2012; Rak and Guha 2012)

PM is composed of two layers of polar lipids facing each other with their hydrophobic heads located inside of the membrane, thus forming hydrophobic core of the membrane. The hydrophilic tails of polar lipids face the intracellular and extracellular space/fluids. Such lipidic bilayer acts as electric insulator which is resistant to movements of electrically charged ions across the membrane. This insulator-like character of the PM is critical for maintaining the voltage differences between the negative interior and positive exterior. As the intracellular fluid is electronegative in comparison to the extracellular fluid, while difference in charge is actively maintained by ATPase-based transporters, living cells resemble a battery maintaining voltage differences between the positive exterior and negative interior of the cell (Veech et al. 2002). Numerous proteins are embedded in lipid bilayer and these proteins determine permeability of membrane to diverse ions, which results in membrane resting potential. In animal cells, this membrane resting potential is typically about -100 mV, being negative within cell and positive outside (Olivotto et al. 1996; Hirsch et al. 1998; Bezanilla 2006, 2008; Pedersen et al. 2012). Surprisingly, plants are superior to animal in this respect due to the PM H^+-ATPase (Pedersen et al. 2012) which allow much larger electrochemical membrane potentials, exceeding −200 mV (Pedersen et al. 2012).

The higher membrane voltage in plants has consequences for several aspects of the PM activities, including endocytosis and endocytic vesicle recycling. Studies on roots revealed that not all root cells have the same membrane potential, but that there is an increasing electrochemical gradient from the meristematic cells at the root tip with lower values, toward elongated cells with the highest values (Illéš et al. 2006). Bioelectric fields generated around PM, cell files, and tissues are critical not only for protein conformation but also alter distributions of proteins within the PM via electrophoresis-like process (Jaffe 1977).

3 Endogenous Electric Fields: From Cell Polarity to Navigation of Pollen Tubes and Roots

Electric fields are associated not only with the PM but permeate the eukaryotic cells both extracellularly and intracellularly (McCaig et al. 2005, 2009). For example, complex wiring of neurons in brains is accomplished via electric field-guided neuron migration (Yao et al. 2008, 2011). Electric fields control polarity of neurons by polarizing organelles and division planes (Yao et al. 2009). Extracellular electric fields direct also cell movements during tissue regeneration of wound healing in animals (McCaig et al. 2005, 2009; Wang and Zhao 2010; Messerli and Graham 2011), and in protozoa and other unicellular organisms which often show electrotaxis (for cells of Dictyostelium, see Gao et al. 2011). In plants, pollen tubes, root hairs and whole roots were used to study electric fields linked with cell polarity, growth, as well as with vesicle trafficking for almost hundred years now (Lund 1923; McAulay and Scott 1954; Scott 1956; Murr 1963; Jaffe and Nuccitelli 1977; Goldsworthy and Rathore 1985; Vissenberg et al. 2001; Goldsworthy 2006). Growing apices of plant roots generate characteristic patterns of electric fields with the inward current around the apical meristem and transition zone (Scott 1967; Weisenseel et al. 1975, 1979; Jaffe and Nuccitelli 1977; Jaffe 1981; Hush and Overall 1989; Iwabuchi et al. 1989; Miller and Gow 1989; Collings et al. 1992; Meyer and Weisenseel 1997). These electric fields, which resemble those measured around brains by electroencephalography (EEC) (Buzsáki et al. 2012), are strong enough to control the cytoskeleton as well as endocytosis followed by endocytic vesicle recycling. Unfortunately, we have no data available on possible impacts of these root apex-inherent electric fields on the endocytosis, as well as on endosomes and endocytic vesicle recycling. However, pioneering study revealed that weak electric fields can affect cell division planes and root apex architecture in maize (Wawrecki and Zagórska-Marek 2007). This report suggests that cellular processes behind polarity, especially endocytosis and endocytic vesicle recycling, will be very sensitive to electric fields. Interestingly in this respect, polar auxin transport *in planta* and orientation of tip-growing pollen tubes, two systems which require endocytosis and endocytic vesicle recycling (Baluška et al. 2003, 2005b, 2009; Mancuso et al. 2007), are sensitive towards electric currents and electric fields (Weisenseel et al. 1979; Miller and Gow 1989;

Medvedev and Markova 1990; Malho et al. 1992; Karcz and Burdach 1995; Morris 2001; Bou Daher and Geitmann 2011). Moreover, electric field around the root apex is sensitive to external auxin (Miller and Gow 1989) as well as to gravistimulation of roots (Iwabuchi et al. 1989; Ishikawa and Evans 1990; Collings et al. 1992). The ability of cells/tissues to generate currents and to respond to external currents indicates that there are endogenous feedbacks between processes which generate electric fields and those which are responsive to such fields. Endocytosis, endosomes, and recycling vesicles are strong candidates in this respect.

In tip-growing cells such as fungal cells, as well as pollen tubes and root hairs, the inward current is regularly scored at the growing tip (Weisenseel et al. 1979; Schreurs and Harold 1988; Takeuchi et al. 1988; Gow 1989; Malho et al. 1992), where endocytosis-dependent vesicle recycling shows the highest activity (Ovecka et al. 2005; Voigt et al. 2005; Šamaj et al. 2006; Campanoni and Blatt 2007; Araujo-Bazán et al. 2008; Zárský and Potocký 2010; Shaw et al. 2011). It is interesting in this respect that the highest values of the inward current are scored in those root apex zones (Iwabuchi et al. 1989; Collings et al. 1992; Meyer and Weisenseel 1997) which show the highest rates of endocytosis and endocytic vesicle recycling (Baluška et al. 2003, 2005a, b, 2009, 2010).

Applied electric fields induce tropistic responses not only in the tip-growing neuronal, fungal, and plant cells (McGillivray and Gow 1986; Bedlack et al. 1992; Malho et al. 1992; Nozue and Wada 1993; Lever et al. 1994; Bou Daher and Geitmann 2011), but also in much larger multicellular plant roots (Stenz and Weisenseel 1993; Wolverton et al. 1999, 2000). Interestigly, pollen tube electrotropism is initiated via the electric field-guided vesicle targeting and is also sensitive to depolymerization of the actin cytoskeleton (Bou Daher and Geitmann 2011).

4 Electric Homeostasis: Resting and Activated States of the Membrane Potential

Charged nature of the PM is associated with formation of electric field around the PM (Olivotto et al. 1996). This electric field affects most of proteins which are embedded within the membrane. Transmembrane proteins represent electric dipoles which often perform conformational changes when exposed to changed electric field. The most sensitive proteins are known as voltage-gated proteins, typically voltage-gated ion channels and pumps (Bezanilla 2008; Swartz 2008) as well as some phosphoinositide phosphatases and kinases (Murata et al. 2005; Okamura 2007; Okamura et al. 2009; Chen et al. 2011; Okamura and Dixon 2011; Sakata et al. 2011), and G-protein-coupled receptors (Ben-Chaim et al. 2003, 2006; Parnas and Parnas 2007, 2010; Bezanilla 2008; Kupchik et al. 2011). It is predicted that voltage-sensitive proteins are much more abundant and new proteins will be discovered and characterized in future studies (Bezanilla 2006, 2008). These voltage-sensitive proteins allow electricity induced modifications of endocytosis-relevant proteins and molecules, including critical phosphoinositides in the PM. For example, PIP2 is regulating endocytosis, vesicle recycling, and exocytosis. In the last part of our chapter, we discuss the electrically stimulated endocytosis (electro-endocytosis) which is potentially important but often underestimated process in electrically active cells such as neurons and plant cells.

4.1 Transient Reversal of the Membrane Potential Polarity During Action Potential: Possible Impacts on Membrane Properties, Endocytosis, and Exocytosis

One of the characteristic features of the action potential is that there is sudden reversal of the membrane potential polarity, with the inner leaflet accomplishing temporarily electropositive switch (Bezanilla 2006; Andersen et al. 2009). This sudden reversal of the membrane charges results in immediate attraction, and subsequent PM fusion of synaptic vesicles (Fig. 1), as these maintain their negative surface electrostatic charge during action potential (Trkanjec 1996; Trkanjec and Demarin 2001). Thus, the action potential has dramatic physical impact on vesicles not only due to rapid increase of cytoplasmic calcium via ligand-activated calcium channels, but also due to purely physical changes inducing electrostatic attraction of synaptic vesicles toward the PM (Trkanjec and Demarin 2001). This rather dramatic change of the PM electricity not only significantly modifies thickness of the PM (increase by about 16 % is reported by Andersen et al. 2009) but also whole neurons thicken and shorten reversibly (Tobias 1960; Iwasa et al. 1980; Tasaki et al. 1989: Tasaki 1999). These physical changes of membranes and neurons must have impacts on endocytosis and vesicle recycling, and also on other endomembranes and cytoskeleton in general sense. Especially, the PM structure and tension stress have immediate impacts on endocytosis and endocytic vesicle recycling.

The transient reversal of the PM potential during action potential suggests that endocytosis is the active process, requiring expenditure of energy and cytoskeletal activities, whereas exocytosis emerges as passive process resulting from subsequent action of two physical forces. First, the electrostatic attraction of oppositely charged lipid layers bring recycling vesicles close to the PM and then surface tension forces open the fusion pore (Trkanjec and Demarin 2001). As plants generate action potentials (Fromm and Lautner 2007; Masi et al. 2009; Stolarz et al. 2010), it should be expected that these may affect endocytosis and endocytic vesicle recycling significantly too. The sudden reversal of electric charges during action potential, resulting in electrostatic attraction of recycling vesicles to the PM, can be expected to be relevant for repair of damaged PM. In accordance with the evolutionary origins of action potentials in ancient membrane repair processes (Goldsworthy 1983), synaptotagmins are evolutionary conserved proteins which control exocytosis based on the endocytic vesicle recycling both in the neurotransmission and the PM repair in animal as well as in plant cells (Steinhardt et al. 1994; Andrews 2005; Andrews and Chakrabarti 2005; Schapire et al. 2008, 2009; Yamazaki et al. 2008).

5 Mechanical and Structural Homeostasis: Endocytosis, Tubulation, and Autophagy

Originally discovered in neurons and plant cells, mechanics and surface area of cellular periphery is actively maintained at optimal tension stress via balanced exo- and endocytosis (Wolfe and Steponkus 1983; Wolfe et al. 1985, 1986; Zorec and Tester 1993; Dai and Sheetz 1995; Dai et al. 1998; Kubitscheck et al. 2000; Morris and Homann 2001; Shope et al. 2003; Meckel et al. 2005; Zonia and Munnik 2009; Gauthier et al. 2011). Besides exocytosis and endocytosis, specialized cells, such as skeletal muscle myocytes, develop contractile tubular outgrows of their PM called T-tubules (Al-Qusairi and Laporte 2011). These tubules allow, via their high structural plasticity, rapid and effective recovery of mechanically and osmotically stressed cells. Besides muscle myocytes, also osmo-mechanically stressed cultured molluscan neurons generate such tubular protrusions from the PM, termed vacuole-like dilations (Reuzeau et al. 1995; Mills and Morris 1998; Herring et al. 1999).

Obviously, besides other similarities (Baluška et al. 2002), neurons and plant cells are similar also from the perspective of their strategies to cope with the osmotic-mechanical stress and to control their volumes and surface areas (Zorec and Tester 1993; Kubitscheck et al. 2000; Morris and Homann 2001; Shope et al. 2003; Meckel et al. 2005; Zonia and Munnik 2009; Staykova and Stone 2011; Staykova et al. 2011). Interestingly in this respect, phosphoinositide phosphatase myotubularin is critical for the T-tubules structural plasticity (Buj-Bello et al. 2008; Al-Qusairi et al. 2009; Dowling et al. 2009, 2010; Gibbs et al. 2010; Al-Qusairi and Laporte 2011), as well as for osmotic adaptations of plant cells (Ding et al. 2012). Studies in myocytes suggest central roles of myotubularins for the excitation–contraction coupling in T-tubules (Dowling et al. 2010) and for the PM homeostasis (Buj-Bello et al. 2008). Myotubularin MTMR14 is required for excitation–contraction coupling of T-tubules as well as for the regulation of autophagy (Dowling et al. 2010). Interestingly, autophagy is induced by mechanical stress in animal cells (King et al. 2011) and autophagosome secretion (Pfeffer 2010) may also be involved in the mechanical and structural homeostasis of the PM.

In plant cells, tubular-like membrane outgrowths are known as plasmatubules. These represent long tubular outgrowths, about 20 nm in diameter (Harris et al. 1982; Harris and Chaffey 1986; Coetzee and Fineran 1987; Kandasamy et al. 1988). Similar tubular modifications are well known from algae Chara where they are called as charasomes (Barton 1965; Franceschi and Lucas 1981; Lucas and Franceschi 1981; Price and Whitecross 1983). Until now, both plasmatubules and charasomes have been discussed only from the perspective of uptake and transport processes. However, these structures are ideal to safeguard the structural integrity of PM of osmotically and mechanically stressed cells. Moreover, our unpublished data show that cortex cells of the maize root apex are extremely sensitive to mechanical stress as their plasma membranes become dramatically wrinkled after few seconds of microgravity exposure (Baluška František, Šamajová Olga, Voigt Boris, Volkmann Dieter, unpublished data). These tubular protrusions are found in the vicinity of plasmodesmata grouped into pit fields (Šamaj et al. 2006). The same domains also accomplish osmotic stress-related fluid phase endocytosis (Baluška et al. 2004a). Plasmodesmata are known to be mechanosensitive (Oparka and Prior 1992), while trafficking through plasmodesmata involves endocytic recycling pathways (Haupt et al. 2005), including plant synaptotagmin SYT1 (Lewis and Lazarowitz 2010). This protein is involved in calcium-mediated maintenance of the PM integrity in plant cells stressed with cold and osmotic stress (Schapire et al. 2008, 2009; Yamazaki et al. 2008).

It is not well known, but a matter of fact, that electrical activities such as action potentials and external electric fields (generated via growing organs such as plant root apices) accomplish minute damages to the PM which are effectively repaired via resealing system resembling neurotransmitter apparatus (Steinhardt et al. 1994; Togo et al. 2003). Similar as in animal cells, nanosecond electric pulses trigger short-term permeabilization (electroporation) of the PM in plants (Berghoefer et al. 2009). Recovery to this PM electric damage is very rapid and effective (Berghoefer et al. 2012), and the actin cytoskeleton is important for effective resealing of electrically permeabilized PM (Berghoefer et al. 2009; Hohenberger et al. 2011). In future, it would be interesting to study the possible roles of endosomes and endocytic vesicle recycling during the PM recovery from such electric damage. In both neurons and plant cells, PM repair is accomplished via synaptotagmin and annexin-supported vesicular mechanisms which allow very rapid resealing of damaged parts (Steinhardt 2005; Schapire et al. 2009). Electrically active cells maintain large reservoirs of secretion-prone endocytic vesicles and endosomes which safeguard the PM integrity. In accordance with the Andrew Goldsworthy's hypothesis of evolutionary origins of the action potentials as ancient means of the PM repair (Goldsworthy 1983), plant (Schapire et al. 2008, 2009; Yamazaki et al. 2008) and animal synaptotagmin (Andrews 2005; Andrews and Chakrabarti 2005) are involved in the PM repair. Interestingly, this process resembles neuronal neurotransmitter release (Steinhardt et al. 1994).

6 Endocytosis and Endosomes: When Outside is Inside

Recent advances clearly reveal the importance of endocytosis and endosomes for the sensory information processing and for signal transduction (Sigismund et al. 2012; for plant cells see Wan et al. 2012). Endosomes are signalling organelles, allowing not only effective processing of sensory information but also rapid intracellular transfer of this information from the PM/synapse to the nucleus (Polo and Di Fiore 2006; Scita and Di Fiore 2010; Sigismund et al. 2012). These new aspects were discovered in animal neurons but are valid also for plant cells (Šamaj et al. 2004; Geldner et al. 2007; Geldner and Robatzek 2008; Koizumi et al. 2011; Wan et al. 2012). As mentioned above, the early endosomes internalized from the cell periphery are surrounded by topologically inverted PM (with external leaflet facing the endosomal interior) which now encloses the previously extracellular space within the cytoplasm (Fig. 1). This topological inversion has very important consequences for processing of sensory information and for signal transduction. It is important to keep in mind that besides the topological inversion ('outside gets inside'), there is also electrical polarity inversion when the electropositive 'outside' transforms into the electropositive 'inside' of the endocytic vesicle (Fig. 1), whereas its surface gets electronegative (Trkanjec 1996).

Some decades ago, we have discovered in roots of maize and wheat that endocytosis internalizes not only PM-associated proteins but also cell wall molecules such pectins, AGPs, and some xyloglucans (Šamaj et al. 2000; Baluška et al. 2002; Yu et al. 2002). This discovery overturned the view about one way trafficking from the cytoplasm to the cell wall of extracellular components and opened new avenues to understand very dynamic aspects of plant cells. It also allowed to propose the concept of plant synapses as this endocytosis of cell wall components turned out to be very active especially at the F-actin and myosin VIII-enriched cross-walls, which are nongrowing due to the effective balance of exocytosis with endocytosis (Baluška et al. 2003, 2005b, 2009; Barlow et al. 2004; Wojtaszek et al. 2004; Baluška and Hlavacka 2005). Endocytosis of cell wall pectins cross-linked with calcium was reported also for Arabidopsis roots (see the Supplementary Figure S1 h in Paciorek et al. 2005). Endosomes filled with internalized wall pectins have turned out to be critical for the cytokinetic cell plate formation (Baluška et al. 2005a; Dhonukshe et al. 2006). They also represent ready-to-use reservoir for the rapid growth are emerging to be involved also in the tip-growing root hairs and pollen tubes (Voigt et al. 2005; Lycett 2008; Moscatelli and Idilli 2009; Zhang et al. 2010; Zhao et al. 2010); as well as in guard cell movements during stomata opening (Shope et al. 2003; Shope and Mott 2006; Tanaka et al. 2007). Especially, pectic arabinans seem to be critical for the high flexibility of stomata and root apex cells (Jones et al. 2003; Gomez et al. 2009). However, folding of the PM is also important for guard cell movements (Li et al. 2010). Interestingly in this respect, arabinans disappear from cell walls during seed germination (Gomez et al. 2009) while endocytosis of boron cross-linked RGII pectins is active and critical during seed germination in Arabidopsis (Pagnussat et al. 2012). Arabinans are used up as a storage reserve (Gomez et al. 2009), whereas internalized RGII pectins seem to be recycling back to cell walls together with recycling proteins (Baluška et al. 2002; Yu et al. 2002; Baluška et al. 2005a; Dhonukshe et al. 2006).

7 Control of Endocytosis by Diverse Physical Factors

It is well known that endocytosis is active at ambient temperatures while at lower temperatures around 15 °C inhibit and at temperature around 4 °C effectively stops all types of endocytosis. Besides temperature, other physical factors such as electro-magnetic fields, gravity, and blue light are also emerging to control endocytosis. It seems that all these physical factors modulate endocytosis via their impacts on the PM structural properties such as rigidity and tension stress. Moreover, they are closely related with the structural/tensional homeostasis of the PM, as already discussed in previous parts of this chapter. We will briefly discuss all these phenomena below.

7.1 Control of Endocytosis by Temperature

One of the most prominent and best understood phenomenons is inhibition of endocytosis by low temperatures (below 15 °C) with complete block of endocytosis at temperatures around 4 °C (Mamdouh et al. 1996; Punnonen et al. 1998; de Figueiredo and Soares 2000). Low temperatures alter physicochemical properties of membranes and their lipids get highly ordered (gel-like). This prevents all kinds of endocytosis due to the very low membrane fluidity (Laroche et al. 2001; Leidy et al. 2004). Such phase transition-like transformations of membranes preclude not only the act of endocytosis itself but also fission–fusion endosomal processes deeper within the cells (Punnonen et al. 1998). If cells are exposed to cold for longer time, they can get adapted (in a process known as cold hardening) by modifications of protein composition of membranes and by increasing amounts of cryoprotective sugars such as sucrose or trehalose (Leslie et al. 1994, 1995; Lee et al. 2006). In plants, cold stress also inhibits or blocks endocytosis (Baluška et al. 2002; Onelli et al. 2008; Shibasaki et al. 2009); and prevents endocytosis and intracellular trafficking of PIN2, as well as polar auxin transport based on PIN2 in root apices of Arabidopsis (Shibasaki et al. 2009). Plants possess effective cold adaptation (hardening) (Orvar et al. 2000) which is mediated by changes in the membrane fluidity (Sangwan et al. 2002). Temperature-responsive mitogen-activated protein kinases (MAPKs) (Sangwan and Dhindsa 2002) and, as discussed above, also plant synaptotagmins are involved in this adaptive cold-hardening process (Yamazaki et al. 2008).

7.2 Endocytosis Induced by Electro-Magnetic Fields

Exposure of cells to low electric fields induces endocytosis also known as electro-endocytosis (Zimmermann et al. 1990; Glogauer et al. 1993; Rosemberg and Korenstein 1997; Antov et al. 2004, 2005; Lin et al. 2011). Unfortunately, it is unclear which type of endocytosis is activated and if this endocytosis has physiological relevance for multicellular organisms as other types of endocytosis. Electro-endocytosis is also blocked by low temperatures (Antov et al. 2005). Even exposure of cells to extremely low electric and magnetic fields, inducing membrane potential differences below 1 mV, may exert impacts on endocytosis and vesicle recycling. But these phenomena are very controversial and highly debated

(Adair 1999; Weaver et al. 1999, 2000; Foster 2003). Nevertheless, it is very interesting that some animals such as insects, birds, fishes, and also plants (Mina and Goldsworthy 1991; Galland and Pazur 2005; Ahmad et al. 2007; Müller and Ahmad 2011) can sense these low electric and magnetic signals (Kalmijn et al. 2002; Sisneros and Tricas 2002; Johnsen and Lohmann 2005; Solov'yov and Schulten 2009; Lohmann 2010; Phillips et al. 2010; Ritz et al., 2010; Stanley et al. 2012; Wiltschko and Wiltschko 2012). In future, it will be important to study effects of these ultra-weak electromagnetic fields on endocytosis and endosomes.

7.3 Control of Endocytosis by Pressure and Gravity

PM tension plays decisive role in coordination of exocytosis and endocytosis in neurons and plant cells (Morris and Homann 2001). This concept, also known as surface area regulation (SAR), states that when the membrane tension exceeds some critical values then new membrane is added (in the form of exocytic vesicles), but when the membrane tension goes too much down then excess of membrane is removed (in the form of endocytic vesicles). This homeostatic mechanism allows to maintain optimal values of the membrane tension which is essential for its proper function (Morris and Homann 2001). The importance of such complex system is obvious from the fact that the PM can expand elastically only by 2–3 % (Wolfe and Steponkus 1983; Nichol and Hutter 1996). Both neurons and plant cells, which often must change surface area on a large scale in short time period, behave as the SAR theory predicted (Dai et al. 1998; Homann 1998; Meckel et al. 2005). For plant guard cells, the large changes in the surface areas are not possible without internalizing and recycling of the cell wall material. In fact, this has been reported for root cells (Baluška et al. 2002, Baluška et al. 2005a; Dhonukshe et al. 2006), and might be valid for dynamic guard cells too (Shope et al. 2003; Baluška et al. 2006; Shope and Mott 2006; Tanaka et al. 2007).

In accordance with the predictions made by the SAR theory for mechanical and structural homeostasis of the PM, pressure applied to plant cells inhibits endocytosis (Jingquan Li, Yinglang Wan, Boris Voigt, Christian Burbach, František Baluška, unpublished data). Interestingly, also the F-actin gets increasingly bundled and less dynamic in plant cells under pressure (Jingquan Li, Yinglang Wan, Boris Voigt, Christian Burbach, František Baluška, unpublished data). It can be expected that both, changes in the physical properties of the PM (e.g. rigidification due to stress-induced ROS species) as well as in the assembly status of the underlying actin cytoskeleton, contribute to the inhibition of endocytosis in plant cells exposed to moderate external pressures. As transgenic Arabidopsis seedlings and especially roots are often used for in vivo microscopy, it is important to be aware that a cover slip as well as microscopic objective in upright microscopic systems are exerting pressure on analyzed cells which might affect endocytosis, as well as endocytic vesicle recycling. This effect is not so dramatic in inverted microscopes and even less with using microscopic chambers and perfusion chambers because in these chambers pressure is exposed on holders and not on plant tissues/cells directly.

Polarity of plant cells is closely related to transcellular polar auxin transport which is typically following the gravity vector (Baluška et al. 2012). It is very difficult to explain the mechanisms behind the polar auxin transport controlled by gravity in a classical, currently dominating, concept of auxin transport solely via PM-resident auxin transporters. However, this problem is easily solved if we consider the cell-to-cell auxin transport in the framework of synaptic model and the SAR theory (Baluška et al. 2003, 2005b; Baluška and Volkmann 2011; Baluška et al. 2012). In the synaptic model, endocytosis (uptake of auxin into cells) is favored at the upper part of plant cells and exocytosis (secretion of auxin out of cells) is then predominant at the lower part (down the gravity vector) of polarized plant cells. In fact, the polar transport of auxin is following the gravity vector in both shoots and roots, via predictions made by the SAR theory (see the Fig. 1 in Baluška et al. 2005b and the Fig. 3 in Baluška et al. 2009).

7.4 Endocytosis Induced by Blue Light in Plants

Breakthrough has been achieved in our understanding of sensing of blue light (BL) by plants. PHOT1 is Arabidopsis BL receptor which resides in the PM in unstimulated form (Sakamoto and Briggs 2002). BL illumination induces conformation change of PHOT1, resulting in unique light-induced and clathrin-based endocytosis (Kaiserli et al. 2009; Wan et al. 2012). Kaiserli et al. (2009) reported direct interaction between PHOT1 and clathrin heavy chain, suggesting that the blue light-induced PHOT1 internalization is clathrin dependent. Moreover, studies on the PHOT2 revealed a similar behavior of this receptor in BL-induced endocytosis (Kong et al. 2006). Recent evidence suggests that these two phototropins may share a similar endosomal trafficking pathway, because intracellular PHOT1 and PHOT2 colocalized with the FM4-64 endosomes sensitive to brefeldin A (Kong et al. 2006; Kaiserli et al. 2009). Furthermore, several important proteins related to the vesicular transport are also interacting with both PHOTs, including several members of ARF and 14-3-3 proteins (Kaiserli et al. 2009; Sullivan et al. 2010). All this indicates that PHOT1 endocytosis is similar to the classic ligand-receptor-mediated endocytosis in animal cells. Interestingly, the level of PHOT1 endocytosis reflects the amount of BL photons arriving at the cell surface (Wan et al. 2008), supporting the hypothesis that photons can act as the physical ligand (Somers and Fujiwara 2009). Prolonged BL exposure of the PHOT1 results in degradation of this photoreceptor (Sakamoto and Briggs 2002).

However, the physiological and functional role of the PHOT1 endocytosis is still under discussion. Sakamoto and Briggs (2002) reported that part of membrane-associated PHOT1 was released into the cytosol after BL exposure, while Han et al. (2008) found that the inhibition of PHOT1 internalization may increase the BL-induced bending curvatures. Later on, Roberts et al. (2011) reported that the NPH3 (non-phototropic hypocotyls), the essential signal transducer for PHOT1, is not internalized into cytoplasm after BL illumination. All this implies that the PM-localized PHOT1 is perhaps more efficient in signalling and photo-transduction as internalized PHOT1. It is possible that endocytosis of PHOT1 is involved in attenuation of BL signal or in the desensitization and resensitization in light and dark conditions, respectively. However, novel evidences support a tight relationship between the PHOT1 signalling and PIN-based polar auxin transport. Lateral illumination with BL caused an asymmetric distribution of PIN3 protein in the hypocotyls (Ding et al. 2011). In roots, PIN2 is essential for a BL-induced phototropic root growth, while BL-depended cellular distribution of PIN2 is based on the PHOT1/NPH3 signalling (Wan et al. 2012). Furthermore, Christie et al. (2011) reported that the localization and distribution of ABCB19 (another auxin transporter) is also determined by the PHOT1-mediated blue light signalling. These new evidences support a role of endocytosis in the PHOT signalling and point to the cross-talk of BL signalling with transcellular polar auxin flow mechanisms in roots and hypocotyls.

Finally, there are also other more general aspects of light impacts on endosomes. Membrane rigidification by light-induced photo-damage affects endosomal permeability (Zhong et al. 2000) and inhibits endocytosis as well as endocytic vesicle recycling (Andrzejak et al. 2011; Kessel 2011; Kessel et al. 2011). As confocal laser scanning microscopy is exerting some light stress on living plant cells (Dixit and Cyr 2003; Hoebe et al. 2007), it is important to keep this in mind when interpreting in vivo data on endocytosis and endosomes.

8 Electricity of the PM and Endosomes: Border Between Living and Non-Living

One of the biggest challenge not only for biologists, but also for philosophers, is the question where is the border line between non-living and living nature. Until now, there is no satisfactory answer to this enduring mystery despite numerous proposals and definitions of life. However, there is no life without electricity (Pokorny et al. 2005). Recently, Andrew Moore has proposed that besides the replication and heredity, it is important to focus also on the thermodynamics (Moore 2012). Living agents perform work using their structures which embody increased order (locally lowered entropy) evolved via long evolution (Kováč 2007, 2008). This biological work performed by living agent results in further decrease of entropy of its structures (in other words, in increased complexity of the biological organization). The most important organismal structure allowing manipulation of energy flows, and generating bioelectricity, is the semipermeable PM which allows harnessing of energy from the environment. The living agent (Cell Body/Energide, see Baluška et al. 2004b, c, 2012) is enjoying a sheltered niche as its surrounding membrane protects it against the second law of thermodynamics via sensing and manipulating environment and its properties. This ability is allowed via electric properties of the PM which transform effectively all perceived sensory events and signals into the downstream biological processes. Here, endosomes are critical as they amplify the border between living (inside) and non-living (outside) domains (Fig. 1). Obviously, the higher number of endosomes cell generates and recycles, the more it is informed about its environment. Recycling of endosomes allows active manipulation of the adjacent environment as endosomes literally internalize the outside space which is first analyzed, then modified, and finally released back into the adjacent environment (Fig. 2). In this respect, early recycling endosomes and especially late endosomes such as multivesicular bodies (MVBs) are important as they can act as unique secretory organelle. In this way, living cells might actively alter their adjacent environment and modify their living niche which allows their survival (Ozansoy and Denizhan 2009).

9 Conclusions and Outlook

Recently, the Nobel Prize winner Sydney Brenner compared the living organisms with Turing machine (Brenner 2012). However, this comparison is not taking into account the cognitive, behavioral, and conscious nature of living organisms (Baluška 2011; Trewavas and Baluška 2011; Witzany and Baluška 2012). Biological systems obtain and store biological information at the expense of physical entropy. Biological information is then stored (embodied) within biological structures (Kováč 2007) nested within each other at numerous levels of the biological complexity (Agnati et al. 2009; Baluška 2009). Importantly, biological organisms (living, behaving and interpreting agents) act not as Turing machines but rather as Maxwell's demons (Kováč 2007, Battail 2009, 2011; Penzlin 2009).

Electrical signals are upstream of biological processes and the PM generates its own electric field which controls cellular biological processes. This electric field of the cell periphery is sensitive to environmental changes and acts as effective sensory device which senses and transforms environmental parameters into biologically relevant information. Among other downstream features sensitive to the membrane electric fields, membrane tension is upstream of numerous molecular polarity factors (Batchelder et al. 2011; Keren 2011). In addition, depolarized PM changes its physical properties as it is less sensitive to detergents (Grossmann et al. 2007). PM fluidity is a critical parameter not only for sensory perceptions (Mikami and Murata 2003; Los and Murata 2004) but, apparently, also for conscious processing of sensory information as diverse anesthetics affect fluidity of the PM (Giocondi et al. 1995; Sedensky et al. 2004; Morrow and Parton 2005; Anbazhagan et al. 2010; Tsuchiya et al. 2010). As lipid rafts represent possible targets of anesthetics (Sedensky et al. 2004; Morrow and Parton 2005) and lipid raft proteins, such as flotillins, accomplishing endocytosis (e.g. Li et al. 2012) and endocytic recycling (Morrow and Parton 2005; Langhorst et al. 2008), it is important to include endocytosis and endosomes into the focus of studies investigating biological and cellular basis of consciousness. Interestingly in this respect, it is well known that anesthetics perturb electric signals as well as actin cytoskeleton-based vesicle recycling, not just in humans and animals (Larsen and Langmoen 1998; Kaech et al. 1999; Sandstrom 2004; Fu et al. 2005; Sonner 2008); but also in several other organisms such as plants, fungi, and protozoa (Wiklund and Allison 1972; Nunn et al. 1974; Taylorson and Hendricks 1979; Morgan et al. 1990; Saltveit 1993; Milne and Beamish 1999; Humphrey et al. 2002; Uesono 2009; Singaram et al. 2011). Moreover, anesthesia in animals can be induced also by DC currents and magnetic fields (Becker and Selden 1985). Importantly in this respect, external electric fields and stimulations affect neuronal activities and entrain cortical neurons in animal brains (Anastassiou et al. 2010; Ozen et al. 2010).

As anesthesia can be reversed by increasing pressure (Johnson and Flagler 1951; Lever et al. 1971; Wlodarczyk et al. 2006), as expected if there are effects of anesthetics on the physical properties of membranes (Heimburg and Jackson 2007; Andersen et al. 2009; Heimburg 2009), it is probable that the PM/endosomal interface is the target of anesthetics action. In future, it will be important to study effects of anesthetics on endocytosis and endosomes, perhaps with plants as emerging new model for such studies.

References

Adair RK (1999) Effects of very weak magnetic fields on radical pair reformation. Bioelectromagnetics 20:255–263

Agnati LF, Baluska F, Barlow PW, Guidolin D (2009) Mosaic, self-similarity logic, and biological attraction principles: three explanatory instruments in biology. Commun Integr Biol 2:552–563

Ahmad M, Galland P, Ritz T, Wiltschko R, Wiltschko W (2007) Magnetic intensity affects cryptochrome-dependent responses in *Arabidopsis thaliana*. Planta 225:615–624

Al-Qusairi L, Laporte J (2011) T-tubule biogenesis and triad formation in skeletal muscle and implication in human diseases. Skelet Muscle 1:26

Al-Qusairi L, Weiss N, Toussaint A, Berbey C, Messaddeq N, Kretz C, Sanoudou D, Beggs AH, Allard B, Mandel JL, Laporte J, Jacquemond V, Buj-Bello A (2009) T-tubule disorganization and defective excitation–contraction coupling in muscle fibers lacking myotubularin lipid phosphatase.Proc Natl Acad Sci U S A 106:18763–18768

An Q, Hückelhoven R, Kogel KH, van Bel AJ (2006) Multivesicular bodies participate in a cell wall-associated defence response in barley leaves attacked by the pathogenic powdery mildew fungus. Cell Microbiol 8:1009–1019

An Q, van Bel AJ, Hückelhoven R (2007) Do plant cells secrete exosomes derived from multivesicular bodies? Plant Signal Behav 2:4–7

Anastassiou CA, Montgomery SM, Barahona M, Buzsáki G, Koch C (2010) The effect of spatially inhomogeneous extracellular electric fields on neurons. J Neurosci 30:1925–1936

Anbazhagan V, Munz C, Tome L, Schneider D (2010) Fluidizing the membrane by a local anesthetic: phenylethanol affects membrane protein oligomerization. J Mol Biol 404:773–777

Andersen SSL, Jackson AD, Heimburg T (2009) Towards a thermodynamic theory of nerve pulse propagation. Prog Neurobiol 88:104–113

Andrews NW (2005) Membrane resealing: synaptotagmin VII keeps running the show. Sci STKE 282:pe19

Andrews NW, Chakrabarti S (2005) There's more to life than neurotransmission: the regulation of exocytosis by synaptotagmin VII. Trends Cell Biol 15:626–631
Andrzejak M, Santiago M, Kessel D (2011) Effects of endosomal photodamage on membrane recycling and endocytosis. Photochem Photobiol 87:699–706
Antov Y, Barbul A, Korenstein R (2004) Electroendocytosis: stimulation of adsorptive and fluid-phase uptake by pulsed low electric fields. Exp Cell Res 297:348–362
Antov Y, Barbul A, Mantsur H, Korenstein R (2005) Electroendocytosis: exposure of cells to pulsed low electric fields enhances adsorption and uptake of macromolecules. Biophys J 88:2206–2223
Araujo-Bazán L, Peñalva MA, Espeso EA (2008) Preferential localization of the endocytic internalization machinery to hyphal tips underlies polarization of the actin cytoskeleton in *Aspergillus nidulans*. Mol Microbiol 67:891–905
Babst M (2011) MVB vesicle formation: ESCRT-dependent, ESCRT-independent and everything in between. Curr Opin Cell Biol 23:452–457
Baluška F (2009) Cell-cell channels, viruses, and evolution: via infection, parasitism and symbiosis towards higher levels of biological complexity. Ann N Y Acad Sci 1178:106–119
Baluška F (2011) Evolution in revolution. Paradigm shift in our understanding of life and biological evolution. Commun Integr Biol 4:521–523
Baluška F, Hlavačka A (2005) Plant formins come to age: something special about cross-walls. New Phytol 168:499–503
Baluška F, Volkmann D (2011) Mechanical aspects of gravity-controlled growth, development and morphogenesis. In: Wojtaszek P (ed) Mechanical integration of plant cells and plants. Springer, Berlin, pp 195–222
Baluška F, Hlavačka A, Šamaj J, Palme K, Robinson DG, Matoh T, McCurdy DW, Menzel D, Volkmann D (2002) F-actin-dependent endocytosis of cell wall pectins in meristematic root cells: insights from brefeldin A-induced compartments. Plant Physiol 130:422–431
Baluška F, Šamaj J, Menzel D (2003) Polar transport of auxin: carrier-mediated flux across the plasma membrane or neurotransmitter-like secretion? Trends Cell Biol 13:282–285
Baluška F, Šamaj J, Hlavačka A, Kendrick-Jones J, Volkmann D (2004a) Myosin VIII and F-actin enriched plasmodesmata in maize root inner cortex cells accomplish fluid-phase endocytosis via an actomyosin-dependent process. J Exp Bot 55:463–473
Baluška F, Volkmann D, Barlow PW (2004b) Cell bodies in a cage. Nature 428:371
Baluška F, Volkmann D, Barlow PW (2004c) Eukaryotic cells and their cell bodies: cell theory revisited. Ann Bot 94:9–32
Baluška F, Liners F, Hlavačka A, Schlicht M, Van Cutsem P, McCurdy D, Menzel D (2005a) Cell wall pectins and xyloglucans are internalized into dividing root cells and accumulate within cell plates during cytokinesis. Protoplasma 225:141–155
Baluška F, Volkmann D, Menzel D (2005b) Plant synapses: actin-based adhesion domains for cell-to-cell communication. Trends Plant Sci 10:106–111
Baluška F, Baroja-Fernandez E, Pozueta-RomeroJ, Hlavacka A, Etxeberria E, Šamaj J (2006) Endocytic uptake of nutrients, cell wall molecules, and fluidized cell wall portions into heterotrophic plant cells. In: Šamaj J, Baluška F, Menzel D (eds) Plant endocytosis. Springer, Berlin, pp 19–35
Baluška F, Schlicht M, Wan Y-L, Burbach C, Volkmann D (2009) Intracellular domains and polarity in root apices: from synaptic domains to plant neurobiology. Nova Acta Leopolda 96:103–122
Baluška F, Mancuso S, Volkmann D, Barlow PW (2010) Root apex transition zone: a signalling-response nexus in the root. Trends Plant Sci 15:402–408
Baluška F, Volkmann D, Menzel D, Barlow PW (2012) Strasburger's legacy to mitosis and cytokinesis and its relevance for the cell theory. Protoplasma 249 (in press)
Barlow PW, Volkmann D, Baluška F (2004) Polarity in roots. In: K Lindsey (ed) Polarity in plants. Blackwell Publishing, Oxford, pp 192–241
Barton R (1965) An unusual organelle in the peripheral cytoplasm of Chara cells. Nature 205:201

Batchelder EL, Hollopeter G, Campillo C, Mezanges X, Jorgensen EM, Nassoy P, Sens P, Plastino J (2011) Membrane tension regulates motility by controlling lamellipodium organization. Proc Natl Acad Sci U S A 108:11429–11431
Battail G (2009) Living versus inanimate: the information border. Biosemiotics 2:321–341
Battail G (2011) An answer to Schrödinger's what is life? Biosemiotics 4:55–67
Becker RO, Selden G (1985) The body electric. Electromagnetism and the foundation of life. Morrow, New York
Bedlack RS, Md Wei, Loew LM (1992) Localized membrane depolarizations, and localized calcium influx during electric field-guided neuritegrowth. Neuron 9:393–403
Ben-Chaim Y, Tour O, Dascal N, Parnas I, Parnas H (2003) The M2 muscarinic G-protein-coupled receptor is voltage-sensitive. J Biol Chem 278:22482–22491
Ben-Chaim Y, Chanda B, Dascal N, Bezanilla F, Parnas I, Parnas H (2006) Movement of 'gating charge' is coupled to ligand binding in a G-protein-coupled receptor. Nature 444:106–109
Berghoefer T, Eing C, Flickinger B, Hohenberger P, Wegner LH, Frey W, Nick P (2009) Nanosecond electric pulses trigger actin responses in plant cells. Biochem Biophys Res Commun 387:590–595
Berghoefer T, Flickinger B, Frey W (2012) Aspects of plant plasmalemma charging induced by external electric field pulses. Plant Signal Behav 7:322–324
Bezanilla F (2006) The action potential: from voltage-gated conductances to molecular structures. Biol Res 39:425–435
Bezanilla F (2008) How membrane proteins sense voltage. Nat Rev Mol Cell Biol 9:323–332
Bou Daher F, Geitmann A (2011) Actin is involved in pollen tube tropism through redefining the spatial targeting of secretory vesicles. Traffic 12:1537–1551
Brenner S (2012) Turing centenary: Life's code script. Nature 482:461
Buj-Bello A, Fougerousse F, Schwab Y, Messaddeq N, Spehner D, Pierson CR, Durand M, Kretz C, Danos O, Douar AM, Beggs AH, Schultz P, Montus M, Denèfle P, Mandel JL (2008) AAV-mediated intramuscular delivery of myotubularin corrects the myotubular myopathy phenotype in targeted murine muscle and suggests a function in plasma membrane homeostasis. Hum Mol Genet 17:2132–2143
Buzsáki G, Anastassiou CA, Koch C (2012) The origin of extracellular fields and currents—EEG, ECoG, LFP and spikes. Nat Rev Neurosci 13:407–420
Campanoni P, Blatt MR (2007) Membrane trafficking and polar growth in root hairs and pollen tubes. J Exp Bot 58:65–74
Chen X, Zhang X, Jia C, Xu J, Gao H, Zhang G, Du X, Zhang H (2011) Membrane depolarization increases membrane PtdIns(4,5)P2 levels through mechanisms involving PKC βII and PI4 kinase. J Biol Chem 286:39760–39767
Christie JM, Yang H, Richter GL et al (2011) phot1 inhibition of ABCB19 primes lateral auxin fluxes in the shoot apex required for phototropism. PLoS Biol 9:1001076
Coetzee J, Fineran BA (1987) The apoplastic continuum, nutrient absorption and plasmatubules in the dwarf mistletoe *Korthalsella lindsaj* (Viscaceae). Protoplasma 136:145–153
Collings DA, White RG, Overall RL (1992) Ionic current changes associated with the gravity-inducing bending response in roots of *Zea mays* L. Plant Physiol 100:1417–1426
Dai J, Sheetz MP (1995) Regulation of endocytosis, exocytosis, and shape by membrane tension. Cold Spring Harb Symp Quant Biol 60:567–571
Dai J, Sheetz MP, Wan X, Morris CE (1998) Membrane tension in swelling and shrinking molluscan neurons. J Neurosci 18:6681–6692
Dhonukshe P, Baluška F, Schlicht M, Hlavačka A, Šamaj J, Friml J, Gadella Jr TWJ (2006) Endocytosis of cell surface material mediates cell plate formation during plant cytokinesis. Dev Cell 10:137–150
Ding ZJ, Galván-Ampudia CS, Dermarsy E et al (2011) Light-mediated polarization of the PIN3 auxin transporter for the phototropic response in Arabidopsis. NatCell Biol 13:447–452
Ding Y, Ndamukong I, Zhao Y, Xia Y, Riethoven JJ, Jones DR, Divecha N, Avramova Z (2012) Divergent functions of the myotubularin (MTM) homologs AtMTM1 and AtMTM2 in *Arabidopsis thaliana*: evolution of the plant MTM family. Plant J 70:866–878

Dixit R, Cyr R (2003) Cell damage and reactive oxygen species production induced by fluorescence microscopy: effect on mitosis and guidelines for non-invasive fluorescence microscopy. Plant J 36:280–290
Dowling JJ, Vreede AP, Low SE, Gibbs EM, Kuwada JY, Bonnemann CG, Feldman EL (2009) Loss of myotubularin function results in T-tubule disorganization in zebrafish and human myotubular myopathy. PLoS Genet 5:e1000372
Dowling JJ, Low SE, Busta AS, Feldman EL (2010) Zebrafish MTMR14 is required for excitation-contraction coupling, developmental motor function and the regulation of autophagy. Hum Mol Genet 19:2668–2681
de Figueiredo RCBQ, Soares MJ (2000) Low temperature blocks fluid-phase pinocytosis and receptor-mediated endocytosis in *Trypanosoma cruzi* epimastigotes. Parasitol Res 86:413–418
Foster KR (2003) Mechanisms of interaction of extremely low frequency electric fields and biological systems. Radiat Prot Dosimetry 106:301–310
Franceschi VR, Lucas WJ (1981) The charasome periplasmic space. Protoplasrna 107:269–284
Fromm J, Lautner S (2007) Electrical signals and their physiological significance in plants. Plant Cell Environm 30:249–257
Fu D, Vissavajjhala P, Hemmings HC Jr (2005) Volatile anaesthetic effects on phospholipid binding to synaptotagmin 1, a presynaptic Ca^{2+} sensor. Br J Anaesth 95:216–221
Galland P, Pazur A (2005) Magnetoreception in plants. J Plant Res 118:371–389
Gao R-C, Zhang X-D, Sun Y-H, Kamimura Y, Mogilner A, Devreotes PN, Zhao M (2011) Different roles of membrane potentials in electrotaxis and chemotaxis of Dictyostelim cells. Eukaryot Cell 10:1251–1256
Gauthier NC, Fardin MA, Roca-Cusachs P, Sheetz MP (2011) Temporary increase in plasma membrane tension coordinates the activation of exocytosis and contraction during cell spreading. Proc Natl Acad Sci U S A 108:14467–14472
Geldner N, Robatzek S (2008) Plant receptors go endosomal: a moving view on signal transduction. Plant Physiol 147:1565–1574
Geldner N, Hyman DL, Wang X, Schumacher K, Chory J (2007) Endosomal signaling of plant steroid receptor kinase BRI1. Genes Dev 21:1598–1602
Gibbs EM, Feldman EL, Dowling JJ (2010) The role of MTMR14 in autophagy and in muscle disease. Autophagy 6:819–820
Giocondi M-C, Mamdouh Z, Le Grimellec C (1995) Benzyl alcohol differently affects fluid phase endocytosis and exocytosis in renal epithelial cells. Biochim Biophys Acta 1234:197–202
Glogauer M, Lee W, McCulloch CA (1993) Induced endocytosis in human fibroblasts by electrical fields. Exp Cell Res 208:232–240
Goldsworthy A (1983) The evolution of plant action potentials. J Theor Biol 103:645–648
Goldsworthy A (2006) Effects of electrical and electromagnetic fields on plants and related topics. In: Volkov A (ed) Plant electrophysiology. Springer, Berlin, pp 247–267
Goldsworthy A, Rathore KS (1985) The electrical control of growth in plant tissue cultures: the polar transport of auxin. J Exp Bot 36:1134–1141
Gomez LD, Steele-King CG, Jones L, Foster JM, Vuttipongchaikij S, McQueen-Mason SJ (2009) Arabinan metabolism during seed development and germination in Arabidopsis. Mol Plant 2:966–976
Gow NAR (1989) Relationship between growth and the electrical current of fungal hyphae. Bio Bul 176:31–35
Grossmann G, Opekarová M, Malinsky J, Weig-Meckl I, Tanner W (2007) Membrane potential governs lateral segregation of plasma membrane proteins and lipids in yeast. EMBO J 26:1–8
Guescini M, Genedani S, Stocchi V, Agnati LF (2010) Astrocytes and glioblastoma cells release exosomes carrying mtDNA. J Neural Transm 117:1–4
Han IS, Tseng TS, Eisinger W, Briggs WR (2008) Phytochrome A regulates the intracellular distribution of Phototropin 1–green fluorescent protein in *Arabidopsis thaliana*. Plant Cell 20:2835–2847
Harris N, Chaffey NJ (1986) Plasmatubules—real modifications of the plasmalemma. Nord J Bot 6:599–607

Harris N, Oparka KJ, Walker-Smith DJ (1982) Plasmatubules: an alternative to transfer cells? Planta 156:461–465

Haupt S, Cowan GH, Ziegler A, Roberts AG, Oparka KJ, Torrance L (2005) Two plant-viral movement proteins traffic in the endocytic recycling pathway. Plant Cell 17:164–181

Heimburg T (2009) The physics of nerves. arXiv:1008.4279v1

Heimburg T, Jackson AD (2007) The thermodynamics of general anesthesia. Biophys J 92:3159–3165

Herring TL, Cohan CS, Welnhofer EA, Mills LR, Morris CE (1999) F-actin at newly invaginated membrane in neurons: implications for surface area regulation. J Membr Biol 171:151–169

Hirsch RE, Lewis BD, Spalding EP, Sussman MR (1998) A role for the AKT1 potassium channel in plant nutrition. Science 280:918–921

Hoebe RA, Van Oven CH, Gadella TW Jr, Dhonukshe PB, Van Noorden CJ, Manders EM (2007) Controlled light-exposure microscopy reduces photobleaching and phototoxicity in fluorescence live-cell imaging. Nat Biotechnol 25:249–253

Hohenberger P, Eing C, Straessner R, Durst S, Frey W, Nick P (2011) Plant actin controls membrane permeability. Biochim Biophys Acta 1808:2304–2312

Homann U (1998) Fusion and fission of plasma-membrane material accommodates for osmotically induced changes in the surface area, of guard-cell protoplasts. Planta 206:329–333

Humphrey JA, Sedensky MM, Morgan PG (2002) Understanding anesthesia: making genetic sense of the absence of senses. Hum Mol Genet 11:1241–1249

Hush JM, Overall RL (1989) Steady ionic currents around pea (*Pisum sativum* L.) root tips: the effect of tissue wounding. Biol Bull 175:56–64

Iero M, Valenti R, Huber V, Filipazzi P, Parmiani G, Fais S, Rivoltini L (2008) Tumour-released exosomes and their implications in cancer immunity. Cell Death Differ 15:80–88

Illéš P, Schlicht M, Pavlovkin J, Lichtscheidl I, Baluška F, Ovečka M (2006) Aluminium toxicity in plants: internalisation of aluminium into cells of the transition zone in ARABIDOPSIS root apices relates to changes in plasma membrane potential, endosomal behaviour, and nitric oxide production. J Exp Bot 57:4201–4213

Ishikawa H, Evans ML (1990) Gravity-induced changes in, intracellular potentials in elongating cortical cells of mungbean roots. Plant Cell Physiol 31:457–462

Iwabuchi A, Yano M, Shimuzu H (1989) Development of, extracellular electric pattern around Lepidium roots: its possible, role in root growth and gravitropism. Protoplasma 148:94–100

Iwasa K, Tasaki I, Gibbons RC (1980) Swelling of nerve fibers associated with action potentials. Science 210:338–339

Jaffe LF (1977) Electrophoresis along cell membranes. Nature 265:600–602

Jaffe LF (1981) The role of ionic currents in establishing developmental pattern. Philos Trans R Soc Lond B Biol Sci 295:553–566

Jaffe LF, Nuccitelli R (1977) Electrical controls of development. Annu Rev Biophys Bioeng 6:445–476

Johnsen S, Lohmann KJ (2005) Physics and neurobiology of magnetoreception. Nat Rev Neurosci 6:703–712

Johnson FH, Flagler EA (1951) Hydrostatic pressure reversal of narcosis in tadpoles. Science 112:91–92

Jones L, Milne JL, Ashford D, McQueen-Mason SJ (2003) Cell wall arabinan is essential for guard cell function. Proc Natl Acad Sci U S A 100:11783–11788

Kaech S, Brinkhaus H, Matus A (1999) Volatile anesthetics block actin-based motility in dendritic spines. Proc Natl Acad Sci U S A 96:10433–10437

Kaiserli E, Sullivan S, Jones MA, Feeney KA, Christie JM (2009) Domain swapping to assess the mechanistic basis of Arabidopsis phototropin 1 receptor kinase activation and endocytosis by blue light. Plant Cell 21:3226–3244

Kalmijn AJ, Gonzalez IF, McClune MC (2002) The physical nature of life. J Physiol (Paris) 96:355–362

Kandasamy MK, Kappler R, Kristen U (1988) Plasmatubules in the pollen tubes of *Nicotiana sylvestris*. Planta 173:35–41
Karcz W, Burdach Z (1995) The effects of electric field on the growth of intact seedlingsand coleoptile segments of *Zea mays* L. Biol Plant 37:391–397
Keren K (2011) Membrane tension leads the way. Proc Natl Acad Sci U S A 108:14379–14380
Kessel D (2011) Inhibition of endocytic processes by photodynamic therapy. Lasers Surg Med 43:542–547
Kessel D, Price M, Caruso J, Reiners J Jr (2011) Effects of photodynamic therapy on the endocytic pathway. Photochem Photobiol Sci 10:491–498
King JS, Veltmann DW, Insall RH (2011) The induction of autophagy by mechanical stress. Autophagy 7:1490–1499
Koizumi K, Wu S, MacRae-Crerar A, Gallagher KL (2011) An essential protein that interacts with endosomes and promotes movement of the SHORT-ROOT transcription factor. Curr Biol 21:1559–1564
Koles K, Nunnari J, Korkut C, Barria R, Brewer C, Li Y, Leszyk J, Zhang B, Budnik V (2012) Mechanism of evenness interrupted (evi)-exosome release at synaptic boutons. J Biol Chem 287:16820–16834
Kong SG, Suzuki T, Tamura K, Mochizuki N, Hara-Nishimura I, Nagatani A (2006) Blue light-induced association of phototropin 2 with the Golgi apparatus. Plant J 45:994–1005
Kováč L (2007) Information and knowledge in biology. Time for reappraisal. Plant Signal Behav 2:65–73
Kováč L (2008) Bioenergetics. A key to brain and mind. Commun Integr Biol 1:114–122
Kubitscheck U, Homann U, Thiel G (2000) Osmotically evoked shrinking of guard-cell protoplasts causes vesicular retrieval of plasma membrane into the cytoplasm. Planta 210:423–431
Kupchik YM, Barchad-Avitzur O, Wess J, Ben-Chaim Y, Parnas I, Parnas H (2011) A novel fast mechanism for GPCR-mediated signal transduction–control of neurotransmitter release. J Cell Biol 192:137–151
Lachenal G, Pernet-Gallay K, Chivet M, Hemming FJ, Belly A, Bodon G, Blot B, Haase G, Goldberg Y, Sadoul R (2011) Release, of exosomes from differentiated neurons and its regulation by synaptic, glutamatergic activity. Mol Cell Neurosci 46:409–418
Langhorst MF, Reuter A, Jaeger FA, Wippich FM, Luxenhofer G, Plattner H, Stuermer CAO (2008) Trafficking of the microdomain scaffolding protein reggie-1/flotillin-2. Eur J Cell Biol 87:211–226
Laroche C, Beney L, Marechal PA, Gervais P (2001) The effect of osmotic pressure on the membrane fluidity of Saccharomyces cerevisiae at different physiological temperatures. Appl Microbiol Biotechnol 56:249–254
Larsen M, Langmoen IA (1998) The effect of volatile anaesthetics on synaptic release and uptake of glutamate. Toxicol Lett 100–101:59–64
Lee RE Jr, Damodaran K, Yi S-X, Lorigan GA (2006) Rapid cold-hardening increases membrane fluidity and cold, tolerance of insect cells. Cryobiology 52:459–463
Lee TH, D'Asti E, Magnus N, Al-Nedawi K, Meehan B, Rak J (2011) Microvesicles as mediators of intercellular communication in cancer–the emerging science of cellular 'debris'. Semin Immunopathol 33:455–467
Leidy C, Gousset K, Ricker J, Wolkers WF, Tsvetkova NM, Tablin F, Crowe JH (2004) Lipid phase behavior and stabilization of domains in membranes of platelets. Cell Biochem Biophys 40:123–148
Leslie SB, Teter SA, Crowe LM, Crowe JH (1994) Trehalose lowersmembrane phase transition in dry yeast cells. Biochim Biophys Acta 1192:7–13
Leslie SB, Israeli E, Lighthart B, Crowe JH, Crowe LM (1995) Trehalose and sucrose protect both membranes and proteins inintact bacteria during drying. Appl Environ Microbiol 61:3592–3597
Lever MJ, Miller KW, Paton WDM, Smith EB (1971) Pressure reversal of anaesthesia. Nature 231:368–371

Lever MC, Robertson BEM, Buchan ADB, Miller PFP, Gooday GW, Gow NAR (1994) pH and Ca^{2+} dependent galvanotropism of filamentous fungi: implications and mechanisms. Mycol Res 98:301–306

Lewis JD, Lazarowitz SG (2010) Arabidopsis synaptotagmin SYTA regulates endocytosis and virus movement protein cell-to-cell transport. Proc Natl Acad Sci U S A 107:2491–2496

Li B, Liu G, Deng Y, Xie M, Feng Z, Sun M, Zhao Y, Liang L, Ding N, Jia W (2010) Excretion and folding of plasmalemma function to accommodate alterations in guard cell volume during stomatal closure in *Vicia faba* L. J Exp Bot 61:3749–3758

Li R, Liu P, Wan Y, Chen T, Wang Q, Mettbach U, Baluška F, Šamaj J, Fang X, Lucas WJ, Lin J (2012) A membrane microdomain-associated protein, Arabidopsis flot1, is involved in a clathrin-independent endocytic pathway and is required for seedling development. Plant Cell 24:2105-2122

Lin R, Chang DC, Lee YK (2011) Single-cell electroendocytosis on a micro chip using in situ fluorescence microscopy. Biomed Microdevices 13:1063–1073

Lohmann KJ (2010) Magnetic-field perception. Nature 464(1140):1142

Los DA, Murata N (2004) Membrane fluidity and the perception of environmental signals. Biochem Biophys Acta 1666:142–157

Lucas WJ, Franceschi VR (1981) Characean charasome-complex and plasmalemma vesicle development. Protoplasma 107:255–267

Lund EJ (1923) Electric control of organic polarity in the egg of Fucus. Bot Gaz 76:288–301

Lycett G (2008) The role of Rab GTPases in cell wall metabolism. J Exp Bot 59:4061–4074

Mamdouh Z, Giocondi M-C, Laprade R, Le Grimellec C (1996) Temperature dependence of endocytosis in renal epithelial cells in culture. Biochim Biophys Acta 1282:171–173

Masi E, Ciszak M, Stefano G, Renna L, Azzarello E, Pandolfi C, Mugnai S, Baluška F, Arecchi FT, Mancuso S (2009) Spatio-temporal dynamics of the electrical network activity in the root apex. Proc Natl Acad Sci U S A 106:4048–4053

Malho R, Feijo JA, Pais MS (1992) Effect of electric fields and external ionic currents on pollen tube orientation. Sex Plant Reprod 5:57–63

Mancuso S, Marras AM, Mugnai S, Schlicht M, Zarsky V, Li G, Song L, Hue HW, Baluška F (2007) Phospholipase Dζ2 drives vesicular secretion of auxin for its polar cell–cell transport in the transition zone of the root apex. Plant Signal Behav 2:240–244

McAulay AL, Scott BI (1954) A new approach to the study of electric fields produced by growing roots. Nature 174:924–925

McCaig CD, Rajnicek AM, Song B, Zhao M (2005) Controlling cell behavior electrically: current views and future potential. Physiol Rev 85:943–978

McCaig CD, Song B, Rajnicek AM (2009) Electrical dimensions in cell science. J Cell Sci 122:4267–4276

McGillivray AM, Gow NAR (1986) Applied electrical fields polarize thegrowth of mycelial fungi. J Gen Microbiol 132:2515–2525

Meckel T, Hurst AC, Thiel G, Homann U (2005) Guard cells undergo constitutive and pressure-driven membrane turnover. Protoplasma 226:23–29

Medvedev SS, Markova IV (1990) How can the electrical polarity of axial organs regulate plant growth and IAA transport? Physiol Plant 78:38–42

Messerli MA, Graham DM (2011) Extracellular electrical fields direct wound healing and regeneration. Biol Bull 221:79–92

Meyer AJ, Weisenseel MH (1997) Wound-induced changes of membrane voltage, endogenous currents, and ion fluxes in primary roots of maize. Plant Physiol 114:989–998

Mikami K, Murata N (2003) Membrane fluidity and the perception of environmental signals in cyanobacteria. Progr Lipid Res 42:527–543

Miller AL, Gow NA (1989) Correlation between root-generated ionic currents, pH, fusicoccin, indoleacetic acid, and growth of the primary root of *Zea mays*. Plant Physiol 89:1198–1206

Milne A, Beamish T (1999) Inhalational and local anesthetics reduce tactile and thermal responses in *Mimosa pudica*. Can J Anesth 46:287–289

Mills LR, Morris CE (1998) Neuronal plasma membrane dynamics evoked by osmomechanical perturbations. J Membr Biol 166:223–235

Mina MG, Goldsworthy A (1991) Changes in the electrical polarity of tobacco following the application of weak external currents. Planta 186:104–108

Moore A (2012) Life defined. BioEssays 34:253–254

Morgan PG, Sedensky M, Meneely PM (1990) Multiple sites of action of volatile anesthetics in *Caenorhabditis elegans*. Proc Natl Acad Sci U S A 87:2965–2969

Morris CE (2001) Mechanosensitive membrane traffic and an optimal strategy for volume and surface area regulation in CNS neurons. Amer Zool 41:721–727

Morris CE, Homann U (2001) Cell surface area regulation and membrane tension. J Membr Biol 179:79–102

Morrow IC, Parton RG (2005) Flotillins and the PHB domain protein family: rafts, worms and anaesthetics. Traffic 6:725–740

Moscatelli A, Idilli AI (2009) Pollen tube growth: a delicate equilibrium between secretory and endocytic pathways. J Integr Plant Biol 51:727–739

Murata Y, Iwasaki H, Sasaki M, Inaba K, Okamura Y (2005) Phosphoinositide phosphatase activity coupled to an intrinsic voltage sensor. Nature 435:1239–1243

Murr LE (1963) Plant growth responses in a stimulated electric field environment. Nature 200:490–491

Müller P, Ahmad M (2011) Light-activated cryptochrome reacts with molecular oxygen to form a flavin-superoxide radical pair consistent with magnetoreception. J Biol Chem 286:21033–21040

Nichol JA, Hutter OF (1996) Tensile strength and dilatational elasticity, of giant sarcolemmal vesicles shed from rabbit muscle. J Physiol (Lond) 493:187–198

Nozue K, Wada M (1993) Electrotropism of Nicotiana pollen tubes. Plant Cell Physiol 34:1291–1296

Nunn JF, Sturrock JE, Wills EJ, Richmond JE, McPherson CK (1974) The effect of inhalational anaesthetics on the swimming velocity of *Tetrahymena pyriformis*. J Cell Sci 15:537–554

Okamura Y (2007) Biodiversity of voltage sensor domain proteins. Pflugers Arch—Eur J Physiol 454:361–371

Okamura Y, Dixon JE (2011) Voltage-sensing phosphatase: its molecular relationship with PTEN. Physiology 26:6–13

Okamura Y, Murata Y, Iwasaki H (2009) Voltage-sensing phosphatase: actions and potentials. J Physiol 587:513–520

Olivotto M, Arcangeli A, Carlà M, Wanke E (1996) Electric fields at the plasma membrane level: a neglected element in the mechanisms of cell signalling. BioEssays 18:495–504

Onelli E, Prescianotto-Baschong C, Caccianiga M, Moscatelli A (2008) Clathrin-dependent and independent endocytic pathways in tobacco protoplasts revealed by labelling with charged nanogold. J Exp Bot 59:3051–3068

Oparka KJ, Prior DAM (1992) Direct evidence for pressure-generated closure of plasmodesmata. Plant J 2:741–750

Orvar BL, Sangwan V, Omann F, Dhindsa RS (2000) Early steps in cold sensing by plant cells: the role of actin cytoskeleton and membrane fluidity. Plant J 23:785–794

Ovečka M, Lang I, Baluška F, Ismail A, Illeš P, Lichtscheidl IK (2005) Endocytosis and vesicle trafficking during tip growth of root hairs. Protoplasma 226:39–54

Ozansoy M, Denizhan Y (2009) The endomembrane system: a representation of the extracellular medium? Biosemiotics 2:255–267

Ozen S, Sirota A, Belluscio MA, Anastassiou CA, Stark E, Koch C, Buzsáki G (2010) Transcranial electric stimulation entrains cortical neuronal populations in rats. J Neurosci 30:11476–11485

Paciorek T, Zažímalová E, Ruthardt N, Petrášek J, Stierhof YD, Kleine-Vehn J, Morris DA, Emans N, Jürgens G, Geldner N, Friml J (2005) Auxin inhibits endocytosis and promotes its own efflux from cells. Nature 435:1251–1256

Pagnussat L, Burbach C, Baluška F, de la Canal (2012) Rapid endocytosis is triggered upon imbibition in Arabidopsis seeds. Plant Signal Behav 7:417–422

Parnas I, Parnas H (2010) Control of neurotransmitter release: from Ca^{2+} to voltage dependent G-protein coupled receptors. Pflugers Arch 460:975–990

Parnas I, Parnas H (2007) The chemical synapse goes electric: Ca^{2+}- and voltage-sensitive GPCRs control neurotransmitter release. Trends Neurosci 30:54–61

Pedersen CNS, Axelsen KB, Harper JF, Palmgren MG (2012) Evolution of plant P-type ATPases. Frontiers Plant Sci 3:31

Penzlin H (2009) The riddle of "life," a biologist's critical view. Naturwissenschaften 96:1–23

Pfeffer SR (2010) Unconventional secretion by autophagosome exocytosis. J Cell Biol 188:451–452

Phillips JB, Jorge PE, Muheim R (2010) Light-dependent magnetic compass orientation in amphibians and insects: candidate receptors and candidate molecular mechanisms. J R Soc Interface 7(Suppl)2:S241–S256

Pokorny J, Hasek J, Jelinek F (2005) Endogenous electric field and organization of living matter. Electromagn Biol Med 24:185–197

Polo S, Di Fiore PP (2006) Endocytosis conducts the cell signaling orchestra. Cell 124:897–900

Price GD, Whitecross MI (1983) Cytochemical localisation of ATPase activity on the plasmalemma of *Chara corallina*. Protoplasma 116:65–74

Punnonen E-L, Kirsi R, Marjommaki VS (1998) At reduced temperature, endocytic membrane is blocked in multivesicular carrier endosomes of cardiac myocytes. Eur J Cell Biol 49:281–294

Rak J, Guha A (2012) Extracellular vesicles—vehicles that spread cancer genes. BioEssays 34:489–497

Reuzeau C, Mills LR, Harris JA, Morris CE (1995) Discrete and reversible vacuole-like dilations induced by osmomechanical perturbation of neurons. J Membr Biol 145:33–47

Roberts D, PedmaleUV MorrowJ et al (2011) Modulation of phototropic responsiveness in Arabidopsis through ubiquitination of phototropin 1 by the CUL3-ring E3 ubiquitin ligase CRL3NPH3. Plant Cell 23:3627–3640

Rosemberg Y, Korenstein R (1997) Incorporation of macromolecules into cells and vesicles by low electirc field: induction of endocytotic-like processes. Biolelectrochem Bionenerg 42:275–281

Sakamoto K, Briggs WR (2002) Cellular and subcellular localization of phototropin 1. Plant Cell 14:1723–1735

Sakata S, Hossain MI, Okamura Y (2011) Coupling of the phosphatase activity of Ci-VSP to itsvoltage sensor activity over the entire range of voltage sensitivity. J Physiol 589:2687–2705

Saltveit ME Jr (1993) Effect of high-pressure gas atmospheres and anaesthetics on chilling injury of plants. J Exp Bot 265:1361–1368

Šamaj J, Šamajová O, Peters M, Baluška F, Lichtscheidl I, Knox JP, Volkmann D (2000) Immunolocalization of LM2 arabinogalactan protein epitope associated with endomembranes of plant cells. Protoplasma 212:186–196

Šamaj J, Baluška F, Voigt B, Schlicht M, Volkmann D, Menzel D (2004) Endocytosis, actin cytoskeleton, and signaling. Plant Physiol 135:1150–1161

Šamaj J, Müller J, Beck M, Böhm N, Menzel D (2006) Vesicular trafficking, cytoskeleton and signalling in root hairs and pollen tubes. Trends Plant Sci 11:594–600

Sandstrom DJ (2004) Isoflurane depresses glutamate release by reducing neuronal excitability at the Drosophila neuromuscular junction. J Physiol 558:489–502

Sangwan V, Dhindsa RS (2002) In vivo and in vitro activation of temperature-responsive plant MAP kinases. FEBS Lett 531:561–564

Sangwan V, Orvar BL, Beyerly J, Hirt H, Dhindsa RS (2002) Opposite changes in membrane fluidity mimic cold and heat stress activation of distinct plant MAP kinase pathways. Plant J 31:629–638

Schapire AL, Voigt B, Jasik J, Rosado A, Lopez-Cobollo R, Menzel D, Salinas J, Mancuso S, Valpuesta V, Baluška F, Botella MA (2008) Arabidopsis synaptotagmin 1 is required for the maintenance of plasma membrane integrity and cell viability. Plant Cell 20:3374–3388

Schapire AL, Valpuesta V, Botella MA (2009) Plasma membrane repair in plants. Trends Plant Sci 14:645–652
Schreurs WJ, Harold FM (1988) Transcellular proton current in *Achlya bisexualis* hyphae: relationship to polarized growth. Proc Natl Acad Sci U S A 85:1534–1538
Scita G, Di Fiore PP (2010) Endocytic matrix. Nature 463:464–473
Scott BIH (1956) Electric oscillations generated by plant roots and a possible feedback mechanism responsible for them. Aust J Biol Sci 10:164–179
Scott BIH (1967) Electric fields in plants. Annu Rev Plant Physiol 18:409–418
Sedensky MM, Siefker JM, Koh JY, Miller DM III, Morgan PG (2004) A stomatin and a degenerin interact in lipid rafts of the nervous systemof *Caenorhabditis elegans*. Am J Physiol Cell Physiol 287:C468–C474
Shaw BD, Chung DW, Wang CL, Quintanilla LA, Upadhyay S (2011) A role for endocytic recycling in hyphal growth. Fungal Biol 115:541–546
Shibasaki K, Uemura M, Tsurumi S, Rahman A (2009) Auxin response in Arabidopsis under cold stress: underlying auxin response in Arabidopsis under cold stress. Plant Cell 21:3823–3838
Shope JC, Mott KA (2006) Membrane trafficking and osmotically induced volumechanges in guard cells. J Exp Bot 57:4123–4131
Shope JC, DeWald DB, Mott KA (2003) Changes in surface area of intact guard cells are correlated with membrane internalization. Plant Physiol 133:1314–1321
Sigismund S, Confalonieri S, Ciliberto A, Polo S, Scita G, Di Fiore PP (2012) Endocytosis and signaling: cell logistics shape the eukaryotic cell plan. Physiol Rev 92:273–366
Simons M, Raposo G (2009) Exosomes—vesicular carriers for intercellular communication. Curr Opin Cell Biol 21:575–581
Singaram VK, Somerlot BH, Falk SA, Falk MJ, Sedensky MM, Morgan PG (2011) Optical reversal of halothane-induced immobility in *C. elegans*. Curr Biol 21:2070–2076
Sisneros JA, Tricas TC (2002) Neuroethology and life history adaptations of the elasmobranch electric sense. J Physiol - Paris 96:379–389
Smalheiser NR (2007) Exosomal transfer of proteins and RNAs at synapses in the nervous system. Biol Direct 2:35
Solovyov IA, Schulten K (2009) Magnetoreception through cryptochrome may involve superoxide. Biophys J 96:4804-4813
Somers DE, Fujiwara S (2009) Thinking outside the F-box: novel ligands for novel receptors. Trends Plant Sci 14:206–213
Sonner JM (2008) A hypothesis on the origin and evolution of the response to inhaled anesthetics. Anesth Analg 107:849–854
Stanley SA, Gagner JE, Damanpour S, Yoshida M, Dordick JS, Friedman JM (2012) Radio-wave heating of iron oxide nanoparticles can regulate plasma glucose in mice. Science 336:604–608
Staykova M, Stone HA (2011) The role of the membrane confinement in the surface area regulation of cells. Commun Integr Biol 4:616–618
Staykova M, Holmes DP, Read C, Stone HA (2011) Mechanics of surface area regulation in cells examined with confined lipid membranes. Proc Natl Acad Sci U S A 108:9084–9088
Steinhardt RA (2005) The mechanisms of cell membrane repair: a tutorial guide to key experiments. Ann N Y Acad Sci 1066:152–165
Steinhardt RA, Bi G, Alderton JM (1994) Cell membrane resealing by a vesicular, mechanism similar to neurotransmitter release. Science 263:390–393
Stenz HG, Weisenseel MH (1993) Electrotropism of maize (*Zea mays* L.) roots. Plant Physiol 101:1107–1111
Stolarz M, El Krol, Dziubinska H, Kurenda A (2010) Glutamate induces series of action potentials and a decrease in circumnutation rate in *Helianthus annuus*. Physiol Plant 138:329–338
Sullivan S, Kaiserli E, Tseng TS, Christie JM (2010) Subcellular localization and turnover of Arabidopsis phototropin 1. Plant Signal Behav 5:184–186
Swartz KJ (2008) Sensing voltage across lipid membranes. Nature 456:891–897
Takeuchi Y, Schmid J, Caldwell JH, Harold FM (1988) Transcellular ion currents and extension of *Neurospora crassa* hyphae. J Membr Biol 101:33–41

Tanaka Y, Kutsuna N, Kanazawa Y, Kondo N, Hasezawa S, Sano T (2007) Intra-vacuolar reserves of membranes during stomatal closure: the possible role of guard cell vacuoles estimated by 3-D reconstruction. Plant Cell Physiol 48:1159–1169

Tasaki I (1999) Evidence for phase transitions in nerve fibers, cells and synapses. Ferroelectrics 220:305–316

Tasaki I, Kusano K, Byrne PM (1989) Rapid mechanical and thermal changes in the garfish olfactory nerve associated with a propagated impulse. Biophys J 55:1033–1040

Taylor DD, Gercel-Taylor C (2011) Exosomes/microvesicles: mediators of cancer-associated immunosuppressive microenvironments. Semin Immunopathol 33:441–454

Taylorson RB, Hendricks SB (1979) Overcoming dormancy in seeds with ethanol and other anesthetics. Planta I45:507–510

Théry C, Ostrowski M, Segura E (2009) Membrane vesicles as, conveyors of immune responses. Nat Rev Immunol 9:581–593

Tobias JM (1960) Further studies on the nature of the excitable system in nerve. I. Voltage-induced axoplasm movement in squid axons. J Gen Physiol 43:57–71

Togo T, Alderton JM, Steinhardt RA (2003) Long-term potentiation of exocytosis, and cell membrane repair in fibroblasts. Mol Biol Cell 14:93–106

Trewavas AJ, Baluška F (2011) The ubiquity of consciousness. EMBO Rep 12:1221–1225

Trkanjec Z (1996) Eletrostatic attraction: the driving force for the presynaptic vesicle-cell membrane fusion. Med Hypoth 47:93–96

Trkanjec Z, Demarin V (2001) Presynaptic vesicle, exocytosis, membrane fusion and basic physical forces. Med Hypoth 56:540–546

Tsuchiya H, Ueno T, Mizogami M, Takakura K (2010) Local anesthetics structure-dependently interact with anionic phospholipid membranes to modify the fluidity. Chem Biol Interact 183:19–24

Uesono Y (2009) Environmental stresses and clinical drugs paralyze a cell. Commun Integr Biol 2:1–4

Valadi H, Ekström K, Bossios A, Sjöstrand M, Lee JJ, Lötvall JO (2007) Exosome-mediated transfer of mRNAs and microRNAs is a novel mechanism of genetic exchange between cells. Nat Cell Biol 9:654–659

Veech RL, Kashiwaya Y, Gates DN, King MT, Clarke K (2002) The energetics of ion distribution: the origin of the resting electric potential of cells. IUBMB Life 54:241–252

Vissenberg K, Feijó JA, Weisenseel MH, Verbelen J-P (2001) Ion fluxes, auxin and the induction of elongation growth in Nicotiana tabacum cells. J Exp Bot 52:2161–2167

Voigt B, Timmers A, Šamaj J, Hlavacka A, Ueda T, Preuss M, Nielsen E, Mathur J, Emans N, Stenmark H, Nakano A, Baluška F, Menzel D (2005) Actin-based motility of endosomes is linked to polar tip-growth of root hairs. Eur J Cell Biol 84:609–621

Wan Y, Eisinger W, Ehrhardt D, Kubitscheck U, Baluška F, Briggs W (2008) The subcellular localization and blue-light-induced movement of phototropin 1-GFP in etiolated seedlings of *Arabidopsis thaliana*. Mol Plant 1:103–117

Wan Y, Jasik J, Wang L, Hao H, Volkmann D, Menzel D, Mancuso S, Baluška F, Lin J (2012) The signal transducer NPH3 integrates the Phototropin1 photosensor with PIN2-based polar auxin transport in Arabidopsis root phototropism. Plant Cell 24:551–565

Wang ET, Zhao M (2010) Regulation of tissue repair and regeneration by electric fields. Chin J Traumatol 13:55–61

Wawrecki W, Zagórska-Marek B (2007) Influence of a weak DC electric field on root meristem architecture. Ann Bot 100:791–796

Weaver JC, Vaughan TE, Martin GT (1999) Biological effects due to weak electric and magnetic fields: the temperature variation threshold. Biophys J 76:3026–3030

Weaver JC, Vaughan TE, Astumian RD (2000) Biological sensing of small field differences by magnetically sensitive chemical reactions. Nature 405:707–709

Wei T, Hibino H, Omura T (2009) Release of rice dwarf virus from insect vector cells involves secretory, exosomes derived from multivesicular bodies. Commun Integr Biol 2:324–326

Weisenseel MH, Nuccitelli R, Jaffe LF (1975) Large electrical currents traverse growing pollen tubes. J Cell Biol 66:556–567
Weisenseel MH, Dorn A, Jaffe LF (1979) Natural H currents traverse growing roots and root hairs of Barley (*Hordeum vulgare L.*). Plant Physiol 64:512–518
Wiklund RA, Allison AC (1972) The effects of anaesthetics on the motility of *Dictyostelium discoideum*: evidence for a possible mechanism of anaesthesia. Br J Anaesth 44:622
Wiltschko R, Wiltschko W (2012) Magnetoreception. Adv Exp Med Biol 739:126–141
Witzany G, Baluška F (2012) Turing: a formal clash of codes. Nature 483:541
Wlodarczyk A, McMillan PF, McMillan SA (2006) High pressure effects in anaesthesia and narcosis. Chem Soc Rev 35:890–898
Wojtaszek P, Volkmann D, Baluška F (2004) Polarity and cell walls. In: Lindsey K (ed) Polarity in plants. Blackwell Publishing, Oxford, pp 72–121
Wolfe J, Steponkus PL (1983) Mechanical properties of the plasma membrane of isolated plant protoplasts. Plant Physiol 71:276–285
Wolfe J, Dowgert MF, Steponkus PL (1985) Dynamics of incorporation of material into the plasma membrane and the lysis of protoplasts during rapid expansions in area. J Membr Biol 86:127–138
Wolfe J, Dowgert MF, Steponkus PL (1986) Mechanical study of the deformation and rupture of the plasma membranes of protoplasts during osmotic expansions. J Membr Biol 93:63–74
Wolverton C, Mullen JL, Aizawa S, Yoshizaki I, Kamigaichi S, Mukai C, Shimazu T, Fukui K, Evans ML, Ishikawa H (1999) Inhibition of root elongation in microgravity by an applied electric field. J Plant Res 112:493–496
Wolverton C, Mullen JL, Ishikawa H, Evans ML (2000) Two distinct regions of response drive differential growth in Vigna root electrotropism. Plant Cell Environ 23:1275–1280
Yamazaki T, Kawamura Y, Minami A, Uemura M (2008) Calcium-dependent freezing tolerance in Arabidopsis involves membrane resealing via synaptotagmin SYT1. Plant Cell 20:3389–3404
Yao L, Shanley L, McCaig C, Zhao M (2008) Small applied electric fields guide migration of hippocampal neurons. J Cell Physiol 216:527–535
Yao L, McCaig CD, Zhao M (2009) Electrical signals polarize neuronal organelles, direct neuron migration, and orient cell division. Hippocampus 19:855–868
Yao L, Pandit A, Yao S, McCaig CD (2011) Electric field-guided neuron migration: a novel approach in neurogenesis. Tissue Eng Part B Rev 17:143–153
Yu Q, Hlavačka A, Matoh T, Volkmann D, Menzel D, Goldbach HE, Baluška F (2002) Short-term boron deprivation inhibits endocytosis of cell wall pectins in meristematic cells of maize and wheat root apices. Plant Physiol 130:415–421
Žárský V, Potocký M (2010) Recycling domains in plant cell morphogenesis: small GTPase effectors, plasma membrane signalling and the exocyst. Biochem Soc Trans 38:723–728
Zhang Y, He J, Lee D, McCormick S (2010) Interdependence of endomembrane trafficking and actin dynamics during polarized growth of Arabidopsis pollen tubes. Plant Physiol 152:2200–2210
Zhao Y, Yan A, Feijó JA, Furutani M, Takenawa T, Hwang I, Fu Y, Yang Z (2010) Phosphoinositides regulate clathrin-dependent endocytosis at the tip of pollen tubes in Arabidopsis and tobacco. Plant Cell 22:4031–4044
Zhong Y-G, Zhang G-J, Yang L, Zheng Y-Z (2000) Effects of photoinduced membrane rigidification on the lysosomal permeability to potassium ions. Photochem Photobiol 71:627–633
Zimmermann U, Schnettler R, Klöck G, Watzka H, Donath E, Glaser RW (1990) Mechanisms of electrostimulated uptake of macromolecules into living cells. Naturwissenschaften 77:543–545
Zonia L, Munnik T (2009) Uncovering hidden treasures in pollen tube growth mechanics. Trends Plant Sci 14:318–327
Zorec R, Tester M (1993) Rapid pressure driven exocytosis–endocytosis cycle in a single plant cell. FEBS Lett 333:283–286

Receptor-Mediated Endocytosis in Plants

Simone Di Rubbo and Eugenia Russinova

Abstract Introducing the concept of receptor-mediated endocytosis (RME) has completely changed the traditional view of endocytosis as a process by which cells simply transport molecules from the plasma membrane (PM) and extracellular space. Internalization of molecules by means of specific cell-surface receptors led to the notion that endocytosis is the master organizer of cellular signalling. RME spatially regulates the signalling outputs of PM receptors by either targeting them for degradation or relocating them to signalling endosomes. Recent studies revealed highly conserved mechanisms behind RME in all eukaryotes, including plants, demonstrating a major role of clathrin as well as post-translational modifications (PTMs) of PM receptors, such as ubiquitination and phosphorylation. In this chapter, we will review the latest data on RME in plants and its function in regulating receptor-mediated signalling. While these recent developments contributed to a better understanding of RME in plants, further work is needed to precisely describe the molecular machinery and to resolve the signalling role of receptor pools localized to different endomembrane compartments.

S. Di Rubbo · E. Russinova (✉)
Department of Plant Systems Biology VIB, Ghent University, 9052 Ghent, Belgium
e-mail: jenny.russinova@psb.vib-ugent.beeurus@psb.vib-ugent.be

S. Di Rubbo · E. Russinova
Department of Plant Biotechnology and Bioinformatics, Ghent University, 9052 Ghent, Belgium

J. Šamaj (ed.), *Endocytosis in Plants*,
DOI: 10.1007/978-3-642-32463-5_7,

1 Introduction

Receptor-mediated endocytosis (RME) represents the uptake of soluble ligands from the extracellular space into the cell, which is mediated by the specific binding between ligands and plasma membrane (PM)-localized receptors (Sorkin and Von Zastrow 2009). The best characterized RME route is clathrin-mediated endocytosis (CME). This pathway involves the recruitment of PM-resident cargo into clathrin-coated pits through the invagination of the PM driven by clathrin polymerization (McMahon and Boucrot 2011). Assembly of clathrin cages at the PM is primed by the binding of FCH domain only (FCHO) proteins to phosphatidylinositol 4,5-bisphosphate in order to establish the initiation point for CME (Henne et al. 2010; Umasankar et al. 2012). Cargo-selective clathrin adaptors, such as the Adaptor Protein-2 (AP-2) complex, bridge the interaction between clathrin and the PM cargo protein (McMahon and Boucrot 2011). Recruitment of cargo proteins into clathrin-coated pits depends on the recognition of sorting signals in the cargo itself by the AP-2 (Traub 2009). Accessory proteins associated with clathrin-coated pits, such as epsins and β-arrestins, can substitute for the AP-2 and bind both cargo and clathrin (Traub 2009). PTMs also regulate RME. Ligand binding to the epidermal growth factor receptor (EGFR) induces its phosphorylation and ubiquitination (Umebayashi et al. 2008). Ubiquitin can act as an endocytic signal as EGFR defective in the sites for ubiquitination is not internalized (Goh et al. 2010) and knock-down of E3 ubiquitin ligase leads to a block in the uptake of EGFR (Bertelsen et al. 2011). Ubiquitination of EGFR ensures its interaction with the accessory clathrin adaptor protein, EGFR substrate 15 (EPS15), and epsin1 in the absence of AP-2 (Hawryluk et al. 2006; Bertelsen et al. 2011). Alternative mechanisms of non-clathrin endocytosis (NCE) have been described for RME. Impairment of CME does not block the endocytosis of transforming growth factor-β receptor (TGFβ-R), which is internalized by lipid raft/caveolar pathways (Di Guglielmo et al. 2003).

RME redistributes active receptors from the PM to different cellular compartments and thus regulates their signalling outputs (Sorkin and Von Zastrow 2009; Sigismund et al. 2012). Removal of active receptors from the PM mainly attenuates signalling (Goh et al. 2010; Brankatschk et al. 2012). Endocytosis can also relocate receptors into signalling endosomes from where active receptors can either initiate or sustain signalling (Murphy et al. 2009). For example, the EGFR-dependent MAPK signalling is negatively regulated by CME (Goh et al. 2010), whereas RNA interference or knock-down of clathrin or AP-2 impairs EGFR-dependent and AKT-mediated signalling (Sigismund et al. 2008; Rappoport and Simon 2009; Goh et al. 2010). RME can therefore be a positive or negative regulator of signalling pathways (Scita and Di Fiore 2010).

2 Receptor-Mediated Endocytosis in Plants

2.1 Plant Plasma Membrane Receptors, Ligands, and Their Endocytosis

Knowledge about ligand-receptor pairs is essential to study RME. More than 600 receptor-like kinases (RLKs) and 1,000 of their potential peptide ligands exist in the genome of the model plant Arabidopsis. These numbers are even higher in monocot plants such as rice (Shiu et al. 2004; Lease and Walker 2006). Plant RLKs share a common domain organization common with the animal receptor tyrosine kinases, consisting of a single-coil transmembrane domain, an extracellular domain responsible for ligand binding, and an intracellular domain that provides the kinase activity to the protein (Shiu et al. 2004). Furthermore, a large number of receptor-like proteins (RLPs) resembling the extracellular domains of RLKs, but lacking functional cytoplasmic domains, are also found in Arabidopsis and other plant species (Kawchuk et al. 2001; Shiu and Bleecker 2003; Wang et al. 2008a). RLKs and RLPs are involved in a wide range of processes, such as root development, floral meristem maintenance, and pollen maturation (Albrecht et al. 2005; De Smet et al. 2008; Nimchuk et al. 2011), as well as in responses to pathogens (Ron and Avni 2004; Robatzek et al. 2006). Direct ligand-receptor binding has been demonstrated only for a few plant RLKs (Kinoshita et al. 2005; Chinchilla et al. 2007; Ogawa et al. 2008; Ohyama et al. 2009; Zipfel 2009; Albert et al. 2010; Lee et al. 2012), although genetic studies have predicted more receptor-ligand couples (Gifford et al. 2005; Krol et al. 2010; Matsuzaki et al. 2010; Yamaguchi et al. 2010; Huffaker et al. 2011; Whitford et al. 2012). For the first time in plants, the receptor-ligand interaction between the brassinosteroid (BR) receptor BRASSINOSTEROID INSENSITIVE 1 (BRI1) and its ligand, the steroid-like plant hormone brassinolide (BL), was recently resolved at the structural level (Hothorn et al. 2011; She et al. 2011).

RME in plants was first demonstrated in 1989 by the saturable and temperature-dependent uptake of a labeled elicitor fraction into a soybean cell suspension (Horn et al. 1989), although neither the identity of the receptor nor the nature of the elicitor molecule was known. Since then, RME has been described for several plant receptors (Ron and Avni 2004; Robatzek et al. 2006; Nimchuk et al. 2011; Irani et al. 2012, see also Chap. 15 in this volume). Among them are the most studied receptor-ligand pairs: the BR receptor BRI1 and its ligand BL (Kinoshita et al. 2005), the elicitor receptor FLAGELLIN-SENSING 2 (FLS2) recognizing the bacterial peptide flagellin (flg22) (Chinchilla et al. 2007), the signalling peptide receptor CLAVATA 1 (CLV1) and its ligand CLV3 (Ohyama et al. 2009), and the fungal elicitor ethylene-inducing xylanase (EIX) receptor LeEIX2 in tomato (Ron and Avni 2004; Bar and Avni 2009).

Until now, endocytosis of plant cell-surface receptors has been studied mainly by live imaging of genetically engineered green fluorescence protein (GFP)-tagged receptors in Arabidopsis root meristem cells, leaf protoplasts, or tobacco leaf epidermis (Russinova et al. 2004; Gifford et al. 2005; Robatzek et al. 2006; Geldner et al. 2007; Ivanov and Gaude 2009; Nimchuk et al. 2011; Sharfman et al. 2011; Irani et al. 2012). Recently, synthesis and imaging of a BR molecule labeled with a small fluorophore, Alexa Fluor 647, have allowed to specifically track the endocytosis of the BR receptor-ligand complexes in the Arabidopsis root epidermis, revealing a trafficking route to the vacuole through *trans*-Golgi network/early endosome (TGN/EE) and multivesicular bodies (Irani et al. 2012).

Similar to animal systems, the RME in plants can either be constitutive or ligand-induced. Exogenous application of the ligand stimulates the endocytosis of FLS2, LeEIX2, S-Receptor Kinase, and CLV1 (Robatzek et al. 2006; Bar and Avni 2009; Ivanov and Gaude 2009; Nimchuk et al. 2011). Further, deletion of the extracellular domain of ARABIDOPSIS CRINKLY 4 (ACR4) can block its internalization (Gifford et al. 2005). On the other hand, no differences in the endosomal localization of BRI1 are observed after a treatment with exogenous BRs, suggesting constitutive endocytosis (Russinova et al. 2004; Geldner et al. 2007; Irani et al. 2012).

2.2 Mechanisms of Receptor-Mediated Endocytosis in Plants

CME is the main mechanism for the internalization of cell surface receptors in animals as well of the Arabidopsis auxin transporter PIN FORMED 2 (PIN2) (Dhonukshe et al. 2007; Chen et al. 2011). Block of endocytosis of LeEIX2 and BRI1 by overexpressing a truncated clathrin or after treatment with tyrphostin A23 (Sharfman et al. 2011; Irani et al. 2012), both described as specific inhibitors of CME (Liu et al. 1998; Banbury et al. 2003; Dhonukshe et al. 2007), demonstrates that clathrin takes part in RME in plants. Although to date, no genetic interference approach for the AP-2 complex has been used in plants, experimental data indicate an evolutionary conserved mechanism for CME. In the animal system, the AP-2 complex mediates the contact between clathrin and PM receptors by recognizing di-leucine ([ED]xxxL[LI]) or tyrosine (YxxΦ, where Φ is a bulky hydrophobic residue) motifs in the receptors (Banbury et al. 2003; Goh et al. 2010). Similarly, mutations in the YxxΦ motif in LeEIX2 block its endocytosis (Bar and Avni 2009). It has also been shown that animal transferrin receptors expressed in Arabidopsis protoplasts mediate the internalization of transferrin (Ortiz-Zapater et al. 2006), a process dependent on the AP-2 complex in animal cells (Motley et al. 2003). Experiments with LeEIX2 demonstrate that other accessory adaptor proteins exist in plants. In particular, the EPSIN HOMOLOGY DOMAIN 2 (EHD2) protein can directly interact with AP-2 and LeEIX2 to modulate the ligand-induced endocytosis of LeEIX2 (Bar and Avni 2009; Bar et al. 2009).

Evidence from genetic and drug studies shows that the guanine-nucleotide exchange factors for ADP-ribosylation factor GTPases (ARF-GEFs) GNOM and GNOM-LIKE1 (GNL1), as well as the Arabidopsis homolog of the animal Rab5 GTPases, ARA7, mediate the endocytosis of the PM-localized auxin transporter

PIN1 in Arabidopsis (Teh and Moore 2007; Dhonukshe et al. 2008; Naramoto et al. 2010). A simultaneous inhibition of the activity of these two ARF-GEFs or the expression of a dominant negative ARA7 can partially impair or fully block the endocytosis of BRI1-BR complexes in living cells (Irani et al. 2012). The ARF-GEF inhibitor brefeldin A (BFA), affecting the activity of GNOM (Naramoto et al. 2010), can also partially impair the endocytosis of LeEIX2 when transiently expressed in tobacco leaf cells (Sharfman et al. 2011). It remains to be investigated if the ARF-GEFs or ARA7-mediated endocytosis depends on clathrin, but as GNOM partially co-localizes with clathrin at the PM, ARF-GEF-mediated endocytosis could be part of CME (Naramoto et al. 2010).

2.3 Post-Translational Modifications and Receptor-Mediated Endocytosis in Plants

2.3.1 Phosphorylation

Animal PM receptors are known to be post-translationally modified upon ligand-binding. PTMs can regulate the endocytosis of receptors, targeting them to different trafficking routes (Traub 2009). In plants, phosphorylation of the auxin transporter PIN2 has been proven essential for its polar PM localization in root epidermis, a process dependent on its intracellular trafficking (Zhang et al. 2010). Studies on phosphorylation and endocytosis of FLS2 suggest that such PTMs might trigger internalization of the receptor. Indeed, mutations in putative phosphorylation sites of FLS2 block its endocytosis (Robatzek et al. 2006; Salomon and Robatzek 2006). Phosphorylation of receptors leads to the recruitment of co-receptors resulting in the activation of the downstream signalling pathway (Wang et al. 2008b). In Arabidopsis, several RLKs, including BRI1, FLS2, and EF-TU RECEPTOR (EFR), share the co-receptor RLK SOMATIC EMBRIOGENESIS RECEPTOR-LIKE KINASE 3 (SERK3)/BRI1 ASSOCIATED KINASE 1 (BAK1) (Russinova et al. 2004; Chinchilla et al. 2007; Heese et al. 2007; Roux et al. 2011; Schwessinger et al. 2011). BAK1 does not participate in ligand binding, but is essential for downstream signalling and seems to have a role in regulating the endocytosis of the main receptors (Kinoshita et al. 2005; Chinchilla et al. 2007). For example, overexpression of BAK1 enhances the rate of BRI1 internalization in leaf protoplasts (Russinova et al. 2004), while FLS2 endocytosis is reduced in the *bak1* mutant (Chinchilla et al. 2007). As in both cases interaction with BAK1 is a prerequisite for activation of the main receptor, it can be speculated that BAK1 facilitates RME by either receptor phosphorylation or by specifically recruiting the endocytic machinery. Interestingly, the tomato BAK1 inhibits elicitor-stimulated endocytosis and signalling of LeEIX2 through its interaction with a decoy receptor LeEIX1 (Bar et al. 2010), but the mechanism of this inhibition is not known yet.

2.3.2 Ubiquitination

Ubiquitination of plant PM proteins is also essential for their endocytosis and trafficking to the vacuole (Pan et al. 2009; Barberon et al. 2011; Kasai et al. 2011; Leitner et al. 2012). Ubiquitination could have a similar role in plant RME, as mutations in the putatively ubiquitinated PEST motif in FLS2 abolish its ligand-induced endocytosis (Salomon and Robatzek 2006). While no experimental mapping of ubiquitination sites of FLS2 exists to date, it has recently been shown that FLS2 is ubiquitinated in a ligand-dependent fashion by the U-box E3 ubiquitin ligases PUB12 and PUB13, proteins that are also necessary for receptor degradation after ligand stimulation (Lu et al. 2011). Ubiquitination and degradation of FLS2 can also be mediated by the bacterial E3 ubiquitin ligase AvrPtoB through its physical interaction with BAK1 (Göhre et al. 2008). These data link ligand-depended phosphorylation of the receptor to its ubiquitination and degradation in the vacuole. It remains to be determined if ubiquitination is necessary for the first steps of RME or is needed for sorting in endosomes.

3 Endocytosis and Receptor-Mediated Signalling in Plants

3.1 Endocytosis Regulates Signalling at the Plasma Membrane

RME has a well-documented biological role in the regulation of receptor signalling (Sorkin and Von Zastrow 2009; Sigismund et al. 2012). In animal systems, RME can attenuate signalling by removing active receptors from the PM (Scita and Di Fiore 2010). This model is corroborated by a recent study showing endocytosis of the receptor-ligand couple, BRI1-BR. Block of CME through genetic or pharmacological interference activates BR signalling. Similar results are obtained by the inhibition of ARF-GEF-dependent endocytosis, as well as by the expression of a dominant-negative ARA7 protein (Irani et al. 2012). Therefore, endocytosis appears to function as a negative regulator of BR signalling. Additional evidence comes from studies associating BRI1 inactivation and degradation with its dephosphorylation by the PROTEIN PHOSPHATASE 2A (PP2A) (Di Rubbo et al. 2011; Wu et al. 2011). Attenuation of BR signalling via RME also implies that the PM is the signalling platform for the BRI1 receptor. This idea is strengthened by previous reports showing that the MEMBRANE STEROID-BINDING PROTEIN 1 (MSBP1) down-regulates BR signalling by inducing endocytosis of BAK1 and consequently making it unavailable for BRI1 (Song et al. 2009). This model also matches the known PM-localization of several downstream BR signalling components (Vert and Chory 2006; Tang et al. 2008).

3.2 Endosomal Signalling in Plants

Apart from signalling attenuation, RME is essential for signalling as it re-localizes active receptors to signalling endosomes from where signalling is sustained or initiated de novo (Sorkin and Von Zastrow 2009; Sigismund et al. 2012). In plants, the genetic or pharmacological block of LeEIX2 endocytosis impairs the EIX-induced signalling (Bar and Avni 2009; Bar et al. 2009; Sharfman et al. 2011). A similar scenario emerges for FLS2. A block of FLS2 endocytosis with wortmannin (inhibitor of phosphoinositide 3-kinase and phosphoinositide 4-kinases) or by mutations in its phosphorylation sites leads to defective pathogen responses due to impaired flagellin signalling (Robatzek et al. 2006; Salomon and Robatzek 2006). Plants defective in the ubiquitination pathway, which is linked to FLS2 internalization and vacuolar targeting, also show a reduced flagellin-induced production of reactive oxygen species (Salomon and Robatzek 2006; Lu et al. 2011). This differs from the RME signalling attenuation model proposed for BRI1 and supports a role for RME in signalling of flagellin and xylanase elicitors, implicating that endosomes might be essential for signalling. Application of the TGN aggregator endosidin 1 (ES1) enhances the EIX-induced signalling which depends on LeEIX2 (Sharfman et al. 2011), confirming that LeEIX2 receptors are still active in TGN/EEs and that EIX signalling is initiated from this compartment. Signalling endosomes have also been investigated in BR signalling. The use of other TGN aggregating drugs, such as concanamycin A, monensin or salinomycin, results in the accumulation of the BRI1-BR complex but does not enhance BR signalling (Irani et al. 2012). The ARF-GEF inhibitor BFA is also commonly used to accumulate some PM proteins in endosomes (BFA compartments) by inhibiting their recycling to the PM. BFA treatment increases the endosomal pool of BRI1 and activates BR signalling (Geldner et al. 2007; Irani et al. 2012). However, no correlation between the amount of endosomal BRI1 and BR signalling was observed and the effect of BFA on BR signalling has been linked to its inhibition of ARF-GEF-mediated endocytosis of BRI1 (Irani et al. 2012). In contrast to FLS2 and LeEIX2, BRI1 does not need re-localization to endosomes in order to perform its signalling function, indicating that the biological role of RME can vary from receptor to receptor (Fig. 1).

4 Conclusions and Future Prospects

Several recent studies highlight similarities between RME in animals and plants (Geldner and Robatzek 2008; Irani and Russinova 2009). Therefore, the basic mechanisms and components behind RME seem to be conserved throughout evolution with CME standing as the major RME pathway. However, subtle differences in this process are present between kingdoms. Although some plant receptors carry canonical motifs for internalization via CME as identified in animals (Geldner and Robatzek 2008), the functionality of such motifs is only

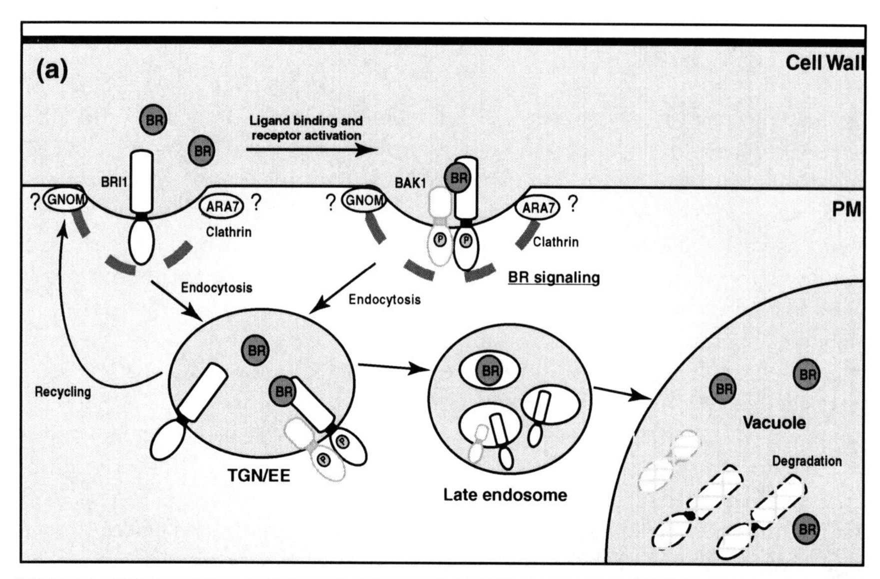

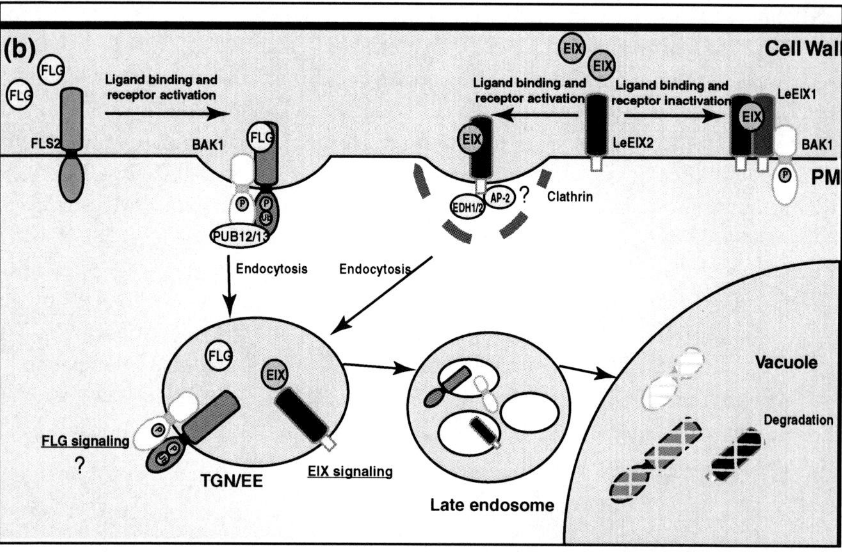

Fig. 1 RME in plants. **a** BRs traffic from the PM to the vacuole via TGN/EEs and multivesicular bodies after binding to BRI1, though endocytosis of this receptor is constitutive and ligand-independent. Endocytosis of BRI1 is clathrin-, GNOM-, and ARA7-dependent and is required to attenuate signalling of the active BRI1-BR complexes from the PM. After endocytosis, BRI1 is either targeted to the vacuole or recycled back to the PM. **b** Elicitors flagellin and EIX and their receptors FLS2 and LeEIX2, respectively, undergo ligand-dependent endocytosis. FLS2 requires phosphorylation and ubiquitination for endocytosis. Internalization of LeEIX2 is clathrin-mediated. Endocytosis of both receptor-ligand couples is required for their signalling activity. *PM* plasma membrane, *TGN/EE trans*-Golgi network/early endosome, *BR* brassinosteroid, *FLG* flagellin

demonstrated for LeEIX2 (Sharfman et al. 2011), suggesting that recruitment of cargos in clathrin-coated vesicles might be different in plants. Also, members of the large ARF-GEF family are involved in CME of receptors in plants (Irani et al. 2012), showing another peculiarity of the plant endomembrane system, as animal counterparts are known to have a role only in endosomes and Golgi (Naramoto et al. 2010).

PTM of receptors has been shown to play a role in plant RME, being linked to trafficking and degradation of FLS2 and BRI1 (Robatzek et al. 2006; Lu et al. 2011; Wu et al. 2011). However, it is still not clear if these modifications are involved in early stages of RME and/or in receptor sorting in endosomes. In animals, different ubiquitination patterns are known to distinguish between recycling and degradation after RME (Mukhopadhyay and Riezman 2007). It is of particular interest to see if similar mechanisms apply to plant receptors such as BRI1 undergoing constitutive endocytosis and recycling.

While no evidence for NCE of plant receptors has been gathered to date, it has already been described that trafficking of animal receptors toward recycling and degradation is driven by, respectively, CME and NCE (Sigismund et al. 2008). In a recent work, NCE has been suggested to be an uptake route of glucose in plants (Bandmann and Homann 2012). Additionally, the detergent-resistant membrane-associated Flotillin1 (Flot1) functions in endocytosis independent of clathrin (Li et al. 2012), giving a first glimpse of the mechanism of NCE in plants and providing a potential tool for the study of NCE of plant receptors.

Here, we propose two different models for RME in plants (Fig. 1). In the first one, endocytosis leads to signal attenuation and receptor degradation (Irani et al. 2012) while in the second one, endocytosis is required for signalling (Robatzek et al. 2006; Lu et al. 2011; Sharfman et al. 2011). This is in accordance with what has been reported for EGFR, where RME appears to be required for both signal attenuation and signalling (Sigismund et al. 2008; Brankatschk et al. 2012). However, while in animal systems RME seems to have two different effects on the signalling of the same receptor, the data in plants suggest that only ligand-stimulated endocytosis might be required for signalling. The two current models of cellular compartmentalization of signalling in plants could be strengthened by identifying additional components of signal transduction and investigating their subcellular localization. This information is missing for flg22 or EIX, while it is only partially resolved for BR signalling. This knowledge will help to design tools for monitoring signalling activities from different subcellular compartments.

In animals, downstream signalling components can be part of the endocytic machinery and thus provide direct feedback on RME (Sorkin and Von Zastrow 2009). In plants, BAK1 has been implicated both in induction and repression of endocytosis and signalling of plant receptors (Russinova et al. 2004; Chinchilla et al. 2007; Bar et al. 2010). However, more studies are required to elucidate evolutionary conserved as well as divergent processes in plant RME.

The growing field of RME in plants will benefit from the creation of new research tools. One of the limiting factors is the absence of biologically, active labeled ligands for plant receptors, which have been routinely used in animal systems during the last decades. To date, only fluorescently—tagged BRs have been developed and used for studies in plants (Irani et al. 2012). Creating new fluorescent ligands will be instrumental in defining the cellular maps of active receptor-ligand complexes and understanding the signalling contribution of different subcellular (e.g., PM versus endosomes) compartments. A better knowledge of the link between endocytosis and receptor-mediated signalling would require a better characterization of the plant endomembrane system. In this regard, chemical genetics has a great potential both to unravel new endocytic components in plants and to supply new drugs for rapid and reversible perturbations of endocytosis. A large-scale chemical genetic screen to identify small molecules affecting the localization of different PM proteins, including BRI1, has recently been carried out (Drakakaki et al. 2011), creating a platform for further development of new research tools that will allow a better understanding of the mechanism and biological function of plant RME.

Acknowledgments We thank N.G. Irani for the critical reading of the manuscript and useful suggestions and A. Bleys for helping with manuscript preparation. This work is supported by the BRAVISSIMO Marie-Curie Initial Training Network (PITN-GA-2008-215118).

References

Albert M, Jehle AK, Mueller K, Eisele C, Lipschis M, Felix G (2010) *Arabidopsis thaliana* pattern recognition receptors for bacterial elongation factor Tu and flagellin can be combined to form functional chimeric receptors. J Biol Chem 285:19035–19042. doi:10.1074/jbc.M110.124800

Albrecht C, Russinova E, Hecht V, Baaijens E, De Vries S (2005) The *Arabidopsis thaliana* somatic embryogenesis receptor-like kinases 1 and 2 control male sporogenesis. Plant Cell 17:3337–3349. doi:10.1105/tpc.105.036814

Banbury AN, Oakley JD, Sessions RB, Banting G (2003) Tyrphostin A23 inhibits internalization of the transferrin receptor by perturbing the interaction between tyrosine motifs and the medium chain subunit of the AP-2 adaptor complex. J Biol Chem 278:12022–12028. doi:10.1074/jbc.M211966200

Bandmann V, Homann U (2012) Clathrin-independent endocytosis contributes to uptake of glucose into BY-2 protoplasts. Plant J 70:578–584. doi:10.1111/j.1365-313X.2011.04892.x

Bar M, Avni A (2009) EHD2 inhibits ligand-induced endocytosis and signaling of the leucine-rich repeat receptor-like protein LeEix2. Plant J 59:600–611. doi:10.1111/j.1365-313X.2009.03897.x

Bar M, Sharfman M, Schuster S, Avni A (2009) The coiled-coil domain of EHD2 mediates inhibition of LeEix2 endocytosis and signaling. PLoS One 4:e7973. doi:10.1371/Journal.Pone.0007973

Bar M, Sharfman M, Ron M, Avni A (2010) BAK1 is required for the attenuation of ethylene-inducing xylanase (Eix)-induced defense responses by the decoy receptor LeEix1. Plant J 63:791–800. doi:10.1111/j.1365-313X.2010.04282.x

Barberon M, Zelazny E, Robert S, Conéjéro G, Curie C, Friml J, Vert G (2011) Monoubiquitin-dependent endocytosis of the iron-regulated transporter 1 (IRT1) transporter controls iron uptake in plants. Proc Natl Acad Sci U.S.A. 108:E450–E458. doi:10.1073/pnas.1100659108

Bertelsen V, Sak MM, Breen K, Rødland MS, Johannessen LE, Traub LM, Stang E, Madshus IH (2011) A chimeric pre-ubiquitinated EGF receptor is constitutively endocytosed in a clathrin-dependent, but kinase-independent manner. Traffic 12:507–520. doi:10.1111/j.1600-0854.2011.01162.x

Brankatschk B, Wichert SP, Johnson SD, Schaad O, Rossner MJ, Gruenberg J (2012) Regulation of the EGF transcriptional response by endocytic sorting. Sci Signal 5:ra21. doi: 10.1126/scisignal.2002351

Chen X, Irani NG, Friml J (2011) Clathrin-mediated endocytosis: the gateway into plant cells. Curr Opin Plant Biol 14:674–682. doi:10.1016/j.pbi.2011.08.006

Chinchilla D, Zipfel C, Robatzek S, Kemmerling B, Nürnberger T, Jones JDG, Felix G, Boller T (2007) A flagellin-induced complex of the receptor FLS2 and BAK1 initiates plant defence. Nature 448:497–500. doi:10.1038/nature05999

De Smet I, Vassileva V, De Rybel B, Levesque MP, Grunewald W, Van Damme D, Van Noorden G, Naudts M, Van Isterdael G, De Clercq R, Wang JY, Meuli N, Vanneste S, Friml J, Hilson P, Jürgens G, Ingram GC, Inzé D, Benfey PN, Beeckman T (2008) Receptor-like kinase ACR4 restricts formative cell divisions in the Arabidopsis root. Science 322:594–597. doi:10.1126/science.1160158

Dhonukshe P, Aniento F, Hwang I, Robinson DG, Mravec J, Stierhof YD, Friml J (2007) Clathrin-mediated constitutive endocytosis of PIN auxin efflux carriers in Arabidopsis. Curr Biol 17:520–527. doi:10.1016/j.cub.2007.01.052

Dhonukshe P, Tanaka H, Goh T, Ebine K, Mähönen AP, Prasad K, Blilou I, Geldner N, Xu J, Uemura T, Chory J, Ueda T, Nakano A, Scheres B, Friml J (2008) Generation of cell polarity in plants links endocytosis, auxin distribution and cell fate decisions. Nature 456:962–966. doi:10.1038/nature07409

Di Guglielmo GM, Le Roy C, Goodfellow AF, Wrana JL (2003) Distinct endocytic pathways regulate TGF-β receptor signaling and turnover. Nat Cell Biol 5:410–421. doi:10.1038/Ncb975

Di Rubbo S, Irani NG, Russinova E (2011) PP2A phosphatases: the "on-off" regulatory switches of brassinosteroid signaling. Sci Signal 4:25. doi:10.1126/scisignal.2002046

Drakakaki G, Robert S, Szatmari AM, Brown MQ, Nagawa S, Van Damme D, Leonard M, Yang Z, Girke T, Schmid SL, Russinova E, Friml J, Raikhel NV, Hicks GR (2011) Clusters of bioactive compounds target dynamic endomembrane networks in vivo. Proc Natl Acad Sci U.S.A. 108:17850–17855. doi:10.1073/pnas.1108581108

Geldner N, Robatzek S (2008) Plant receptors go endosomal: a moving view on signal transduction. Plant Physiol 147:1565–1574. doi:10.1104/pp.108.120287

Geldner N, Hyman DL, Wang X, Schumacher K, Chory J (2007) Endosomal signaling of plant steroid receptor kinase BRI1. Genes Dev 21:1598–1602. doi:10.1101/gad.1561307

Gifford ML, Robertson FC, Soares DC, Ingram GC (2005) ARABIDOPSIS CRINKLY4 function, internalization, and turnover are dependent on the extracellular crinkly repeat domain. Plant Cell 17:1154–1166. doi:10.1105/tpc.104.029975

Goh LK, Huang F, Kim W, Gygi S, Sorkin A (2010) Multiple mechanisms collectively regulate clathrin-mediated endocytosis of the epidermal growth factor receptor. J Cell Biol 189:871–883. doi:10.1083/jcb.201001008

Göhre V, Spallek T, Häweker H, Mersmann S, Mentzel T, Boller T, De Torres M, Mansfield JW, Robatzek S (2008) Plant pattern-recognition receptor FLS2 is directed for degradation by the bacterial ubiquitin ligase AvrPtoB. Curr Biol 18:1824–1832. doi:10.1016/j.cub.2008.10.063

Hawryluk MJ, Keyel PA, Mishra SK, Watkins SC, Heuser JE, Traub LM (2006) Epsin 1 is a polyubiquitin-selective clathrin-associated sorting protein. Traffic 7:262–281. doi:10.1111/j.1600-0854.2006.00383.x

Heese A, Hann DR, Gimenez-Ibanez S, Jones AME, He K, Li J, Schroeder JI, Peck SC, Rathjen JP (2007) The receptor-like kinase SERK3/BAK1 is a central regulator of innate immunity in plants. Proc Natl Acad Sci U.S.A. 104:12217–12222. doi:10.1073/pnas.0705306104

Henne WM, Boucrot E, Meinecke M, Evergren E, Vallis Y, Mittal R, McMahon HT (2010) FCHo proteins are nucleators of clathrin-mediated endocytosis. Science 328:1281–1284. doi:10.1126/science.1188462

Horn MA, Heinstein PF, Low PS (1989) Receptor-mediated endocytosis in plant cells. Plant Cell 1:1003–1009. doi:10.1105/tpc.1.10.1003

Hothorn M, Belkhadir Y, Dreux M, Dabi T, Noel JP, Wilson IA, Chory J (2011) Structural basis of steroid hormone perception by the receptor kinase BRI1. Nature 474:467–471. doi:10.1038/Nature10153

Huffaker A, Dafoe NJ, Schmelz EA (2011) ZmPep1, an ortholog of Arabidopsis elicitor peptide 1, regulates maize innate immunity and enhances disease resistance. Plant Physiol 155:1325–1338. doi:10.1104/pp.110.166710

Irani NG, Russinova E (2009) Receptor endocytosis and signaling in plants. Curr Opin Plant Biol 12:653–659. doi:10.1016/j.pbi.2009.09.011

Irani NG, Di Rubbo S, Mylle E, Van Den Begin J, Schneider-Pizoń J, Hniliková J, Šíša M, Buyst D, Vilarrasa-Blasi J, Szatmári AM, Van Damme D, Mishev K, Codreanu MC, Kohout L, Strnad M, Caño-Delgado AI, Friml J, Madder A, Russinova E (2012) Fluorescent castasterone reveals BRI1 signaling from the plasma membrane. Nat Chem Biol 8:583–589. doi:10.1038/nchembio.958

Ivanov R, Gaude T (2009) Endocytosis and endosomal regulation of the *S*-receptor kinase during the self-incompatibility response in *Brassica oleracea*. Plant Cell 21:2107–2117. doi:10.1105/tpc.108.063479

Kasai K, Takano J, Miwa K, Toyoda A, Fujiwara T (2011) High boron-induced ubiquitination regulates vacuolar sorting of the BOR1 borate transporter in *Arabidopsis thaliana*. J Biol Chem 286:6175–6183. doi:10.1074/jbc.M110.184929

Kawchuk LM, Hachey J, Lynch DR, Kulcsar F, Van Rooijen G, Waterer DR, Robertson A, Kokko E, Byers R, Howard RJ, Fischer R, Prüfer D (2001) Tomato *Ve* disease resistance genes encode cell surface-like receptors. Proc Natl Acad Sci U.S.A. 98:6511–6515. doi:10.1073/pnas.091114198

Kinoshita T, Caño-Delgado A, Seto H, Hiranuma S, Fujioka S, Yoshida S, Chory J (2005) Binding of brassinosteroids to the extracellular domain of plant receptor kinase BRI1. Nature 433:167–171. doi:10.1038/Nature.03227

Krol E, Mentzel T, Chinchilla D, Boller T, Felix G, Kemmerling B, Postel S, Arents M, Jeworutzki E, Al-Rasheid KaS, Becker D, Hedrich R (2010) Perception of the Arabidopsis danger signal peptide 1 involves the pattern recognition receptor *At*PEPR1 and its close homologue *At*PEPR2. J Biol Chem 285:13471–13479. doi:10.1074/jbc.M109.097394

Lease KA, Walker JC (2006) The Arabidopsis unannotated secreted peptide database, a resource for plant peptidomics. Plant Physiol 142:831–838. doi:10.1104/pp.106.086041

Lee JS, Kuroha T, Hnilova M, Khatayevich D, Kanaoka MM, Mcabee JM, Sarikaya M, Tamerler C, Torii KU (2012) Direct interaction of ligand-receptor pairs specifying stomatal patterning. Genes Dev 26:126–136. doi:10.1101/gad.179895.111

Leitner J, Petrášek J, Tomanov K, Retzer K, Pařezová M, Korbei B, Bachmair A, Zažímalová E, Luschnig C (2012) Lysine^{63}-linked ubiquitylation of PIN2 auxin carrier protein governs hormonally controlled adaptation of Arabidopsis root growth. Proc Natl Acad Sci U.S.A. 109:8322–8327. doi:10.1073/pnas.1200824109

Li R, Liu P, Wan Y, Chen T, Wang Q, Mettbach U, Baluška F, Šamaj J, Fang X, Lucas WJ, Lin JA (2012) A membrane microdomain-associated protein, Arabidopsis Flot1, is involved in a clathrin-independent endocytic pathway and is required for seedling development. Plant Cell. 24: 2105–2122. doi:10.1105/tpc.112.095695

Liu S-H, Marks MS, Brodsky FM (1998) A dominant-negative clathrin mutant differentially affects trafficking of molecules with distinct sorting motifs in the class II major histocompatibility complex (MHC) pathway. J Cell Biol 140:1023–1037. doi:10.1083/jcb.140.5.1023

Lu D, Lin W, Gao X, Wu S, Cheng C, Avila J, Heese A, Devarenne TP, He P, Shan L (2011) Direct ubiquitination of pattern recognition receptor FLS2 attenuates plant innate immunity. Science 332:1439–1442. doi:10.1126/science.1204903

Matsuzaki Y, Ogawa-Ohnishi M, Mori A, Matsubayashi Y (2010) Secreted peptide signals required for maintenance of root stem cell niche in Arabidopsis. Science 329:1065–1067. doi:10.1126/science.1191132

McMahon HT, Boucrot E (2011) Molecular mechanism and physiological functions of clathrin-mediated endocytosis. Nat Rev Mol Cell Biol 12:517–533. doi:10.1038/nrm3151

Motley A, Bright NA, Seaman MNJ, Robinson MS (2003) Clathrin-mediated endocytosis in AP-2-depleted cells. J Cell Biol 162:909–918. doi:10.1083/jcb.200305145

Mukhopadhyay D, Riezman H (2007) Proteasome-independent functions of ubiquitin in endocytosis and signaling. Science 315:201–205. doi:10.1126/science.1127085

Murphy JE, Padilla BE, Hasdemir B, Cottrell GS, Bunnett NW (2009) Endosomes: a legitimate platform for the signaling train. Proc Natl Acad Sci U.S.A. 106:17615–17622. doi:10.1073/pnas.0906541106

Naramoto S, Kleine-Vehn J, Robert S, Fujimoto M, Dainobu T, Paciorek T, Ueda T, Nakano A, Van Montagu MCE, Fukuda H, Friml J (2010) ADP-ribosylation factor machinery mediates endocytosis in plant cells. Proc Natl Acad Sci U.S.A. 107:21890–21895. doi:10.1073/pnas.1016260107

Nimchuk ZL, Tarr PT, Ohno C, Qu X, Meyerowitz EM (2011) Plant stem cell signaling involves ligand-dependent trafficking of the CLAVATA1 receptor kinase. Curr Biol 21:345–352. doi:10.1016/j.cub.2011.01.039

Ogawa M, Shinohara H, Sakagami Y, Matsubayashi Y (2008) *Arabidopsis* CLV3 peptide directly binds CLV1 ectodomain. Science 319:294. doi:10.1126/science.1150083

Ohyama K, Shinohara H, Ogawa-Ohnishi M, Matsubayashi Y (2009) A glycopeptide regulating stem cell fate in *Arabidopsis thaliana*. Nat Chem Biol 5:578–580. doi:10.1038/nchembio.182

Ortiz-Zapater E, Soriano-Ortega E, Marcote MJ, Ortiz-Masiá D, Aniento F (2006) Trafficking of the human transferrin receptor in plant cells: effects of tyrphostin A23 and brefeldin A. Plant J 48:757–770. doi:10.1111/j.1365-313X.2006.02909.x

Pan JW, Fujioka S, Peng JL, Chen JH, Li GM, Chen RJ (2009) The E3 ubiquitin ligase $SCF^{TIR1/AFB}$ and membrane sterols play key roles in auxin regulation of endocytosis, recycling, and plasma membrane accumulation of the auxin efflux transporter PIN2 in *Arabidopsis thaliana*. Plant Cell 21:568–580. doi:10.1105/tpc.108.061465

Rappoport JZ, Simon SM (2009) Endocytic trafficking of activated EGFR is AP-2 dependent and occurs through preformed clathrin spots. J Cell Sci 122:1301–1305. doi:10.1242/Jcs.040030

Robatzek S, Chinchilla D, Boller T (2006) Ligand-induced endocytosis of the pattern recognition receptor FLS2 in Arabidopsis. Genes Dev 20:537–542. doi:10.1101/Gad.366506

Ron M, Avni A (2004) The receptor for the fungal elicitor ethylene-inducing xylanase is a member of a resistance-like gene family in tomato. Plant Cell 16:1604–1615. doi:10.1105/Tpc.022475

Roux M, Schwessinger B, Albrecht C, Chinchilla D, Jones A, Holton N, Malinovsky FG, Tör M, De Vries S, Zipfel C (2011) The Arabidopsis leucine-rich repeat receptor-like kinases BAK1/SERK3 and BKK1/SERK4 are required for innate immunity to hemibiotrophic and biotrophic pathogens. Plant Cell 23:2440–2455. doi:10.1105/tpc.111.084301

Russinova E, Borst J-W, Kwaaitaal M, Caño-Delgado A, Yin Y, Chory J, De Vries SC (2004) Heterodimerization and endocytosis of Arabidopsis brassinosteroid receptors BRI1 and AtSERK3 (BAK1). Plant Cell 16:3216–3229. doi:10.1105/tpc.104.025387

Salomon S, Robatzek S (2006) Induced endocytosis of the receptor kinase FLS2. Plant Signal Behav 1:293–295. doi:10.4161/psb.1.6.3594

Schwessinger B, Roux M, Kadota Y, Ntoukakis V, Sklenar J, Jones A, Zipfel C (2011) Phosphorylation-dependent differential regulation of plant growth, cell death, and innate immunity by the regulatory receptor-like kinase BAK1. PLoS Genet 7:e1002046. doi:10.1371/journal.pgen.1002046

Scita G, Di Fiore PP (2010) The endocytic matrix. Nature 463:464–473. doi:10.1038/nature08910

Sharfman M, Bar M, Ehrlich M, Schuster S, Melech-Bonfil S, Ezer R, Sessa G, Avni A (2011) Endosomal signaling of the tomato leucine-rich repeat receptor-like protein LeEix2. Plant J 68:413–423. doi:10.1111/j.1365-313X.2011.04696.x

She J, Han Z, Kim T-W, Wang J, Cheng W, Chang J, Shi S, Wang J, Yang M, Wang Z-Y, Chai J (2011) Structural insight into brassinosteroid perception by BRI1. Nature 474:472–476. doi:10.1038/Nature10178

Shiu S-H, Bleecker AB (2003) Expansion of the receptor-like kinase/pelle gene family and receptor-like proteins in Arabidopsis. Plant Physiol 132:530–543. doi:10.1104/pp.103.021964
Shiu S-H, Karlowski WM, Pan R, Tzeng Y-H, Mayer KFX, Li W-H (2004) Comparative analysis of the receptor-like kinase family in Arabidopsis and rice. Plant Cell 16:1220–1234. doi:10.1105/tpc.020834
Sigismund S, Argenzio E, Tosoni D, Cavallaro E, Polo S, Di Fiore PP (2008) Clathrin-mediated internalization is essential for sustained EGFR signaling but dispensable for degradation. Dev Cell 15:209–219. doi:10.1016/j.devcel.2008.06.012
Sigismund S, Confalonieri S, Ciliberto A, Polo S, Scita G, Di Fiore PP (2012) Endocytosis and signaling: cell logistics shape the eukaryotic cell plan. Physiol Rev 92:273–366. doi:10.1152/physrev.00005.2011
Song L, Shi Q-M, Yang X-H, Xu Z-H, Xue H-W (2009) Membrane steroid-binding protein 1 (MSBP1) negatively regulates brassinosteroid signaling by enhancing the endocytosis of BAK1. Cell Res 19:864–876. doi:10.1038/cr.2009.66
Sorkin A, Von Zastrow M (2009) Endocytosis and signaling: intertwining molecular networks. Nat Rev Mol Cell Biol 10:609–622. doi:10.1038/Nrm2748
Tang W, Kim T-W, Oses-Prieto JA, Sun Y, Deng Z, Zhu S, Wang R, Burlingame AL, Wang Z-Y (2008) BSKs mediate signal transduction from the receptor kinase BRI1 in Arabidopsis. Science 321:557–560. doi:10.1126/science.1156973
Teh O-K, Moore I (2007) An ARF-GEF acting at the golgi and in selective endocytosis in polarized plant cells. Nature 448:493–496. doi:10.1038/nature06023
Traub LM (2009) Tickets to ride: selecting cargo for clathrin-regulated internalization. Nat Rev Mol Cell Biol 10:583–596. doi:10.1038/Nrm2751
Umasankar PK, Sanker S, Thieman JR, Chakraborty S, Wendland B, Tsang M, Traub LM (2012) Distinct and separable activities of the endocytic clathrin-coat components Fcho1/2 and AP-2 in developmental patterning. Nat Cell Biol 14:488–501. doi:10.1038/ncb2473
Umebayashi K, Stenmark H, Yoshimori T (2008) Ubc4/5 and c-Cbl continue to ubiquitinate EGF receptor after internalization to facilitate polyubiquitination and degradation. Mol Biol Cell 19:3454–3462. doi:10.1091/mbc.E07-10-0988
Vert G, Chory J (2006) Downstream nuclear events in brassinosteroid signaling. Nature 441:96–100. doi:10.1038/nature04681
Wang G, Ellendorff U, Kemp B, Mansfield JW, Forsyth A, Mitchell K, Bastas K, Liu CM, Woods-Tōr A, Zipfel C, De Wit PJGM, Jones JDG, Tōr M, Thomma BPHJ (2008a) A genome-wide functional investigation into the roles of receptor-like proteins in Arabidopsis. Plant Physiol 147:503–517. doi:10.1104/pp.108.119487
Wang X, Kota U, He K, Blackburn K, Li J, Goshe MB, Huber SC, Clouse SD (2008b) Sequential transphosphorylation of the BRI1/BAK1 receptor kinase complex impacts early events in brassinosteroid signaling. Dev Cell 15:220–235. doi:10.1016/j.devcel.2008.06.011
Whitford R, Fernandez A, Tejos R, Cuéllar Pérez A, Kleine-Vehn J, Vanneste S, Drozdzecki A, Leitner J, Abas L, Aerts M, Hoogewijs K, Baster P, De Groodt R, Lin YC, Storme V, Van De Peer Y, Beeckman T, Madder A, Devreese B, Luschnig C, Friml J, Hilson P (2012) GOLVEN secretory peptides regulate auxin carrier turnover during plant gravitropic responses. Dev Cell 22:678–685. doi:10.1016/j.devcel.2012.02.002
Wu G, Wang X, Li X, Kamiya Y, Otegui MS, Chory J (2011) Methylation of a phosphatase specifies dephosphorylation and degradation of activated brassinosteroid receptors. Sci Signal 4:ra29. doi: 10.1126/scisignal.2001258
Yamaguchi Y, Huffaker A, Bryan AC, Tax FE, Ryan CA (2010) PEPR2 is a second receptor for the pep1 and pep2 peptides and contributes to defense responses in Arabidopsis. Plant Cell 22:508–522. doi:10.1105/tpc.109.068874
Zhang J, Nodzyński T, Pěnčík A, Rolčík J, Friml J (2010) PIN phosphorylation is sufficient to mediate PIN polarity and direct auxin transport. Proc Natl Acad Sci U.S.A. 107:918–922. doi:10.1073/pnas.0909460107
Zipfel C (2009) Early molecular events in PAMP-triggered immunity. Curr Opin Plant Biol 12:414–420. doi:10.1016/j.pbi.2009.06.003

Endocytic Trafficking of PIN Proteins and Auxin Transport

Tomasz Nodzyński, Steffen Vanneste and Jiří Friml

Abstract Plants display an amazing developmental plasticity which compensates for a sessile lifestyle and inability to move away from unfavorable conditions as in the case of animals. Many aspects of this adaptability inflict changing of cell fate, de novo organ formation, and rearrangement of the plant body plan. Coordination of those processes is facilitated by the hormone auxin, which through its directional flow and local accumulation patterns provides the spatial information linking cellular and developmental modifications. This tight control of auxin distribution is in a major extent facilitated by the activity of PIN-FORMED (PIN) auxin transporters mediating export of auxin from cells. Members of the PIN protein family can display polar localization at the plasma membrane which enables the directional auxin transit through cells. The local levels of PINs in the membrane are dynamically, controlled by subcellular vesicular trafficking events encompassing secretion, recycling, degradation, and most prominently endocytosis. This well-characterized process also provides entry points for different signals that engage the endocytic machinery forging PINs into different downstream trafficking pathways, in accordance with ontogenetic programs and environmental stimuli, thus facilitating plant development.

T. Nodzyński · S. Vanneste · J. Friml (✉)
Department of Plant Systems Biology, VIB, 9052, Ghent, Belgium
e-mail: jifri@psb.vib-ugent.be

T. Nodzyński
e-mail: tonod@psb.vib-ugent.be

S. Vanneste
e-mail: stnes@psb.vib-ugent.be

T. Nodzyński · S. Vanneste · J. Friml
Department of Plant Biotechnology and Bioinformatics, Ghent University, 9052, Ghent, Belgium

J. Šamaj (ed.), *Endocytosis in Plants*,
DOI: 10.1007/978-3-642-32463-5_8,

Keywords Endocytosis · Recycling · Trafficking · PIN proteins · Polar auxin transport

1 Auxin Transport Machinery for Differential Auxin Distribution

The evolution of multicellularity in living organisms gave the opportunity for specialization of different cell types in distinct structures that could more efficiently perform tasks vital to every life form such as reproduction, nutrient uptake, or light perception. However, development of complex cellular systems inadvertently created the need for means of commutation between the basic building blocks enabling a coordinated functioning that conforms with the overall plan of the organism that they build. In plants this organisation is mediated, among others, by the phytohormone auxin which via its differential distribution within plant tissues sketches the basic plan of plant body guiding the embryonic (Friml et al. 2003; Schlereth et al. 2010) and post-embryonic development (Benková et al. 2003; Berleth et al. 2007; Benjamins and Scheres 2008; Dubrovsky et al. 2008; Swarup et al. 2008; Vanneste and Friml 2009). Apart from local biosynthesis or release of active forms from inactive precursors (Ruiz Rosquete et al. 2011), the carrier based directional cell-to-cell transport contributes mostly to the generation of cellular auxin gradients providing instructive cues for plant development (Petrášek and Friml 2009).

The necessity of carrier-mediated auxin transport was highlighted by the chemiosmotic polar diffusion model (Goldsmith 1977) which considered the physical and chemical properties of this molecule. It had been reasoned that auxin (IAA—indole-3-acetic acid) residing in the acidic apoplast (pH 5.5) is mostly in an undissociated lipophilic form; thus, it can only slowly penetrate the lipid bilayer. However, upon entering the more basic cytoplasm (pH 7) the molecule dissociates to a proton and an IAA anion which cannot penetrate the plasma membrane (PM) and must be transported out of the cellular environment. The asymmetric distribution of carriers in the cells would facilitate the directionality of auxin transport. In addition, experimental data hinted at an active uptake of auxin anions facilitated by $2H^+$ symporters (Rubery and Sheldrake 1974). Consequently, following investigations have proven the existence of a group of PM carriers (auxin resistant 1/Like AUX1—AUX1/LAX family) that facilitate auxin influx (Bennett et al. 1996; Kramer 2004).

Auxin efflux is carried out by the family of PIN proteins encompassing eight members which contribute mostly to the directionality of auxin transport (Vieten et al. 2007, Zažímalová et al. 2007). The transporters PIN1, 2, 3, 4, and 7 are localized in the PM where they mediate auxin efflux from the cells (Petrášek et al. 2006). In contrast, PIN5, 6, and 8 are presumably localized to the endoplasmic reticulum

(ER) membrane where they mediate auxin exchange between the cytosol and the ER lumen (Mravec et al. 2009). The auxin transport capacity of PINs was tested both in plant and heterologous systems (Petrášek et al. 2006; Yang and Murphy 2009). These results are also supported by the fact that genetic interference with PIN proteins led to reduced auxin transport and diverse developmental phenotypes (Okada et al. 1991; Gälweiler et al. 1998) similar to the ones observed after application of auxin transport inhibitors (Okada et al. 1991; Vieten et al. 2007).

Auxin efflux is also facilitated by members of the multidrug resistance/phosphoglycoprotein belonging to the group of ATP-binding cassette sub-family B (ABCB) previously abbreviated in the literature as MDR/PGP (Noh et al. 2001; Verrier et al. 2008). These auxin translocators, in most of the cells, do not have a clearly defined polarity at the PM and rather seem to facilitate auxin availability for the directional PIN-mediated auxin efflux (Geisler et al. 2005; Mravec et al. 2008; Kubeš et al. 2011, Christie et al. 2011).

Recently, a new group of putative auxin transport facilitators, called PIN-LIKES (PILS), was shown to regulate auxin levels by sequestration in ER (Barbez et al. 2012). This important finding has once again highlighted the significance of compartmentalization and metabolism as mechanisms regulating auxin levels (Mravec et al. 2009; Wabnik et al. 2011b; Barbez et al. 2012).

2 Entering the Intracellular Trafficking Pathway by Clathrin-Dependent Endocytosis

The abundance of PINs in the PM is the product of exocytosis and endocytosis coupled with polar recycling of these proteins (Kleine-Vehn et al. 2011). On the endocytic route from PM the first compartments reached by internalized proteins are called the early endosomes (EE) which in plants are a apart or derive from the *trans*-Golgi network (TGN/EE) (Dettmer et al. 2006; Lam et al. 2007). From the TGN cargos can be trafficked to the recycling endosomes (RE), from where they are redirected to PM (polar recycling), or travel onwards to the late endosomes/prevacuolar compartment (PVC), and finally to the vacuole (Robinson et al. 2008), where they are degraded (Fig. 1). The combined action of these processes contributes to the regulation of abundance and polar localization of auxin efflux carriers.

PINs engage in their journey from the PM via endocytosis, a process in which proteins integrated in the lipid bilayer or extracellular components are internalized by pinching off vesicles from the PM. Some decades ago, the existence of endocytosis in plants has been controversial reflecting the high turgor pressure inside the cells. Nonetheless, available data supported the existence of clathrin-mediated endocytosis (Holstein et al. 1994; Low and Chandra 1994; Holstein 2002). Subsequently, endocytosis was described for many proteins (Meckel et al. 2004;

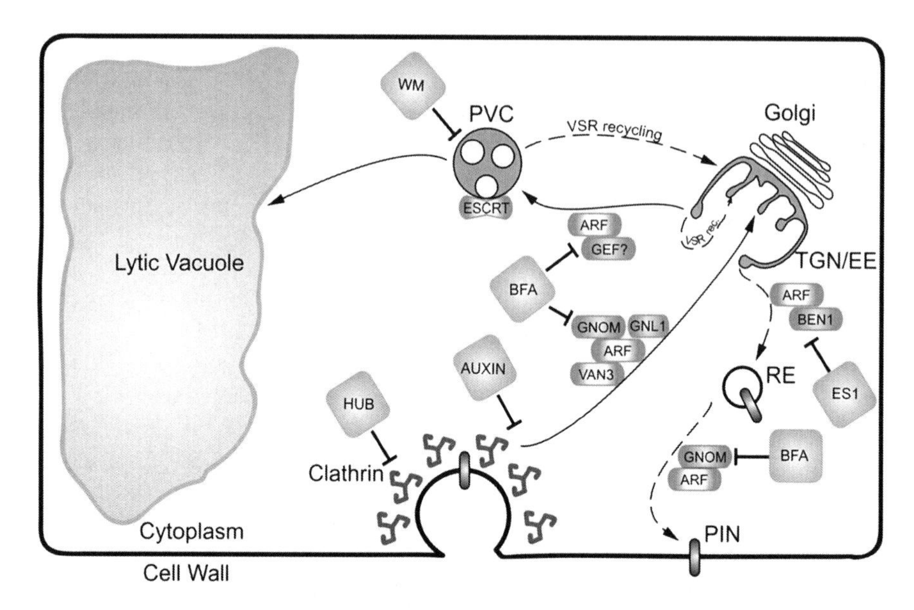

Fig. 1 Subcellular trafficking pathways of PIN proteins. The endocytic route is indicated in *full-line arrows*. *HUB* truncated C-terminal part of clathrin heavy chain interfering with clathrin-coated pit formation; *BFA* brefeldin A; *ES1* endosidin1; *TGN/EE trans*-Golgi network/early endosome; *RE recycling* endosome; *PVC* pre-vacuolar compartment; *WM* wortmannin; *VSR recycling* retromer-mediated retrieval of vacuolar sorting receptors (VSR)

Šamaj et al. 2005; Robinson et al. 2008; Reyes et al. 2011) including PIN-auxin carriers (Geldner et al. 2001; Dhonukshe et al. 2007). Indeed, conserved genes for the clathrin mediated endocytic (CME) machinery are encoded in plant genomes and a core element, clathrin has been localized to the endomembrane structures and at the PM (Dhonukshe et al. 2007). In support of the functionality of CME pathway it has been shown that genetic and pharmacological interference with clathrin function impairs PIN endocytosis (Fig. 1; Dhonukshe et al. 2007; Kitakura et al. 2011), strengthening the evidence that PINs are endocytosed via CME. In addition, the dynamin-related proteins (DRPs) necessary for scission of endocytic vesicles during CME have been shown to associate with PIN proteins at the cell plate. Genetic interference with DRP1 resulted in altered PIN distribution, especially regarding the repositioning of carriers after cell division. Consequently, *drp*1 mutants displayed phenotypes associated with auxin transport defects (Mravec et al. 2011).

Besides the endocytic machineries related to clathrin, the existence of other mechanisms controlling PIN endocytosis cannot be excluded. Interestingly, several reports demonstrated that membrane composition affects PIN endocytosis and re-establishment of PIN2 polarity after cytokinesis is impaired in the *cyclopropylsterol isomerase1-1* (*cpi1-1*) mutant (Men et al. 2008). Similarly, disruption of the molecular machinery required for very long fatty acid biosynthesis also results in aberration in PIN endocytic recycling and polarity leading to developmental defects (Roudier et al. 2010). The cytoskeleton which in general is guiding endocytosed vesicles containing multiple cargos toward their subcellular destinations has also a role in PIN trafficking. The depolymerization of actin filaments by cytochalasin D or latrunculin B (LatB) interferes with internalization and recycling of PINs (Geldner et al. 2001; Kleine-Vehn et al. 2006; Kleine-Vehn et al. 2008b). Also, actin hyperstabilization by 2,3,5-triiodobenzoic acid (TIBA) and 2-(1-pyrenoyl) benzoic acid (PBA) interferes with PIN cycling (Dhonukshe et al. 2008a, see also chapter by Baral and Dhonukshe in this volume). Both modes of disrupting actin dynamics lead to auxin transport defects, implicating that not only polar localization but also constitutive trafficking of PINs are necessary for their function in auxin transport (Geldner et al. 2001). In maize roots, auxin-like molecules could be detected in brefeldin A (BFA)-sensitive endosomes via antibodies against IAA. These endosomes localize close to PIN polar domains at the PM, suggesting that they might serve as vessels for secretion of the hormone contributing to its transport (Schlicht et al. 2006). On the other hand, auxin transport was enhanced by stabilization of PINs in the PM after interference with CME (Kitakura et al. 2011). This indicates that the polar localization of PINs in the PM is sufficient to mediate auxin transport even during inhibition of endosomal cycling. In summary, the existence of a neuro-transmitter-like mode of auxin transport and/or its contribution to polar auxin transport remains a matter of debate that has to be further experimentally supported.

3 PIN Early Endocytic Recycling

After endocytosis, PINs quickly reach the TGN/EE (Geldner et al. 2001; Dettmer et al. 2006) where they are presumably sorted for their recycling or degradation (Fig. 1). The accumulation of PINs in TGN/EE and RE can be visualized by inhibition of exocytosis via the fungal toxin BFA (Geldner et al. 2001). This toxin inhibits the ARF-GEF (ARF-Guanine nucleotide Exchange Factor for ADP-Ribosylation Factor GTPases—ARFs) called GNOM and subsequently leads to the formation of TGN aggregates which become surrounded by Golgi stacks often designated as "BFA-bodies" or "BFA-compartments" in Arabidopsis roots. Subsequent studies demonstrated that BFA acts mostly on the recycling step of PINs hence leading to the depletion of these transporters from the PM by the ongoing endocytosis (Geldner et al. 2003). In addition, GNOM-like1 (GNL1), a close homolog of GNOM, has also been implicated in the control of PIN2 trafficking as in *gnl1* mutants the internalization of this auxin carrier was impaired (Teh and Moore 2007). Recently, it has been reported that, besides their function on intracellular organelles, ARF-GEFs GNOM and GNL1 as well as ARF-GAP (ARF-GTPase-Activating Protein) VAN3 (Vascular Network Defective 3) control also the endocytic uptake at the PM (Fig. 1), and genetic interference with either of these players significantly decreases PIN1-GFP internalization as visualised by BFA treatment (Naramoto et al. 2010).

Besides the previously discussed GNOM and GNL1 assigned to the GBF class (Golgi-associated BFA-resistant guanine nucleotide exchange Factor), the *Arabidopsis thaliana* genome encodes also a second group of ARF-GEFs constituting the BIG (BFA-Inhibited Guanine nucleotide-exchange protein) family but their functional characterization is largely incomplete. A member of this group has been isolated from a forward genetic screen and designated BEN1 (BFA-visualized endocytic trafficking defective 1 also known as BIG5/MIN7). It has been shown to regulate the BFA-induced internalization of basally localized PIN1 in the stele and PIN2 in the cortex cells which accumulated significantly less in *ben1* genetic background after BFA application (Tanaka et al. 2009). In epidermal root cells of *ben1* mutant, the apically localized PIN2 showed comparable BFA sensitivity to wild type indicating that only basal cargos were affected by the mutation (the upper side of the cell facing the shoot pole is termed apical while the lower cell site closer to the root pole is termed basal, those designation for cellular polarity will be used through the entire text; see also designations in Figs. 2a, 2b, 2c). BEN1 shows TGN/EE localization, distinct from the GNOM-positive RE, and seems to control the EE to RE transit of constitutively cycling PM proteins (Fig. 1), including PINs recycling to the basal cell site (Tanaka et al. 2009). Conversely, the trafficking of apically localized cargos, including PIN2, was affected by endosidin1 (ES1) suggesting that this chemical compound might affect specifically the recycling step to the apical cell site (Robert et al. 2008). Other PM markers including PIN1 and PIN7 were not affected by ES1. These results hint at the existence of separate endocytic routes for apically and basally localized cargos. This is also in accordance with the experimental data showing differences in recycling pathways of PIN1 versus AUX1 where the PM pool of apically localized AUX1 is not affected by BFA treatment, while basally positioned PIN1 is (Kleine-Vehn et al. 2006), and by different BFA sensitivities of basal and apical targeting pathways (Kleine-Vehn et al. 2008a), suggesting that different sets of ARF GEFs are involved in these recycling routes.

4 PINs on the Late Endocytic Route

Another mechanism by which PM abundance of auxin carriers can be controlled is by their rate of degradation. PINs that have been trafficked to the TGN/EE can be re-directed back to PM via RE or become targeted toward the vacuole. This degradation route goes through the PVCs/multivesicualar bodies (MVBs) which represent late endosomes in plants (Fig. 1; Müller et al. 2007; Scheuring et al. 2011). Wortmannin (WM) treatment was shown to result in PVC swelling and consumption of TGN by PVC, subsequently, leading to accumulation of cargos in the dilated PVC/MVB compartment (Tse et al. 2004; Takáč et al. 2012). PIN2 vacuolar trafficking was also inhibited by WM and the protein seemed to accumulate at the PVC (Kleine-Vehn et al. 2008b). BFA, apart from affecting preferentially the endocytic recycling at lower concentrations (<25 μM), has also been

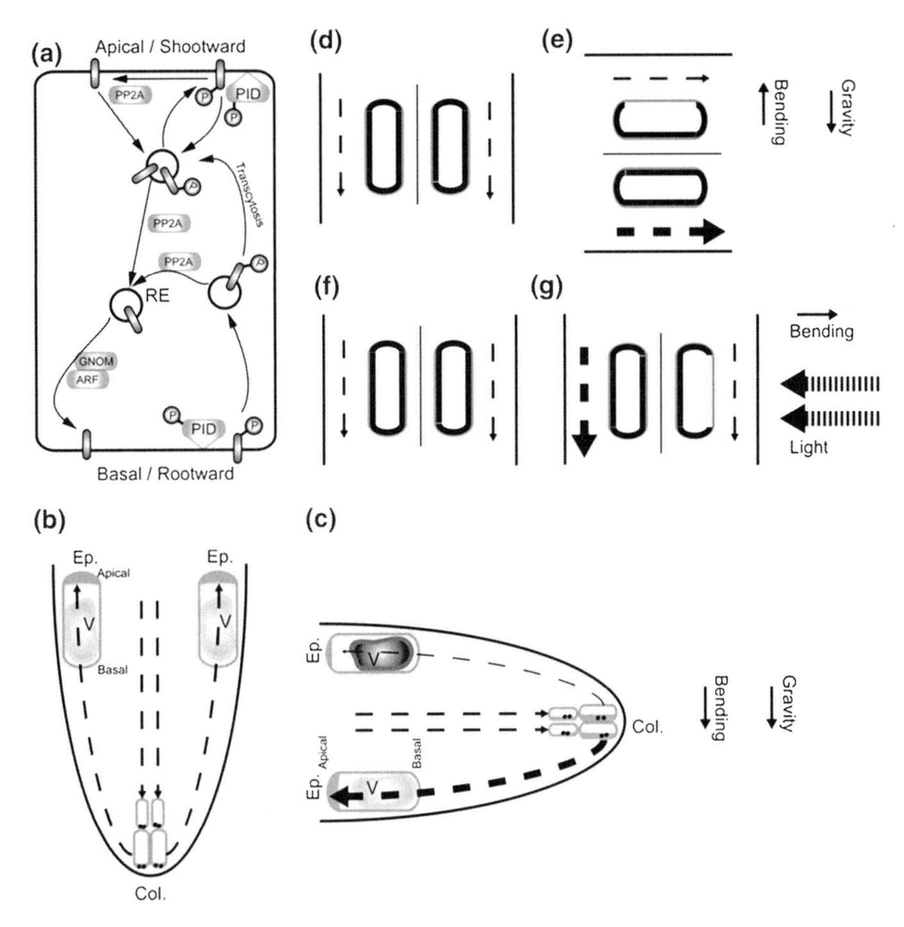

Fig. 2 Regulation of PIN trafficking by phosphorylation translates environmental signals into differential auxin distribution. **a** Recruitment of PIN auxin carrier into distinct intracellular trafficking pathways depending on the phosphorylation status of the protein. **b**, **c** Maintenance of auxin flow during gravitropic growth **b** and its redirectioning **c** facilitating the gravitropic response of the root. The vascular acropetal flux of auxin toward the root tip is differentially reoriented by PIN3 and 7 polarity shift in columella cells of the root tip upon statolith sedimentation during gravistimulation **c**. **b**, **c** Differential PM levels of PIN2 in epidermal cells are supported by the vacuolar degradation rate of the carrier resulting in auxin flux asymmetry triggering gravitropic bending of the root; *Ep* epidermal cells; *Col* columella root cap cells; *V* lytic vacuole. **d–f** Schematic representation of a hypocotyl fragment where PIN3 polarization in endodermis cells leads to changes in auxin distribution and differential growth in response to gravity **d**, **e** and light **f**, **g** stimulus. Auxin flux directions are depicted in the figures as *dashed arrows*, thickness of the lines corresponds to relative auxin concentrations

shown to inhibit vacuolar trafficking of PIN2 and other PM proteins at higher concentrations (>50 μM) (Geldner et al. 2007; Kleine-Vehn et al. 2008b; Robert et al. 2010; Takano et al. 2010).

PINs on the path to the vacuole can evade degradation via retromer-mediated retrograde trafficking (Kleine-Vehn et al. 2008b) that has been shown to influence vacuolar sorting (Seaman 2005; Shimada et al. 2006). In higher eukaryotes, this complex consists of vacuolar sorting protein VPS29, 26, and 35 trimer which interacts with sorting nexin proteins (SNX) (Robinson et al. 2012). Interestingly, the retromer constituents SNX1 and VPS29 have been proposed to control early endocytic recycling of PINs to the PM (Jaillais et al. 2006; Jaillais et al. 2007, see also chapter by Zelazny et al. in this volume). However, a following report revealed that SNX1 and VPS29 might not directly affect PIN polarity but rather control the rate of vacuolar trafficking of those auxin carriers by promoting their retrograde transport to the early endocytic pathways (Kleine-Vehn et al. 2008b). These data are in accordance with the conserved function of retromer in recycling of vacuolar sorting receptors (VSRs) and thus saving them from lytic degradation for further rounds of sorting between TGN and PVC/MVB (Arighi et al. 2004; Seaman 2005).

Within the PVC/MVB, internal vesicles are formed that will be released in the vacuolar lumen upon fusion with the vacuole (Scheuring et al. 2011). Formation of these intraluminal structures is possible, thanks to another molecular machinery termed the endosomal-sorting complexes required for transport (ESCRT; Winter and Hauser 2006; Wollert et al. 2009). It has been shown that disruption of the ESCRT-related CHMP1A and CHMP1B (Charged Multivesicular Body Protein/ Chromatin Modifying Protein1) components in Arabidopsis leads to accumulation of PINs and other PM proteins on the surface of the PVC/MVB and vacuole, thus hampering the initiation of the degradation process (Spitzer et al. 2009, see also chapter by Otegui et al. in this volume).

The rapid turnover of PIN proteins facilitated the isolation of mutants in Adaptor Protein (AP) complex 3 subunits AP-3 β (*pat2*) and AP-3 δ (*pat4*) defective in morphology and function of lytic and protein storage vacuoles (Feraru et al. 2010; Zwiewka et al. 2011). Biochemical and genetic investigations of *pat2* and *pat4* mutants have shown a physical and functional association of AP-3β and AP-3δ with the other putative μ and σ subunits of the AP-3 complex and several components of clathrin and dynamin assemblies (Zwiewka et al. 2011). Genetic lesion of the AP-3 machinery led to the accumulation of PINs and other cargos in abnormal lytic vacuoles (Feraru et al. 2010; Zwiewka et al. 2011). Regarding the fact the *pat3* and *pat4* mutants show an aberrant pattern of PVC/MVB markers, it is tempting to speculate that the fusion of this compartment with the vacuole is disrupted in these mutants. The examples of endocytic trafficking summarized above demonstrate the value of PIN proteins that undergo such elaborate subcellular dynamics, in understanding the complexity of endocytic pathways which still needs to be fully unravelled.

5 PIN-Specific Tags for Distinct Polar Destinations

First insights into ways how the intracellular machinery can distinguish between different cargos was provided by the identification of sequences within the central loop of the PIN1 that when disrupted caused a basal-to-apical localization shift of this protein (Wiśniewska et al. 2006). Subsequent evidence indicated that these amino acid sequences are related to the phosphorylation sites found in the PINs (Friml et al. 2004; Michniewicz et al. 2007; Huang et al. 2010, Zhang et al. 2010), even before two regulators related to phosphorylation were identified, a Ser/Thr protein kinase PINOID (PID) that phosphorylates the hydrophilic loop of PINs and its counteractor, a phosphatase 2A (PP2A) (Muday and DeLong 2001; Michniewicz et al. 2007). Enhanced phosphorylation of PIN1 by overexpression of PID (Benjamins et al. 2001; Friml et al. 2004) or genetic interference with PP2A (Michniewicz et al. 2007) resulted in apicalization of this protein, leading to severe embryo and root defects. In *pid* mutant a converse localization of PIN was observed with the protein targeted preferentially to the basal cell site. In this case, the switched basal PIN polarity resulted in defective organogenesis of the shoot apical meristem (Benjamins et al. 2001; Friml et al. 2004). Collectively, these results show that PIN phosphorylation status acts as a binary switch directing the auxin efflux carrier to the apical or basal cell site (Fig. 2a). Accordingly, a direct disruption of the PID phosphorylated sites within PIN1 hydrophilic loop resulted in basal targeting of this carrier and increased auxin accumulation in the root tip. Consequently, a converse scenario was observed when phosphorylation-mimicking mutations were introduced (Huang et al. 2010).

Apart from PID-specific phosphorylation sites, additional ones have also been identified that are sufficient to alter PIN polarity and the concomitant auxin flow. However, in vitro phosphorylation studies indicate that those amino acids are not direct targets of PID (Zhang et al. 2010). This opens up a possibility for other kinases to regulate the directional PIN-mediated auxin transport as indicated by some reports (Zourelidou et al. 2009). As discussed previously, GNOM mediates preferentially basal polar delivery of PIN1 and genetic or pharmacological interferences with this ARF-GEF cause basal-to-apical PIN polarity shift resembling the situation observed in *pp2a* mutants and PID overexpressors (Friml et al. 2004; Michniewicz et al. 2007). Therefore, it has been suggested that dephosphorylated PIN1 is recruited preferential to the basal recycling pathway controlled by GNOM, whereas phosphorylated PINs traffic by a GNOM-independent apical pathway (Fig. 2a; Kleine-Vehn et al. 2009). This example shows how phosphorylation can decide about distinct trafficking pathways resulting in different polar localization of PIN1, causing auxin flux redirecting and subsequent developmental output (Kleine-Vehn et al. 2009). However, it remains unclear what controls the activity of PID and PP2A.

Moreover, it has been shown that phosphatidylinositol-(3,4,5)trisphosphate (PtdIns(3,4,5)P_3)-dependent calcium signalling regulates PID activity resulting in an apical-basal PIN1 polarity shift (Benjamins et al. 2003; Zegzouti et al. 2006;

Zhang et al. 2011). It will be interesting to see whether these signalling molecules are able to spatio-temporally regulate phosphotransfer enzymes and coordinate the localization of auxin efflux transporters. The available data on PIN phosphorylation and polarity are sufficient to conceive a model where this post-translational modification (PTM) regulates PIN differential entry into apical/basal endocytic pathways.

6 Regulation of PIN Trafficking in Answer to Developmental Stimuli

PM localized PIN auxin carriers display an asymmetric localization in cells (Wiśniewska et al. 2006) and undergo complex subcellular dynamics encompassing secretion that might not initially establish the polarity (Dhonukshe et al. 2008), clathrin-mediated endocytosis (Dhonukshe et al. 2007; Kitakura et al. 2011), polar recycling (Geldner et al. 2001,), and transcytosis (Kleine-Vehn et al. 2008a) as well as controlled vacuolar trafficking (Fig. 1 and 2a; Abas et al. 2006; Kleine-Vehn et al. 2008b). Apart from transcriptional regulation of *PIN* genes (Blilou et al. 2005; Vieten et al. 2007) it is mainly the action of the intracellular trafficking machinery, at the above-mentioned steps, that controls the abundance of PIN auxin transporters at distinct polar domains of PM. How plants integrate multi-level cues into an optimized growth via regulation of auxin transport is slowly being unraveled.

One of the auxin transport system features is that it has the ability to integrate developmental and environmental cues. Auxin flow directions are dynamically changed during organ patterning and regeneration (Benková et al. 2003; Friml et al. 2003, Reinhardt et al. 2003, Sauer et al. 2006, Scarpella et al. 2006; Heisler et al. 2010). Similarly, photo- and gravitropic growth responses require PIN-mediated auxin flow regulations (Friml et al. 2002). Interestingly, interference with subcellular dynamics of auxin efflux carriers strongly affects these processes, suggesting that regulated PIN trafficking is a central theme in many plant growth responses (Kleine-Vehn et al. 2010; Ding et al. 2011, Rakusová et al. 2011).

One of the long known features of auxin is the ability to shape its own transport (Sachs 1981; Sachs 1991), but the underlying molecular mechanisms are still not fully understood. It has been shown that auxin-induced transcriptional changes affect PIN polarization and therefore provide means by which auxin regulates its transport directionality (Sauer et al. 2006). However, it was also reported that auxin can inhibit endocytosis, thereby promoting the retention of its transporters at the PM (Paciorek et al. 2005). This mechanism involves auxin binding protein 1 (ABP1). It is still unclear how ABP1 regulates endocytosis, but presumably in auxin-bound state it inhibits clathrin-mediated endocytosis of PINs via a downstream ROP-dependent mechanism (Robert et al. 2010, Xu et al. 2010; Nagawa et al. 2012; see also chapter by Tejos and Friml in this volume). Thus, auxin by regulating the endocytic machinery is able to control the abundance of PIN efflux carriers promoting its own transport. This provides a rationale for the self-organizing role of auxin in guiding vasculature regeneration after wounding (Sauer et al. 2006; Wabnik et al. 2011a) and patterning of leaf venation (Scarpella et al. 2006). However, it is still not fully known how action of the trafficking machinery controlled by auxin leads to PIN polarity switches occurring in those processes.

Auxin has been also shown to play a pivotal role in gravitropic responses of roots and shoots (Friml et al. 2002). It has been reported that gravity perception is facilitated by specialized amyloplasts called statoliths that are present in the root cap and shoot endodermis, which due to their high density are able to sediment according to the gravity vector (Morita 2010), thus informing the particular organ about its orientation. Despite that it has not been fully elucidated how the movement of statoliths is detected by the perceptive cells; it was shown that following gravistimulation, PIN3 and PIN7 in the columella root cap cells polarize accordingly with the gravity vector, reorienting the auxin flux toward the bottom side of the root and eliciting gravitropic bending (Kleine-Vehn et al. 2010). In the root, PIN3 acts redundantly with PIN7 to redirect auxin flux (Fig. 2b and 2c). In the hypocotyl, gravity-induced growth responses also depend on PIN-mediated auxin transport to the bottom of the organ, but here flow reorientation is facilitated more exclusively by relocation of PIN3 (Rakusová et al. 2011; Fig. 2d, 2e). During hypocotyl and root gravitropic responses the relocation of PIN proteins is mediated by the GNOM-dependent endocytic recycling. Additionally, PIN3 relocation seems to be regulated also by the phosphorylation activity of PID kinase and its homologs WAG1 and WAG2 in the hypocotyl (Rakusová et al. 2011). As for gravitropism, also in phototropic response the light-induced PIN3 relocation and resulting auxin accumulation on the shaded side of the hypocotyl triggers its asymmetric growth and also involves PID activity and the trafficking pathway controlled by ARF-GEF GNOM (Ding et al. 2011; Fig. 2f and 2g). Similarly, the importance of PIN3-mediated auxin relocation has been shown in shade avoidance (Keuskamp et al. 2010).

In contrast to the positive phototropism of Arabidopsis hypocotyl, the root bends away from light. This tropic response depends on phot1 and phot2 blue light photoreceptors. The asymmetric distribution of the auxin, subsequently triggering root bending away from light, is facilitated, by differential PM levels of PIN2 protein controlled presumably by the rate of its degradation and endocytic recycling (Wan et al. 2012).

PIN2-mediated auxin transport has also been shown to play an important role in root gravitropism. It has been reported that gravistimulation induces a retromer-dependent vacuolar degradation of PIN2 (Kleine-Vehn et al. 2008b). This mechanism assists in the differential regulation of PIN2 levels during gravitropic bending by enhanced vacuole trafficking of the transporter at the upper site of the root, contributing indirectly to its depletion from the PM and facilitating differential auxin flux necessary for gravitropic growth (Fig 2b and 2c; Müller et al. 1998; Abas et al. 2006; Kleine-Vehn et al. 2008b). It is still unclear whether this asymmetric PIN2 degradation is caused by the original auxin asymmetry or by an auxin-independent mechanism.

The plant hormone cytokinin is an example of another endogenous signal that regulates PIN trafficking, namely PIN degradation. It has been shown to affect specifically the endocytic recycling of PIN1 by redirecting it toward the vacuole (Marhavý et al. 2011). This effect on PIN1 trafficking requires functional cytokinin signalling but is mediated by virtue of the endocytic machinery. Accordingly, both *ben1* and *ben2* endocytic mutants (Tanaka et al. 2009) were resistant to cytokinin-induced PIN1 vacuolar trafficking and concomitantly were not responsive to effect of the hormone in lateral root primordia development and root meristem differentiation (Marhavý et al. 2011).

Recently, a group of peptides designated GOLVEN has been shown to affect PIN2 trafficking, temporally increasing the pool of PIN2 in the PM presumably by inhibiting the endocytic step involved in constitutive cycling of these proteins (Whitford et al. 2012).

It is possible, but has not been demonstrated yet, that also other endogenous and environmental signals regulate PIN endocytic trafficking and thus auxin transport-mediated development.

7 Conclusions and Future Prospects

As briefly discussed in the beginning of this chapter, plants display an amazing developmental plasticity encompassing cell fate respecification and de novo organ formation in post-embryonic life. At large, these processes are guided by the plant hormone auxin. The developmentally instructive, differential accumulation patterns of this signalling molecule are mainly established via directional transport through tissues, mediated by PIN transporters. It seems that the elaborate subcellular dynamics of these auxin efflux carriers is linked to the adaptive potential of plants to the dynamically changing environment. Since in plants, unlike in animals, many responses to environmental changes are growth mediated and involve a considerable investment of energy it is not surprising that the abundance and localization of transporters facilitating auxin-controlled growth is also under a tight and elaborate control by the endocytic machinery. The examples summarized above illustrate that disruption of early endocytic events affects the PM localization of PINs and concomitantly auxin transport, leading to detrimental consequences in development as exemplified by embryo development, lateral root formation and vasculature patterning defects (Dhonukshe et al. 2008; Naramoto et al. 2010; Kitakura et al. 2011). The late endocytic machinery is also involved in governing PIN-mediated auxin efflux. The significance of controlled PIN degradation is visible at circumstances when rapid depletion of these proteins from the PM is necessary to maintain asymmetric growth during gravitropism (Abas et al. 2006; Kleine-Vehn et al. 2008b; Marhavý et al. 2011). These early and late endocytic events are targets of regulation by different endogenous signals including hormones and signalling peptides, as well as environmental cues. By regulating auxin transport via control over PIN subcellular localization these signals can co-ordinately control asymmetric auxin distribution and multiple facets of plant growth and development. It seems that all this elaborate subcellular dynamic of PINs strives to ensure the precise control of auxin-orchestrated growth facilitating the totipotency of plants and also minimizing costly developmental errors.

Acknowledgments This work was supported by the Odysseus Programme of the Research Foundation-Flanders. SV is a postdoctoral fellow of the Research Foundation-Flanders (FWO09/PDO/196); TN is a predoctoral fellow of Jiří Friml's research group which is supported by the Odysseus Program of the Research Foundation-Flanders (Grant no. G091608).

References

Abas L, Benjamins R, Malenica N, Paciorek T, Wiśniewska J, Moulinier–Anzola JC, Sieberer T, Friml j, Luschnig c (2006) Intracellular trafficking and proteolysis of the Arabidopsis auxin-efflux facilitator PIN2 are involved in root gravitropism. Nat Cell Biol 8:249–256. doi:10.1038/ncb1369

Arighi CN, Hartnell LM, Aguilar RC, Haft CR, Bonifacino JS (2004) Role of the mammalian retromer in sorting of the cation-independent mannose 6-phosphate receptor. J Cell Biol 165:123–133. doi:10.1083/jcb200312055

Barbez E, Kubeš M, Rolčík J, Béziat C, Pěnčík A, Wang B, Rosquete MR, Zhu J, Dobrev PI, Lee Y, Zažímalovà E, Petrášek J, Geisler M, Friml J, Kleine-Vehn J (2012) A novel putative auxin carrier family regulates intracellular auxin homeostasis in plants. Nature. doi:10.1038/nature11001

Benjamins R, Ampudia CSG, Hooykaas PJJ, Offringa R (2003) PINOID-mediated signaling involves calcium-binding proteins. Plant Physiol 132:1623–1630. doi:10.1104/pp103.019943

Benjamins R, Quint A, Weijers D, Hooykaas P, Offringa R (2001) The PINOID protein kinase regulates organ development in Arabidopsis by enhancing polar auxin transport. Development 128:4057–4067

Benjamins R, Scheres B (2008) Auxin: the looping star in plant development. Annu Rev Plant Biol 59:443–465. doi:10.1146/annurev.arplant.58.032806.103805

Benková E, Michniewicz M, Sauer M, Teichmann T, Seifertová D, Jürgens G, Friml J (2003) Local, efflux-dependent auxin gradients as a common module for plant organ formation. Cell 115:591–602. doi:10.1016/S0092-8674(03)00924-3

Bennett MJ, Marchant A, Green HG, May ST, Ward SP, Millner PA, Walker AR, Schulz B, Feldmann KA (1996) Arabidopsis AUX1 gene: a permease-like regulator of root gravitropism. Science 273:948–950. doi:10.1126/science.273.5277.948

Berleth T, Scarpella E, Prusinkiewicz P (2007) toward the systems biology of auxin-transport-mediated patterning. Trends Plant Sci 12:151–159. doi:10.1016/j.tplants.2007.03.005

Blilou I, Xu J, Wildwater M, Willemsen V, Paponov I, Friml J, Heidstra R, Aida M, Palme K, Scheres B (2005) The PIN auxin efflux facilitator network controls growth and patterning in Arabidopsis roots. Nature 433:39–44. doi:10.1038/nature03184

Christie JM, Yang H, Richter GL, Sullivan S, Thomson CE, Lin J, Titapiwatanakun B, Ennis M, Kaiserli E, Lee OR, Adamec J, Peer WA, Murphy AS (2011) phot1 inhibition of ABCB19 primes lateral auxin fluxes in the shoot apex required for phototropism. PLoS Biol. doi:10.1371/journal.pbio.1001076

Dettmer J, Hong-Hermesdorf A, Stierhof Y-D, Schumacher K (2006) Vacuolar H^{+}-ATPase activity is required for endocytic and secretory trafficking in Arabidopsis. Plant Cell 18: 715–730. doi:10.1105/tpc.105.037978

Dhonukshe P, Aniento F, Hwang I, Robinson DG, Mravec J, Stierhof Y-D, Friml J (2007) Clathrin-mediated constitutive endocytosis of PIN auxin efflux carriers in Arabidopsis. Curr Biol 17:520–527. doi:10.1016/j.cub.2007.01.052

Dhonukshe P, Tanaka H, Goh T, Ebine K, Mähönen AP, Prasad K, Blilou I, Geldner N, Xu J, Uemura T, Chory J, Ueda T, Nakano A, Scheres B, Friml J (2008) Generation of cell polarity in plants links endocytosis, auxin distribution and cell fate decisions. Nature 456:962–966. doi:10.1038/nature07409

Ding Z, Galván-Ampudia CS, Demarsy E, Łangowski Ł, Kleine-Vehn J, Fan Y, Morita MT, Tasaka M, Fankhauser C, Offringa R, Friml J (2011) Light-mediated polarization of the PIN3 auxin transporter for the phototropic response in Arabidopsis. Nat Cell Biol 13:447–452. doi:10.1038/ncb2208

Dubrovsky JG, Sauer M, Napsucialy-Mendivil S, Ivanchenko MG, Friml J, Shishkova S, Celenza J, Benková E (2008) Auxin acts as a local morphogenetic trigger to specify lateral root founder cells. Proc Natl Acad Sci U S A 105:8790–8794. doi:10.1073/pnas.0712307105

Feraru E, Paciorek T, Feraru MI, Zwiewka M, De Groodt R, De Rycke R, Kleine-Vehn J, Friml J (2010) The AP-3 B adaptin mediates the biogenesis and function of lytic vacuoles in Arabidopsis. Plant Cell 22:2812–2824. doi:10.1105/tpc.110.075424

Friml J, Vieten A, Sauer M, Weijers D, Schwarz H, Hamann T, Offringa R, Jürgens G (2003) Efflux-dependent auxin gradients establish the apical-basal axis of Arabidopsis. Nature 426:147–153. doi:10.1038/nature02085

Friml J, Wiśniewska J, Benková E, Mendgen K, Palme K (2002) Lateral relocation of auxin efflux regulator PIN3 mediates tropism in Arabidopsis. Nature 415:806–809. doi:10.1038/415806a

Friml J, Yang X, Michniewicz M, Weijers D, Quint A, Tietz O, Benjamins R, Ouwerkerk PBF, Ljung K, Sandberg G, Hooykaas PJJ, Palme K, Offringa R (2004) A PINOID-dependent binary switch in apical-basal PIN polar targeting directs auxin efflux. Science 306:862–865. doi:10.1126/science.1100618

Gälweiler L, Guan C, Müller A, Wisman E, Mendgen K, Yephremov A, Palme K (1998) Regulation of polar auxin transport by AtPIN1 in Arabidopsis vascular tissue. Science 282:2226–2230. doi:10.1126/science.282.5397.2226

Geisler M, Blakeslee JJ, Bouchard R, Lee OR, Vincenzetti V, Bandyopadhyay A, Titapiwatanakun B, Peer WA, Bailly A, Richards EL, Ejendal KFK, Smith AP, Baroux C, Grossniklaus U, Müller A, Hrycyna CA, Dudler R, Murphy AS, Martinoia E (2005) Cellular efflux of auxin catalyzed by the Arabidopsis MDR/PGP transporter AtPGP1. Plant J 44:179–194. doi:10.1111/j.1365-313X.2005.02519.x

Geldner N, Anders N, Wolters H, Keicher J, Kornberger W, Muller P, Delbarre A, Ueda T, Nakano A, Jürgens G (2003) The Arabidopsis GNOM ARF-GEF mediates endosomal recycling, auxin transport, and auxin-dependent plant growth. Cell 112:219–230. doi:10.1016/S0092-8674(03)00003-5

Geldner N, Friml J, Stierhof Y-D, Jürgens G, Palme K (2001) Auxin transport inhibitors block PIN1 cycling and vesicle trafficking. Nature 413:425–428. doi:10.1038/35096571

Geldner N, Hyman DL, Wang X, Schumacher K, Chory J (2007) Endosomal signaling of plant steroid receptor kinase BRI1. Genes Dev 21:1598–1602. doi:10.1101/gad.1561307

Goldsmith MHM (1977) The polar transport of auxin. Annu Rev Plant Physiol 28:439–478. doi:10.1146/annurev.pp.28.060177.002255

Heisler MG, Hamant O, Krupinski P, Uyttewaal M, Ohno C, Jönsson H, Traas J, Meyerowitz EM (2010) Alignment between PIN1 polarity and microtubule orientation in the shoot apical meristem reveals a tight coupling between morphogenesis and auxin transport. PLoS Biol. doi:10.1371/journal.pbio.1000516

Holstein SE, Drucker M, Robinson DG (1994) Identification of a beta-type adaptin in plant clathrin-coated vesicles. J Cell Sci 107:945–953

Holstein SEH (2002) Clathrin and plant endocytosis. Traffic 3:614–620. doi:10.1034/j.1600-0854.2002.30903.x

Huang F, Kemel Zago M, Abas L, van Marion A, Galván-Ampudia CS, Offringa R (2010) Phosphorylation of conserved PIN motifs directs Arabidopsis PIN1 polarity and auxin transport. Plant Cell 22:1129–1142. doi:10.1105/tpc.109.072678

Jaillais Y, Fobis-Loisy I, Miège C, Rollin C, Gaude T (2006) AtSNX1 defines an endosome for auxin-carrier trafficking in Arabidopsis. Nature 443:106–109. doi:10.1038/nature05046

Jaillais Y, Santambrogio M, Rozier F, Fobis-Loisy I, Miège C, Gaude T (2007) The retromer protein VPS29 links cell polarity and organ initiation in plants. Cell 130:1057–1070. doi:10.1016/j.cell.2007.08.040

Keuskamp DH, Pollmann S, Voesenek LACJ, Peeters AJM, Pierik R (2010) Auxin transport through PIN-FORMED 3 (PIN3) controls shade avoidance and fitness during competition. PNAS 107(52):22740–22744. doi:10.1073/pnas.1013457108

Kitakura S, Vanneste S, Robert S, Löfke C, Teichmann T, Tanaka H, Friml J (2011) Clathrin mediates endocytosis and polar distribution of PIN auxin transporters in Arabidopsis. Plant Cell 23:1920–1931. doi:10.1105/tpc.111.083030

Kleine-Vehn J, Dhonukshe P, Sauer M, Brewer PB, Wiśniewska J, Paciorek T, Benková E, Friml J (2008a) ARF GEF-dependent transcytosis and polar delivery of PIN auxin carriers in Arabidopsis. Curr Biol 18:526–531. doi:10.1016/j.cub.2008.03.021

Kleine-Vehn J, Dhonukshe P, Swarup R, Bennett M, Friml J (2006) Subcellular trafficking of the Arabidopsis auxin influx carrier AUX1 uses a novel pathway distinct from PIN1. Plant Cell 18:3171–3181. doi:10.1105/tpc.106.042770

Kleine-Vehn J, Ding Z, Jones AR, Tasaka M, Morita MT, Friml J (2010) Gravity-induced PIN transcytosis for polarization of auxin fluxes in gravity-sensing root cells. PNAS 107:22344–22349. doi:10.1073/pnas.1013145107

Kleine-Vehn J, Huang F, Naramoto S, Zhang J, Michniewicz M, Offringa R, Friml J (2009) PIN auxin efflux carrier polarity is regulated by PINOID kinase-mediated recruitment into GNOM-independent trafficking in Arabidopsis. Plant Cell 21:3839–3849. doi:10.1105/tpc.109.071639

Kleine-Vehn J, Leitner J, Zwiewka M, Sauer M, Abas L, Luschnig C, Friml J (2008b) Differential degradation of PIN2 auxin efflux carrier by retromer-dependent vacuolar targeting. PNAS 105:17812–17817. doi:10.1073/pnas.0808073105

Kleine-Vehn J, Wabnik K, Martinière A, Łangowski Ł, Willig K, Naramoto S, Leitner J, Tanaka H, Jakobs S, Robert S, Luschnig C, Govaerts W, Hell W, Runions S, Friml J (2011) Recycling, clustering, and endocytosis jointly maintain PIN auxin carrier polarity at the plasma membrane. Mol Syst Biol 7:540. doi:10.1038/msb.2011.72

Kramer EM (2004) PIN and AUX/LAX proteins: their role in auxin accumulation. Trends Plant Sci 9:578–582. doi:10.1016/j.tplants.2004.10.010

Kubeš M, Yang H, Richter GL, Cheng Y, Młodzińska E, Wang X, Blakeslee JJ, Carraro N, Petrášek J, Zažímalová E, Hoyerová K, Peer WA, Murphy AS (2011) The Arabidopsis concentration-dependent influx/efflux transporter ABCB4 regulates cellular auxin levels in the root epidermis. Plant J 69:640–654. doi:10.1111/j.1365-313X.2011.04818.x

Lam SK, Siu CL, Hillmer S, Jang S, An G, Robinson DG, Jiang L (2007) Rice SCAMP1 defines clathrin-coated, trans-golgi–located tubular-vesicular structures as an early endosome in tobacco BY-2 cells. Plant Cell 19:296–319. doi:10.1105/tpc.106.045708

Low PS, Chandra S (1994) Endocytosis in plants. Annu Rev Plant Physiol Plant Mol Biol 45:609–631. doi:10.1146/annurev.pp.45.060194.003141

Marhavý P, Bielach A, Abas L, Abuzeineh A, Duclercq J, Tanaka H, Pařezová M, Petrášek J, Friml J, Kleine-Vehn J, Benková E (2011) Cytokinin modulates endocytic trafficking of PIN1 auxin efflux carrier to control plant organogenesis. Dev Cell 21:796–804. doi:10.1016/j.devcel.2011.08.014

Meckel T, Hurst AC, Thiel G, Homann U (2004) Endocytosis against high turgor: intact guard cells of *Vicia faba* constitutively endocytose fluorescently labelled plasma membrane and GFP-tagged K^+-channel KAT1. Plant J 39:182–193. doi:10.1111/j.1365-313X.2004.02119.x

Men S, Boutté Y, Ikeda Y, Li X, Palme K, Stierhof Y-D, Hartmann M-A, Moritz T, Grebe M (2008) Sterol-dependent endocytosis mediates post-cytokinetic acquisition of PIN2 auxin efflux carrier polarity. Nat Cell Biol 10:237–244. doi:10.1038/ncb1686

Michniewicz M, Zago MK, Abas L, Weijers D, Schweighofer A, Meskiene I, Heisler MG, Ohno C, Zhang J, Huang F, Schwab R, Weigel D, Meyerowitz EM, Luschnig C, Offringa R, Friml J (2007) Antagonistic regulation of PIN phosphorylation by PP2A and PINOID directs auxin flux. Cell 130:1044–1056. doi:10.1016/j.cell.2007.07.033

Morita MT (2010) Directional gravity sensing in gravitropism. Annu Rev Plant Biol 61:705–720. doi:10.1146/annurev.arplant.043008.092042

Mravec J, Kubeš M, Bielach A, Gaykova V, Petrášek J, Skůpa P, Chand S, Benková E, Zažímalová E, Friml J (2008) Interaction of PIN and PGP transport mechanisms in auxin distribution-dependent development. Development 135:3345–3354. doi:10.1242/dev.021071

Mravec J, Petrášek J, Li N, Boeren S, Karlova R, Kitakura S, Pařezová M, Naramoto S, Nodzyński T, Dhonukshe P, Bednarek SY, Zažímalová E, de Vries S, Friml J (2011) Cell plate restricted association of DRP1A and PIN proteins is required for cell polarity establishment in Arabidopsis. Curr Biol 21:1055–1060. doi:10.1016/j.cub.2011.05.018

Mravec J, Skůpa P, Bailly A, Hoyerová K, Křeček P, Bielach A, Petrášek J, Zhang J, Gaykova V, Stierhof Y-D, Dobrev PI, Schwarzerová K, Rolčík J, Seifertová D, Luschnig C, Benková E, Zažímalovà E, Geisler M, Friml J (2009) Subcellular homeostasis of phytohormone auxin is mediated by the ER-localized PIN5 transporter. Nature 459:1136–1140. doi:10.1038/nature08066

Muday GK, DeLong A (2001) Polar auxin transport: controlling where and how much. Trends Plant Sci 6:535–542. doi:10.1016/S1360-1385(01)02101-X

Müller A, Guan C, Gälweiler L, Tänzler P, Huijser P, Marchant A, Parry G, Bennett M, Wisman E, Palme K (1998) AtPIN2 defines a locus of Arabidopsis for root gravitropism control. EMBO J 17:6903–6911. doi:10.1093/emboj/17.23.6903

Müller J, Mettbach U, Menzel D, Šamaj J (2007) Molecular dissection of endosomal compartments in plants. Plant Physiol 145:293–304. doi:10.1104/pp.107.102863

Nagawa S, Xu T, Lin D, Dhonukshe P, Zhang X, Friml J, Scheres B, Fu Y, Yang Z (2012) ROP GTPase-dependent actin microfilaments promote PIN1 polarization by localized inhibition of clathrin-dependent endocytosis. PLoS Biol. doi:10.1371/journal.pbio.1001299

Naramoto S, Kleine-Vehn J, Robert S, Fujimoto M, Dainobu T, Paciorek T, Ueda T, Nakano A, Van Montagu MCE, Fukuda H, Friml J (2010) ADP-ribosylation factor machinery mediates endocytosis in plant cells. Proc Natl Acad Sci U S A 107:21890–21895. doi:10.1073/pnas.1016260107

Noh B, Murphy AS, Spalding EP (2001) Multidrug resistance–like genes of Arabidopsis required for auxin transport and auxin-mediated development. Plant Cell 13:2441–2454. doi:10.1105/tpc.010350

Okada K, Ueda J, Komaki MK, Bell CJ, Shimura Y (1991) Requirement of the auxin polar transport system in early stages of Arabidopsis floral bud formation. Plant Cell 3:677–684. doi:10.1105/tpc.3.7.677

Paciorek T, Zazímalová E, Ruthardt N, Petrásek J, Stierhof Y-D, Kleine-Vehn J, Morris DA, Emans N, Jürgens G, Geldner N, Friml J (2005) Auxin inhibits endocytosis and promotes its own efflux from cells. Nature 435:1251–1256. doi:10.1038/nature03633

Petrášek J, Friml J (2009) Auxin transport routes in plant development. Development 136:2675–2688. doi:10.1242/dev.030353

Petrášek J, Mravec J, Bouchard R, Blakeslee JJ, Abas M, Seifertová D, Wiśniewska J, Tadele Z, Kubeš M, Čovanová M, Dhonukshe P, Skůpa P, Benková E, Perry L, Křeček P, Lee OR, Fink GR, Geisler M, Murphy AS, Luschnig C, Zažímalová E, Friml J (2006) PIN proteins perform a rate-limiting function in cellular auxin efflux. Science 312:914–918. doi:10.1126/science.1123542

Rakusová H, Gallego-Bartolomé J, Vanstraelen M, Robert HS, Alabadí D, Blázquez MA, Benková E, Friml J (2011) Polarization of PIN3-dependent auxin transport for hypocotyl gravitropic response in *Arabidopsis thaliana*. Plant J 67:817–826. doi:10.1111/j.1365-313X.2011.04636.x

Reinhardt D, Pesce E-R, Stieger P, Mandel T, Baltensperger K, Bennett M, Traas J, Friml J, Kuhlemeier C (2003) Regulation of phyllotaxis by polar auxin transport. Nature 426:255–260. doi:10.1038/nature02081

Reyes FC, Buono R, Otegui MS (2011) Plant endosomal trafficking pathways. Curr Opin Plant Biol 14:666–673. doi:10.1016/j.pbi.2011.07.009

Robert S, Chary SN, Drakakaki G, Li S, Yang Z, Raikhel NV, Hicks GR (2008) Endosidin1 defines a compartment involved in endocytosis of the brassinosteroid receptor BRI1 and the auxin transporters PIN2 and AUX1. Proc Natl Acad Sci U S A 105:8464–8469. doi:10.1073/pnas.0711650105

Robert S, Kleine-Vehn J, Barbez E, Sauer M, Paciorek T, Baster P, Vanneste S, Zhang J, Simon S, Čovanová M, Hayashi K, Dhonukshe P, Yang Z, Bednarek SY, Jones AM, Luschnig C, Aniento F, Zažímalová E, Friml J (2010) ABP1 mediates auxin inhibition of clathrin-dependent endocytosis in Arabidopsis. Cell 143:111–121. doi:10.1016/j.cell.2010.09.027

Robinson DG, Jiang L, Schumacher K (2008) The endosomal system of plants: charting new and familiar territories. Plant Physiol 147:1482–1492. doi:10.1104/pp.108.120105

Robinson DG, Pimpl P, Scheuring D, Stierhof Y-D, Sturm S, Viotti C (2012) Trying to make sense of retromer. Trends Plant Sci. doi:10.1016/j.tplants.2012.03.005

Roudier F, Gissot L, Beaudoin F, Haslam R, Michaelson L, Marion J, Molino D, Lima A, Bach L, Morin H, Tellier F, Palauqui J-C, Bellec Y, Renne C, Miquel M, DaCosta M, Vignard J, Rochat C, Markham JE, Moreau P, Napier J, Faure J-D (2010) Very-long-chain fatty acids are involved in polar auxin transport and developmental patterning in Arabidopsis. Plant Cell 22:364–375. doi:10.1105/tpc.109.071209

Rubery PH, Sheldrake AR (1974) Carrier-mediated auxin transport. Planta 118:101–121. doi:10.1007/BF00388387

Ruiz Rosquete M, Barbez E, Kleine-Vehn J (2011) Cellular auxin homeostasis: gatekeeping is housekeeping. Mol Plant. doi:10.1093/mp/ssr109

Sachs T (1981) The control of the patterned differentiation of vascular tissues. Adv Botanical Res 9:151–262

Sachs T (1991) Cell polarity and tissue patterning in plants. Development 113:83–93

Šamaj J, Read ND, Volkmann D, Menzel D, Baluška F (2005) The endocytic network in plants. Trends Cell Biol 15:425–433. doi:10.1016/j.tcb.2005.06.006

Sauer M, Balla J, Luschnig C, Wiśniewska J, Reinöhl V, Friml J, Benková E (2006) Canalization of auxin flow by Aux/IAA-ARF-dependent feedback regulation of PIN polarity. Genes Dev 20:2902–2911. doi:10.1101/gad.390806

Scarpella E, Marcos D, Friml J, Berleth T (2006) Control of leaf vascular patterning by polar auxin transport. Genes Dev 20:1015–1027. doi:10.1101/gad.1402406

Scheuring D, Viotti C, Krüger F, Künzl F, Sturm S, Bubeck J, Hillmer S, Frigerio L, Robinson DG, Pimpl P, Schumacher K (2011) Multivesicular bodies mature from the *trans*-Golgi network/early endosome in Arabidopsis. Plant Cell 23:3463–3481. doi:10.1105/tpc.111.086918

Schlereth A, Möller B, Liu W, Kientz M, Flipse J, Rademacher EH, Schmid M, Jürgens G, Weijers D (2010) MONOPTEROS controls embryonic root initiation by regulating a mobile transcription factor. Nature 464:913–916. doi:10.1038/nature08836

Schlicht M, Strnad M, Scanlon MJ, Mancuso S, Hochholdinger F, Palme K, Volkmann D, Menzel D, Baluška F (2006) Auxin immunolocalization implicates vesicular neurotransmitter-like mode of polar auxin transport in root apices. Plant Signal Behav 1:122–133

Seaman MNJ (2005) Recycle your receptors with retromer. Trends Cell Biol 15:68–75. doi:10.1016/j.tcb.2004.12.004

Shimada T, Koumoto Y, Li L, Yamazaki M, Kondo M, Nishimura M, Hara-Nishimura I (2006) AtVPS29, a putative component of a retromer complex, is required for the efficient sorting of seed storage proteins. Plant Cell Physiol 47:1187–1194. doi:10.1093/pcp/pcj103

Spitzer C, Reyes FC, Buono R, Sliwinski MK, Haas TJ, Otegui MS (2009) The ESCRT-related CHMP1A and B proteins mediate multivesicular body sorting of auxin carriers in Arabidopsis and are required for plant development. Plant Cell 21:749–766. doi:10.1105/tpc.108.064865

Swarup K, Benková E, Swarup R, Casimiro I, Péret B, Yang Y, Parry G, Nielsen E, De Smet I, Vanneste S, Levesque MP, Carrier D, James N, Calvo V, Ljung K, Kramer E, Roberts R, Graham N, Marillonnet S, Patel K, Jones JDG, Taylor CG, Schachtman DP, May S, Sandberg G, Benfey P, Friml J, Kerr I, Beeckman T, Laplaze L, Bennett MJ (2008) The auxin influx carrier LAX3 promotes lateral root emergence. Nat Cell Biol 10:946–954. doi:10.1038/ncb1754

Takáč T, Pechan T, Šamajová O, Ovečka M, Richter H, Eck C, Niehaus K, Šamaj J (2012) Wortmannin treatment induces changes in Arabidopsis root proteome and post-Golgi compartments. J Proteome Res. doi:10.1021/pr201111n

Takano J, Tanaka M, Toyoda A, Miwa K, Kasai K, Fuji K, Onouchi H, Naito S, Fujiwara T (2010) Polar localization and degradation of Arabidopsis boron transporters through distinct trafficking pathways. PNAS. doi:10.1073/pnas.0910744107

Tanaka H, Kitakura S, De Rycke R, De Groodt R, Friml J (2009) Fluorescence imaging-based screen identifies ARF GEF component of early endosomal trafficking. Curr Biol 19:391–397. doi:10.1016/j.cub.2009.01.057

Teh O-K, Moore I (2007) An ARF-GEF acting at the golgi and in selective endocytosis in polarized plant cells. Nature 448:493–496. doi:10.1038/nature06023

Tse YC, Mo B, Hillmer S, Zhao M, Lo SW, Robinson DG, Jiang L (2004) Identification of multivesicular bodies as prevacuolar compartments in *Nicotiana tabacum* BY-2 cells. Plant Cell 16:672–693. doi:10.1105/tpc.019703

Vanneste S, Friml J (2009) Auxin: a trigger for change in plant development. Cell 136:1005–1016. doi:10.1016/j.cell.2009.03.001

Verrier PJ, Bird D, Burla B, Dassa E, Forestier C, Geisler M, Klein M, Kolukisaoglu Ü, Lee Y, Martinoia E, Murphy A, Rea PA, Samuels L, Schulz B, Spalding EP, Yazaki K, Theodoulou FL (2008) Plant ABC proteins—a unified nomenclature and updated inventory. Trends Plant Sci 13:151–159. doi:10.1016/j.tplants.2008.02.001

Vieten A, Sauer M, Brewer PB, Friml J (2007) Molecular and cellular aspects of auxin-transport-mediated development. Trends Plant Sci 12:160–168. doi:10.1016/j.tplants.2007.03.006

Wabnik K, Govaerts W, Friml J, Kleine-Vehn J (2011a) Feedback models for polarized auxin transport: an emerging trend. Mol BioSyst 7:2352. doi:10.1039/c1mb05109a

Wabnik K, Kleine-Vehn J, Govaerts W, Friml J (2011b) Prototype cell-to-cell auxin transport mechanism by intracellular auxin compartmentalization. Trends Plant Sci 16:468–475. doi:10.1016/j.tplants.2011.05.002

Wan Y, Jasik J, Wang L, Hao H, Volkmann D, Menzel D, Mancuso S, Baluška F, Lin J (2012) The signal transducer NPH3 integrates the phototropin1 photosensor with PIN2-based polar auxin transport in Arabidopsis root phototropism. Plant Cell 24:551–565. doi:10.1105/tpc.111.094284

Whitford R, Fernandez A, Tejos R, Pérez AC, Kleine-Vehn J, Vanneste S, Drozdzecki A, Leitner J, Abas L, Aerts M, Hoogewijs K, Baster P, De Groodt R, Lin Y-C, Storme V, Van de Peer Y, Beeckman T, Madder A, Devreese B, Luschnig C, Friml J, Hilson P (2012) GOLVEN secretory peptides regulate auxin carrier turnover during plant gravitropic responses. Dev Cell 22:678–685. doi:10.1016/j.devcel.2012.02.002

Winter V, Hauser M-T (2006) Exploring the ESCRTing machinery in eukaryotes. Trends Plant Sci 11:115–123. doi:10.1016/j.tplants.2006.01.008

Wiśniewska J, Xu J, Seifertová D, Brewer PB, Růžička K, Blilou I, Rouquié D, Benková E, Scheres B, Friml J (2006) Polar PIN localization directs auxin flow in plants. Science 312:883–884. doi:10.1126/science.1121356

Wollert T, Wunder C, Lippincott-Schwartz J, Hurley JH (2009) Membrane scission by the ESCRT-III complex. Nature 458:172–177. doi:10.1038/nature07836

Xu T, Wen M, Nagawa S, Fu Y, Chen J-G, Wu M-J, Perrot-Rechenmann C, Friml J, Jones AM, Yang Z (2010) Cell surface- and Rho GTPase-based auxin signaling controls cellular interdigitation in Arabidopsis. Cell 143:99–110. doi:10.1016/j.cell.2010.09.003

Yang H, Murphy AS (2009) Functional expression and characterization of Arabidopsis ABCB, AUX 1 and PIN auxin transporters in *Schizosaccharomyces pombe*. Plant J 59:179–191. doi:10.1111/j.1365-313X.2009.03856.x

Zažímalová E, Krecek P, Skůpa P, Hoyerová K, Petrášek J (2007) Polar transport of the plant hormone auxin—the role of PIN-FORMED (PIN) proteins. Cell Mol Life Sci 64:1621–1637. doi:10.1007/s00018-007-6566-4

Zegzouti H, Anthony RG, Jahchan N, Bögre L, Christensen SK (2006) Phosphorylation and activation of PINOID by the phospholipid signaling kinase 3-phosphoinositide-dependent protein kinase 1 (PDK1) in Arabidopsis. PNAS 103:6404–6409. doi:10.1073/pnas.0510283103

Zhang J, Nodzyński T, Pěnčík A, Rolčík J, Friml J (2010) PIN phosphorylation is sufficient to mediate PIN polarity and direct auxin transport. PNAS 107:918–922. doi:10.1073/pnas.0909460107

Zhang J, Vanneste S, Brewer PB, Michniewicz M, Grones P, Kleine-Vehn J, Löfke C, Teichmann T, Bielach A, Cannoot B, Hoyerová K, Chen X, Xue H-W, Benková E, Zažímalová E, Friml J (2011) Inositol trisphosphate-induced Ca^{2+} signaling modulates auxin transport and PIN polarity. Dev Cell 20:855–866. doi:10.1016/j.devcel.2011.05.013

Zourelidou M, Müller I, Willige BC, Nill C, Jikumaru Y, Li H, Schwechheimer C (2009) The polarly localized D6 PROTEIN KINASE is required for efficient auxin transport in *Arabidopsis thaliana*. Development 136:627–636. doi:10.1242/dev.028365

Zwiewka M, Feraru E, Müller B, Hwang I, Feraru MI, Kleine-Vehn J, Weijers D, Friml J (2011) The AP-3 adaptor complex is required for vacuolar function in Arabidopsis. Cell Res 21:1711–1722. doi:10.1038/cr.2011.99

Dynamic Behavior and Internalization of Aquaporins at the Surface of Plant Cells

Doan-Trung Luu and Christophe Maurel

Abstract Aquaporins, channel proteins that facilitate water transport across the tonoplast and the plasma membrane of plant cells, have classically been used as reference markers for these two membranes. Yet, recent studies have shown that the subcellular localization of aquaporins is constantly adjusted in response to environmental stimuli. This chapter addresses the mechanisms that determine the density at the cell surface of aquaporins of the Plasma Membrane Intrinsic Protein (PIP) subclass. While pharmacological interference coupled to confocal imaging was extensively used in initial studies, single particle tracking of PIPs fused to GFP, fluorescence correlation spectroscopy (FCS), and a novel fluorescence recovery after photobleaching approach have provided unique insights into the peculiarity of PIP cellular dynamics. It was shown in particular that, while endocytosis of PIPs is predominantly clathrin-dependent under standard conditions, PIP cycling is enhanced under salt stress possibly involving a clathrin- and membrane raft-mediated endocytosis. Future research will address the genetic bases of these pathways and their possible control by the plant hormone auxin.

D.-T. Luu · C. Maurel (✉)
Biochimie et Physiologie Moléculaire des Plantes, Institut de Biologie Intégrative des Plantes, Centre National de la Recherche Scientifique, UMR 5004 CNRS/UMR 0386 INRA/ Montpellier SupAgro/Université Montpellier 2, 2 place Viala, 34060 Montpellier Cedex 2, France
e-mail: maurel@supagro.inra.fr

J. Šamaj (ed.), *Endocytosis in Plants*,
DOI: 10.1007/978-3-642-32463-5_9,

1 Aquaporin Functions and Regulations and the Importance of Aquaporin Trafficking

Aquaporins are water channel proteins that are present in virtually all living organisms and contribute to water transport and equilibration throughout the body of terrestrial plants. The plant aquaporin family is subdivided into seven homology classes (Anderberg et al. 2011). The four classes that are present in all higher plant species can be associated with distinct subcellular localizations. Members of the Tonoplast Intrinsic Protein (TIP) and the Plasma membrane Intrinsic Protein (PIP) classes represent the most abundant aquaporins in the tonoplast and plasma membrane (PM), respectively. The PIP class can be further divided in the PIP1 and PIP2 subclasses. Some isoforms of the Nodulin-26-like Intrinsic Protein (NIP) class have also been found in the PM (Ma et al. 2006; Takano et al. 2006). However, other NIP homologs, similar to members of the fourth class (Small basic Intrinsic Proteins; SIPs), are localized to endoplasmic reticulum (ER) membranes (Ishikawa et al. 2005; Mizutani et al. 2006).

Plants are subjected to multiple biotic and abiotic stresses that impact on their water status and aquaporin-mediated water transport plays an important role in the maintenance of this status (for a review see Maurel et al. 2008). In Arabidopsis roots, aquaporins of the PIP class (*At*PIP) have been shown to be differentially regulated in response to oxygen deprivation, oxidative stress, salinity, and treatments with salicylic acid. These regulations can occur at the transcript level, but more generally at the protein level with changes in pore opening and closing (gating) or changes in aquaporin subcellular localization (Tournaire-Roux et al. 2003; Boursiac et al. 2005; Boursiac et al. 2008; Prak et al. 2008). Elucidating the modes of aquaporin trafficking within the cell or at the cell surface is therefore of prime importance for understanding aquaporin function.

Most of current knowledge on the intracellular trafficking (endocytosis and exocytosis) of plant PM proteins comes from studies on the PIN-FORMED (PIN) auxin transport proteins (see also chapter by Nodzynski et al. in this volume). Other molecular models have been punctually investigated such as the flagellin receptor FLS2 (Robatzek 2007), the boron transporter BOR1 (Takano et al. 2005), the brassinosteroid receptor BRI1 (Vert et al. 2005), the endo-1-4-beta-D-glucanase KOR1 (Robert et al. 2005), and the potassium channel KAT1 (Sutter et al. 2007). Although PIP aquaporins have long been used as canonical PM protein markers, data on their trafficking properties are still scarce. Recent studies (Boursiac et al. 2005, 2008; Zelazny et al. 2007, 2009; Prak et al. 2008; Sorieul et al. 2011; Luu et al. 2012) have shown, however, that PIPs are excellent models for studying the cell biology of plant PM proteins. As an example, data have emerged on PIP export from the ER which is considered as the first step, following synthesis, for transit within the secretory pathway of PM proteins. In the case of maize and Arabidopsis PIP2s, this process was shown to be dependent on canonical diacidic ER-export motifs (Zelazny et al. 2009; Sorieul et al. 2011). Using heterologous and transient expression systems, it was also established that hetero-oligomerization between maize *Zm*PIP2s and *Zm*PIP1s is required for proper export of the latter from the ER (Zelazny et al. 2007). While these studies are crucial for understanding the mechanisms that govern PIP aquaporin biogenesis and proper targeting to the PM, additional mechanisms allow to constantly adjust the density of PIPs in the PM, thereby regulating cell water permeability. This chapter focuses on this topic and addresses the generality and peculiarity of PIP aquaporin dynamics at the cell surface.

2 Constitutive Cycling of PIPs

2.1 Protein Components Involved in the Cycling of PM Cargoes in Plant Cells

Animal cells exhibit various uptake mechanisms for surface proteins, such as clathrin-dependent endocytosis, membrane raft-mediated endocytosis, caveolae-mediated endocytosis, and phagocytosis. In plants, the functional importance of endocytosis has been pioneerly demonstrated in the case of cell wall pectins (Baluska et al. 2002) and PIN auxin efflux transporters, using developmental mutant and pharmacological interference analyses (Geldner et al. 2001, 2003). Polarity of auxin transport is achieved by the coordinated localization of PINs to the one end of cells that mediates auxin efflux. In these and other studies, the fungal toxin brefeldin A (BFA), which inhibits the adenosine ribosylation factor (ARF)-guanine nucleotide exchange factor (GEF) GNOM and thus supposedly blocks exocytosis, was crucial to demonstrate the existence and importance of a continuous cycling of PM proteins between the PM and endosomal compartments. Although its precise localization is still unknown, GNOM is assumed to sit in a recycling endosome compartment and thus should act in the interface between the *trans*-Golgi network (TGN) and the PM (Šamaj et al. 2004; Robinson et al. 2008).

Further evidence that internalization mechanisms similar to those in animals exist in plant cells comes from pharmacological interference using the tyrosine (Tyr) analog, Tyrphostin A23 (A23). One of the effects of A23 in mammalian cells is to prevent interaction between the μ2 subunit of the clathrin-binding Adaptor Protein (AP)-2 complex and cytosolic motifs of cargo PM proteins (Banbury et al. 2003). Tyr-based YXXΦ motifs (where Y is Tyr, X is any amino acid, and Φ is an amino acid with a bulky hydrophobic side chain) mediate cargo recruitment into clathrin-coated vesicles by binding to the μ-subunit of the AP complex and have been identified in animal as well as plant proteins (Happel et al. 2004; Ron and Avni 2004; daSilva et al. 2006; Takano et al. 2010). Although A23 side effects cannot be excluded (see Robinson et al. 2008), several laboratories have shown that inhibition by A23 can be used to dissect membrane trafficking processes in plants (Ortiz-Zapater et al. 2006; Dhonukshe et al. 2007; Konopka et al. 2008; Leborgne-Castel et al. 2008; Fujimoto et al. 2010). More specifically, A23 treatment prevented the labeling of BFA compartments by several PM proteins including PINs, H^+-ATPase, low temperature inducible protein 6b (LTi6b), and most importantly *At*PIP2;1, suggesting an inhibition of their endocytosis (Dhonukshe et al. 2007). The synthetic auxin analog, naphthalene-1-acetic acid (NAA), has also been shown to inhibit endocytosis of PINs, *At*PIP2;1 and PM H^+-ATPase (Paciorek et al. 2005; Dhonukshe et al. 2008; Pan et al. 2009). Wortmannin (Wm), an inhibitor of phosphatidylinositol-3-phosphate (PI-3-P) and phosphatidylinositol-4-phosphate (PI-4-P) kinases, induces in Arabidopsis root cells an enlargement of endosomes labeled with the *At*SORTING NEXIN1 (*At*SNX1), late endosomal (RabF2a) as well as TGN (RabA1d, RabA1e, RabA4b and VTI12) markers (Jaillais et al. 2006; Takáč et al. 2012). Such endosomes may be assimilated to an intermediate compartment between TGN and the pre-vacuolar compartment (PVC) (Jaillais et al. 2006; Niemes et al. 2010; Takáč et al. 2012). Wm-induced enlarged endosomes (Wm compartment) are also labeled by *At*PIP2;1, suggesting that this aquaporin may cycle between the PM and the endosomes labeled by *At*SNX1 or may have been detected in the PVC during its route toward vacuolar degradation (Jaillais et al. 2008). Finally, a role for membrane sterols in the endocytosis of PIN proteins and other PM proteins such as *At*PIP2;1 has been proposed (Kleine-Vehn et al. 2006; Men et al. 2008; Pan et al. 2009).

As a complement of the above-mentioned pharmacological approaches, reverse genetics has allowed to identify specific protein components involved in the cycling of plant PM cargoes. For instance, the involvement of clathrin in *At*PIP2;1 endocytosis was demonstrated by means of a dominant-negative mutant strategy (Dhonukshe et al. 2007). When overexpressed in plant cells, the C-terminal part of clathrin heavy chain (CHC) termed "clathrin Hub" binds to and titers away the clathrin light chains, thus making them unavailable for clathrin cage formation. This expression strategy was shown to impair the labeling of BFA compartments by *At*PIP2;1, indicating a disruption of its endocytosis. More recently, a single *chc*2 loss-of-function mutant and dominant-negative *CHC*1 (HUB) transgenic lines were shown to be defective in bulk lipid membrane endocytosis as well as in PIN1 subcellular localization (Kitakura et al. 2011). Yet, the phenotypic effects on PIP trafficking remain unknown.

Genetic approaches were also used to dissect the dependency of endocytosis on auxin. In a study using auxin signalling $SCF^{TIR1/AFB}$ mutants, PIN2 and LTI6a but not *At*PIP2;1 were able to label BFA compartments in the presence of NAA (Pan et al. 2009). It was concluded that the corresponding nuclear auxin receptor complex is required for inhibition by auxin of PIN2 and LTi6a endocytosis, but not that of *At*PIP2;1. In another study, auxin-binding protein 1 (ABP1) was shown to act as a positive factor during clathrin recruitment to the PM (Robert et al. 2010). In the absence of auxin, expression of ABP1 at the plant PM promotes endocytosis of PINs as well as internalization of an ectopically expressed human transferrin receptor. By contrast, binding of auxin to ABP1 impaired the recruitment of clathrin to the PM, and therefore inhibited endocytosis, at least transiently. It remains unknown whether ABP1 affects PIP endocytosis constitutively or only after binding to auxin.

Finally, a plant ortholog of the yeast and mammalian vacuolar protein sorting 29 (VPS29), a member of the retromer complex, has been shown to be required for PIN trafficking (Jaillais et al. 2007). In roots of an Arabidopsis *vps*29 mutant, PIN proteins were polarly localized but also showed an additional accumulation in abnormally enlarged compartments. Noticeably, *At*PIP2;1 remained correctly localized at the PM of *vps*29 and did not label the enlarged intracellular compartments. This indicates that, in contrast to PINs, the trafficking of PIPs does not involve VSP29 (Jaillais et al. 2007, see also chapter by Zelazny et al. in this volume).

Altogether, these data indicate that PIPs share with other plant PM proteins, and PINs in particular, main subcellular trafficking routes. The data also point to molecular and functional specificities that possibly underlie regulation mechanisms restricted to PIPs (Fig. 1a).

2.2 Novel Approaches for Exploring the Cycling of PIPs

Variable-angle evanescent wave microscopy, also named variable-angle epifluorescence microscopy, was first applied in plants for the visualization of vesicles and proteins within or near the plane of the PM of pollen tubes and epidermal cells (Wang et al. 2006; Konopka and Bednarek 2008; Konopka et al. 2008). In a recent study, this technique was extended to analyzing the lateral mobility of plant PM proteins at a high spatio-temporal resolution. This was achieved by tracking single particles of *At*PIP2;1-fluorescent protein constructs expressed in epidermal cells of Arabidopsis roots (Li et al. 2011). Single particle tracking (SPT) allowed to distinguish for *At*PIP2;1 four types of trajectories and modes of diffusion either Brownian (33.7 $\pm$ 3.3 %), directed (27.5 $\pm$ 2.4 %), restricted (17.5 $\pm$ 2.1 %) or with mixed trajectories (21.2 $\pm$ 3.1 %). These analyses also revealed a broad distribution of individual *At*PIP2;1 diffusion coefficients. However, the modal value (2.46×10^{-3} $\mu m^2 s^{-1}$) was ten times lower than the one for LTi6a (2.37×10^{-2} $\mu m^2 s^{-1}$), indicating an extremely low lateral diffusion for the aquaporin. In this study, an independent technique, fluorescence correlation spectroscopy (FCS), was used to measure the density of *At*PIP2;1-fluorescent protein constructs in the PM of root epidermal cells (Li et al. 2011). A value of 30.3 $\pm$ 5.1 molecules.μm^{-2} was found for *At*PIP2;1 in cells under resting conditions.

The role in *At*PIP2;1 behavior of various membrane components, such as membrane-raft microdomains or clathrin coats, was first addressed using drug interference (Li et al. 2011). Treatment of root cells with methyl-β-cyclodextrin (MβCD), a sterol disrupting reagent, induced a subpopulation of lowly mobile *At*PIP2;1 particles (diffusion coefficient reduced by 20-fold) and modified the lateral mobility characteristics of *At*PIP2;1 by increasing by 64 % the proportion of particles with a restricted diffusion. These important changes were not accompanied by any change in *At*PIP2;1 density in the PM. When cells were treated by A23, a marked (10-fold) reduction of the diffusion coefficient value was observed (1.59×10^{-3} $\mu m^2.s^{-1}$), accompanied by a twofold increase of the

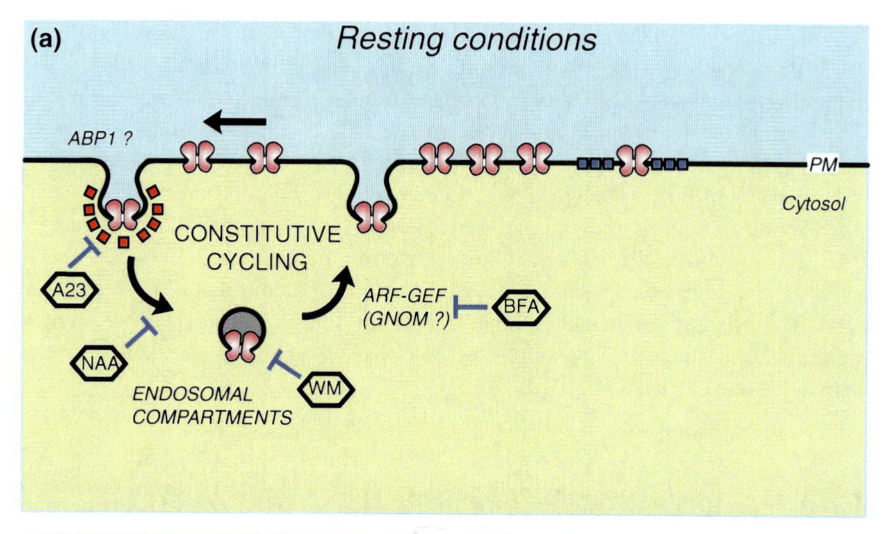

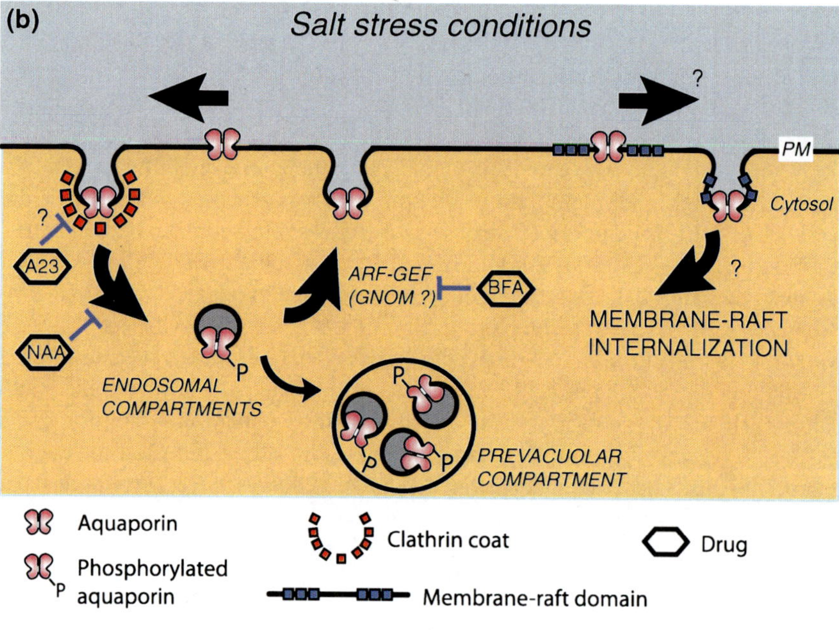

Fig. 1 Tentative model showing the subcellular trafficking mechanisms controlling PIP cycling under resting and stress conditions. **a** In resting conditions, PIP aquaporins exhibit a reduced lateral diffusion and may in part be associated with membrane-raft microdomains. PIPs constitutively cycle between the plasma membrane (*PM*) and endosomal compartments. Endocytosis is clathrin-mediated and sensitive to *A*23. Auxin (*NAA*) or *BFA* can antagonize PIP cycling at the indicated steps. Functional importance of putative auxin receptor ABP1 in PIP endocytosis is unknown. Enlarged endosomes labeled with PIPs are formed after *Wm* treatments. **b** Under salt stress, the proportion of PIPs with restricted trajectories at the cell surface and the rate of PIP cycling are increased by comparison to resting conditions. Functional importance of clathrin in the salt-induced endocytosis needs to be clarified. A membrane-raft-associated pathway is possibly involved in the internalization of PIPs. Prevacuolar compartments are enriched in phosphorylated forms of *At*PIP2;1 whereas unphosphorylated forms would be confined to undetermined endosomes

proportion of particles with a mixed diffusion mode. Importantly, and consistent with an inhibition of endocytosis, the A23 treatment increased by 32 % the density of GFP-tagged *At*PIP2;1 in the PM. The differential effects of these drugs suggest that both clathrin-dependent endocytosis and membrane microdomains determine the dynamic characteristics of *At*PIP2;1, but through independent modes. The influence of these two components on *At*PIP2;1 surface behavior was also suggested by co-localizations of the aquaporin with both Flotillin 1 (*At*Flot1), a marker protein of membrane rafts, and the Clathrin Light Chain (*At*CLC). These data also confirm previous studies showing that PIPs co-purified with detergent-resistant membranes (Mongrand et al. 2004; Morel et al. 2006).

The low lateral diffusion of *At*PIP2;1 in the plane of the PM described above is reminiscent of the behavior described for the PM K^+ channel, KAT1, and the H^+-ATPase, PMA2 (Sutter et al. 2006; Sutter et al. 2007). It contrasts with the behavior of proteins resident or retained in the ER, which exhibit an extremely high lateral motion (Runions et al. 2006; Luu et al. 2012). In this kind of studies, fluorescence recovery after photobleaching (FRAP) techniques, and the use of a photo-activable version of GFP (*pa*GFP) have been instrumental for monitoring the lateral mobility of membrane proteins. The low lateral diffusion of PIPs was recently examined in Arabidopsis root epidermal cells expressing fusions of GFP with PM (*At*PIP1;2, *At*PIP2;1, LTi6a) or tonoplast (*At*TIP1;1) proteins by using the same set of techniques (Luu et al. 2012). The fluorescence recovery responses were similar for the two *At*PIP isoforms, fused with GFP either at the N- or C-terminus, suggesting that the position of the fused GFP did not affect the lateral mobility of *At*PIPs. While the fluorescence signal of GFP-LTi6a had recovered by ~55 % at 50 s and almost completely at 7 min after photobleaching, the relative amplitude of signal recovery of *At*PIP fusions was below 60 % even after 30 min. A kymographic analysis, allowing to represent as a function of time the recovery of the fluorescence signal along a line which crosses the bleached region, also confirmed the low dynamics of *At*PIPs in the plane of the PM. Complementary experiments were performed by monitoring the decrease in fluorescent signal, after photo-activation in a region of interest of *pa*GFP fused to LTi6a or *At*PIP2;1. Here again, the decrease in signal was found slower for *At*PIP2;1 than for LTi6a. This series of data indicate a minute, if any, lateral diffusion of GFP-tagged *At*PIPs, over the 30 min of monitoring in the PM of Arabidopsis root epidermal cells. Therefore, the slow FRAP response observed over the same period of time cannot be explained by the lateral diffusion of the GFP-tagged *At*PIPs and other processes must be invoked. These findings suggested that the slow FRAP response of the *At*PIP constructs could be used to dissect their constitutive cycling. A close inspection of the corresponding fluorescence recovery curves showed them to be biphasic, with a fast process completed at 60 s and a slower process that developed for up to 30 min and beyond. It was suggested that the first process is due to a fast cytoplasmic streaming which drags intracellular compartments containing GFP-tagged *At*PIPs into the initially bleached region. Because of a lack of lateral diffusion of the GFP-tagged *At*PIPs in the PM, the slower recovery observed during the next 30 min would, by contrast, be due to their recycling from the endosomal compartments to the PM. As a consequence, the amplitude of the early recovery phase can be used to estimate the labeling intensity of endomembrane compartments (in other words, the abundance of GFP-tagged *At*PIP in these compartments). The slow recovery phase would report on the cycling of GFP-tagged *At*PIPs between the endomembrane and PM compartments, including sequential steps in endocytosis, sorting in the endosomes, and exocytosis. This tentative model was validated using a pharmacological dissection of FRAP responses. Consistent with its inhibiting effects on endocytosis, A23 was found to reduce the labeling of endomembrane compartments by GFP-tagged *At*PIPs. A23 and also NAA and BFA reduced the recovery of fluorescence at 30 min and beyond suggesting an inhibition of the recycling to the PM. Surprisingly, NAA did not change the intensity of endosomal labeling in these experiments. Thus, the site of NAA inhibition may be downstream of endocytic uptake and upstream of the TGN. Nevertheless, the whole set of data validates FRAP as a new technique for studying the cycling of PM proteins, with a higher quantitative resolution than standard confocal imaging (Fig. 1a).

3 Stress-Induced Trafficking of PIPs

3.1 *Physiological Effects of Salinity and Oxidative Stress on Root Water Transport*

Aquaporins contribute to a significant part of the root hydraulic conductivity (Lp_r) in numerous plant species including Arabidopsis, rice, maize, grapevine, or legumes (for review, see Maurel et al. 2008). Therefore, the regulation of Lp_r in response to multiple abiotic stimuli such as drought, salinity, temperature extremes as well as oxygen and nutrient deprivation provides interesting physiological contexts in which to investigate the molecular and cellular mechanisms of aquaporin regulation. For instance, a treatment with 100 mM NaCl was found to provoke a fast (halftime = 45 min) and strong (−70 %) inhibition of Arabidopsis Lp_r (Boursiac et al. 2005). Salinity was perceived as an osmotic stimulus, since an equivalent osmotic challenge with 200 mM mannitol induced a similar reduction in Lp_r. A strong reduction (60–75 %) in aquaporin transcript abundance was observed, but at a rather late stage (2–4 h) after exposure to salt. By contrast, *At*PIP1 protein abundance was decreased by 40 % as early as 30 min after salt exposure, whereas *At*PIP2 levels remained constant during the first 6 h of the salt treatment. In addition, salt enhanced the intracellular labeling of root cells expressing *At*PIP1;2-GFP and *At*PIP2;1-GFP fusions, as early as 2 h after salt treatment. Internalization of *At*PIPs and the resulting reduction of aquaporin density at the PM, would therefore contribute to salt-induced reduction of Lp_r.

In a more general context, exposure of cucumber roots to cold stress (4–8 °C) (Lee et al. 2004), or exposure of Arabidopsis to salt (150 mM NaCl) or salicylic acid

(0.5 mM SA) (Boursiac et al. 2008) were all shown to induce a concomitant accumulation of reactive oxygen species (ROS) and inhibition of Lp_r. Exogenous application of hydrogen peroxide (H_2O_2), one of major ROS, was also able to reduce Lp_r by up to 90 % in less than 15 min (Boursiac et al. 2008), suggesting that the effects of cold, salt or SA on Lp_r were mediated through production of ROS. This was partly demonstrated in the case of Arabidopsis by showing that the salt- or SA-induced internalization of *At*PIPs could be counteracted by ROS scavenging using a catalase treatment. Functional expression of *At*PIPs in *Xenopus* oocytes showed, in addition, that the intrinsic water transport activity of these aquaporins was insensitive to ROS. Thus, these experiments establish the central role of a ROS-induced cellular redistribution of aquaporins during the regulation of root water transport in response to multiple hormonal and environmental signals.

3.2 Cell Surface Dynamics and Cycling of PIPs in Response to Salinity and Oxidative Stress

Under biotic or abiotic stresses, regulation of endocytosis is essential in plant cells (Leborgne-Castel and Luu 2009), as it allows a rapid adjustment of membrane transport or stress perception and signalling. For instance, when rice cells were subjected to a saline stress, the endocytosis of biotinylated bovine serum albumin was initially inhibited but activated after a 24 h-extended period (Bahaji et al. 2003). By contrast, a study in Arabidopsis roots (Leshem et al. 2007) demonstrated an enhancement of bulk-flow endocytosis shortly after a massive salt treatment (0.2 M NaCl). More specifically, the salt-induced internalization of aquaporins was found to be strongly dose dependent. Intracellular labeling with *At*PIP-GFP was occasionally observed after 2 h treatment with 100 mM NaCl, whereas a treatment with 150 mM NaCl resulted in a pronounced labeling of fuzzy structures and of spherical bodies, tentatively identified as endomembrane compartments and small vacuoles, respectively (Boursiac et al. 2005, 2008). The same intracellular structures were labeled following SA or H_2O_2 application. While some PM cargoes, such as BOR1 or the iron transporter IRT1 undergo stimulus-induced internalization prior to being targeted to vacuolar degradation (Takano et al. 2005; Barberon et al. 2011), the K^+-channel *At*KAT1 was shown to be reversibly sequestered upon ABA-induced internalization (Sutter et al. 2007). Whether stress-induced *At*PIP internalization plays a role in degradation and/or intracellular sequestration remains as yet unknown.

The two recent studies partially described in the previous section have brought significant insights into the fundamental effects of salt on the membrane dynamics of PIPs. First, SPT in root epidermal cells expressing *At*PIP2;1-GFP showed that a short-term salt stress (100 mM NaCl for 10 min) increased by twofold the diffusion coefficient of *At*PIP2;1 particles and by 60 % the proportion of particles with a restricted diffusion mode (Li et al. 2011). In addition, FCS showed that these effects were accompanied by a 46 % decrease of the *At*PIP2;1 density in the PM. The latter decrease could be partly antagonized by both A23 and M*β*CD. Thus, while endocytosis is predominantly clathrin-dependent under standard conditions, it is enhanced under salt stress and involves both clathrin- and membrane microdomain-mediated endocytic pathways (Fig 1b).

In the second study (Luu et al. 2012), FRAP and photo-activation approaches were used to dissect the cycling properties of *At*PIPs under salt stress. Kymographic analyses and photo-activation experiments showed that although the diffusion coefficient of *At*PIP2;1 particles was increased by twofold under salt stress (Li et al. 2011), the contribution of lateral diffusion to the recovery of fluorescence was negligible, similar to control conditions (Luu et al. 2012). Yet, the amplitude of fluorescence recovery was increased by twofold by salt treatment, suggesting that salt induced an enhanced cycling of *At*PIPs. This idea was validated by showing that the kinetics of labeling by GFP-tagged *At*PIPs of BFA compartments, as well as the reversal of labeling after washout experiment, were accelerated under salt stress conditions. The overall data show that salt stress enhanced both endocytosis and exocytosis of *At*PIPs, and therefore their cycling. A pharmacological dissection of the FRAP response, similar to the one performed under control conditions, showed that PIP cycling under salt stress was blocked by NAA but had become insensitive to A23. This was interpreted to mean that an enhanced clathrin-mediated endocytosis could overrun the effect of this drug or that, contrary to the standard conditions, an endocytic mechanism, which is insensitive to A23, specifically operates in salt stress conditions. The latter hypothesis is consistent with involvement of a membrane microdomain-dependent pathway, as suggested by FCS measurements (Li et al. 2011).

In mammalian cells, early endosomes comprise two distinct populations: a dynamic population that is highly mobile on microtubules and matures rapidly toward late endosomes, and a static population that matures much more slowly (Lakadamyali et al. 2006). The dynamic population was found to be linked to an AP-2-independent endocytic machinery and transports cargoes such as low-density lipoprotein receptors. Here, we suggest that a similar pathway may be activated in plant cells under salt stress. Interestingly, evidence for a clathrin-independent endocytic pathway has recently been presented in the context of glucose uptake by BY-2 tobacco cells (Bandmann and Homann 2012). Flotillin-dependent endocytosis was also recently described in plant cells (Li et al. 2012). Nevertheless, there is an urging need to know further molecular components involved in these emerging pathway(s) of plant endocytosis.

Phosphorylation has well-identified effects on plant aquaporin gating, but was recently shown to interfere with the subcellular trafficking of PIPs (Prak et al. 2008). Mass spectrometry analyses of membrane protein extracts from Arabidopsis roots allowed the identification of two phosphorylation sites at Ser280 and Ser283 in the C-terminal tail of *At*PIP2;1. The functional role of these sites was addressed by introducing into a GFP-*At*PIP2;1 fusion, mutations of Ser280 or Ser283 to an Alanine or an Aspartate, to mimic a constitutive dephosphorylation or phosphorylation, respectively, of these two sites. The overall data indicated that phosphorylation of Ser283 but not of Ser280 is necessary for correct targeting of *At*PIP2;1 to the PM of epidermal root cells. The former site is also involved in the intracellular sorting of *At*PIP2;1 following salt-induced internalization. In brief, the dephosphorylated form (Ser283Ala mutant) was associated with the labeling of "fuzzy" structures whereas the phosphorylated form (Ser283Asp mutant) labeled spherical bodies (Fig. 1b). The protein kinases and phosphatases acting on the two phosphorylation sites and the protein partners involved in the phosphorylation-dependent sorting of *At*PIP2;1 under salt stress are as yet unknown.

4 Conclusions and Future Prospects

While PIPs have been extensively used as canonical and stable markers of the PM, the trafficking of these proteins is now the matter of intense investigations. Data gathered over the last few years have pointed to similarities with the PIN transporters, another intensively studied model for protein subcellular trafficking. For instance, the involvement of clathrin-dependent endocytosis and the similar effects of BFA indicate common constitutive cycling pathways for PIP and PIN homologs (Fig. 1). However, a possible role of ABP1 in the recruitment of clathrin for subsequent endocytosis of PIP remains to be defined. This is a totally open question as recent pharmacological studies (Luu et al. 2012, Wudick et al. 2012 in preparation) indicate that auxin may interfere with PIP trafficking downstream of clathrin-mediated endocytosis. Another important line of research will concern the activation of both clathrin-dependent and -independent endocytic pathways, in cells under oxidative and salt stress. A very useful strategy is to combine pharmacological and genetic interference approaches. Such a strategy has proved to be very informative for the dissection of clathrin-dependent endocytosis of PINs and of course needs to be extended to PIPs. Another component which potentially functions in PIP trafficking is the exocyst complex. This complex tethers secretory vesicles to the PM in yeast and animals (Munson and Novick 2006; He and Guo 2009). By contrast to their mammalian and yeast counterparts, the majority of the eight predicted plant exocyst subunits exist as multiple copies with only a few of these genes having assigned functions in cell–cell interactions, apical cell growth, and development (Synek et al. 2006; Hala et al. 2008; Samuel et al. 2009). Their primary role in exocytosis of PIPs and other PM cargos remains to be determined.

Although providing a very partial knowledge, studies on the trafficking pathways of PIPs have already pointed to significant differences to those described for PINs. These differences might be related to specific PM distributions of the two classes of proteins. Polar targeting of PIN proteins is of fundamental importance for directional auxin transport from cell to cell and for maintaining the auxin gradients that are required for organ development. By contrast, very few aquaporins have been described to be polarly localized in plant cells. For instance, *At*NIP5;1 serves as a boric acid channel and preferentially localizes at the exofacial side of root cells under boron limiting conditions (Takano et al. 2010).

Comparative studies of the trafficking pathways of *At*NIP5;1 and of the non-polar *At*PIP2;1 might reveal specific and common mechanisms. Finally, homologs of the PIP1 and PIP2 subclasses are known to have distinct trafficking abilities within the secretory pathway (Fetter et al. 2004). Yet, these homologs can assemble in heterotetramers and, while PIP1 homotetramers are retained intracellularly, PIP2 homotetramers or PIP1-PIP2 heterotetramers exhibit a facilitated trafficking to the PM (Zelazny et al. 2007). Thus, we anticipate that the PIP1-PIP2 molecular interactions may have specific impacts on the dynamics at the cell surface and endocytosis of each of these PIPs. Complex cellular dynamics of aquaporins would therefore depend on the combinatorial co-expression of numerous PIP isoforms within the same plant cell type.

References

Anderberg H, Danielson J, Johanson U (2011) Algal MIPs, high diversity and conserved motifs. BMC Evol Biol 11:110

Bahaji A, Aniento F, Cornejo MJ (2003) Uptake of an endocytic marker by rice cells: variations related to osmotic and saline stress. Plant Cell Physiol 44:1100–1111

Baluška F, Hlavacka A, Šamaj J, Palme K, Robinson DG, Matoh T, McCurdy DW, Menzel D, Volkmann D (2002) F-actin-dependent endocytosis of cell wall pectins in meristematic root cells. Insights from brefeldin A-induced compartments. Plant Physiol 130:422–431

Banbury DN, Oakley JD, Sessions RB, Banting G (2003) Tyrphostin A23 inhibits internalization of the transferrin receptor by perturbing the interaction between tyrosine motifs and the medium chain subunit of the AP-2 adaptor complex. J Biol Chem 278:12022–12028

Bandmann V, Homann U (2012) Clathrin-independent endocytosis contributes to uptake of glucose into BY-2 protoplasts. Plant J (in press)

Barberon M, Zelazny E, Robert S, Conejero G, Curie C, Friml J, Vert G (2011) Monoubiquitin-dependent endocytosis of the iron-regulated transporter 1 (IRT1) transporter controls iron uptake in plants. Proc Natl Acad Sci U S A 108:E450–E458

Boursiac Y, Boudet J, Postaire O, Luu DT, Tournaire-Roux C, Maurel C (2008) Stimulus-induced downregulation of root water transport involves reactive oxygen species-activated cell signalling and plasma membrane intrinsic protein internalization. Plant J 56:207–218

Boursiac Y, Chen S, Luu DT, Sorieul M, van den Dries N, Maurel C (2005) Early effects of salinity on water transport in Arabidopsis roots. Molecular and cellular features of aquaporin expression. Plant Physiol 139:790–805

daSilva LL, Foresti O, Denecke J (2006) Targeting of the plant vacuolar sorting receptor BP80 is dependent on multiple sorting signals in the cytosolic tail. Plant Cell 18:1477–1497

Dhonukshe P, Aniento F, Hwang I, Robinson DG, Mravec J, Stierhof YD, Friml J (2007) Clathrin-mediated constitutive endocytosis of PIN auxin efflux carriers in Arabidopsis. Curr Biol 17:520–527

Dhonukshe P, Grigoriev I, Fischer R, Tominaga M, Robinson DG, Hasek J, Paciorek T, Petrasek J, Seifertova D, Tejos R, Meisel LA, Zazimalova E, Gadella TWJ Jr, Stierhof Y-D, Ueda T, Oiwa K, Akhmanova A, Brock R, Spang A, Friml J (2008) Auxin transport inhibitors impair vesicle motility and actin cytoskeleton dynamics in diverse eukaryotes. Proc Natl Acad Sci U S A 105:4489–4494

Fetter K, Van Wilder V, Moshelion M, Chaumont F (2004) Interactions between plasma membrane aquaporins modulate their water channel activity. Plant Cell 16:215–228

Fujimoto M, Arimura S-i, Ueda T, Takanashi H, Hayashi Y, Nakano A, Tsutsumi N (2010) Arabidopsis dynamin-related proteins DRP2B and DRP1A participate together in clathrin-coated vesicle formation during endocytosis. Proc Natl Acad Sci U S A 107:6094–6099
Geldner N, Anders N, Wolters H, Keicher J, Kornberger W, Muller P, Delbarre A, Ueda T, Nakano A, Jurgens G (2003) The Arabidopsis GNOM ARF-GEF mediates endosomal recycling, auxin transport, and auxin-dependent plant growth. Cell 112:219–230
Geldner N, Friml J, Stierhof Y-D, Jurgens G, Palme K (2001) Auxin transport inhibitors block PIN1 cycling and vesicle trafficking. Nature 413:425–428
Hala M, Cole R, Synek L, Drdova E, Pecenkova T, Nordheim A, Lamkemeyer T, Madlung J, Hochholdinger F, Fowler JE, Zarsky V (2008) An exocyst complex functions in plant cell growth in Arabidopsis and tobacco. Plant Cell 20:1330–1345
Happel N, Honing S, Neuhaus JM, Paris N, Robinson DG, Holstein SE (2004) Arabidopsis μA-adaptin interacts with the tyrosine motif of the vacuolar sorting receptor VSR-PS1. Plant J 37:678–693
He B, Guo W (2009) The exocyst complex in polarized exocytosis. Curr Opin Cell Biol 21:537–542
Ishikawa F, Suga S, Uemura T, Sato MH, Maeshima M (2005) Novel type aquaporin SIPs are mainly localized to the ER membrane and show cell-specific expression in *Arabidopsis thaliana*. FEBS Lett 579:5814–5820
Jaillais Y, Fobis-Loisy I, Miege C, Gaude T (2008) Evidence for a sorting endosome in Arabidopsis root cells. Plant J 53:237–247
Jaillais Y, Fobis-Loisy I, Miege C, Rollin C, Gaude T (2006) *At*SNX1 defines an endosome for auxin-carrier trafficking in Arabidopsis. Nature 443:106–109
Jaillais Y, Santambrogio M, Rozier F, Fobis-Loisy I, Miege C, Gaude T (2007) The retromer protein VPS29 links cell polarity and organ initiation in plants. Cell 130:1057–1070
Kitakura S, Vanneste S, Robert S, Lofke C, Teichmann T, Tanaka H, Friml J (2011) Clathrin mediates endocytosis and polar distribution of PIN auxin transporters in Arabidopsis. Plant Cell 23:1920–1931
Kleine-Vehn J, Dhonukshe P, Swarup R, Bennett M, Friml J (2006) Subcellular trafficking of the Arabidopsis auxin influx carrier AUX1 uses a novel pathway distinct from PIN1. Plant Cell 18:3171–3181
Konopka CA, Backues SK, Bednarek SY (2008) Dynamics of Arabidopsis dynamin-related protein 1C and a clathrin light chain at the plasma membrane. Plant Cell 20:1363–1380
Konopka CA, Bednarek SY (2008) Variable-angle epifluorescence microscopy: a new way to look at protein dynamics in the plant cell cortex. Plant J 53:186–196
Lakadamyali M, Rust MJ, Zhuang X (2006) Ligands for clathrin-mediated endocytosis are differentially sorted into distinct populations of early endosomes. Cell 124:997–1009
Leborgne-Castel N, Lherminier J, Der C, Fromentin J, Houot V, Simon-Plas F (2008) The plant defense elicitor cryptogein stimulates clathrin-mediated endocytosis correlated with reactive oxygen species production in bright yellow-2 tobacco cells. Plant Physiol 146:1255–1266
Leborgne-Castel N, Luu D-T (2009) Regulation of endocytosis by external stimuli in plant cells. Plant Biosyst 143:630–635
Lee SH, Singh AP, Chung GC (2004) Rapid accumulation of hydrogen peroxide in cucumber roots due to exposure to low temperature appears to mediate decreases in water transport. J Exp Bot 55:1733–1741
Leshem Y, Seri L, Levine A (2007) Induction of phosphatidylinositol 3-kinase-mediated endocytosis by salt stress leads to intracellular production of reactive oxygen species and salt tolerance. Plant J 51:185–197
Li X, Wang X, Yang Y, Li R, He Q, Fang X, Luu DT, Maurel C, Lin J (2011) Single-molecule analysis of PIP2;1 dynamics and partitioning reveals multiple modes of Arabidopsis plasma membrane aquaporin regulation. Plant Cell 23:3780–3797
Li R, Liu P, Wan Y, Chen T, Wang Q, Mettbach U, Baluška F, Šamaj J, Fang X, Lucas WJ, Lin J (2012) A membrane microdomain-associated protein, Arabidopsis Flot1, is involved in a

clathrin-independent endocytic pathway and is required for seedling development. Plant Cell, (in press, published on line May 15)

Luu D-T, Martinière A, Sorieul M, Runions J, Maurel C (2012) Fluorescence recovery after photobleaching reveals high cycling dynamics of plasma membrane aquaporins in Arabidopsis roots under salt stress. Plant J 69:894–905

Ma JF, Tamai K, Yamaji N, Mitani N, Konishi S, Katsuhara M, Ishiguro M, Murata Y, Yano M (2006) A silicon transporter in rice. Nature 440:688–691

Maurel C, Verdoucq L, Luu DT, Santoni V (2008) Plant aquaporins: membrane channels with multiple integrated functions. Annu Rev Plant Biol 59:595–624

Men S, Boutte Y, Ikeda Y, Li X, Palme K, Stierhof YD, Hartmann MA, Moritz T, Grebe M (2008) Sterol-dependent endocytosis mediates post-cytokinetic acquisition of PIN2 auxin efflux carrier polarity. Nat Cell Biol 10:237–244

Mizutani M, Watanabe S, Nakagawa T, Maeshima M (2006) Aquaporin NIP2;1 is mainly localized to the ER membrane and shows root-specific accumulation in *Arabidopsis thaliana*. Plant Cell Physiol 47:1420–1426

Mongrand S, Morel J, Laroche J, Claverol S, Carde JP, Hartmann MA, Bonneu M, Simon-Plas F, Lessire R, Bessoule JJ (2004) Lipid rafts in higher plant cells: purification and characterization of Triton X-100-insoluble microdomains from tobacco plasma membrane. J Biol Chem 279:36277–36286

Morel J, Claverol S, Mongrand S, Furt F, Fromentin J, Bessoule JJ, Blein JP, Simon-Plas F (2006) Proteomics of plant detergent-resistant membranes. Mol Cell Proteomics 5:1396–1411

Munson M, Novick P (2006) The exocyst defrocked, a framework of rods revealed. Nat Struct Mol Biol 13:577–581

Niemes S, Langhans M, Viotti C, Scheuring D, San Wan Yan M, Jiang L, Hillmer S, Robinson DG, Pimpl P (2010) Retromer recycles vacuolar sorting receptors from the *trans*-Golgi network. Plant J 61:107–121

Ortiz-Zapater E, Soriano-Ortega E, Marcote MJ, Ortiz-Masia D, Aniento F (2006) Trafficking of the human transferrin receptor in plant cells: effects of tyrphostin A23 and brefeldin A. Plant J 48:757–770

Paciorek T, Zažímalová E, Ruthardt N, Petrasek J, Stierhof YD, Kleine-Vehn J, Morris DA, Emans N, Jurgens G, Geldner N, Friml J (2005) Auxin inhibits endocytosis and promotes its own efflux from cells. Nature 435:1251–1256

Pan J, Fujioka S, Peng J, Chen J, Li G, Chen R (2009) The E3 ubiquitin ligase SCFTIR1/AFB and membrane sterols play key roles in auxin regulation of endocytosis, recycling, and plasma membrane accumulation of the auxin efflux transporter PIN2 in *Arabidopsis thaliana*. Plant Cell 21:568–580

Prak S, Hem S, Boudet J, Viennois G, Sommerer N, Rossignol M, Maurel C, Santoni V (2008) Multiple phosphorylations in the C-terminal tail of plant plasma membrane aquaporins: role in subcellular trafficking of *At*PIP2;1 in response to salt stress. Mol Cell Proteomics 7:1019–1030

Robatzek S (2007) Vesicle trafficking in plant immune responses. Cell Microbiol 9:1–8

Robert S, Bichet A, Grandjean O, Kierzkowski D, Satiat-Jeunemaitre B, Pelletier S, Hauser MT, Höfte H, Vernhettes S (2005) An Arabidopsis endo-1,4-beta-D-glucanase involved in cellulose synthesis undergoes regulated intracellular cycling. Plant Cell 17:3378–3389

Robert S, Kleine-Vehn J, Barbez E, Sauer M, Paciorek T, Baster P, Vanneste S, Zhang J, Simon S, Covanova M, Hayashi K, Dhonukshe P, Yang Z, Bednarek SY, Jones AM, Luschnig C, Aniento F, Zazimalova E, Friml J (2010) ABP1 mediates auxin inhibition of clathrin-dependent endocytosis in Arabidopsis. Cell 143:111–121

Robinson DG, Jiang L, Schumacher K (2008) The endosomal system of plants: charting new and familiar territories. Plant Physiol 147:1482–1492

Ron M, Avni A (2004) The receptor for the fungal elicitor ethylene-inducing xylanase is a member of a resistance-like gene family in tomato. Plant Cell 16:1604–1615

Runions J, Brach T, Kuhner S, Hawes C (2006) Photoactivation of GFP reveals protein dynamics within the endoplasmic reticulum membrane. J Exp Bot 57:43–50

Šamaj J, Baluška F, Voigt B, Schlicht M, Volkmann D, Menzel D (2004) Endocytosis, actin cytoskeleton, and signaling. Plant Physiol 135:1150–1161

Samuel MA, Chong YT, Haasen KE, Aldea-Brydges MG, Stone SL, Goring DR (2009) Cellular pathways regulating responses to compatible and self-incompatible pollen in Brassica and Arabidopsis stigmas intersect at Exo70A1, a putative component of the exocyst complex. Plant Cell 21:2655–2671

Sorieul M, Santoni V, Maurel C, Luu DT (2011) Mechanisms and effects of retention of over-expressed aquaporin *At*PIP2;1 in the endoplasmic reticulum. Traffic 12:473–482

Sutter JU, Campanoni P, Tyrrell M, Blatt MR (2006) Selective mobility and sensitivity to SNAREs is exhibited by the Arabidopsis KAT1 K^+ channel at the plasma membrane. Plant Cell 18:935–954

Sutter JU, Sieben C, Hartel A, Eisenach C, Thiel G, Blatt MR (2007) Abscisic acid triggers the endocytosis of the Arabidopsis KAT1 K^+ channel and its recycling to the plasma membrane. Curr Biol 17:1396–1402

Synek L, Schlager N, Elias M, Quentin M, Hauser MT, Zarsky V (2006) AtEXO70A1, a member of a family of putative exocyst subunits specifically expanded in land plants, is important for polar growth and plant development. Plant J 48:54–72

Takáč T, Pechan T, Šamajová O, Ovečka M, Richter H, Eck C, Niehaus K, Šamaj J (2012) Wortmannin treatment induces changes in Arabidopsis root proteome and post-Golgi compartments. J Proteome Res, (in press, published on line May 10)

Takano J, Miwa K, Yuan L, von Wiren N, Fujiwara T (2005) Endocytosis and degradation of BOR1, a boron transporter of *Arabidopsis thaliana*, regulated by boron availability. Proc Natl Acad Sci U S A 102:12276–12281

Takano J, Tanaka M, Toyoda A, Miwa K, Kasai K, Fuji K, Onouchi H, Naito S, Fujiwara T (2010) Polar localization and degradation of Arabidopsis boron transporters through distinct trafficking pathways. Proc Natl Acad Sci U S A 107:5220–5225

Takano J, Wada M, Ludewig U, Schaaf G, von Wiren N, Fujiwara T (2006) The Arabidopsis major intrinsic protein NIP5;1 is essential for efficient boron uptake and plant development under boron limitation. Plant Cell 18:1498–1509

Tournaire-Roux C, Sutka M, Javot H, Gout E, Gerbeau P, Luu DT, Bligny R, Maurel C (2003) Cytosolic pH regulates root water transport during anoxic stress through gating of aquaporins. Nature 425:393–397

Vert G, Nemhauser JL, Geldner N, Hong F, Chory J (2005) Molecular mechanisms of steroid hormone signaling in plants. Annu Rev Cell Dev Biol 21:177–201

Wang X, Teng Y, Wang Q, Li X, Sheng X, Zheng M, Šamaj J, Baluška F, Lin J (2006) Imaging of dynamic secretory vesicles in living pollen tubes of *Picea meyeri* using evanescent wave microscopy. Plant Physiol 141:1591–1603

Wudick MM, Geldner N, Chory J, Maurel C Luu DT (2012) Sub-cellular pathway and role of hydrogen peroxide-induced redistribution of root aquaporins (in preparation)

Zelazny E, Borst JW, Muylaert M, Batoko H, Hemminga MA, Chaumont F (2007) FRET imaging in living maize cells reveals that plasma membrane aquaporins interact to regulate their subcellular localization. Proc Natl Acad Sci U S A 104:12359–12364

Zelazny E, Miecielica U, Borst JW, Hemminga MA, Chaumont F (2009) An N-terminal diacidic motif is required for the trafficking of maize aquaporins ZmPIP2;4 and ZmPIP2;5 to the plasma membrane. Plant J 57:346–355

The Role of RAB GTPases and SNARE Proteins in Plant Endocytosis and Post-Golgi Trafficking

Takashi Ueda, Masa H. Sato and Tomohiro Uemura

Abstract Each membrane trafficking pathway involves several evolutionarily conserved key molecules, including RAB GTPases and SNARE proteins. Distinct sets of RAB and SNARE molecules regulate tethering and fusion of the carrier membrane to the target membrane for different trafficking pathways. These proteins are thought to control the specificity of directional targeting and membrane fusion to the correct target organelles. These molecules also exhibit distinctive subcellular localizations and are, therefore, regarded as earmarks for organelles. Several subgroups of RAB and SNARE are widely conserved among eukaryotic lineages, indicating ancient origins and conserved functions. In contrast, recent comparative genomics indicated RAB and SNARE members have expanded in a lineage-specific manner. This finding suggests novel trafficking routes, which are unique to each lineage, developed during evolution. Plant-unique sets of RAB and SNARE proteins have been identified and characterized in recent years. In this chapter, we summarize the conserved and unique features of plant RAB and SNARE proteins with a special focus on post-Golgi trafficking pathways, including the endocytic pathway.

T. Ueda (✉) · T. Uemura
Department of Biological Sciences, Graduate School of Sciences,
University of Tokyo, 7-3-1 Hongo, Bunkyo-ku Tokyo, 113-0033, Japan
e-mail: tueda@biol.s.u-tokyo.ac.jp

T. Ueda
Japan Science and Technology Agency (JST), PRESTO,
4-1-8 Honcho Kawaguchi, Saitama, 332-0012, Japan

M. H. Sato
Graduate School of Life and Environmental Sciences, Kyoto Prefectural University,
1-5, Shimogamo-nakaragi-cho, Sakyo-ku, Kyoto, 606-8522, Japan

J. Šamaj (ed.), *Endocytosis in Plants*,
DOI: 10.1007/978-3-642-32463-5_10,

1 Conserved Actions of RAB and SNARE Proteins

The Rab/Ypt protein is a member of the Ras super family. It acts as a molecular switch in membrane trafficking by cycling between active GTP-bound and inactive GDP-bound states (Saito and Ueda 2009). RAB is activated by the guanine nucleotide exchange factor (GEF) at the organelle membrane, which triggers the tethering of membrane carriers or organelles to the target membrane (Fig. 1). Tethering between the two membranes is mediated by multi-subunit complexes such as HOPS, TRAPP, and Exocyst and monomeric long fibrous proteins such as EEA1 and p115/Uso1p (Warren and Mellman 2006; Markgraf et al. 2007). In yeast and animal systems, many of these tethering molecules (tethers) act as effector proteins of RAB GTPases, which specifically bind to GTP-bound RAB to evoke downstream reactions (Novick et al. 2006; Cai et al. 2007). The tethering of membrane carriers to the target membranes appears to be a conserved (thus ancient) function of RAB GTPases, as RAB and most tethers are highly conserved among animals, yeasts, and plants. Although RABs and tethers have not been linked directly in plants, the components of some conserved tethering complexes function in membrane trafficking and organelle biogenesis in plants (Rojo et al. 2001; Ishikawa et al. 2008), suggesting homologous systems are employed in a wide range of eukaryotic lineages. On the other hand, lineage-specific effector proteins have been reported; for example, most animal RAB5 effectors, including Rabaptin-5 and EEA1, do not have homologs in plants and interactions between plant RABA/RAB11 or RABE/RAB8 and their effectors seem to be unique to the plant system (Preuss et al. 2006; Camacho et al. 2009). The RAB GTPases have been diversified in a unique way in plants; thus, the identification and characterization of plant-unique effectors and the study of conserved tethering factors are important to determine how plants have added unique regulatory systems to the conserved frameworks of membrane trafficking.

Once membrane carriers are tethered to the target membranes, another group of evolutionary conserved molecules, soluble N-ethylmaleimide-sensitive factor attachment protein receptor (SNARE) proteins, carry out membrane fusion between membrane carriers and target organelles (Jahn and Scheller 2006; Wickner and Schekman 2008; Saito and Ueda 2009) (Fig. 1). The SNARE protein family consists of four subgroups, Qa-, Qb-, Qc-, and R-SNAREs, which are classified according to sequence similarity in a typical helical SNARE domain. In many cases, Q- and R-SNAREs reside on distinct membrane compartments, and three Q-SNAREs (Qa, Qb, and Qc) and one R-SNARE assemble into a tight complex in specific combinations, which lead to membrane fusion between two membrane compartments. Most SNARE proteins contain one SNARE domain in their polypeptides, but some, such as SNAP25-like proteins, contain two (Qb+Qc) (Jahn and Scheller 2006).

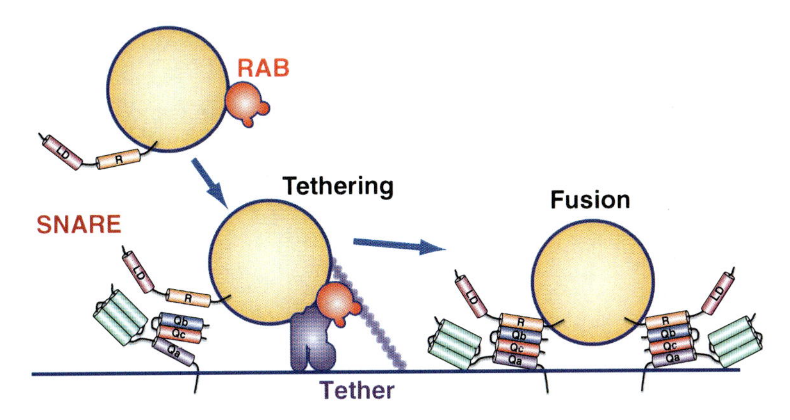

Fig. 1 Model which provides general mechanism of vesicle tethering and fusion. Transport vesicles are attached to the target membranes by some RAB effectors, which are called "tethers." Once the tethering step is accomplished, Qa-, Qb-, Qc- (or Qb+Qc- in the case of SNAP-25–like SNAREs), and R-SNAREs assemble into a tight complex to execute membrane fusion. *LD*, longin domain

2 Unique Aspects of Plant RAB and SNARE Proteins

Recent comparative genomic studies suggested that the acquisition of new organelles and diversification of membrane trafficking pathways are tightly associated with increases in the net number of genes encoding key regulators of membrane trafficking, including RAB and SNARE proteins (Dacks and Field 2007). After branching from other eukaryotic lineages, RABs and SNAREs successively expanded in the plant lineages during evolution, resulting in the current unique organization of plant RAB and SNARE molecules (Rutherford and Moore 2002; Vernoud et al. 2003; Sanderfoot 2007; Ebine and Ueda 2009; Ebine et al. 2012). In addition to conserved sets of RABs and SNAREs (core RABs and SNAREs), plants have acquired plant-unique RAB and SNARE proteins with distinctive structural features. The ARA6/RABF1 group (plant-unique RAB5 members) and VAMP727 members (R-SNARE with a unique structure) are obvious examples, demonstrating plant-unique innovations in organelle functions and membrane trafficking pathways. On the other hand, plants seem to recruit core RABs and SNAREs for higher order functions that are unique to plants. A distinctive expansion of several RAB and SNARE molecules, including RAB11 and SYP1, is evident in plant lineages, which probably reflects functional diversification in core RAB and SNARE groups (Rutherford and Moore 2002; Sanderfoot 2007). In the following sections, we discuss examples of the conserved and unique features of plant RAB and SNARE molecules that act in the post-Golgi trafficking network.

3 Conventional RAB Functioning in Diverse Plant Activities

In the genome of *Arabidopsis thaliana*, 57 RAB GTPases are encoded, which are classified into eight groups (RABA ~ RABH) (Rutherford and Moore 2002; Vernoud et al. 2003). Each of these eight groups exhibits high similarity to animal RAB1, RAB2, RAB5, RAB6, RAB7, RAB8, RAB11, or RAB18. All the genomes of land plants sequenced thus far essentially consist of these eight groups, with a few additions with unknown functions in basal plants (Rensing et al. 2008). There are numerous studies on RAB GTPases carried out using various plant species; thus, our knowledge of their functions is rapidly increasing. There are already comprehensive reviews on plant RAB functions (Nielsen et al. 2008, Woollard and Moore 2008; Saito and Ueda 2009); thus, in this chapter, we focus our attention on some post-Golgi RAB groups. A remarkable feature of plant RAB GTPases is an expanded RABA/RAB11 group; 26 of 57 RAB GTPases in *A. thaliana* belong to this group, which is further divided into six subgroups (Rutherford and Moore 2002). On the other hand, animals harbor only a few RAB11 members (two or three). The diversity in plant RABA members suggests plants assigned diverse functions to this group in a unique way during evolution. Recent studies have demonstrated this notion, as some RABA subgroups were shown to regulate several important biological processes unique to plants. For example, RABA2 and RABA3 are involved in cell plate formation, a plant unique mechanism of cytokinesis, and RABA1 and RABA4 groups are required for normal tip growth of pollen tubes and root hairs (Preuss et al. 2004; Graaf et al. 2005; Szumlanski and Nielsen 2009; Ovecka et al. 2010). Members of these RABA subgroups localize on the Golgi apparatus, *trans*-Golgi network (TGN), and small vesicles, and constitutively active mutants of these members sometimes target the plasma membrane (PM) (Ueda et al. 1996; Preuss et al. 2004; Graaf et al. 2005; Chow et al. 2008; Szumlanski and Nielsen 2009). Thus, RABA members appear to act on the trafficking pathway from the TGN to the PM. On the other hand, overexpression of the nucleotide free form of a RABA2 member in tobacco leaf cells perturbed transport of soluble proteins to vacuoles (Bottanelli et al. 2011); thus, some RABA members may act during the early secretory process, which is involved in both exocytic and vacuolar trafficking pathways. The plant TGN also acts as an early endosome (Dettmer et al. 2006; Viotti et al. 2010). Further studies are needed to determine if RABA members are involved in the endocytic pathway and to better understand the precise functions of RABA members. RABA groups may also have roles in fruit development, such as in the biogenesis and degradation of the cell wall during fruit softening (Zainal et al. 1996; Lu et al. 2001; Abbal et al. 2008; Lycett 2008).

4 ARA6/RABF1 Group, a Subfamily of RAB GTPases Unique to Plants

In addition to the remarkable expansion of RAB11 members, another distinctive characteristic of plant RAB GTPases is the existence of two distinct RAB5 groups. RAB5 is a member of the widely conserved core RAB GTPases, which regulate a wide variety of early endosomal trafficking events in animal cells (Somsel Rodman and Wandinger-Ness 2000; Benmerah 2004). Currently, all sequenced plant genomes contain *RAB5* genes orthologous to animal *RAB5*, with the exception of a unicellular rhodophyte, *Cyanidioschyzon merolae* (Matsuzaki et al. 2004). In addition, plants harbor unique RAB5 with a characteristic structure, the ARA6/RABF1 group. ARA6 group members are well conserved among land plants and were thought to be absent in more basal lineages of plants because the *ARA6*-like gene does not exist in the genomes of *Chlamydomonas reinhardtii* and *C. merolae*. However, a more ancient origin of ARA6 members is recently suggested based on dramatic progress in genome and expressed sequence tag (EST) sequencing (unpublished result).

Three *RAB5*-related genes exist in the *A. thaliana* genome: two orthologs of animal *RAB5* (*RHA1*/*RABF2a* and *ARA7*/*RABF2b*) and one plant-unique RAB5, *ARA6*/*RABF1* (Ueda et al. 2001). All products of these three genes are localized on multivesicular endosomes (MVEs) (Haas et al. 2007) and they are all activated by the same GEF, VPS9a (Goh et al. 2007). Thus, the two plant RAB5 groups appear to have redundant functions. However, localization patterns of ARA6 and conventional RAB5 (ARA7 and RHA1) do not overlap completely (Ueda et al. 2004). In addition, the constitutively active forms of ARA6 and ARA7 exert distinct effects on the partial loss-of-function mutant of *VPS9a* (Goh et al. 2007) when overexpressed, suggesting these two RAB5 groups have different functions. Moreover, inconsistent results have been reported for the effects of overexpression of dominant-negative ARA6 in protoplasts, while the function of conventional RAB5 in transporting soluble cargos to vacuoles was convincingly demonstrated in a similar system (Sohn et al. 2003; Bolte et al. 2004). Thus, it is unclear how the function of ARA6 diverges from conventional RAB5 and why plants possess the two types of RAB5.

Recently, a function of ARA6 was determined, which differs from conventional RAB5, using loss-of-function mutants of *A. thaliana*. Although the single mutants of each RAB5 member exhibited phenotypes that were indistinguishable from wild-type plants, mutations in *ARA6* and conventional *RAB5* resulted in totally different genetic interactions with a *syp22-1*/*vam3-1* mutant. The SYP22/VAM3 protein is an endosomal/vacuolar Qa-SNARE, whose loss-of-function mutant exhibits pleiotropic phenotypes including semi-dwarfism, late flowering, excessive accumulation of myrosinase, and delayed endocytosis (Ohtomo et al. 2005; Ueda et al. 2006; Ebine et al. 2011, 2012b;). Interestingly, the *ara6* mutation suppressed the phenotypes of *syp22-1*, while *ara7* and *rha1* exaggerated them (Ebine et al. 2011). This result clearly indicates that the two types of plant RAB5s have distinct functions in the trafficking pathways mediated by MVEs. Biochemical and imaging analyses further determined that ARA6 is required for efficient SNARE complex formation (and membrane fusion) at the PM, while conventional RAB5 is required for the trafficking pathway between MVEs and vacuoles (Ebine et al. 2011). These results indicate the plant-unique RAB5, ARA6, acts in the trafficking pathway from endosomes to the PM. Interestingly, this trafficking pathway also involves the R-SNARE molecule VAMP727, with a unique structural feature (Ebine and Ueda 2009). The *ara6* mutant is hypersensitive to salinity stress, and overexpression of constitutively active ARA6 confers salinity and osmotic stress tolerance to *A. thaliana* plants (Ebine et al. 2011, 2012a). The possible link between ARA6 group members and stress responses was also reported in other plant species (Bolte et al. 2000; Zhang et al. 2009); however, the precise functions of ARA6 homologs in plants other than *A. thaliana* have not yet been explored. It would be an interesting future project to examine whether the ARA6 function is shared by all ARA6 group members throughout plant lineages or if their functions are diversified during plant evolution.

5 Function of the RAB7/RABG Group in the Vacuolar Trafficking Pathway

RAB7 and its yeast homolog Ypt7 are known to act in the late endosomal/vacuolar trafficking pathway (Epp et al. 2011, Wang et al. 2011). In animal and yeast cells, RAB5 and RAB7 sequentially act in the same trafficking pathway, which is mediated by the conserved endosomal complexes, CORVET and HOPS (Rink et al. 2005; Epp et al. 2011). On the other hand, our knowledge of the plant RAB7 function is still sporadic, and the functional relevance between RAB5- and RAB7-dependent trafficking pathways remains largely unknown. RAB7 is localized mainly on vacuolar membranes in *A. thaliana* leaf epidermal cells (Saito et al. 2002) and substantial amounts were observed in *Medicago truncatula* root cells on MVE membranes (Limpens et al. 2009). RAB5 and RAB7 are localized on different endosomes with partial overlap in *M. truncatula*, which might reflect the sequential action of these RAB GTPases, as suggested by Bottanelli et al. (2012). However, it is also reported that overexpression of nucleotide-free RAB5 and RAB7 in tobacco leaf cells confer different effects on the transport of SYP22 to vacuolar membranes (Bottanelli et al. 2011). Furthermore, the phenotypes of two embryonic lethal mutants, *vps9a-1* and *vcl1*, of an activating factor for RAB5 and a shared subunit of HOPS and CORVET complexes, respectively, were different when observed by TEM (Rojo et al. 2001; Goh et al. 2007). These results strongly indicate that RAB5 and RAB7 act, at least partly, in different trafficking events in vacuolar transport.

6 SNAREs Regulating the Vacuolar Transport Pathway

The functions of SNARE proteins in the vacuolar trafficking pathway seem to be clearer than for RAB GTPases. A Qa-SNARE group, SYP2, acts between prevacuolar compartments (PVCs) and vacuoles. In *A. thaliana*, the SYP2 group consists of three genes: *SYP21/PEP12*, *SYP22/VAM3*, and *SYP23/PLP*. Immunogold electron microscopy and fluorescent protein tagging revealed that SYP21/PEP12 and SYP22/VAM3 are predominantly localized on the PVC and vacuolar membranes, respectively, with partial overlap. Initially, SYP21 and SYP22 were believed to have non-redundant functions (Sanderfoot et al. 2001b); however, recent studies indicated these two proteins can replace the function of each other (Shirakawa et al. 2010, Uemura et al. 2010). Interestingly, the *syp*22-3 mutation is also suppressed by overexpression of SYP23, which has no transmembrane domain. These results suggest that SYP2 members possess redundant functions, although distinct effects of overexpression are also reported for SYP21 and SYP22 (Foresti et al. 2006). Thus, each SYP2 member in *A. thaliana* could have an additional specialized function.

SYP22 forms a complex with VTI11 (Qb SNARE), SYP5 (Qc-SNARE), and VAMP727 (R-SNARE). This complex plays critical roles in protein trafficking between the PVC and vacuoles and in vacuole morphogenesis (Ebine et al. 2008). SYP21 is also reported to form a complex with VTI11 and SYP5 (Sanderfoot et al. 2001a). Interestingly, *SYP*22 and *VTI*11 were identified as the genes responsible for *sgr*3 and *zig/sgr*4 mutations, respectively, which result in defective shoot gravitropism (Morita et al. 2002; Yano et al. 2003). Thus, vacuolar traffic and/or vacuole integrity are essential for normal shoot gravitropism and future studies should reveal the precise mechanism. Another example in higher ordered plant functions is the R-SNARE VAMP711, which resides on the vacuolar membrane. Genetic suppression of *VAMP*711 gene resulted in elevated salinity stress tolerance, which could be attributed to a defective fusion between H_2O_2-containing vesicles and vacuoles (Leshem et al. 2006).

7 SNAREs in the Endocytic Pathway

As described above, the TGN acts as an early endosome in plants (Dettmer et al. 2006). RAB5-positive MVEs also play a role in the endocytic pathway, namely *en route* to the final destination for cargo degradation in vacuoles (Ueda et al. 2001; Takano et al. 2005). Thus, SNARE proteins residing on the TGN, MVE, and vacuoles appear to have roles in the endocytic pathway. In a consistent manner, the endocytic trafficking of PIN2-GFP and BOR1-GFP are defective in *syp*42*syp*43, a double mutant of TGN-resident Qa-SNAREs, and *syp*22, a mutant of a vacuolar Qa-SNARE (Ebine et al. 2011; Uemura et al. 2012). However, it is unclear whether other SNARE molecules localized to endocytic compartments are involved in endocytosis. Further studies involving bioimaging as well as biochemical and genetic analyses of other vacuolar and endosomal SNAREs will hopefully reveal their functions in the endocytic pathway.

8 SNAREs on the TGN Regulate Multiple Trafficking Pathways

In addition to the endocytic pathway, SNARE molecules on the TGN regulate other transport pathways. In the *syp*42*syp*43 double mutant, secGFP (GFP with a secretory signal peptide) and 12S globulin are not transported to their correct destinations, indicating impairments in secretory and vacuolar trafficking pathways (Uemura et al. 2012). In addition, the morphologies of the Golgi apparatus and TGN are also altered. These results suggest the SYP4 group is required in multiple trafficking pathways mediated by the post-Golgi network of membrane compartments. Consistent with its secretory role, SYP4 proteins are required for extracellular resistance responses to a fungal pathogen (Uemura et al. 2012). Moreover, SYP4 also functions in the protection of chloroplasts from salicylic acid-dependent biotic stress, but detailed mechanism should be elucidated in future studies.

In vivo SYP41 and SYP42 form SNARE complexes with Qb-VTI12 and Qc-SYP61 (Sanderfoot et al. 2001a) while R-YKT6 mediates membrane fusion with SYP41 and SYP61 in vitro (Chen et al. 2005). These results suggest SYP4, VTI12, SYP61, and YKT6 form a complex at the TGN to execute membrane fusion. However, loss-of-function mutants of *VTI12* and *SYP*61 exhibited phenotypes different from the *syp*4 mutant described above (Zhu et al. 2002; Surpin et al. 2003), suggesting these SNARE molecules have distinct roles in post-Golgi trafficking pathways. A recent proteomic study of the SYP61-positive TGN identified many PM-localized proteins (Drakakaki et al. 2012). Thus, SYP61 might be involved in vesicle fusions at the PM, in addition to vesicle fusions at the TGN.

9 Various Functions of SNAREs at the Plasma Membrane

In plant cells, vesicle-mediated secretion plays important roles in various biological events including generating a cell plate during cytokinesis, primary cell wall deposition, tip growth of pollen tubes and root hairs, and immune responses. Therefore, vesicular trafficking to the PM should be highly organized. The SYP1 group (Qa-SNARE), consisting of nine members, can localize to the PM in *A. thaliana* (Ueda et al. 2004), with various expression patterns (Enami et al. 2009; Silva et al. 2010; Ul-Rehman et al. 2011). This could reflect the functional diversity in this group. One SYP1 member, KNOLLE/SYP111, is responsible for membrane fusions in forming cell plates in coordination with SNAP33, a SNAP25-like Qb+Qc SNARE, and KEULE, a Sec1-Munc18 protein (Lukowitz et al. 1996; Lauber et al. 1997; Assaad et al. 2001, Heese et al. 2001). The R-SNARE involved in this process has not yet been identified, although VAMP721 and VAMP722, which localize preferentially to the expanding cell plates in dividing cells, could be candidates because the double mutant of these R-SNAREs exhibits a cytokinesis defect (Zhang et al. 2011). It is not known how KNOLLE is targeted to forming cell plates. KNOLLE is expressed in mitotic cells and predominantly localizes to developing cell plates during cytokinesis. When expressed constitutively, KNOLLE is targeted to the PM (Volker et al. 2001). Analyses using chimeric and truncated versions of KNOLLE indicated there is no specific targeting sequence in the KNOLLE polypeptide, suggesting the targeting of KNOLLE to forming cell plates is accomplished by a default sorting pathway without active sorting (Touihri et al. 2011).

SYP121, another SYP1 member, also provides important insights into SNARE functions in higher ordered plant functions. NtSYR1, an ortholog of *A. thaliana* SYP121 in tobacco, was identified during screening for abscisic acid-related signalling components (Leyman et al. 1999; Leyman et al. 2000). Overexpression of a cytosolic truncated fragment of NtSYR1 (Sp2) induced severe growth inhibition (Geelen et al. 2002), while distribution and trafficking dynamics of the KAT1 K^+ channel were perturbed by this fragment (Sutter et al. 2006). Thus, the distribution of KAT1 at subdomains of the PM appears to be mediated by the SYP121/SYR1-dependent trafficking pathway. SYP121 also interacts directly with the K^+ channel subunit KC1 at the unique FxRF motif within the SNARE domain and forms a tripartite complex with a second K^+ channel subunit, AKT1, to control channel gating and K^+ transport (Honsbein et al. 2009; Grefen et al. 2010).

SYP121 is also responsible for plant defense response against fungal pathogens. *SYP*121 was identified as a causal gene of the *pen*1 mutation, which results in elevated penetration of barley powdery mildew *Blumeria graminis* f. sp. *hordei* (*Bgh*) into non-host *A. thaliana* plants (Collins et al. 2003). In addition, SYP121 accumulates in papilla formed at the PM penetration sites. Although a mutant of *SYP*122 (the closest paralog of *SYP*121/*PEN*1) had more pronounced primary cell wall defects than the *pen*1 mutant, this mutant exhibited subtle or no defect in penetration resistance. However, the *pen1syp*122 double mutant demonstrated dwarf and necrotic phenotypes, suggesting overlapping functions of these genes encoding syntaxins (Assaad et al. 2004). VAMP721 and VAMP722 also focally accumulate at the fungal entry site and form ternary SNARE complexes with SNAP33 and SYP121/PEN1 (Kwon et al. 2008).

Orthologs of SYP132 seem to play pivotal roles in bacterial infection and symbiosis (Catalano et al. 2007, Kalde et al. 2007). Silencing of NbSYP132 in *Nicotiana benthamiana* impaired both effector-triggered and basal and salicylic acid-associated defenses, indicating SYP132-dependent secretion is responsible for multiple defense responses against bacterial pathogens (Kalde et al. 2007). In *M. truncatula*, MtSYP132 localizes to the PM surrounding infection threads and the infection droplet membrane, suggesting it acts in infection thread development and symbiosome formation (Catalano et al. 2007).

10 Functions of the Plant-Unique R-SNARE, VAMP727

Expansion of the R-SNARE group, VAMP7, is also a unique characteristic of the SNARE composition in plants (Sanderfoot 2007; Ebine and Ueda 2009). VAMP7 group consists of three subgroups, VAMP71, VAMP72, and VAMP727. Phylogenic analysis suggests VAMP71 is a prototype, from which VAMP72 and VAMP727 were derived (Sanderfoot 2007). VAMP727 harbors a unique structural characteristic: an acidic amino acid-rich insertion in the N-terminal longin domain (Ebine and Ueda 2009, Vedovato et al. 2009). This type of VAMP7 member has only been found in seed plants; thus, this is another example, in addition to ARA6, of the innovations of plants in membrane trafficking. VAMP727 localizes on the MVE with RAB5 members, and mediates membrane fusion between the PVC and vacuolar membranes by forming a complex with Qa-SYP22, Qb-VTI11, and Qc-SYP51 (Ebine et al. 2008). This complex is essential for efficient transport of storage proteins to protein storage vacuoles in maturating seeds. The double loss-of-function mutation of *VAMP*727 and *SYP*22 results in defective secretion of storage proteins and impaired desiccation tolerance.

VAMP727 also mediates membrane fusion at the PM by forming a complex with Qa-SYP121 (Ebine et al. 2011). Intriguingly, assembly of this SNARE complex is under regulation of ARA6. Plants might have developed a novel trafficking pathway from the MVE to the PM through the acquisition of these two plant-unique molecules, as proposed by Dacks and Field (2007).

11 Conclusions and Perspectives

As summarized above, the plant post-Golgi trafficking network is organized differently from animals and yeasts, although, homologous molecular components are employed. Recent studies revealed unique trafficking pathways and molecular machineries that take part in various higher order plant functions, where RAB and SNARE proteins play fundamental roles. Nevertheless, most of the upstream and downstream regulations of these molecules remain to be elucidated. It is also still unclear whether specific cargos are transported via specific pathways and how cargo transport leads to the specific output. Moreover, the functional relevance between RABs and SNARE complexes is still largely unexamined. Further genetic, biochemical, and bioimaging studies as well as systems biology approaches (e.g., genomics and proteomics) are needed to gain a better understanding of post-Golgi trafficking pathways. In this respect, recent combined proteomic and cell biology study by Takáč et al. (2012) revealed wortmannin induced consumption of TGN by MVB accompanied by downregulation of RABA1d while the same protein was upregulated by brefeldin A treatment (Takáč et al. 2011).

References

Abbal P, Pradal M, Muniz L, Sauvage FX, Chatelet P, Ueda T, Tesniere C (2008) Molecular characterization and expression analysis of the Rab GTPase family in *Vitis vinifera* reveal the specific expression of a VvRabA protein. J Exp Bot 59(9):2403–2416. doi:10.1093/jxb/ern132

Assaad FF, Huet Y, Mayer U, Jurgens G (2001) The cytokinesis gene KEULE encodes a Sec1 protein that binds the syntaxin KNOLLE. J cell biol 152(3):531–543

Assaad FF, Qiu JL, Youngs H, Ehrhardt D, Zimmerli L, Kalde M, Wanner G, Peck SC, Edwards H, Ramonell K, Somerville CR, Thordal-Christensen H (2004) The PEN1 syntaxin defines a novel cellular compartment upon fungal attack and is required for the timely assembly of papillae. Mol Biol Cell 15(11):5118–5129. doi:10.1091/mbc.E04-02-0140

Benmerah A (2004) Endocytosis: signaling from endocytic membranes to the nucleus. Curr Biol 14(8):R314–R316

Bolte S, Brown S, Satiat-Jeunemaitre B (2004) The N-myristoylated Rab-GTPase m-Rabmc is involved in post-Golgi trafficking events to the lytic vacuole in plant cells. J Cell Sci 117(Pt 6):943–954

Bolte S, Schiene K, Dietz KJ (2000) Characterization of a small GTP-binding protein of the rab 5 family in *Mesembryanthemum crystallinum* with increased level of expression during early salt stress. Plant Mol Biol 42(6):923–936

Bottanelli F, Foresti O, Hanton S, Denecke J (2011) Vacuolar transport in tobacco leaf epidermis cells involves a single route for soluble cargo and multiple routes for membrane cargo. Plant Cell 23(8):3007–3025. doi:10.1105/tpc.111.085480

Bottanelli F, Gershlick DC, Denecke J (2012) Evidence for sequential action of Rab5 and Rab7 GTPases in prevacuolar organelle partitioning. Traffic 13(2):338-354

Cai H, Reinisch K, Ferro-Novick S (2007) Coats, tethers, Rabs, and SNAREs work together to mediate the intracellular destination of a transport vesicle. Dev Cell 12(5):671–682

Camacho L, Smertenko AP, Perez-Gomez J, Hussey PJ, Moore I (2009) Arabidopsis Rab-E GTPases exhibit a novel interaction with a plasma-membrane phosphatidylinositol-4-phosphate 5-kinase. J Cell Sci 122(Pt 23):4383–4392. doi:10.1242/jcs.053488

Catalano CM, Czymmek KJ, Gann JG, Sherrier DJ (2007) *Medicago truncatula* syntaxin SYP132 defines the symbiosome membrane and infection droplet membrane in root nodules. Planta 225(3):541–550. doi:10.1007/s00425-006-0369-y

Chen Y, Shin YK, Bassham DC (2005) YKT6 is a core constituent of membrane fusion machineries at the Arabidopsis *trans*-Golgi network. J Mol Biol 350(1):92–101. doi:10.1016/j.jmb.2005.04.061

Chow CM, Neto H, Foucart C, Moore I (2008) Rab-A2 and Rab-A3 GTPases define a *trans*-Golgi endosomal membrane domain in Arabidopsis that contributes substantially to the cell plate. Plant Cell 20(1):101–123

Collins NC, Thordal-Christensen H, Lipka V, Bau S, Kombrink E, Qiu JL, Huckelhoven R, Stein M, Freialdenhoven A, Somerville SC, Schulze-Lefert P (2003) SNARE-protein-mediated disease resistance at the plant cell wall. Nature 425(6961):973–977. doi:10.1038/nature02076

Dacks JB, Field MC (2007) Evolution of the eukaryotic membrane-trafficking system: origin, tempo and mode. J Cell Sci 120(Pt 17):2977–2985

de Graaf BH, Cheung AY, Andreyeva T, Levasseur K, Kieliszewski M, Wu HM (2005) Rab11 GTPase-regulated membrane trafficking is crucial for tip-focused pollen tube growth in tobacco. Plant Cell 17(9):2564–2579

Dettmer J, Hong-Hermesdorf A, Stierhof YD, Schumacher K (2006) Vacuolar H+-ATPase activity is required for endocytic and secretory trafficking in Arabidopsis. Plant Cell 18(3):715–730. doi:10.1105/tpc.105.037978

Drakakaki G, van de Ven W, Pan S, Miao Y, Wang J, Keinath NF, Weatherly B, Jiang L, Schumacher K, Hicks G, Raikhel N (2012) Isolation and proteomic analysis of the SYP61 compartment reveal its role in exocytic trafficking in Arabidopsis. Cell Res 22(2):413–424. doi:10.1038/cr.2011.129

Ebine K, Fujimoto M, Okatani Y, Nishiyama T, Goh T, Ito E, Dainobu T, Nishitani A, Uemura T, Sato MH, Thordal-Christensen H, Tsutsumi N, Nakano A, Ueda T (2011) A membrane trafficking pathway regulated by the plant-specific RAB GTPase ARA6. Nat Cell Biol 13(7):853–859. doi:10.1038/ncb2270

Ebine K, Miyakawa N, Fujimoto M, Uemura T, Nakano A, Ueda T (2012a) An endosomal trafficking pathway regulated by the plant-unique RAB5, ARA6. Small GTPases 3(1):1–5. doi:10.1038/ncb2270

Ebine K, Uemura T, Nakano A, Ueda T (2012b) Flowering time modulation by a vacuolar SNARE via *FLOWERING LOCUS C* in *Arabidopsis thaliana*. PLoS ONE 7(7):e42239

Ebine K, Okatani Y, Uemura T, Goh T, Shoda K, Niihama M, Morita MT, Spitzer C, Otegui MS, Nakano A, Ueda T (2008) A SNARE complex unique to seed plants is required for protein storage vacuole biogenesis and seed development of *Arabidopsis thaliana*. Plant Cell 20(11):3006–3021. doi:10.1105/tpc.107.057711

Ebine K, Ueda T (2009) Unique mechanism of plant endocytic/vacuolar transport pathways. J Plant Res 122(1):21–30. doi:10.1007/s10265-008-0200-x

Enami K, Ichikawa M, Uemura T, Kutsuna N, Hasezawa S, Nakagawa T, Nakano A, Sato MH (2009) Differential expression control and polarized distribution of plasma membrane-resident SYP1 SNAREs in *Arabidopsis thaliana*. Plant Cell Physiol 50(2):280–289. doi:10.1093/pcp/pcn197

Epp N, Rethmeier R, Kramer L, Ungermann C (2011) Membrane dynamics and fusion at late endosomes and vacuoles–Rab regulation, multisubunit tethering complexes and SNAREs. Eur J Cell Biol 90(9):779–785. doi:10.1016/j.ejcb.2011.04.007

Foresti O, daSilva LL, Denecke J (2006) Overexpression of the Arabidopsis syntaxin PEP12/SYP21 inhibits transport from the prevacuolar compartment to the lytic vacuole in vivo. Plant Cell 18(9):2275–2293. doi:10.1105/tpc.105.040279

Geelen D, Leyman B, Batoko H, Di Sansebastiano GP, Moore I, Blatt MR (2002) The abscisic acid-related SNARE homolog NtSyr1 contributes to secretion and growth: evidence from competition with its cytosolic domain. Plant Cell 14(2):387–406

Goh T, Uchida W, Arakawa S, Ito E, Dainobu T, Ebine K, Takeuchi M, Sato K, Ueda T, Nakano A (2007) VPS9a, the common activator for two distinct types of Rab5 GTPases, is essential for the development of *Arabidopsis thaliana*. Plant Cell 19(11):3504–3515. doi:10.1105/tpc.107.053876

Grefen C, Chen Z, Honsbein A, Donald N, Hills A, Blatt MR (2010) A novel motif essential for SNARE interaction with the K(+) channel KC1 and channel gating in Arabidopsis. Plant Cell 22(9):3076–3092. doi:10.1105/tpc.110.077768

Haas TJ, Sliwinski MK, Martinez DE, Preuss M, Ebine K, Ueda T, Nielsen E, Odorizzi G, Otegui MS (2007) The Arabidopsis AAA ATPase SKD1 is involved in multivesicular endosome function and interacts with its positive regulator LYST-INTERACTING PROTEIN5. Plant Cell 19(4):1295–1312. doi:10.1105/tpc.106.049346

Heese M, Gansel X, Sticher L, Wick P, Grebe M, Granier F, Jurgens G (2001) Functional characterization of the KNOLLE-interacting t-SNARE AtSNAP33 and its role in plant cytokinesis. J Cell Biol 155(2):239–249. doi:10.1083/jcb.200107126

Honsbein A, Sokolovski S, Grefen C, Campanoni P, Pratelli R, Paneque M, Chen Z, Johansson I, Blatt MR (2009) A tripartite SNARE-K+ channel complex mediates in channel-dependent K+ nutrition in Arabidopsis. Plant Cell 21(9):2859–2877. doi:10.1105/tpc.109.066118

Ishikawa T, Machida C, Yoshioka Y, Ueda T, Nakano A, Machida Y (2008) EMBRYO YELLOW gene, encoding a subunit of the conserved oligomeric Golgi complex, is required for appropriate cell expansion and meristem organization in *Arabidopsis thaliana*. Genes to cells: devoted to molecular and cellular mechanisms 13 (6):521–535. doi:10.1111/j.1365-2443.2008.01186.x

Jahn R, Scheller RH (2006) SNAREs–engines for membrane fusion. Nat Rev Mol Cell Biol 7(9):631–643

Kalde M, Nuhse TS, Findlay K, Peck SC (2007) The syntaxin SYP132 contributes to plant resistance against bacteria and secretion of pathogenesis-related protein 1. Proc Nat Acad Sci USA 104(28):11850–11855. doi:10.1073/pnas.0701083104

Kwon C, Neu C, Pajonk S, Yun HS, Lipka U, Humphry M, Bau S, Straus M, Kwaaitaal M, Rampelt H, El Kasmi F, Jurgens G, Parker J, Panstruga R, Lipka V, Schulze-Lefert P (2008) Co-option of a default secretory pathway for plant immune responses. Nature 451(7180):835–840. doi:10.1038/nature06545

Lauber MH, Waizenegger I, Steinmann T, Schwarz H, Mayer U, Hwang I, Lukowitz W, Jurgens G (1997) The Arabidopsis KNOLLE protein is a cytokinesis-specific syntaxin. J cell biol 139(6):1485–1493

Leshem Y, Melamed-Book N, Cagnac O, Ronen G, Nishri Y, Solomon M, Cohen G, Levine A (2006) Suppression of Arabidopsis vesicle-SNARE expression inhibited fusion of H2O2-containing vesicles with tonoplast and increased salt tolerance. Proc Nat Acad Sci U S A 103(47):18008–18013. doi:10.1073/pnas.0604421103

Leyman B, Geelen D, Blatt MR (2000) Localization and control of expression of Nt-Syr1, a tobacco SNARE protein. Plant J 24(3):369–381

Leyman B, Geelen D, Quintero FJ, Blatt MR (1999) A tobacco syntaxin with a role in hormonal control of guard cell ion channels. Science 283(5401):537–540

Limpens E, Ivanov S, van Esse W, Voets G, Fedorova E, Bisseling T (2009) Medicago N2-fixing symbiosomes acquire the endocytic identity marker Rab7 but delay the acquisition of vacuolar identity. Plant Cell 21(9):2811–2828. doi:10.1105/tpc.108.064410

Lu C, Zainal Z, Tucker GA, Lycett GW (2001) Developmental abnormalities and reduced fruit softening in tomato plants expressing an antisense Rab11 GTPase gene. Plant Cell 13(8):1819–1833

Lukowitz W, Mayer U, Jurgens G (1996) Cytokinesis in the Arabidopsis embryo involves the syntaxin-related KNOLLE gene product. Cell 84(1):61–71

Lycett G (2008) The role of Rab GTPases in cell wall metabolism. J Exp Bot 59(15):4061–4074. doi:10.1093/jxb/ern255

Markgraf DF, Peplowska K, Ungermann C (2007) Rab cascades and tethering factors in the endomembrane system. FEBS Lett 581(11):2125–2130

Matsuzaki M, Misumi O, Shin IT, Maruyama S, Takahara M, Miyagishima SY, Mori T, Nishida K, Yagisawa F, Yoshida Y, Nishimura Y, Nakao S, Kobayashi T, Momoyama Y, Higashiyama T, Minoda A, Sano M, Nomoto H, Oishi K, Hayashi H, Ohta F, Nishizaka S, Haga S, Miura S, Morishita T, Kabeya Y, Terasawa K, Suzuki Y, Ishii Y, Asakawa S, Takano H, Ohta N, Kuroiwa H, Tanaka K, Shimizu N, Sugano S, Sato N, Nozaki H, Ogasawara N, Kohara Y, Kuroiwa T (2004) Genome sequence of the ultrasmall unicellular red alga *Cyanidioschyzon merolae* 10D. Nature 428(6983):653–657. doi:10.1038/nature02398

Morita MT, Kato T, Nagafusa K, Saito C, Ueda T, Nakano A, Tasaka M (2002) Involvement of the vacuoles of the endodermis in the early process of shoot gravitropism in Arabidopsis. Plant Cell 14(1):47–56

Nielsen E, Cheung AY, Ueda T (2008) The regulatory RAB and ARF GTPases for vesicular trafficking. Plant Physiol 147(4):1516–1526. doi:10.1104/pp.108.121798

Novick P, Medkova M, Dong G, Hutagalung A, Reinisch K, Grosshans B (2006) Interactions between Rabs, tethers, SNAREs and their regulators in exocytosis. Biochem Soc Trans 34(Pt 5):683–686. doi:10.1042/BST0340683

Ohtomo I, Ueda H, Shimada T, Nishiyama C, Komoto Y, Hara-Nishimura I, Takahashi T (2005) Identification of an allele of VAM3/SYP22 that confers a semi-dwarf phenotype in *Arabidopsis thaliana*. Plant Cell Physiol 46(8):1358–1365. doi:10.1093/pcp/pci146

Ovečka M, Berson T, Beck M, Derksen J, Šamaj J, Baluška F, Lichtscheidl IK (2010) Structural sterols are involved in both the initiation and tip growth of root hairs in *Arabidopsis thaliana*. Plant Cell 22(9):2999–3019. doi:10.1105/tpc.109.069880

Preuss ML, Schmitz AJ, Thole JM, Bonner HK, Otegui MS, Nielsen E (2006) A role for the RabA4b effector protein PI-4Kbeta1 in polarized expansion of root hair cells in *Arabidopsis thaliana*. J Cell Biol 172(7):991–998

Preuss ML, Serna J, Falbel TG, Bednarek SY, Nielsen E (2004) The Arabidopsis Rab GTPase RabA4b localizes to the tips of growing root hair cells. Plant Cell 16(6):1589–1603. doi:10.1105/tpc.021634

Rensing SA, Lang D, Zimmer AD, Terry A, Salamov A, Shapiro H, Nishiyama T, Perroud PF, Lindquist EA, Kamisugi Y, Tanahashi T, Sakakibara K, Fujita T, Oishi K, Shin IT, Kuroki Y, Toyoda A, Suzuki Y, Hashimoto S, Yamaguchi K, Sugano S, Kohara Y, Fujiyama A, Anterola A, Aoki S, Ashton N, Barbazuk WB, Barker E, Bennetzen JL, Blankenship R, Cho SH, Dutcher SK, Estelle M, Fawcett JA, Gundlach H, Hanada K, Heyl A, Hicks KA, Hughes J, Lohr M, Mayer K, Melkozernov A, Murata T, Nelson DR, Pils B, Prigge M, Reiss B, Renner T, Rombauts S, Rushton PJ, Sanderfoot A, Schween G, Shiu SH, Stueber K, Theodoulou FL, Tu H, Van de Peer Y, Verrier PJ, Waters E, Wood A, Yang L, Cove D, Cuming AC, Hasebe M, Lucas S, Mishler BD, Reski R, Grigoriev IV, Quatrano RS, Boore JL (2008) The Physcomitrella genome reveals evolutionary insights into the conquest of land by plants. Science 319(5859):64–69. doi:10.1126/science.1150646

Rink J, Ghigo E, Kalaidzidis Y, Zerial M (2005) Rab conversion as a mechanism of progression from early to late endosomes. Cell 122(5):735–749. doi:10.1016/j.cell.2005.06.043

Rojo E, Gillmor CS, Kovaleva V, Somerville CR, Raikhel NV (2001) VACUOLELESS1 is an essential gene required for vacuole formation and morphogenesis in Arabidopsis. Dev Cell 1(2):303–310

Rutherford S, Moore I (2002) The Arabidopsis Rab GTPase family: another enigma variation. Curr Opin Plant Biol 5(6):518–528

Saito C, Ueda T (2009) Functions of RAB and SNARE proteins in plant life. Int rev cell mol biol 274:183–233. doi:10.1016/S1937-6448(08)02004-2

Saito C, Ueda T, Abe H, Wada Y, Kuroiwa T, Hisada A, Furuya M, Nakano A (2002) A complex and mobile structure forms a distinct subregion within the continuous vacuolar membrane in young cotyledons of Arabidopsis. Plant J 29(3):245–255

Sanderfoot A (2007) Increases in the number of SNARE genes parallels the rise of multicellularity among the green plants. Plant Physiol 144(1):6–17. doi:10.1104/pp.106.092973

Sanderfoot AA, Kovaleva V, Bassham DC, Raikhel NV (2001a) Interactions between syntaxins identify at least five SNARE complexes within the Golgi/prevacuolar system of the Arabidopsis cell. Mol Biol Cell 12(12):3733–3743

Sanderfoot AA, Pilgrim M, Adam L, Raikhel NV (2001b) Disruption of individual members of Arabidopsis syntaxin gene families indicates each has essential functions. Plant Cell 13(3):659–666

Shirakawa M, Ueda H, Shimada T, Koumoto Y, Shimada TL, Kondo M, Takahashi T, Okuyama Y, Nishimura M, Hara-Nishimura I (2010) Arabidopsis Qa-SNARE SYP2 proteins localized to different subcellular regions function redundantly in vacuolar protein sorting and plant development. Plant J 64(6):924–935. doi:10.1111/j.1365-313X.2010.04394.x

Silva PA, Ul-Rehman R, Rato C, Di Sansebastiano GP, Malho R (2010) Asymmetric localization of Arabidopsis SYP124 syntaxin at the pollen tube apical and sub-apical zones is involved in tip growth. BMC Plant Biol 10:179. doi:10.1186/1471-2229-10-179

Sohn EJ, Kim ES, Zhao M, Kim SJ, Kim H, Kim YW, Lee YJ, Hillmer S, Sohn U, Jiang L, Hwang I (2003) Rha1, an Arabidopsis Rab5 homolog, plays a critical role in the vacuolar trafficking of soluble cargo proteins. Plant Cell 15(5):1057–1070

Somsel Rodman J, Wandinger-Ness A (2000) Rab GTPases coordinate endocytosis. J Cell Sci 113(Pt 2):183–192

Surpin M, Zheng H, Morita MT, Saito C, Avila E, Blakeslee JJ, Bandyopadhyay A, Kovaleva V, Carter D, Murphy A, Tasaka M, Raikhel N (2003) The VTI family of SNARE proteins is necessary for plant viability and mediates different protein transport pathways. Plant Cell 15(12):2885–2899. doi:10.1105/tpc.016121

Sutter JU, Campanoni P, Tyrrell M, Blatt MR (2006) Selective mobility and sensitivity to SNAREs is exhibited by the Arabidopsis KAT1 K+ channel at the plasma membrane. Plant Cell 18(4):935–954. doi:10.1105/tpc.105.038950

Szumlanski AL, Nielsen E (2009) The Rab GTPase RabA4d regulates pollen tube tip growth in *Arabidopsis thaliana*. Plant Cell 21(2):526–544. doi:10.1105/tpc.108.060277

Takáč T, Pechan T, Richter H, Muller J, Eck C, Bohm N, Obert B, Ren H, Niehaus K, Šamaj J (2011) Proteomics on brefeldin A-treated Arabidopsis roots reveals profilin 2 as a new protein involved in the cross-talk between vesicular trafficking and the actin cytoskeleton. J Proteome Res 10(2):488–501. doi:10.1021/pr100690f

Takáč T, Pechan T, Šamajová O, Ovečka M, Richter H, Eck C, Niehaus K, Šamaj J (2012) Wortmannin treatment induces changes in Arabidopsis root proteome and post-Golgi compartments. J Proteome Res. doi:10.1021/pr201111n

Takano J, Miwa K, Yuan L, von Wiren N, Fujiwara T (2005) Endocytosis and degradation of BOR1, a boron transporter of *Arabidopsis thaliana*, regulated by boron availability. Proc Nat Acad Sci U S A 102(34):12276–12281

Touihri S, Knoll C, Stierhof YD, Muller I, Mayer U, Jurgens G (2011) Functional anatomy of the Arabidopsis cytokinesis-specific syntaxin KNOLLE. Plant J 68(5):755–764. doi:10.1111/j.1365-313X.2011.04736.x

Ueda H, Nishiyama C, Shimada T, Koumoto Y, Hayashi Y, Kondo M, Takahashi T, Ohtomo I, Nishimura M, Hara-Nishimura I (2006) AtVAM3 is required for normal specification of idioblasts, myrosin cells. Plant Cell Physiol 47(1):164–175

Ueda T, Anai T, Tsukaya H, Hirata A, Uchimiya H (1996) Characterization and subcellular localization of a small GTP-binding protein (Ara-4) from Arabidopsis: conditional expression under control of the promoter of the gene for heat-shock protein HSP81-1. Mol Gen Genet 250(5):533–539

Ueda T, Uemura T, Sato MH, Nakano A (2004) Functional differentiation of endosomes in Arabidopsis cells. Plant J 40(5):783–789. doi:10.1111/j.1365-313X.2004.02249.x

Ueda T, Yamaguchi M, Uchimiya H, Nakano A (2001) Ara6, a plant-unique novel type Rab GTPase, functions in the endocytic pathway of *Arabidopsis thaliana*. EMBO J 20(17):4730–4741

Uemura T, Kim H, Saito C, Ebine K, Ueda T, Schulze-Lefert P, Nakano A (2012) Qa-SNAREs localized to the *trans*-Golgi network regulate multiple transport pathways and extracellular disease resistance in plants. Proc Nat Acad Sci U S A 109(5):1784–1789. doi:10.1073/pnas.1115146109

Uemura T, Morita MT, Ebine K, Okatani Y, Yano D, Saito C, Ueda T, Nakano A (2010) Vacuolar/pre-vacuolar compartment Qa-SNAREs VAM3/SYP22 and PEP12/SYP21 have interchangeable functions in Arabidopsis. Plant J 64(5):864–873. doi:10.1111/j.1365-313X.2010.04372.x

Ul-Rehman R, Silva PA, Malho R (2011) Localization of Arabidopsis SYP125 syntaxin in the plasma membrane sub-apical and distal zones of growing pollen tubes. Plant Sign Behav 6(5):665–670

Vedovato M, Rossi V, Dacks JB, Filippini F (2009) Comparative analysis of plant genomes allows the definition of the "Phytolongins": a novel non-SNARE longin domain protein family. BMC Genomics 10:510. doi:10.1186/1471-2164-10-510

Vernoud V, Horton AC, Yang Z, Nielsen E (2003) Analysis of the small GTPase gene superfamily of Arabidopsis. Plant Physiol 131(3):1191–1208. doi:10.1104/pp.013052

Viotti C, Bubeck J, Stierhof YD, Krebs M, Langhans M, van den Berg W, van Dongen W, Richter S, Geldner N, Takano J, Jurgens G, de Vries SC, Robinson DG, Schumacher K (2010) Endocytic and secretory traffic in Arabidopsis merge in the *trans*-Golgi network/early endosome, an independent and highly dynamic organelle. Plant Cell 22(4):1344–1357. doi:10.1105/tpc.109.072637

Volker A, Stierhof YD, Jurgens G (2001) Cell cycle-independent expression of the Arabidopsis cytokinesis-specific syntaxin KNOLLE results in mistargeting to the plasma membrane and is not sufficient for cytokinesis. J Cell Sci 114(Pt 16):3001–3012

Wang T, Ming Z, Xiaochun W, Hong W (2011) Rab7: role of its protein interaction cascades in endo-lysosomal traffic. Cell Signal 23(3):516–521. doi:10.1016/j.cellsig.2010.09.012

Warren G, Mellman I (2006) Protein trafficking between membranes. In: Lewin B (ed) Cell, 1st edn. Jones & Bartlett Pub, Sudbury, pp 153–204

Wickner W, Schekman R (2008) Membrane fusion. Nat Struct Mol Biol 15(7):658–664

Woollard AA, Moore I (2008) The functions of Rab GTPases in plant membrane traffic. Curr Opin Plant Biol 11(6):610–619. doi:10.1016/j.pbi.2008.09.010

Yano D, Sato M, Saito C, Sato MH, Morita MT, Tasaka M (2003) A SNARE complex containing SGR3/AtVAM3 and ZIG/VTI11 in gravity-sensing cells is important for Arabidopsis shoot gravitropism. Proc Nat Acad Sci U S A 100(14):8589–8594. doi:10.1073/pnas.1430749100

Zainal Z, Tucker GA, Lycett GW (1996) A rab11-like gene is developmentally regulated in ripening mango (*Mangifera indica* L.) fruit. Biochim Biophys Acta 1314(3):187–190

Zhang L, Tian LH, Zhao JF, Song Y, Zhang CJ, Guo Y (2009) Identification of an apoplastic protein involved in the initial phase of salt stress response in rice root by two-dimensional electrophoresis. Plant Physiol 149(2):916–928

Zhang L, Zhang H, Liu P, Hao H, Jin JB, Lin J (2011) Arabidopsis R-SNARE proteins VAMP721 and VAMP722 are required for cell plate formation. PLoS One 6(10):e26129. doi:10.1371/journal.pone.0026129

Zhu J, Gong Z, Zhang C, Song CP, Damsz B, Inan G, Koiwa H, Zhu JK, Hasegawa PM, Bressan RA (2002) OSM1/SYP61: a syntaxin protein in Arabidopsis controls abscisic acid-mediated and non-abscisic acid-mediated responses to abiotic stress. Plant Cell 14(12):3009–3028

SCAMP, VSR, and Plant Endocytosis

Angus Ho Yin Law, Jinbo Shen and Liwen Jiang

Abstract In plants, secretory carrier membrane proteins (SCAMPs) are integral membrane proteins localized to plasma membrane (PM) and *trans*-Golgi network (TGN)/early endosome, whereas vacuolar sorting receptors (VSRs) are type I integral membrane proteins localized to TGN and prevacuolar compartment (PVC)/late endosome in plant cells. Recent studies demonstrated that SCAMPs and VSRs exhibited distinct patterns of distribution in growing pollen tube and both proteins are found to be essential for pollen tube growth. Interestingly, similar to SCAMPs, VSRs are also found to locate in PM in pollen tube, indicating a possible role in endocytosis. This chapter summarizes the recent findings on plant SCAMPs and VSRs in respect to their possible roles in endocytic trafficking within plant cells.

Keywords Endocytosis · Endosome · SCAMP · VSR · *Trans*-Golgi network · Plasma membrane

Abbreviations

Arf	ADP-ribosylation factor
AP	Adaptor protein
ARF-GEF	ADP-ribosylation factor guanine nucleotide exchange factors
BFA	Brefeldin A
CCV	Clathrin-coated vesicle
CD3G	T cell surface glycoprotein CD3 gamma chain
ConcA	Concanamycin A

A. H. Y. Law and J. Shen contributed equally to this work.

A. H. Y. Law · J. Shen · L. Jiang (✉)
School of Life Sciences, Centre for Cell and Developmental Biology, The Chinese University of Hong Kong, Shatin, New Territories, Hong Kong, China
e-mail: ljiang@cuhk.edu.hk

J. Šamaj (ed.), *Endocytosis in Plants*,
DOI: 10.1007/978-3-642-32463-5_11,

CT	Cytoplasmic tail
EH	Eps15-homology
EMP12	Endomembrane protein 12
ER	Endoplasmic reticulum
Immuno-EM	Immuno-electron microscopy
MPR	Mannose-6-phosphate receptor
NPF	Asn-Pro-Phe
PM	Plasma membrane
PVC	Prevacuolar compartment
SCAMP	Secretory carrier membrane protein
SP	Signal peptide
TMD	Transmembrane domain
TGN	*Trans*-Golgi network
VSR	Vacuolar sorting receptor

1 Introduction

Endocytosis is a fundamental and complex process supporting many essential cellular functions in eukaryotes, including nutrient uptake, receptor signalling and downregulation, and defense against pathogens (Miaczynska and Stenmark 2008). This process consists of membrane invagination, budding, and formation of transport vesicles directed to an intracellular membrane-bound organelle specialized in receiving internalized materials, known as early endosome (Lam et al. 2007a, b; Traub 2009). In mammalian cells, tubular membrane of the early endosome is recycled back to the plasma membrane (PM), while the remaining vesicular membrane matures into a late endosome with increased acidification prior to its fusion with lysosome (Geldner and Jurgens 2006; Grant and Donaldson 2009).

Plant cells share a similar endomembrane system as in animal cells and the operation of endocytosis in plant cells has been already well recognized (Šamaj et al. 2005; Lam et al. 2007a, b; Otegui and Spitzer 2008; Robinson et al. 2008; Cai et al. 2012). Over the past years, significant progress has been achieved in our understanding of molecular components regulating plant endocytosis as well as the nature and identity of endosomes in plant cells (Fig. 1) (Tse et al. 2004; Dettmer et al. 2006; Lam et al. 2007a; Müller et al. 2007; Richter et al. 2007). Vesicular trafficking and organelle maturation of the endosomal system have been usually studied with organelle markers that are key players of vesicular transport, such as ADP-ribosylation factor (Arf) and Rab (Ueda et al. 2001; Preuss et al. 2004; Chow et al. 2008). In their GTP-bound stage, these proteins are membrane bound and function in regulating vesicular transport, while in GDP-bound stage they are recycled back to the cytosol (Zerial and McBride 2001; Behnia and Munro 2005). Other studies utilized integral membrane proteins as stable and reliable organelle

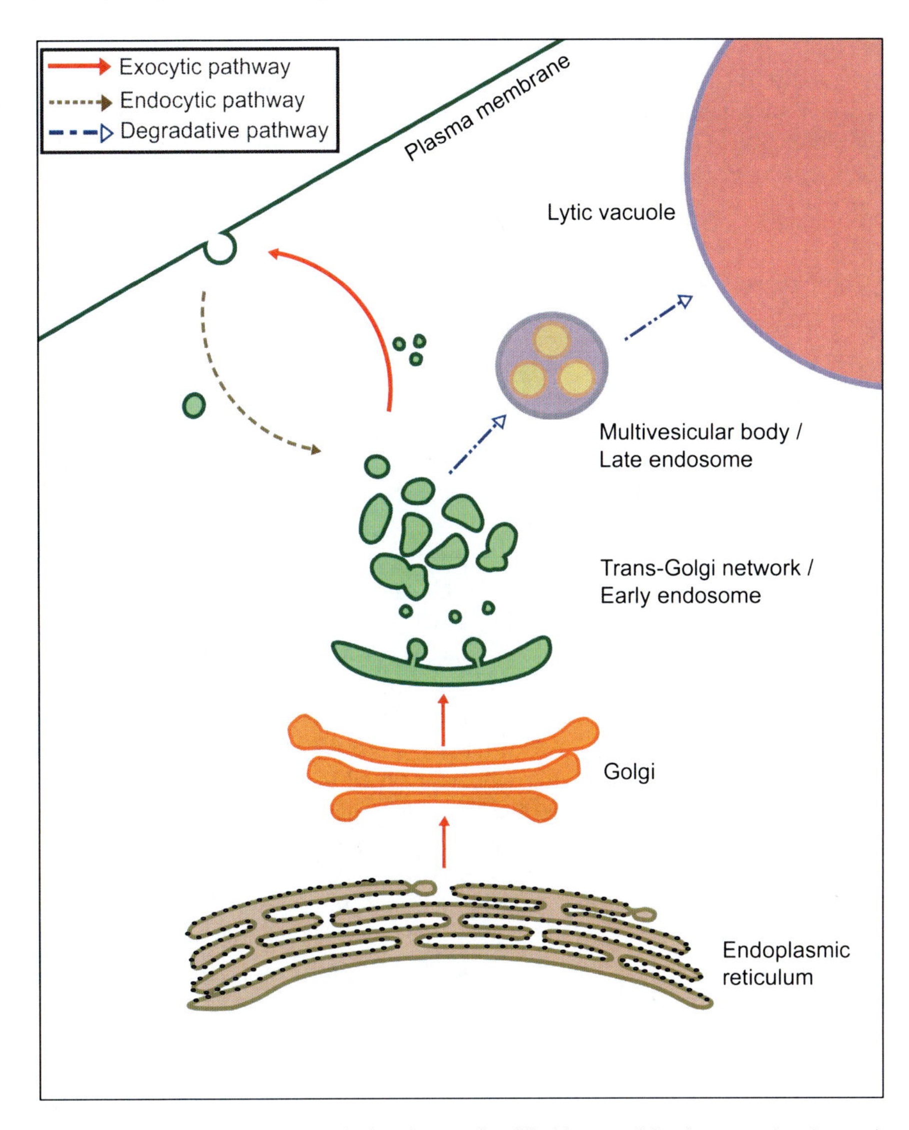

Fig. 1 Endocytosis and exocytosis in plant cells. Working model of conventional protein transport pathways in the plant endomembrane system. In exocytic pathways (*red solid arrow*), proteins synthesized in rough endoplasmic reticulum (*ER*) are transported via the Golgi apparatus and *trans*-Golgi network (*TGN*) secretory vesicles toward the plasma membrane or via prevacuolar compartments (*PVC*)/multivesicular bodies (*MVBs*) toward vacuoles. In the endocytic pathways (*brown dashed arrow*), proteins are internalized from the plasma membrane or extracellular space and first reach the early endosomes/TGNs. From there they are either recycled back to the PM or they move to the late endosome/MVB (*blue line*-and-*dashed arrow*) for further transport to the lytic vacuole for degradation

markers to follow the dynamics of membrane flow in endocytic pathways. Here, we will discuss two examples: secretory carrier membrane proteins (SCAMPs) and vacuolar sorting receptors (VSRs) (Jiang and Rogers 1998, 1999a, b; Jiang and Sun 2002; Tse et al. 2004; Lam et al. 2007a; Min et al. 2007; Wang et al. 2007, 2009, 2010a; Cai et al. 2011).

2 SCAMPs and Endocytosis

Secretory carrier membrane proteins (SCAMPs) are represented by family of ubiquitously expressed tetra-spinning membrane proteins that are enriched in secretory membranes of mammalian cells (Brand et al. 1991; Laurie et al. 1993). They are structurally conserved, having four transmembrane domains (TMDs), three inter-TMD hydrophilic loops, and cytosolic N- and C-termini (Hubbard et al. 2000). Studies on both mammalian and plant SCAMPs indicated that these TMDs and cytosolic termini are essential for their proper subcellular localization and trafficking (Fernandez-Chacon et al. 2000; Cai et al. 2011).

The first study using the rice SCAMP1 (OsSCAMP1) stably expressed in transgenic tobacco BY-2 cells showed that OsSCAMP1-YFP localized to both the PM and cytosolic organelles that are sensitive to wortmannin and brefeldin A (BFA) treatments (Lam et al. 2007a). These SCAMP1-positive cytosolic organelles, identified as *trans*-Golgi network (TGN) vesicles via immunogold EM, turn out to be an early endosome as the endocytic dye FM4–64 reached them first prior to reaching the late endosome and finally the tonoplast (Tse et al. 2004; Lam et al. 2007a). Thus, endocytic and secretory pathways in plant cells are merged at TGN/early enodosome (Dettmer et al. 2006; Lam et al. 2007a, b; Robinson et al. 2008). Interestingly, overexpression of OsSCAMP1 resulted in a shift to preferential PM localization instead of cytosolic TGN puncta, likely owing to an inefficient transport from PM to TGN caused by saturation of the endocytic machinery, or by retention of OsSCAMP1 in the PM (Lam et al. 2007a).

OsSCAMP1 lacks a typical signal peptide (SP) sequence at its N-terminus. Further study on the trafficking of OsSCAMP1 using transient expression system in BY-2 protoplasts (Miao and Jiang 2007) demonstrated that OsSCAMP1 reached the PM via the conventional endoplasmic reticulum (ER)-Golgi-TGN-PM pathway (Fig. 2) (Cai et al. 2011). For example, when OsSCAMP1-YFP was co-expressed with sec12p, the guanine nucleotide exchange factor of Sar1, or with constitutive active ARF1 mutant (Q71L, locked in GTP-bound active state), OsSCAMP1 was trapped within the ER (Fig. 2b), a result indicating that SCAMP1 is synthesized in the ER. Similarly, concanamycin A (ConcA), a post-Golgi trafficking inhibitor, trapped OsSCAMP1-YFP within the TGN and thus blocked its delivery to the PM (Fig. 2b). Taken together, these results support an ER-Golgi-TGN-PM pathway of OsSCAMP1 traffic in plant cells (Cai et al. 2011).

Brefeldin A (BFA), a fungal metabolite which targets some Arf guanine nucleotide exchange factors (ARF-GEFs), induced both Golgi and TGN to form

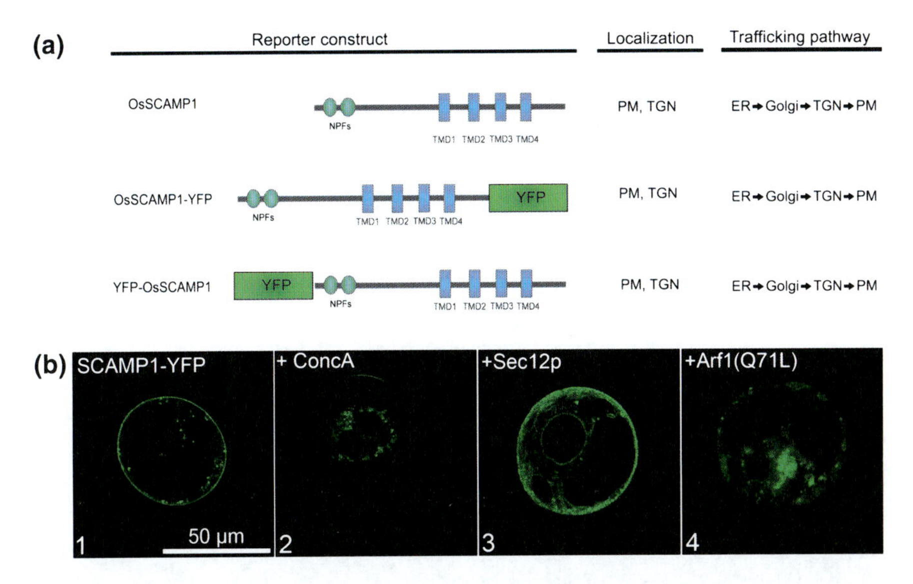

Fig. 2 Subcellular localization and trafficking pathway of SCAMP1 and YFP-tagged SCAMP1 in suspension culture cells. **a** Gene constructs, their subcellular localization, and transport pathways upon transient expression in plant cells. **b** CLSM images of tobacco BY-2 protoplasts transiently expressing single OsSCAMP1-YFP (*panel 1*), or expressing single OsSCAMP1-YFP in the presence of post-Golgi transport inhibitor ConcA (*panel 2*), or co-expressing OsSCAMP1-YFP with Sec12p (*panel 3*) or with Arf1(Q71L) mutant (*panel 4*)

aggregates (Tse et al. 2004, 2006; Lam et al. 2007a). Interestingly, in plant cells co-expressing GFP-OsSCAMP1 with the Golgi marker Man1-RFP, the BFA-induced aggregates derived from the GFP-tagged TGN and the RFP-tagged Golgi are morphologically and physically distinct from each other (Lam et al. 2009). However, when these cells were treated with the endocytic inhibitor tyrphostin A23 together with the BFA, Man1-RFP but not GFP-OsSCAMP1 was trapped in the BFA-induced aggregates, suggesting that the formation of these BFA-induced aggregates is not dependent on endocytosed materials from the PM (Lam et al. 2009). Figure 2 summarizes the subcellular localization, trafficking, and transport pathways of OsSCAMP1 and its reporters in plant cells.

3 VSRs and Secretory Trafficking

Vacuolar sorting receptors (VSRs) are type I integral membrane proteins containing a single TMD and a short cytoplasmic tail (CT) (Fig. 3a; Paris et al. 1997; Neuhaus and Rogers 1998). The N-terminus of a VSR is responsible for sorting cargo proteins at TGN via specific cargo-receptor interaction (Cao et al. 2000) for further delivery via clathrin-coated vesicle (CCV) to PVC (Tse et al. 2004). The N-terminus of the pea BP-80, the first identified VSR in plants (Kirsch et al. 1994), is

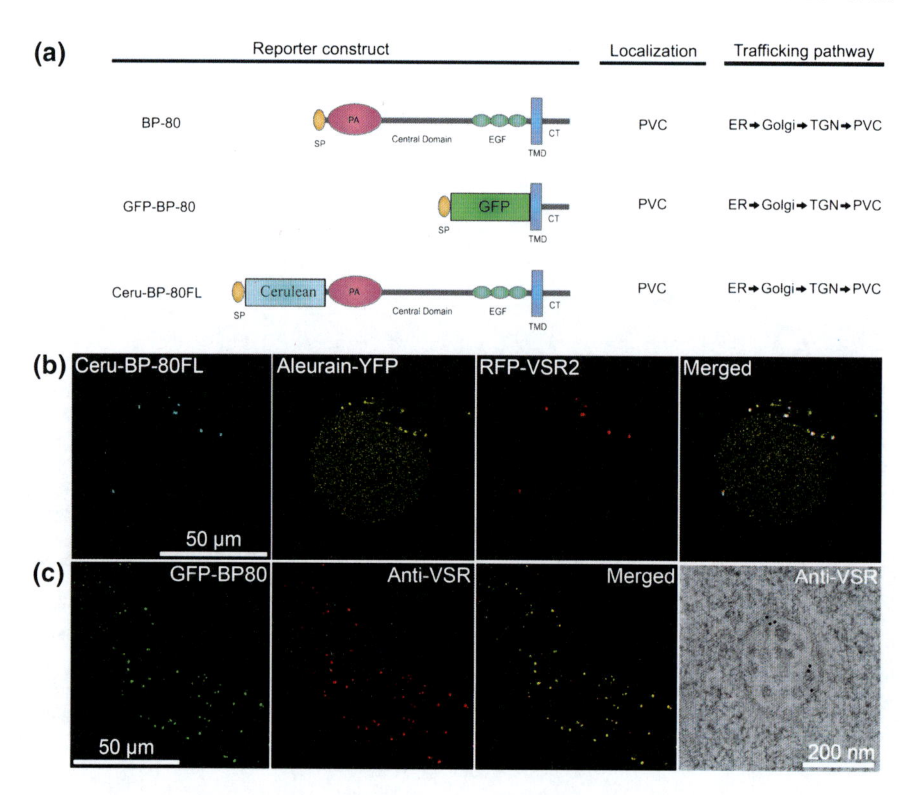

Fig. 3 Subcellular localization and trafficking pathway of VSR and VSR reporters in suspension culture cells. **a** Gene constructs, their subcellular localization, and transport pathways upon transient expression in plant cells. **b** CLSM images of Arabidopsis PSB-D protoplast co-expressing ceru-BP-80FL (cerulean), aleurain-YFP (*yellow*), and RFP-AtVSR2 (*red*). **c** CLSM images of transgenic BY-2 cells expressing the PVC marker GFP-BP-80 (*green*) colocalized with VSR (*yellow color* in merged image) using anti-VSR antibodies (*red*), the most right image represents immunogold EM localization of VSR in the MVB

responsible for binding the proaleurain NPIR motif (Suen et al. 2010). Interestingly, recent studies using transient expression system in tobacco protoplasts suggested that such cargo-receptor interaction important for protein sorting could initiate already in the ER (Niemes et al. 2010a, b).

Immunofluorescence study using CLSM indicated that VSRs were enriched in PVCs of BY-2 cells (Fig. 3b; Li et al. 2002), and these were subsequently identified as MVBs via immunogold EM study (Fig. 3c; Tse et al. 2004). In addition, using transient expression in Arabidopsis and BY-2 culture cells, the TMD and CT of a VSR were shown to be essential and sufficient for PVC targeting in plant cells (Jiang and Rogers 1998; Tse et al. 2004; Miao et al. 2006), whereas these PVCs mediated the vacuolar transport of both hydrolytic enzymes and storage protein reporters (Miao et al. 2008).

The seven Arabidopsis VSRs showed different spatial and temporal expression profiles (Paris and Neuhaus 2002; Laval et al. 2003; Shimada et al. 2003), suggesting their possible functional diversity and redundancy in plant development. For example, AtVSR1 knockout severely affected the correct sorting of 2S albumin and 12S globulin in Arabidopsis seeds (Shimada et al. 2003) whereas the antisense AtVSR1 mutants failed to germinate (Laval et al. 2003). A more recent genetic study using various Arabidopsis T-DNA insertional mutants showed that AtVSR1, AtVSR3, and AtVSR4 are the major sorting receptors for seed storage proteins, whereas AtVSR1 and AtVSR4 have redundant functions in sorting aleurain in leaves (Zouhar et al. 2010).

4 Essential Roles of SCAMPs and VSRs in Pollen Tube Growth

Growing pollen tube is an excellent single cell system for studying membrane dynamics in the endocytic and secretory pathways (Cheung and Wu 2008) because tube growth requires large amounts of materials being transported by secretory vesicles to the right place at the right time while surplus material is retrieved by compensatory endocytosis (Malho et al. 2006; Šamaj et al. 2006; Krichevsky et al. 2007).

The dynamics and possible roles of SCAMPs and VSRs in pollen tubes were investigated recently. Microinjection of either SCAMP1 or VSR antibodies into growing lily pollen tube inhibited its growth (Wang et al. 2010a), indicating that protein transport mediated by SCAMPs and VSRs is essential for pollen tube growth. Interestingly, growing lily pollen tube expressing GFP-tagged SCAMP or VSR showed distinct patterns of distribution: GFP-LISCAMP (lily SCAMP) is enriched in the tip region (Fig. 4a) colocalizing with the internalized endocytic marker FM4–64 (Wang et al. 2010a). In contrast, GFP-tagged VSR was missing in the apical inverted-cone region representing clear zone (Fig. 4b) (Wang et al. 2010a). Consistent with previous results obtained in BY-2 cells (Lam et al. 2007a), SCAMP1 antibodies also labeled PM or vesicles close to the PM, TGN, and vacuole in growing pollen tube as revealed by immunogold EM (Fig. 4a; Wang et al. 2010a, 2011). In addition, BFA treatment disrupted the tip-focused localization of GFP-LISCAMP and FM4–64 dye (Wang et al. 2010a). Collectively, these data support the roles of SCAMPs in endocytosis in pollen tube, but their possible roles in exocytosis remain elusive because the apical tip region of a pollen tube is enriched with both endocytic and exocytic vesicles (Šamaj et al. 2006; Zonia and Munnik 2008). Interestingly, immunogold EM studies with VSR antibodies showed that VSRs were localized to both PVC/MVB and PM in lily and tobacco pollen tubes (Fig. 4b) (Wang et al. 2011), indicating likely an additional role of VSR in regulating protein trafficking in both secretory and endocytic pathways (Wang et al. 2010a, 2011) (see also discussion below).

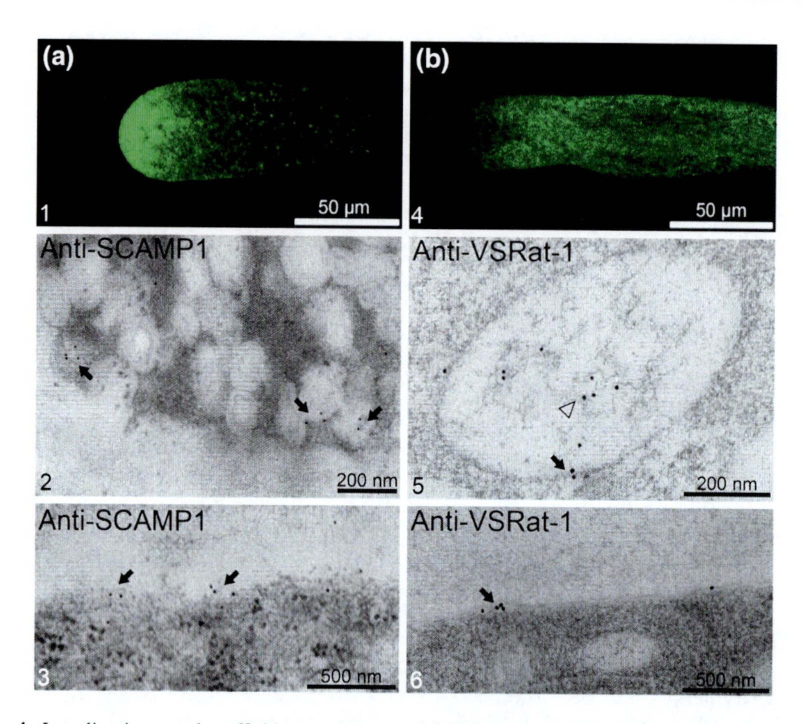

Fig. 4 Localizations and trafficking pathways of SCAMP and VSR reporters in pollen tube. **a** CLSM image of lily pollen tube transiently expressing GFP-LISCAMP (*panel 1*) and immunogold EM images with anti-SCAMP1 showing the subcellular localization of SCAMP1 in both endocytic vesicles in pollen tip region (*arrows*, *panel 2*) and at the plasma membrane (*arrows*, *panel 3*). **b** CLSM image of lily pollen tube transiently expressing GFP-LIVSR (*panel 4*) and immunogold EM images with anti-VSRat-1 showing the subcellular localization of VSRs in tonoplast (*arrow*) and vacuolar lumen (*arrowhead*) (*panel 5*) and at the plasma membrane (*arrow*, *panel 6*)

5 PM Localization of VSR and its Implication in Endocytosis

In addition to functional roles in vacuolar sorting, VSRs may somehow contribute to endocytosis due to the presence at the PM in plant cells, as recently demonstrated in pollen tube (Wang et al. 2011). Another study also supports this notion. A functional full-length AtVSR4 receptor was first generated by placing the fluorescent reporter citrine between the SP and the rest of the coding sequence of AtVSR4. When the resulting SP-citrine-AtVSR4 was either transiently expressed in tobacco epidermal cells or in transgenic Arabidopsis plants, the reporter protein was also found partially in the PM (Saint-Jean et al. 2010). These PM-localized VSRs may undergo endocytosis. In transgenic roots treated with BFA, a drug leading to the intracellular accumulation of TGN/endocytic vesicles in BFA compartments, the PM-localized citrine-AtVSR4 was reduced dramatically, indicating that the receptor undergoes endocytosis. Upon washing out the BFA in the presence of cycloheximide, an inhibitor of protein synthesis, citrine-AtVSR4 positive BFA compartments disappeared which was accompanied with the recovery of PM signal. Thus, VSRs may recycle between an endomembrane compartment and PM, and have functions both in endocytosis and exocytosis (Saint-Jean et al. 2010). However, it remains elusive whether VSRs are able to transport cargos outside of the cells via exocytosis and/or receive ligands from the PM for internalization to vacuole via endocytic pathway in plant cells.

6 Targeting Mechanisms of SCAMPs

All plant SCAMPs contain a conserved tripeptide Asn-Pro-Phe (NPF) motif in the cytosolic N-terminus, capable of interacting with the eps15 homology (EH) domain proteins (Fig. 5a; Law et al. 2011). EH domain proteins are key players of endocytosis and vesicular trafficking in yeast and mammalian cells (Salcini et al. 1997; Paoluzi et al. 1998; Xiao et al. 2009). Two plant proteins containing EH domain have been shown to play role in endocytosis of EIX receptor (Bar et al. 2008). Thus, better understanding of targeting mechanisms of SCAMPs in plant cells may have implications for their biological functions in plants.

OsSCAMP1 was previously shown to localize to TGN and PM (Lam et al. 2007a) as well as to the cell plate (Lam et al. 2008) in tobacco BY-2 cells. The targeting and degradation mechanisms of OsSCAMP1 were further illustrated by two recent studies using a transient expression system in BY-2 protoplasts (Cai et al. 2011, 2012). Although without having a predicted SP at its N-terminus, OsSCAMP1 was shown to reach the PM via typical ER-Golgi-TGN-PM pathway in plant cells (Cai et al. 2011), which is distinct from the unconventional secretion pathway (Wang et al. 2010b). Deletion analysis showed that a short N-terminal sequence (119–144aa) adjacent to the first TMD of OsSCAMP1 served as a conformational ER export signal, whereas TMD1 was essential for its TGN-to-PM targeting. Truncated OsSCAMP1 lacking its N-terminus and the first TMD was directed to the vacuole via the Golgi-TGN-PVC-LV pathway, without reaching the PM (Cai et al. 2011). Interestingly, when both TMD2 and TMD3 were deleted, the resulting OsSCAMP1 mutant was trapped in Golgi, without reaching post-Golgi compartments TGN and PM, indicating that TMD2 and TMD3 are important for Golgi export of OsSCAMP1 in plant cells (Cai et al. 2011). Such Golgi-retention mechanism is different in the case of OsSCAMP1 from that of a Golgi-localized integral membrane protein termed EMP12 (endomembrane protein 12). Short cytosolic tail of EMP12 was shown to contain both ER export and Golgi retention signals in Arabidopsis cells (Gao et al. 2012). OsSCAMP1 was internalized from PM and reached vacuole via PVC for degradation via typical endocytic pathway in plant cells (Cai et al. 2012). Interestingly, internalization of OsSCAMP1 from the PM for vacuolar degradation was achieved via internal vesicles of PVC/MVB in a cargo ubiquitination-independent manner (Cai et al. 2012). In future study, it will be of interest to find out the possible relationship between the trafficking and function of SCAMPs in plants.

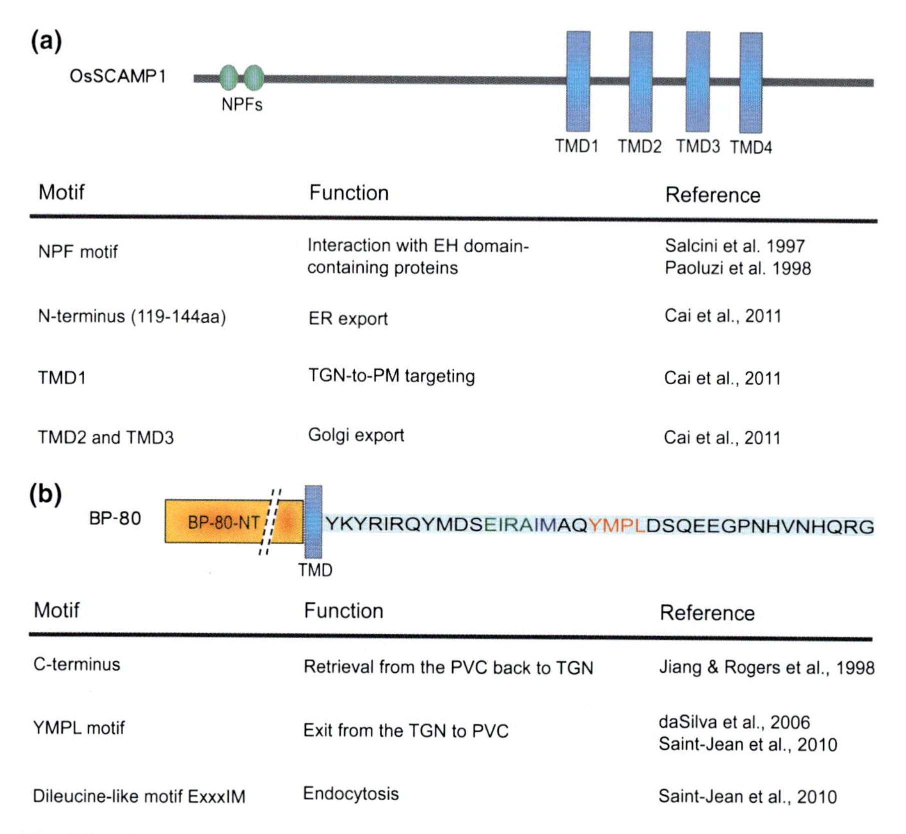

Fig. 5 Possible roles of SCAMPs and VSRs in plant endocytosis and exocytosis. **a** Topology and functional domains of OsSCAMP1. OsSCAMP1 contains four TMDs and cytosolic N-terminus and C-terminus. A tripeptide Asn-Pro-Phe (*NPF*) motif in the cytosolic N-terminus is believed to interact with EH-domain-containing proteins as reported in animals and yeast. In plant cells, the N-terminus (119–144aa) of OsSCAMP1 is found to be essential for ER export, whereas the TMD1 of OsSCAMP1 is essential for TGN to PM targeting while TMD2 and TMD3 are essential for Golgi export. **b** Topology of BP-80 and its amino acid sequence of cytosolic tail (*CT*). Both TMD and CT (with YMPL motif) of BP-80 are essential for its correct targeting to PVC. At the plasma membrane, the receptor might bind a ligand and is endocytosed using the IM dipeptide as part of a dileucine-like motif ExxxIM, possibly with the help of the Tyr motif

7 Targeting Mechanisms of VSRs

Both TMD and CT of a VSR were shown to be sufficient and essential for its correct targeting to the PVC/MVB in plant cells (Fig. 5b; Jiang and Rogers 1998; Tse et al. 2004; Miao et al. 2006). A Tyr motif (YMPL/YIPL) is highly conserved in the CT of all seven VSRs in *Arabidopsis thaliana* (Hadlington and Denecke 2000) and this motif was shown to be important for in vitro interaction with the Golgi-localized μA-adaptin (Happel et al. 2004). Indeed, detailed mutagenesis analysis on the pea BP-80 CT using an in vivo competition assay for vacuolar transport showed that substitution of Tyr-612 for Ala caused its partial retention in the Golgi apparatus and mistargeting to PM (daSilva et al. 2006). BP-80 was first purified from a CCV–enriched fraction (Kirsch et al. 1994) and it could recruit adaptor protein (AP) complexes to the Golgi membranes and integrate clathrin coat assembly with its inclusion into the nascent vesicle (Robinson et al. 2005). These results suggest an active role of BP-80 CT in the clathrin-mediated anterograde transport from the Golgi apparatus to the PVC. However, the PM-localized Tyr motif mutant may play a role in mediating BP-80 traffic in the exocytic pathway.

In addition to the Tyr motif in the BP-80 CT, the IM dipeptide as part of a dileucine-like motif ExxxIM may also play a role in endocytosis (Fig. 5b), as double mutation in Tyr and IM dipeptide caused significant increase in the fluorescence of this mutated receptor at the PM (Saint-Jean et al. 2010). The context of the IM dipeptide in VSRs, ExxxIM, resembles the canonic sequence of a dileucine signal [D/E]xxxL[L/I/M]. In mammalian cells, such signals have been identified as endocytotic signals in several membrane proteins including T-cell surface glycoprotein CD3 gamma chain (CD3G) and Mannose-6-phosphate receptor (MPR). The ExxxIM motif found in the CT of VSRs may well serve as such endocytic dileucine-like motif (Sandoval et al. 2000).

8 Conclusions and Future Perspectives

In recent years, integral membrane proteins have been used as markers for defining endocytic trafficking pathways and the molecular nature of endosomes (Delhaize et al. 2007; Lam et al. 2007a, b; Müller et al. 2007; Robinson et al. 2008; Cai et al. 2011, 2012; Gao et al. 2012). Indeed, their stability and defined trafficking along the secretory and endocytic pathways have made them to be reliable markers in this research field. Although the biosynthetic and degradation pathways of SCAMPs have been well-defined in plant cells (Law et al. 2011; Cai et al. 2011, 2012), the functional roles of SCAMP family proteins remain elusive in plants. Similarly, very little is known about native cargo proteins for the seven VSRs in Arabidopsis. Recent discoveries on the abundance of SCAMPs in the apical region of pollen tube and the PM localization of VSRs have thus opened a new direction of future research focused on functional characterization of these proteins in plants. For example, PM localization of sorting receptors for acid hydrolases is common in eukaryotic cells. In mammals, in addition to mediating lysosomal transport of hydrolytic enzymes from the Golgi apparatus, the MPR also functions in recovery of acid hydrolases from the PM (Braulke and Bonifacino 2009). Therefore, it would be interesting to find out if VSRs have functions in recovery of acid hydrolases at the PM in plants. Future studies should reveal identity of putative PM-localized cargoes of VSRs and shed more light on cargo selection, sorting, and subcellular trafficking.

Acknowledgments Our research has been supported by grants from the Research Grants Council of Hong Kong (CUHK466309, CUHK466610, CUHK466011, CUHK2/CRF/11G and HKBU1/CRF/10) and CUHK Schemes to L. Jiang.

Conflict of Interest The authors declare that they have no conflict of interest.

References

Bar M, Aharon M, Benjamin S, Rotblat B, Horowitz M, Avni A (2008) AtEHDs, novel Arabidopsis EH-domain-containing proteins involved in endocytosis. Plant J 55:1025–1038
Behnia R, Munro S (2005) Organelle identity and the signposts for membrane traffic. Nature 438:597–604
Brand SH, Laurie SM, Mixon MB, Castle JD (1991) Secretory carrier membrane proteins 31–35 define a common protein composition among secretory carrier membranes. J Bio Chem 266:18949–18957
Braulke T, Bonifacino JS (2009) Sorting of lysosomal proteins. Biochim Biophys Acta 1793: 605–614
Cai Y, Jia T, Lam SK, Ding Y, Gao C, San MWY, Pimpl P, Jiang L (2011) Multiple cytosolic and transmembrane determinants are required for the trafficking of SCAMP1 via an ER-Golgi-TGN-PM pathway. Plant J 65:882–896
Cai Y, Zhuang X, Wang J, Wang H, Lam SK, Gao X, Wang X, Jiang L (2012) Vacuolar degradation of two integral plasma membrane proteins, AtLRR84A and OsSCAMP1, is cargo ubiquitination-independent and prevacuolar compartment-mediated in plant cells. Traffic 13:1023–1040
Cao X, Rogers SW, Butler J, Beevers L, Rogers JC (2000) Structural requirements for ligand binding by a probable plant vacuolar sorting receptor. Plant Cell 12:493–506
Cheung AY, Wu HM (2008) Structural and signaling networks for the polar cell growth machinery in pollen tubes. Annu Rev Plant Biol 59:547–572
Chow CM, Neto H, Foucart C, Moore I (2008) Rab-A2 and Rab-A3 GTPases define a *trans*-golgi endosomal membrane domain in Arabidopsis that contributes substantially to the cell plate. Plant Cell 20:101–123
daSilva LLP, Foresti O, Denecke J (2006) Targeting of the plant vacuolar sorting receptor BP80 is dependent on multiple sorting signals in the cytosolic tail. Plant Cell 18:1477–1497
Delhaize E, Gruber BD, Pittman JK, White RG, Leung H, Miao Y, Jiang L, Ryan PR, Richardson AE (2007) A role for the AtMTP11 gene of Arabidopsis in manganese transport and tolerance. Plant J 51:198–210
Dettmer J, Hong-Hermesdorf A, Stierhof YD, Schumacher K (2006) Vacuolar H+-ATPase activity is required for endocytic and secretory trafficking in Arabidopsis. Plant Cell 18:715–730
Fernandez-Chacon R, Sudhof TC (2000) Novel SCAMPs lacking NPF repeats: ubiquitous and synaptic vesicle-specific forms implicate SCAMPs in multiple membrane-trafficking functions. J Neurosci 20:7941–7950
Gao C, Yu CKY, Qu S, San MWY, Li KY, Lo SW, Jiang L (2012) The Golgi-localized Arabidopsis Endomembrane Protein12 contains both endoplasmic reticulum export and Golgi retention signals at its C Terminus. Plant Cell 24:2086–2104
Geldner N, Jurgens G (2006) Endocytosis in signalling and development. Curr Opin Plant Biol 9: 589–594
Grant BD, Donaldson JG (2009) Pathways and mechanisms of endocytic recycling. Nature Rev Mol Cell Biol 10:597–608

Hadlington JL, Denecke J (2000) Sorting of soluble proteins in the secretory pathway of plants. Curr Opin Plant Biol 3:461–468
Happel N, Honing S, Neuhaus JM, Paris N, Robinson DG, Holstein SEH (2004) Arabidopsis mu A-adaptin interacts with the tyrosine motif of the vacuolar sorting receptor VSR-PS1. Plant J 37:678–693
Hubbard C, Singleton D, Rauch M, Jayasinghe S, Cafiso D, Castle D (2000) The secretory carrier membrane protein family: structure and membrane topology. Mol Biol Cell 11:2933–2947
Jiang L, Rogers JC (1998) Integral membrane protein sorting to vacuoles in plant cells: evidence for two pathways. J Cell Biol 143:1183–1199
Jiang L, Rogers JC (1999a) Sorting of membrane proteins to vacuoles in plant cells. Plant Sci 146:55–67
Jiang L, Rogers JC (1999b) Functional analysis of a Golgi-localized Kex2p-like protease in tobacco suspension culture cells. Plant J 18:23–32
Jiang L, Sun SSM (2002) Membrane anchors for vacuolar targeting: application in plant bioreactors. Trends Biotech 20:99–102
Kirsch T, Paris N, Butler JM, Beevers L, Rogers JC (1994) Purification and initial characterization of a potential plant vacuolar targeting receptor. Proc Natl Acad Sci U S A 91:3403–3407
Krichevsky A, Kozlovsky SV, Tian GW, Chen MH, Zaltsman A, Citovsky V (2007) How pollen tubes grow. Dev Biol 303:405–420
Lam SK, Siu CL, Hillmer S, Jang S, An G, Robinson DG, Jiang L (2007a) Rice SCAMP1 defines clathrin-coated, *trans*-Golgi-located tubular-vesicular structures as an early endosome in tobacco BY-2 cells. Plant Cell 19:296–319
Lam SK, Tse YC, Robinson DG, Jiang L (2007b) Tracking down the elusive early endosome. Trends Plant Sci 12:497–505
Lam SK, Cai Y, Hillmer S, Robinson DG, Jiang L (2008) SCAMPs highlight the developing cell plate during cytokinesis in tobacco BY-2 Cells. Plant Physiol 147:1637–1645
Lam SK, Cai Y, Tse YC, Wang J, Law AHY, Pimpl P, Chan HY, Xia J, Jiang L (2009) BFA-induced compartments from the Golgi apparatus and *trans*-Golgi network/early endosome are distinct in plant cells. Plant J 60:865–881
Laurie SM, Cain CC, Lienhard GE, Castle JD (1993) The glucose transporter GluT4 and secretory carrier membrane proteins (SCAMPs) colocalize in rat adipocytes and partially segregate during insulin stimulation. J Biol Chem 268:19110–19117
Laval V, Masclaux F, Serin A, Carriere M, Roldan C, Devic M, Pont-Lezica RF, Galaud JP (2003) Seed germination is blocked in Arabidopsis putative vacuolar sorting receptor (atbp80) antisense transformants. J Exp Bot 54:213–221
Law AHY, Chow CM, Jiang L (2011) Secretory carrier membrane proteins. Protoplasma 249:269–283
Li Y-B, Rogers SW, Tse YC, Lo SW, Sun SSM, Jauh G-Y, Jiang L (2002) BP-80 and homologs are concentrated on post-Golgi, probable lytic prevacuolar compartments. Plant Cell Physiol 43:726–742
Malho R, Liu Q, Monteiro D, Rato C, Camacho L, Dinis A (2006) Signalling pathways in pollen germination and tube growth. Protoplasma 228:21–30
Miaczynska M, Stenmark H (2008) Mechanisms and functions of endocytosis. J Cell Biol 180:7–11
Miao Y, Yan PK, Kim H, Hwang I, Jiang L (2006) Localization of green fluorescent protein fusions with the seven Arabidopsis vacuolar sorting receptors to prevacuolar compartments in tobacco BY-2 cells. Plant Physiol 142:945–962
Miao Y, Jiang L (2007) Transient expression of fluorescent fusion proteins in protoplasts of suspension cultured cells. Nature Protoc 2:2348–2353
Miao Y, Li KY, Li HY, Yao XQ, Jiang LW (2008) The vacuolar transport of aleurain-GFP and 2S albumin-GFP fusions is mediated by the same pre-vacuolar compartments in tobacco BY-2 and Arabidopsis suspension cultured cells. Plant J 56:824–839

Min MY, Kim SJ, Miao Y, Shin J, Jiang L, Hwang I (2007) Overexpression of Arabidopsis AGD7 causes relocation of Golgi-localized proteins to the ER and inhibits protein trafficking in plant cells. Plant Physiol 143:1601–1614

Müller J, Mettbach U, Menzel D, Šamaj J (2007) Molecular dissection of endosomal compartments in plants. Plant Physiol 145:293–304

Neuhaus JM, Rogers JC (1998) Sorting of proteins to vacuoles in plant cells. Plant Mol Biol 38:127–144

Niemes S, Labs M, Scheuring D, Krueger F, Langhans M, Jesenofsky B, Robinson DG, Pimpl P (2010a) Sorting of plant vacuolar proteins is initiated in the ER. Plant J 62:601–614

Niemes S, Langhans M, Viotti C, Scheuring D, Yan MSW, Jiang L, Hillmer S, Robinson DG, Pimpl P (2010b) Retromer recycles vacuolar sorting receptors from the *trans*-Golgi network. Plant J 61:107–121

Otegui MS, Spitzer C (2008) Endosomal functions in plants. Traffic 9:1589–1598

Paoluzi S, Castagnoli L, Lauro I, Salcini AE, Coda L, Fre S, Confalonieri S, Pelicci PG, Di Fiore PP, Cesareni G (1998) Recognition specificity of individual EH domains of mammals and yeast. EMBO J 17:6541–6550

Paris N, Neuhaus JM (2002) BP-80 as a vacuolar sorting receptor. Plant Mol Biol 50:903–914

Paris N, Rogers SW, Jiang LW, Kirsch T, Beevers L, Phillips TE, Rogers JC (1997) Molecular cloning and further characterization of a probable plant vacuolar sorting receptor. Plant Physiol 115:29–39

Preuss ML, Serna J, Falbel TG, Bednarek SY, Nielsen E (2004) The Arabidopsis Rab GTPase RabA4b localizes to the tips of growing root hair cells. Plant Cell 16:1589–1603

Richter S, Geldner N, Schrader J, Wolters H, Stierhof YD, Rios G, Koncz C, Robinson DG, Jurgens G (2007) Functional diversification of closely related ARF-GEFs in protein secretion and recycling. Nature 448:488–492

Robinson DG, Oliviusson P, Hinz G (2005) Protein sorting to the storage vacuoles of plants: a critical appraisal. Traffic 6:615–625

Robinson DG, Jiang L, Schumacher K (2008) The endosomal system of plants: charting new and familiar territories. Plant Physiol 147:1482–1492

Saint-Jean B, Seveno-Carpentier E, Alcon C, Neuhaus J-M, Paris N (2010) The cytosolic tail dipeptide Ile-Met of the *Pea* receptor BP80 is required for recycling from the prevacuole and for endocytosis. Plant Cell 22:2825–2837

Salcini AE, Confalonieri S, Doria M, Santolini E, Tassi E, Minenkova O, Cesareni G, Pelicci PG, Di Fiore PP (1997) Binding specificity and in vivo targets of the EH domain, a novel protein–protein interaction module. Genes Dev 11:2239–2249

Sandoval IV, Martinez-Arca S, Valdueza J, Palacios S, Holman GD (2000) Distinct reading of different structural determinants modulates the dileucine-mediated transport steps of the lysosomal membrane protein LIMPII and the insulin-sensitive glucose transporter GLUT4. J Biol Chem 275:39874–39885

Shimada T, Fuji K, Tamura K, Kondo M, Nishimura M, Hara-Nishimura I (2003) Vacuolar sorting receptor for seed storage proteins in *Arabidopsis thaliana*. Proc Natl Acad Sci U S A 100:16095–16100

Suen PK, Shen JB, Sun SSM, Jiang L (2010) Expression and characterization of two functional vacuolar sorting receptor (VSR) proteins, BP-80 and AtVSR4 from culture media of transgenic tobacco BY-2 cells. Plant Sci 179:68–76

Šamaj J, Read ND, Volkmann D, Menzel D, Baluška F (2005) The endocytic network in plants. Trends Cell Biol 15:425–433

Šamaj J, Müller J, Beck M, Böhm N, Menzel D (2006) Vesicular trafficking, cytoskeleton and signalling in root hairs and pollen tubes. Trends Plant Sci 11:594–600

Traub LM (2009) Tickets to ride: selecting cargo for clathrin-regulated internalization. Nature Rev Mol Cell Biol 10:583–596

Tse YC, Mo B, Hillmer S, Zhao M, Lo SW, Robinson DG, Jiang L (2004) Identification of multivesicular bodies as prevacuolar compartments in *Nicotiana tabacum* BY-2 Cells. Plant Cell 16:672–693

Tse YC, Lo SW, Hillmer S, Dupree P, Jiang L (2006) Dynamic response of prevacuolar compartments to brefeldin A in plant cells. Plant Physiol 142:1442–1459

Ueda T, Yamaguchi M, Uchimiya H, Nakano A (2001) Ara6, a plant-unique novel type Rab GTPase, functions in the endocytic pathway of *Arabidopsis thaliana*. EMBO J 20:4730–4741

Wang H, Tse YC, Law AHY, Sun SSM, Sun YB, Xu ZF, Hillmer S, Robinson DG, Jiang LW (2010a) Vacuolar sorting receptors (VSRs) and secretory carrier membrane proteins (SCAMPs) are essential for pollen tube growth. Plant J 61:826–838

Wang J, Ding Y, Wang JQ, Hillmer S, Miao YS, Lo SW, Wang XF, Robinson DG, Jiang L (2010b) EXPO: an exocyst-positive organelle distinct from multivesicular endosomes and autophagosomes, mediates cytosol to cell wall exocytosis in plant cells. Plant Cell 22:4009–4030

Wang H, Zhuang XH, Hillmer S, Robinson DG, Jiang LW (2011) Vacuolar sorting receptor (VSR) proteins reach the plasma membrane in germinating pollen tubes. Mol Plant 4:845–853

Wang JQ, Li YB, Lo SW, Hillmer S, Sun SSM, Robinson DG, Jiang L (2007) Protein mobilization in germinating mung bean seeds involves vacuolar sorting receptors and multivesicular bodies. Plant Physiol 143:1628–1639

Wang JQ, Miao YS, Cai Yi, Jiang L (2009) Wortmannin induced homotypic fusion of prevacuolar compartment in plant cells. J Exp Bot 60:3075–3083

Xiao N, Kam C, Shen C, Jin W, Wang J, Lee KM, Jiang L, Xia J (2009) PICK1 deficiency causes male infertility in mice by disrupting acrosome formation. J Clin Investig 119:802–812

Zerial M, McBride H (2001) Rab proteins as membrane organizers. Nature Rev Mol Cell Biol 2:107–117

Zonia L, Munnik T (2008) Vesicle trafficking dynamics and visualization of zones of exocytosis and endocytosis in tobacco pollen tubes. J Exp Bot 59:861–873

Zouhar J, Muñoz A, Rojo E (2010) Functional specialization within the vacuolar sorting receptor family: VSR1, VSR3 and VSR4 sort vacuolar storage cargo in seeds and vegetative tissues. Plant J 64:577–588

The Plant SNX Family and Its Role in Endocytosis

Enric Zelazny, Rumen Ivanov and Thierry Gaude

Abstract Separation of functions within the cytoplasm of the eukaryotic cell has resulted in the development of a highly dynamic network of membranous compartments. Among them, the endosomes represent a crossing point for the major cellular trafficking pathways. Indeed, they are responsible for the communication between the cellular compartments, as well as between the cell and its environment, through the regulation of signalling networks and transport of material. The proteins from the sorting nexin (SNX) family have an established role as major regulators of endosomal protein transport in mammals, insects, worms, and yeast. By contrast, research on plant SNXs has been initiated relatively recently. Despite that, the accumulated knowledge suggests that plant SNXs may have evolved different mechanisms of action and might work in a different protein complex environment, compared to their yeast and animal counterparts. In this chapter, we highlight both common and specific characteristics of plant SNX protein function and regulation in comparison to the well-studied mammalian SNX machinery.

Abbreviations

BAR Bin/Amphiphysin/Rvs domain
C-ter Nexin C-terminal domain

Enric Zelazny and Rumen Ivanov have contributed equally to this work

---

E. Zelazny · T. Gaude (✉)
Centre National de la Recherche Scientifique, Institut National de la Recherche Agronomique, Unité Mixte de Service 3444 BioSciences Gerland-Lyon Sud, Université Claude Bernard Lyon I, Ecole Normale Supérieure de Lyon, Lyon, France
e-mail: thierry.gaude@ens-lyon.fr

R. Ivanov
Department of Plant Biology, Saarland University, Campus A 2.4, 66123 Saarbruecken, Germany

J. Šamaj (ed.), *Endocytosis in Plants*,
DOI: 10.1007/978-3-642-32463-5_12,

MIT	Microtubule interacting domain
F	Forkhead-associated domain
FERM	4.1/Ezrin/Radixin/Moesin domain
Kinesin	Kinesin motor domain
PDZ	PSD95/Dlg1/Zo-1 domain
PIN	Pin-formed
PX	PHOX homology domain
PXA	PHOX associated domain
RAB	Rab5-binding domain
RAS	Ras-association domain
RGS	Regulator of G-protein signalling domain
RHO	RhoGAP domain
SH3	Src Homology 3 domain
SNX	Sorting nexin
SRK	S locus receptor kinase
VPS	Vacuolar protein sorting

1 SNX Proteins in Eukaryotes: Variations on the Same Theme

The endosomes provide a major distribution network, serving as intermediate sorting stations for cargo proteins coming from the endoplasmic reticulum (ER)/Golgi apparatus and the plasma membrane (PM). Depending on a variety of signals, such as post-translational modifications, proteins destined for secretion, endocytosis, or transfer to the vacuole/lysosome are grouped and allocated to their destination compartments with the help of multi-subunit complexes (Cullen 2008; Reyes et al. 2011). After internalization from the PM, proteins such as receptor kinases (Geldner et al. 2007; Traer et al. 2007, Roepstorff et al. 2009), transporters (Takano et al. 2010; Barberon et al. 2011; Kasai et al. 2011) and others may either be retrieved and rerouted back to the PM, or sent to the lytic vacuole for degradation. Another pathway is the backwards targeting of proteins, such as the yeast Vps10 receptor, mammalian cation-independent mannose 6-phosphate receptor (CI-MPR), or the plant vacuolar sorting receptor (VSR) from multivesicular bodies (MVBs) back to the *trans*-Golgi network (TGN) in an event called retrograde transport (Seaman et al. 1997; Seaman 2004; Yamazaki et al. 2008). Sorting nexin (SNX) proteins, often associated with other cytoplasmic proteins to form a multiprotein complex, were shown to play a key function in mediating some of these intracellular protein trafficking pathways.

SNXs were identified in human cells through the ability of the HsSNX1 to bind the kinase domain of the epidermal growth factor receptor (EGFR) (Kurten et al. 1996, Haft et al. 1998). Because HsSNX1 overexpression led to an increased EGFR degradation rate, HsSNX1 was believed to regulate EGFR levels at the cell surface (Kurten et al. 1996, 2001; Cozier et al. 2002). Similar effects were observed for HsSNX2, where inactivation through site-directed mutagenesis compromised receptor degradation. However, the corresponding mutation in SNX1 did not affect EGFR stability (Gullapalli et al. 2004), an observation later confirmed by loss of HsSNX1 function analyses (Carlton et al. 2004, 2005a). Despite this discrepancy, it is clear that SNX proteins are involved in the trafficking and stability of plasma membrane receptors. In yeast, two members of the SNX family, Vps5 and Vps17, were shown to regulate the late endosome to Golgi trafficking of the Vps10 receptor (Horazdovsky et al. 1997; Nothwehr and Hindes 1997). This retrograde transport is mediated by a multimeric complex, named retromer, which contains at least three other proteins, Vps26, Vps29, and Vps35 (Seaman et al. 1997; Reddy and Seaman 2001). The retromer exists in all known eukaryotes and transports many diverse proteins (Cullen and Korswagen 2012). Mammals possess two Vps5 homologs, HsSNX1 and HsSNX2, and potentially three homologs of Vps17, namely HsSNX5, HsSNX6, and possibly HsSNX32. One Vps5 and one Vps17 homolog are used within the complex with the exception of organisms that lack the Vps17 equivalent (Koumandou et al. 2011).

In plants, SNX proteins are involved in environmental responses and development through regulating the homeostasis of the phytohormone auxin (Jaillais et al. 2006; Kleine-Vehn et al. 2008). In addition, SNXs can also interact with the kinase domains of plant receptor kinases, suggesting a more complex role in cellular signalling (Vanoosthuyse et al. 2003).

From the above selected examples, it is evident that the SNX protein family plays a key role in the regulation of protein trafficking on the crossroad between the secretory and endocytic pathways in all eukaryotes. Because these proteins are present in evolutionary very distant groups of organisms, which have evolved different life strategies, the specifics of SNX action and regulation may vary significantly. Therefore, the study of SNX family in different organisms reveals new aspects of SNX function and regulation.

1.1 The PX Domain: Phox Homology Domain Common to All SNXs

SNXs are endosomal proteins characterized by the presence of a specific form of the lipid-binding PHOX (NADPH Phagocyte Oxidase, PX) homology domain. This domain is required for the interaction with cellular membranes and is present in a large group of proteins, including NADPH oxidase subunits, phospholipases D, and others. The PX domain of SNXs differs significantly from those in other proteins and forms a phylogenetically distinct subgroup which is often referred as PX^{SNX} (Ponting 1996; Teasdale et al. 2001).

In addition, SNX proteins may contain additional domains, such as the BAR (Bin/Amphiphysin/Rvs) domain, which is involved in protein–protein interactions and membrane curvature sensing (Peter et al. 2004). PHOX-associated (PHOX A),

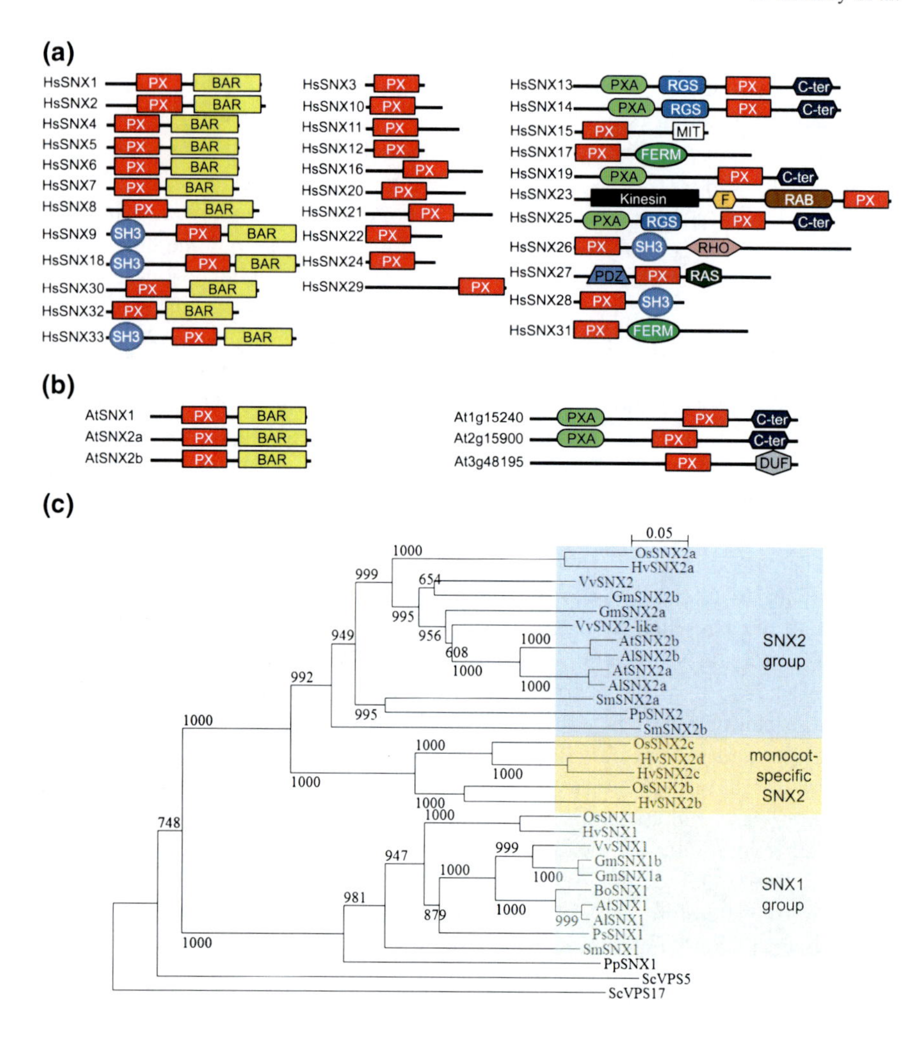

Src homology 3 (SH3), kinesin motor, regulator of G-protein signalling (RGS) are examples of other domains also found in SNX proteins (Fig. 1a).

SNX family members have been identified in all groups of eukaryotes but the number of members within each group varies greatly. Yeast contains approximately 10 SNXs, while mammals possess over 30 (Worby and Dixon 2002). The plant SNX family is relatively small having only three members in the model plant *Arabidopsis thaliana* (Vanoosthuyse et al. 2003; van Leeuwen et al. 2004). Mammalian SNX proteins have been classified into three groups according to their domain structure (Teasdale et al. 2001; Cullen 2008) (Fig. 1a). BAR and PX domain-containing SNX proteins form the SNX-BAR class. The second group consists of proteins that appear to contain only the PX domain. The third class is more heterogeneous and includes SNXs that contain PX and various other domains but do not contain a recognizable BAR. In comparison, the plant SNX family seems to be composed of only members of the first class (Fig. 1b).

◀ **Fig. 1** The SNX protein family in mammals and plants. **a** schematic representation of the mammalian SNX family. Three classes can be distinguished based on the domain structure of SNX proteins. SNX-BAR class (*left*) comprises the SNX proteins that contain both PX and BAR domains. SNXs of the second class (*middle*) contain only PX domain. The third class of SNXs (*right*) have no BAR domain but contain PX and one or more additional domains. *PX* PHOX homology domain; *BAR* Bin/Amphiphysin/Rvs domain; *SH3* Src Homology 3 domain; *PXA* PHOX associated domain, *RGS* Regulator of G-protein signalling domain, *C-ter* Nexin C-terminal domain, *MIT* Microtubule interacting domain, *FERM* 4.1/Ezrin/Radixin/Moesin domain, Kinesin Kinesin motor domain, *F* Forkhead-associated domain, *RAB* Rab5-binding domain, *RHO* RhoGAP domain, *PDZ* PSD95/Dlg1/Zo-1 domain, *RAS* Ras-association domain. Modified from Cullen (2008). **b** schematic representation of the Arabidopsis SNX protein family. The three PX-containing proteins with versatile domains and clustering independently from BAR domain-containing SNXs are also represented due to their predicted homology to human HsSNX13. *DUF*: domain of unknown function. Note that alternative splicing is predicted for the product of the At1g15240 gene (www.arabidopsis.org). The protein represented in the scheme corresponds to gene models 2 and 3. **c** the SNX-BAR class of proteins in plants. SNX-BAR protein sequences of different plant species were identified based on their homology to the yeast Vps5. Further homology searches were made for each species using the identified proteins to ensure that all homologs were found. The sequences, including the yeast Vps5 and Vps17, were aligned using the Clustal X2 algorithm (Larkin et al. 2007) and the resulting neighbor-joining tree was visualized using NJplot (Perrière and Gouy 1996). Vps17 sequence was used to root the tree and bootstrap values based on 1000 repeats are presented upon the branches. Sc *Saccharomyces cerevisiae*, Pp *Physcomitrella patens*, Sm *Selaginella moellendorffii*, Ps *Picea sitchensis*, Al *Arabidopsis lyrata*, At *Arabidopsis thaliana*, Bo *Brassica oleracea*, Gm *Glycine max*, Vv *Vitis vinifera*, Hv *Hordeum vulgare*, Os *Oryza sativa*. The GenBank accession numbers of the used sequences are as follows: PpSNX1: EDQ65698, PpSNX2: EDQ69692, SmSNX1: EFJ32690, SmSNX2a: EFJ20905, SmSNX2b: EFJ35882, PsSNX1: ABK25396, AlSNX1: EFH49499, AlSNX2a: EFH42525, AlSNX2b: EFH49556, AtSNX1: AED90975, AtSNX2a: AED97053, AtSNX2b: AED91112, BoSNX1: CAD29576, GmSNX1a: XP_003518207, GmSNX1b: ACU20395, GmSNX2a: XP_003519271, GmSNX2b: XP_003544040, VvSNX1: XP_002282010, VvSNX2: XP_002273943, VvSNX2-like: XP_002277656, HvSNX1: BAK01506, HvSNX2a: BAJ96230, HvSNX2b: BAJ94950, HvSNX2c: BAK03183, HvSNX2d: BAJ97936, OsSNX1: EAY76579, OsSNX2a: EEC71731, OsSNX2b: EAY97622, OsSNX2c: EEC79392. The scale bar represents the number of amino acid substitutions per position

1.2 The SNX Protein Family in Plants

The first plant SNX protein, (BoSNX1), was isolated as an interactor of the kinase domain of the *Brassica oleracea S*-RECEPTOR KINASE$_{29}$ (BoSRK$_{29}$). It was reported that BoSNX1 can interact in vitro with two other plant receptor kinases, the S-LOCUS FAMILY RECEPTOR1 (SFR1) and CLAVATA3 (CLV3) (Vanoosthuyse et al. 2003). BoSNX1 is a homolog of the mammalian SNX1 and SNX2 proteins, as well as the yeast Vps5, and also possesses both PX and BAR domains. Arabidopsis contains three BoSNX1 homologs, AtSNX1, AtSNX2a, and AtSNX2b (Fig. 1b). AtSNX1 shares 97 % amino acid identity with BoSNX1 and about 25 % with AtSNX2a and AtSNX2b. The identity between AtSNX2a and AtSNX2b is 80 % (Pourcher et al. 2010). The roles of these three Arabidopsis SNXs in protein trafficking have been investigated in more detail, and these results are further discussed in this chapter.

In order to investigate the SNX-BAR protein family and its evolution in the plant kingdom, we made searches for homologs of the yeast Vps5 and Vps17 proteins in selected plant species. Only homologs of Vps5 could be identified (Fig. 1c). In the majority of investigated species, at least two types of homologs exist, which we generally named SNX1 and SNX2. This division could be observed already in mosses (*Physcomitrella patens*) and early vascular plants (the lycophyte *Selaginella moellendorffii*), suggesting that the separation may have occurred early in the evolution of the land plants. No homologs of SNX2 proteins could be identified in the gymnosperm *Picea sitchensis*, probably due to the still incomplete information available for this species. In most cases, the SNX1 group contained a unique member. Exceptions were the two proteins found in *Glycine max*, namely GmSNX1a and GmSNX1b. On the other hand, the SNX2 group was more diverse, often with two members present in each organism. Interestingly, barley (*Hordeum vulgare*) and rice (*Oryza sativa*) possess a separate subgroup of SNX2 proteins, which seem to be unique to monocotyledonous plants (Fig. 1c). Because of a general lack of functional data from plants other than Arabidopsis, it is difficult to speculate on the significance of this divergence in monocots.

Arabidopsis genome contains eight additional genes that encode proteins with PX domain, including two phospholipases D and six proteins of unknown functions. They differ greatly from the SNX1 homologs and in a phylogenetic tree, they form three separate clusters (van Leeuwen et al. 2004). Interestingly, At1g15240, At2g15900 and At3g48195, which are in the same cluster, lack a BAR domain but contain a PHOX A domain and BLAST searches using the whole protein sequence against the human database reveal a slight similarity to HsSNX13 (the heterogeneous SNX class). Notably, the three Arabidopsis proteins lack a recognizable RGS domain. The potential subcellular roles of these three proteins have not been addressed so far. Further analysis should be carried out to determine whether these proteins might belong to the SNX family.

2 Binding and Tubulation of Biological Membranes by SNXs

SNX proteins have a dual distribution in the cytosol and at the endomembrane compartments. Recruitment to endomembranes involves the PX domains that are capable of specific interactions with various phosphoinositides (PIs). As the membranes of the different intracellular compartments and vesicles have distinct PI composition, the PX domains will allow SNX targeting to defined cellular compartments (Teasdale and Collins 2012). Although there is relatively weak primary sequence identity among the different PX domains, their two-dimensional structure consists of a positively charged PI-binding pocket that is remarkably conserved (Seet and Hong 2006). As revealed by crystal structures, binding to phosphatidylinositol 3-phosphate (PtdIns(3)P), for instance, is mediated by hydrogen bonds between arginine residues located in this pocket and the negative phosphate group of PtdIns(3)P (Bravo et al. 2001; Zhou et al. 2003). Depending on their PX domain, HsSNX proteins display different binding specificities to lipids. In mammals, while HsSNX3 binds exclusively to PtdIns(3)P (Xu et al. 2001; Zhong et al. 2005), HsSNX1 and HsSNX2 interact with PtdIns(3)P and phosphatidylinositol 3,5-bisphosphate (PtdIns(3,5)P_2) in liposome-based assays (Cozier et al. 2002; Carlton et al. 2005b). Using the same approach, HsSNX9 was found to be less specific in phospholipid recognition since it binds PtdIns(3)P, phosphatidylinositol 3,4-bisphosphate (PtdIns(3,4)P_2), phosphatidylinositol 4,5-bisphosphate (PtdIns(4,5)P_2), and phosphatidylinositol 3,4,5-trisphosphate (PtdIns(3,4,5)P_3) (Lundmark and Carlsson 2003). This range of phosphoinositide binding suggests diverse roles of SNX proteins in membrane trafficking or signalling from different subcellular compartments. In plants, lipid-binding properties of SNXs are still poorly documented. PtdIns(3)P was demonstrated to be important for the endosomal localization of AtSNX1, since wortmannin, an inhibitor of PtdIns(3)P production at low concentration, induced the release of AtSNX1 to the cytosol (Pourcher et al. 2010). In addition, the PX domain of AtSNX2b heterologously expressed in *E. coli* was shown to specifically bind to PtdIns(3)P in protein-lipid overlay assays (Phan et al. 2008). However, caution should be taken when interpreting this result since subsequent analysis carried out in planta revealed that AtSNX2b fused to a fluorescent reporter was not associated with endosomal membranes in the absence of AtSNX1 (Pourcher et al. 2010). These observations argue for use multiple methods to analyze lipid-binding properties of SNX proteins.

Interaction of SNXs with membrane lipids is also achieved through the BAR domain. Indeed, BAR domains are known to dimerize and to form a rigid banana-shaped structure with concave surface containing basic residues that allow association with the phospholipid bilayer through electrostatic interactions (van Weering et al. 2010). It is worth noting that BAR domains alone are not capable of interacting with endosomal membranes, underscoring the mutual action of PX and BAR domains in completing SNX-BAR protein association with membranes (Zhong et al. 2002). BAR dimers preferentially bind to highly curved negatively charged membranes and hence act as sensors of membrane curvature (Peter et al. 2004; Zimmerberg and McLaughlin 2004). For instance, mammalian HsSNX1 was shown to bind to microdomains of early endosomes defined by high curvature and the presence of PtdIns(3)P (Carlton et al. 2004). In addition, BAR domains revealed to induce membrane remodeling leading to membrane tubulation. So far, several mammalian SNX proteins such as HsSNX1, HsSNX4, HsSNX8, HsSNX9, and HsSNX18 were reported to tubulate membrane in vitro, using liposomes, and in vivo by overexpressing SNX in cultured cells (Carlton et al. 2004; Pylypenko et al. 2007; Haberg et al. 2008; van Weering et al. 2012). Van Weering et al. (2012) showed that HsSNX1 can elicit the formation of endosomal tubules in HeLa cells in the absence of VPS26-VPS29-VPS35 subcomplex, indicating that tubulation is independent of a functional retromer, which is consistent with the ability of recombinant HsSNX1 to induce membrane tubulation in vitro. Intriguingly, these functional properties seem not to be common to all mammalian SNXs as HsSNX2 is able to bind to membranes, however, without inducing membrane tubulation (Carlton et al. 2005a). Structural analysis performed on mammalian BAR modules showed that membrane tubules are formed when BAR dimers polymerize into helical lattices that are held together by lateral and tip-to-tip interactions (Frost et al. 2008). This oligomerization model is also supported by the crystal structure of the PX-BAR unit from HsSNX9, in which tip-to-tip interactions between adjacent BAR domain dimers have been observed (Wang et al. 2008).

In plants, the function of SNXs in membrane remodeling remains an open question and both in vitro and in vivo tubulation experiments will be required to determine whether plant SNXs have conserved tubulation activity.

3 SNXs as Major Actors of Endosomal Sorting

3.1 SNXs in Yeast and Mammalian Cells

In yeast and mammals, the function of the retromer complex is inseparable from the function of SNX proteins. Depletion of SNXs induces membrane dissociation of VPS proteins and hence leads to the loss of retromer activity, underlying the crucial role played by SNXs in the assembly and the function of the entire retromer (Rojas et al. 2007). A model explaining how the retromer mediates endosomal sorting in mammals has been proposed by Cullen (2008). In this model, SNX proteins generate and define a tubular domain on the endosome, which acts as a binding site for the core retromer (VPS26-VPS29-VPS35), the latter being the cargo-selective unit. Hence, the retromer couples membrane deformation with cargo sorting. The retromer complex has been involved in an expending number of physiological processes by regulating the trafficking of diverse cargos including acid hydrolase receptors such as the yeast Vps10 protein (Seaman et al. 1997) and the human CI-MPR (Arighi et al. 2004), the *Drosophila* Wingless morphogen receptor Wntless (Franch-Marro et al. 2008), or the yeast Fet3p-Ftr1p iron transporter (Strochlic et al. 2007).

Interestingly, SNX proteins are able to regulate endosomal sorting independently of the core retromer subcomplex. Indeed, depletion of SNX1 by siRNA in mammalian cells was shown to disturb trafficking of the $P2Y_1$ receptor and the protease-activated receptor-1 (PAR1), whereas inhibition of the retromer subunits VPS26 and VPS35 had no effect (Gullapalli et al. 2006; Nisar et al. 2010). Surprisingly, in both cases, no direct interaction between SNX1 and the cargo proteins was detected. In yeast, Snx4, in association with Snx41 and Snx42, binds to the endosomal SNARE Snc1 and promotes its retrieval from post-Golgi endosomes through a pathway that is genetically distinct from that of the retromer (Hettema et al. 2003). On the other hand, some SNX proteins were reported to directly act as cargo recruiter. For instance, in yeast, Grd19/Snx3 protein interacts with the Fet3p-Ftr1p iron transporter and mediates its recycling to the PM in an iron-dependent manner (Strochlic et al. 2007). In that case, Grd19/Snx3 functions as a cargo-specific adapter for the retromer complex, the latter being required for iron transporter recycling.

In addition to endosomal sorting, a role for SNX proteins in endocytosis was reported through the study of HsSNX9 (Lundmark and Carlsson 2003). This work revealed that HsSNX9 is involved in clathrin-mediated endocytosis at the plasma membrane. Indeed, HsSNX9 interacts with several key proteins of this endocytic pathway, such as clathrin, adaptor protein complex 2 (AP-2), and dynamins (Lundmark and Carlsson 2003), and is probably recruited to clathrin-coated pits during late stage of vesicle formation (Soulet et al. 2005). Morever, HsSNX9 knockdown in HeLa cells was shown to inhibit transferrin receptor internalization (Soulet et al. 2005), whereas its overexpression caused severe defects in synaptic vesicle endocytosis in neurons (Shin et al. 2007).

3.2 SNXs in Plants: Emergence of Specificities

Compared to yeast and mammals, the functions of sorting nexins in plants remain poorly described but it already emerges that they display specific features. For instance, contrary to yeast and mammals, plant SNXs are not necessary for the recruitment of the retromer complex to membranes and its proper functioning. Indeed, in a recent study, *snx1 snx2a snx2b* triple mutants were found to exhibit rather weak developmental phenotypes compared to the strong defects found in *vps29* single mutant, which lacks a functional retromer complex (Pourcher et al. 2010). Moreover, although AtSNX1 and AtVPS29 localize to the same population of endosomes (Jaillais et al. 2007), in the absence of the three Arabidopsis SNX proteins, i.e., in the context of *snx* triple mutant, the retromer subunit AtVPS29 was still associated with endosomes (Pourcher et al. 2010). Multiple attempts failed to demonstrate any physical interaction between Arabidopsis SNX proteins and the core retromer, which supports the model where SNX and the retromer mostly work independently in plants. However, genetic analyses established that SNX and VPS26/VPS29/VPS35 proteins can also work together in some common developmental pathways such as embryo development and seedling growth (Jaillais et al. 2007; Pourcher et al. 2010).

As in yeast and mammals, plant SNXs are involved in diverse physiological functions by mediating the trafficking of membrane proteins. In *B. oleracea*, BoSNX1 was shown to interact with the S locus receptor kinase (SRK) that is essential for self-incompatibility (SI) response. Thus, it has been proposed to play a role in this mechanism; however, the exact role of this interaction remains unknown (Vanoosthuyse et al. 2003). In *A. thaliana*, AtSNX1 and AtSNX2a were shown to participate in the recycling of VSRs, which are involved in the transport of soluble cargos to the lytic vacuole (Niemes et al. 2010a; Niemes et al. 2010b).

The study of *A. thaliana* AtSNX1 revealed a role for AtSNX1 in plant development by mediating proper trafficking of auxin carriers (Jaillais et al. 2006). Auxin efflux carriers of the PIN family are major regulators of auxin transport between cells and they respond dynamically to physiological and environmental changes (Paciorek et al. 2005; Laxmi et al. 2008; Shibasaki et al. 2009; Rahman et al. 2010). In roots, PIN1 and PIN2 are plasma membrane proteins that show distinct patterns of polar localization depending on cell layers. This polar targeting of PINs to one side of root cells is dependent on the constitutive cycling of PIN proteins between plasma membrane and endosomal compartments (Geldner et al. 2001, see also chapter by Nodzynski et al. in this volume). Newly synthesized PINs are initially symmetrically distributed around the cell and their polarity is established following successive rounds of endocytosis/recycling (Dhonukshe et al. 2008). The final polarity of PINs involves ADP ribosylation factor GTP-exchanging factors (ARF-GEFs), such as GNOM, and the phosphorylation status of the proteins, which is regulated by the counterbalancing activities of PINOID Kinase and protein phosphatase 2A (Geldner et al. 2001; Michniewicz et al. 2007; Dhonukshe et al. 2008). Interestingly, in *snx1* mutants, Jaillais et al. (2006) reported that PIN2, but not PIN1, accumulates in endosomal compartments distinct from endosomes possessing GNOM (Jaillais et al. 2006), and were still properly polarized in epidermal cells. The loss of AtSNX1 function was accompanied with multiple auxin-related defects such as shorter primary roots and impaired response to gravity. Stimulation by gravity of root tips results in internalization of PIN2 proteins at the upper side of the root associated with degradation of PIN2, whereas endocytosis is inhibited at the lower side (Paciorek et al. 2005; Abas et al. 2006). In wild-type plants, following root gravistimulation, PIN2 was found to accumulate in SNX1-positive endosomes, indicating a possible function for SNX1 in routing PIN2 for degradation (Jaillais et al. 2006). This was confirmed later by the work of Kleine-Vehn et al. (2008), which suggests that SNX1 has a gating function for endocytic trafficking of PIN2 to the lytic vacuole. Upon gravistimulation, PIN2 is presumably translocated to the vacuole via a SNX1-dependent pathway.

Subsequently, SNX1-compartment has been proposed to correspond to a sorting endosome at the crossroads between the endocytic and secretory pathways. Such sorting endosome is able to redirect cargo proteins toward various destinations such as the plasma membrane for recycling PM proteins, vacuole for protein degradation, or TGN for retrograde transport of proteins (Jaillais et al. 2008). In addition to SNX1, SNX2b was hypothesized to be involved in trafficking from endosomes to the vacuole since overexpression of SNX2b inhibits the trafficking of the membrane FM4-64 dye between these two compartments (Phan et al. 2008).

Until now, the subcellular localization of Arabidopsis AtSNX1 is still a matter of debate. Initially, AtSNX1 fused to the green fluorescent protein (GFP) was demonstrated to colocalize with markers of MVBs, such as RABF2b and the vacuolar sorting receptor BP80, in Arabidopsis root tips, whereas no colocalization was observed with markers of the TGN (Jaillais et al. 2006, 2008). Colocalization between AtSNX1 and the yellow fluorescent protein (YFP)-FYVE, which labels MVBs (Voigt et al. 2005), also supports that AtSNX1-containing compartments correspond to MVBs (Kleine-Vehn et al. 2008). However, based on immunogold electron microscopy, recent work suggests that AtSNX1 is predominantly, if not exclusively, at the TGN in Arabidopsis root cells (Robinson et al. 2012; Stierhof et al. 2012).

In mammals, while some SNX proteins, such as HsSNX9 and HsSNX33, have been reported to solely homodimerize (Haberg et al. 2008; Dislich et al. 2011), other SNXs including HsSNX1, HsSNX2, HsSNX5, HsSNX6 are able to form a variety of heterodimers such as the pairs HsSNX1-HsSNX2, HsSNX1-HsSNX5, HsSNX1-HsSNX6, HsSNX2-HsSNX5, and HsSNX2-HsSNX6 (Rojas et al. 2007; Wassmer et al. 2009). SNX heterodimers likely display distinct properties compared to homodimers, but the cellular functions of these different combinations of SNXs remain unclear. Recently, a very interesting role of heterodimerization in the regulation of SNX subcellular localization has been described in plants. Although all Arabidopsis SNXs possess a PX and a BAR domain (Fig. 1b), Pourcher et al. (2010) showed that AtSNX2a and AtSNX2b, in the absence of AtSNX1 (i.e., in the *snx1* mutant), remain cytosolic, while AtSNX1 alone can associate with endosomal membranes. However, following heterodimerization with AtSNX1, AtSNX2a and AtSNX2b are able to bind to endosomes (Pourcher et al. 2010).

4 Conclusions

During the past decade, a growing interest in deciphering the molecular mechanims that control intracellular trafficking pathways has led to the discovery of unexpected functions for endosomes and endocytic trafficking in a variety of organisms. In particular, SNX proteins, alone or most often associated with the retromer VPS proteins, play a crucial role in diverse physiological and developmental processes in mammals and plants. Work in Arabidopsis uncovered some unique features of these proteins with regard to their distinct binding ability to endosomal membranes and role in endocytic recycling of certain PM proteins. It remains to be investigated whether the other PX-domain-containing proteins found in Arabidopsis genome may also have functions in plant development similar to SNXs. Although yeast two-hybrid assays and pull-down analyses provided evidence for protein interactions between SNXs and a few membrane receptors, cargos of SNX and VPS proteins remain still largely unidentified in plants. Identification of proteins operating through the SNX and/or retromer transport machinery as well as determination of the developmental pathways regulated by these proteins will be undoubtedly the next challenge for the coming years.

References

Abas L, Benjamins R, Malenica N, Paciorek T, Wirniewska J, Moulinier-Anzola JC, Sieberer T, Friml J, Luschnig C (2006) Intracellular trafficking and proteolysis of the Arabidopsis auxin-efflux facilitator PIN2 are involved in root gravitropism. Nat Cell Biol 8:249–256

Arighi CN, Hartnell LM, Aguilar RC, Haft CR, Bonifacino JS (2004) Role of the mammalian retromer in sorting of the cation-independent mannose 6-phosphate receptor. J Cell Biol 165:123–133

Barberon M, Zelazny E, Robert S, Conejero G, Curie C, Friml J, Vert G (2011) Monoubiquitin-dependent endocytosis of the iron-regulated transporter 1 (IRT1) transporter controls iron uptake in plants. Proc Natl Acad Sci U S A 108:E450–E458

Bravo J, Karathanassis D, Pacold CM, Pacold ME, Ellson CD, Anderson KE, Butler PJ, Lavenir I, Perisic O, Hawkins PT, Stephens L, Williams RL (2001) The crystal structure of the PX domain from p40(phox) bound to phosphatidylinositol 3-phosphate. Mol Cell 8:829–839

Carlton J, Bujny M, Peter BJ, Oorschot VM, Rutherford A, Mellor H, Klumperman J, McMahon HT, Cullen PJ (2004) Sorting nexin-1 mediates tubular endosome-to-TGN transport through coincidence sensing of high-curvature membranes and 3-phosphoinositides. Curr Biol 14:1791–1800

Carlton J, Bujny M, Rutherford A, Cullen P (2005a) Sorting nexins–unifying trends and new perspectives. Traffic 6:75–82

Carlton JG, Bujny MV, Peter BJ, Oorschot VM, Rutherford A, Arkell RS, Klumperman J, McMahon HT, Cullen PJ (2005b) Sorting nexin-2 is associated with tubular elements of the early endosome, but is not essential for retromer-mediated endosome-to-TGN transport. J Cell Sci 118:4527–4539

Cozier GE, Carlton J, McGregor AH, Gleeson PA, Teasdale RD, Mellor H, Cullen PJ (2002) The phox homology (PX) domain-dependent, 3-phosphoinositide-mediated association of sorting nexin-1 with an early sorting endosomal compartment is required for its ability to regulate epidermal growth factor receptor degradation. J Biol Chem 277:48730–48736

Cullen PJ (2008) Endosomal sorting and signalling: an emerging role for sorting nexins. Nat Rev Mol Cell Biol 9:574–582

Cullen PJ, Korswagen HC (2012) Sorting nexins provide diversity for retromer-dependent trafficking events. Nat Cell Biol 14:29–37

Dhonukshe P, Tanaka H, Goh T, Ebine K, Mahonen AP, Prasad K, Blilou I, Geldner N, Xu J, Uemura T, Chory J, Ueda T, Nakano A, Scheres B, Friml J (2008) Generation of cell polarity in plants links endocytosis, auxin distribution and cell fate decisions. Nature 456:962–966

Dislich B, Than ME, Lichtenthaler SF (2011) Specific amino acids in the BAR domain allow homodimerization and prevent heterodimerization of sorting nexin 33. Biochem J 433:75–83

Franch-Marro X, Wendler F, Guidato S, Griffith J, Baena-Lopez A, Itasaki N, Maurice MM, Vincent JP (2008) Wingless secretion requires endosome-to-golgi retrieval of Wntless/Evi/Sprinter by the retromer complex. Nat Cell Biol 10:170–177

Frost A, Perera R, Roux A, Spasov K, Destaing O, Egelman EH, De Camilli P, Unger VM (2008) Structural basis of membrane invagination by F-BAR domains. Cell 132:807–817

Geldner N, Friml J, Stierhof YD, Jurgens G, Palme K (2001) Auxin transport inhibitors block PIN1 cycling and vesicle trafficking. Nature 413:425–428

Geldner N, Hyman DL, Wang X, Schumacher K, Chory J (2007) Endosomal signaling of plant steroid receptor kinase BRI1. Genes Dev 21:1598–1602

Gullapalli A, Garrett TA, Paing MM, Griffin CT, Yang Y, Trejo J (2004) A role for sorting nexin 2 in epidermal growth factor receptor down-regulation: evidence for distinct functions of sorting nexin 1 and 2 in protein trafficking. Mol Biol Cell 15:2143–2155

Gullapalli A, Wolfe BL, Griffin CT, Magnuson T, Trejo J (2006) An essential role for SNX1 in lysosomal sorting of protease-activated receptor-1: evidence for retromer-, Hrs-, and Tsg101-independent functions of sorting nexins. Mol Biol Cell 17:1228–1238

Haberg K, Lundmark R, Carlsson SR (2008) SNX18 is an SNX9 paralog that acts as a membrane tubulator in AP-1-positive endosomal trafficking. J Cell Sci 121:1495–1505

Haft CR, de la Luz Sierra M, Barr VA, Haft DH, Taylor SI (1998) Identification of a family of sorting nexin molecules and characterization of their association with receptors. Mol Cell Biol 18:7278–7287

Hettema EH, Lewis MJ, Black MW, Pelham HR (2003) Retromer and the sorting nexins Snx4/41/42 mediate distinct retrieval pathways from yeast endosomes. EMBO J 22:548–557

Horazdovsky BF, Davies BA, Seaman MN, McLaughlin SA, Yoon S, Emr SD (1997) A sorting nexin-1 homologue, Vps5p, forms a complex with Vps17p and is required for recycling the vacuolar protein-sorting receptor. Mol Biol Cell 8:1529–1541

Jaillais Y, Fobis-Loisy I, Miege C, Gaude T (2008) Evidence for a sorting endosome in Arabidopsis root cells. Plant J 53:237–247

Jaillais Y, Fobis-Loisy I, Miege C, Rollin C, Gaude T (2006) AtSNX1 defines an endosome for auxin-carrier trafficking in Arabidopsis. Nature 443:106–109

Jaillais Y, Santambrogio M, Rozier F, Fobis-Loisy I, Miege C, Gaude T (2007) The retromer protein VPS29 links cell polarity and organ initiation in plants. Cell 130:1057–1070

Kasai K, Takano J, Miwa K, Toyoda A, Fujiwara T (2011) High boron-induced ubiquitination regulates vacuolar sorting of the BOR1 borate transporter in *Arabidopsis thaliana*. J Biol Chem 286:6175–6183

Kleine-Vehn J, Leitner J, Zwiewka M, Sauer M, Abas L, Luschnig C, Friml J (2008) Differential degradation of PIN2 auxin efflux carrier by retromer-dependent vacuolar targeting. Proc Natl Acad Sci USA 105:17812–17817

Koumandou VL, Klute MJ, Herman EK, Nunez-Miguel R, Dacks JB, Field MC (2011) Evolutionary reconstruction of the retromer complex and its function in Trypanosoma brucei. J Cell Sci 124:1496–1509

Kurten RC, Cadena DL, Gill GN (1996) Enhanced degradation of EGF receptors by a sorting nexin, SNX1. Science 272:1008–1010

Kurten RC, Eddington AD, Chowdhury P, Smith RD, Davidson AD, Shank BB (2001) Self-assembly and binding of a sorting nexin to sorting endosomes. J Cell Sci 114:1743–1756

Larkin MA, Blackshields G, Brown NP, Chenna R, McGettigan PA, McWilliam H, Valentin F, Wallace IM, Wilm A, Lopez R, Thompson JD, Gibson TJ, Higgins DG (2007) Clustal W and clustal X version 2.0. Bioinformatics 23:2947–2948

Laxmi A, Pan J, Morsy M, Chen R (2008) Light plays an essential role in intracellular distribution of auxin efflux carrier PIN2 in Arabidopsis thaliana. PLoS One 3:e1510

Lundmark R, Carlsson SR (2003) Sorting nexin 9 participates in clathrin-mediated endocytosis through interactions with the core components. J Biol Chem 278:46772–46781

Michniewicz M, Zago MK, Abas L, Weijers D, Schweighofer A, Meskiene I, Heisler MG, Ohno C, Zhang J, Huang F, Schwab R, Weigel D, Meyerowitz EM, Luschnig C, Offringa R, Friml J (2007) Antagonistic regulation of PIN phosphorylation by PP2A and PINOID directs auxin flux. Cell 130:1044–1056

Niemes S, Labs M, Scheuring D, Krueger F, Langhans M, Jesenofsky B, Robinson DG, Pimpl P (2010a) Sorting of plant vacuolar proteins is initiated in the ER. Plant J 62:601–614

Niemes S, Langhans M, Viotti C, Scheuring D, San Wan Yan M, Jiang L, Hillmer S, Robinson DG, Pimpl P (2010b) Retromer recycles vacuolar sorting receptors from the *trans*-Golgi network. Plant J 61:107–121

Nisar S, Kelly E, Cullen PJ, Mundell SJ (2010) Regulation of P2Y1 receptor traffic by sorting Nexin 1 is retromer independent. Traffic 11:508–519

Nothwehr SF, Hindes AE (1997) The yeast VPS5/GRD2 gene encodes a sorting nexin-1-like protein required for localizing membrane proteins to the late golgi. J Cell Sci 110(Pt 9):1063–1072

Paciorek T, Zazimalova E, Ruthardt N, Petrasek J, Stierhof YD, Kleine-Vehn J, Morris DA, Emans N, Jurgens G, Geldner N, Friml J (2005) Auxin inhibits endocytosis and promotes its own efflux from cells. Nature 435:1251–1256

Perrière G, Gouy M (1996) WWW-query: an on-line retrieval system for biological sequence banks. Biochimie 78:364–369

Peter BJ, Kent HM, Mills IG, Vallis Y, Butler PJ, Evans PR, McMahon HT (2004) BAR domains as sensors of membrane curvature: the amphiphysin BAR structure. Science 303:495–499

Phan NQ, Kim SJ, Bassham DC (2008) overexpression of Arabidopsis sorting nexin AtSNX2b inhibits endocytic trafficking to the vacuole. Mol Plant 1:961–976

Ponting CP (1996) Novel domains in NADPH oxidase subunits, sorting nexins, and PtdIns 3-kinases: binding partners of SH3 domains? Protein Sci 5:2353–2357

Pourcher M, Santambrogio M, Thazar N, Thierry AM, Fobis-Loisy I, Miege C, Jaillais Y, Gaude T (2010) Analyses of sorting nexins reveal distinct retromer-subcomplex functions in development and protein sorting in *Arabidopsis thaliana*. Plant Cell 22:3980–3991

Pylypenko O, Lundmark R, Rasmuson E, Carlsson SR, Rak A (2007) The PX-BAR membrane-remodeling unit of sorting nexin 9. EMBO J 26:4788–4800

Rahman A, Takahashi M, Shibasaki K, Wu S, Inaba T, Tsurumi S, Baskin TI (2010) Gravitropism of Arabidopsis thaliana roots requires the polarization of PIN2 toward the root tip in meristematic cortical cells. Plant Cell 22:1762–1776

Reddy JV, Seaman MN (2001) Vps26p, a component of retromer, directs the interactions of Vps35p in endosome-to-golgi retrieval. Mol Biol Cell 12:3242–3256

Reyes FC, Buono R, Otegui MS (2011) Plant endosomal trafficking pathways. Curr Opin Plant Biol 14:666–673

Robinson DG, Pimpl P, Scheuring D, Stierhof YD, Sturm S, Viotti C (2012) Trying to make sense of retromer. Trends Plant Sci 17:431–439

Roepstorff K, Grandal MV, Henriksen L, Knudsen SL, Lerdrup M, Grovdal L, Willumsen BM, van Deurs B (2009) Differential effects of EGFR ligands on endocytic sorting of the receptor. Traffic 10:1115–1127

Rojas R, Kametaka S, Haft CR, Bonifacino JS (2007) Interchangeable but essential functions of SNX1 and SNX2 in the association of retromer with endosomes and the trafficking of mannose 6-phosphate receptors. Mol Cell Biol 27:1112–1124

Seaman MN (2004) Cargo-selective endosomal sorting for retrieval to the Golgi requires retromer. J Cell Biol 165:111–122

Seaman MN, Marcusson EG, Cereghino JL, Emr SD (1997) Endosome to Golgi retrieval of the vacuolar protein sorting receptor, Vps10p, requires the function of the VPS29, VPS30, and VPS35 gene products. J Cell Biol 137:79–92

Seet LF, Hong W (2006) The phox (PX) domain proteins and membrane traffic. Biochim Biophys Acta 1761:878–896

Shibasaki K, Uemura M, Tsurumi S, Rahman A (2009) Auxin response in Arabidopsis under cold stress: underlying molecular mechanisms. Plant Cell 21:3823–3838

Shin N, Lee S, Ahn N, Kim SA, Ahn SG, YongPark Z, Chang S (2007) Sorting nexin 9 interacts with dynamin 1 and N-WASP and coordinates synaptic vesicle endocytosis. J Biol Chem 282:28939–28950

Soulet F, Yarar D, Leonard M, Schmid SL (2005) SNX9 regulates dynamin assembly and is required for efficient clathrin-mediated endocytosis. Mol Biol Cell 16:2058–2067

Stierhof YD, Viotti C, Scheuring D, Sturm S, Robinson DG (2012) Sorting nexins 1 and 2a locate mainly to the TGN. Protoplasma. doi:10.1007/s00709-012-0399-1

Strochlic TI, Setty TG, Sitaram A, Burd CG (2007) Grd19/Snx3p functions as a cargo-specific adapter for retromer-dependent endocytic recycling. J Cell Biol 177:115–125

Takano J, Tanaka M, Toyoda A, Miwa K, Kasai K, Fuji K, Onouchi H, Naito S, Fujiwara T (2010) Polar localization and degradation of Arabidopsis boron transporters through distinct trafficking pathways. Proc Natl Acad Sci USA 107:5220–5225

Teasdale RD, Collins BM (2012) Insights into the PX (phox-homology) domain and SNX (sorting nexin) protein families: structures, functions and roles in disease. Biochem J 441:39–59

Teasdale RD, Loci D, Houghton F, Karlsson L, Gleeson PA (2001) A large family of endosome-localized proteins related to sorting nexin 1. Biochem J 358:7–16

Traer CJ, Rutherford AC, Palmer KJ, Wassmer T, Oakley J, Attar N, Carlton JG, Kremerskothen J, Stephens DJ, Cullen PJ (2007) SNX4 coordinates endosomal sorting of TfnR with dynein-mediated transport into the endocytic recycling compartment. Nat Cell Biol 9:1370–1380

van Leeuwen W, Okresz L, Bogre L, Munnik T (2004) Learning the lipid language of plant signalling. Trends Plant Sci 9:378–384

van Weering JR, Verkade P, Cullen PJ (2010) SNX-BAR proteins in phosphoinositide-mediated, tubular-based endosomal sorting. Semin Cell Dev Biol 21:371–380

van Weering JR, Verkade P, Cullen PJ (2012) SNX-BAR-mediated endosome tubulation is co-ordinated with endosome maturation. Traffic 13:94–107

Vanoosthuyse V, Tichtinsky G, Dumas C, Gaude T, Cock JM (2003) Interaction of calmodulin, a sorting nexin and kinase-associated protein phosphatase with the *Brassica oleracea* S locus receptor kinase. Plant Physiol 133:919–929

Voigt B, Timmers AC, Šamaj J, Hlavacka A, Ueda T, Preuss M, Nielsen E, Mathur J, Emans N, Stenmark H, Nakano A, Baluška F, Menzel D (2005) Actin-based motility of endosomes is linked to the polar tip growth of root hairs. Eur J Cell Biol 84:609–621

Wang Q, Kaan HY, Hooda RN, Goh SL, Sondermann H (2008) Structure and plasticity of endophilin and sorting nexin 9. Structure 16:1574–1587

Wassmer T, Attar N, Harterink M, van Weering JR, Traer CJ, Oakley J, Goud B, Stephens DJ, Verkade P, Korswagen HC, Cullen PJ (2009) The retromer coat complex coordinates endosomal sorting and dynein-mediated transport, with carrier recognition by the *trans*-Golgi network. Dev Cell 17:110–122

Worby CA, Dixon JE (2002) Sorting out the cellular functions of sorting nexins. Nat Rev Mol Cell Biol 3:919–931

Xu Y, Hortsman H, Seet L, Wong SH, Hong W (2001) SNX3 regulates endosomal function through its PX-domain-mediated interaction with PtdIns(3)P. Nat Cell Biol 3:658–666

Yamazaki M, Shimada T, Takahashi H, Tamura K, Kondo M, Nishimura M, Hara-Nishimura I (2008) Arabidopsis VPS35, a retromer component, is required for vacuolar protein sorting and involved in plant growth and leaf senescence. Plant Cell Physiol 49:142–156

Zhong Q, Lazar CS, Tronchere H, Sato T, Meerloo T, Yeo M, Songyang Z, Emr SD, Gill GN (2002) Endosomal localization and function of sorting nexin 1. Proc Natl Acad Sci USA 99:6767–6772

Zhong Q, Watson MJ, Lazar CS, Hounslow AM, Waltho JP, Gill GN (2005) Determinants of the endosomal localization of sorting nexin 1. Mol Biol Cell 16:2049–2057

Zhou CZ, de La Sierra-Gallay IL, Quevillon-Cheruel S, Collinet B, Minard P, Blondeau K, Henckes G, Aufrere R, Leulliot N, Graille M, Sorel I, Savarin P, de la Torre F, Poupon A, Janin J, van Tilbeurgh H (2003) Crystal structure of the yeast phox homology (PX) domain protein Grd19p complexed to phosphatidylinositol-3-phosphate. J Biol Chem 278:50371–50376

Zimmerberg J, McLaughlin S (2004) Membrane curvature: how BAR domains bend bilayers. Curr Biol 14:R250–R252

ESCRT-Dependent Sorting in Late Endosomes

Marisa S. Otegui, Rafael Buono, Francisca C. Reyes and Hannetz Roschzttardtz

Abstract The sorting of plasma membrane (PM) proteins for degradation involves their internalization by endocytosis, delivery to endosomes, and sorting into intraluminal vesicles (ILV) of late endosomes/multivesicular bodies (MVBs). The sorting of cargo proteins into ILVs depends on protein complexes named ESCRT-0 to III (Endosomal Sorting Complex Required for Transport) and the SKD1/Vps4p (Suppressor of K^+ Transport Growth Defect 1/Vacuolar protein sorting (vps) 4). These complexes associate with ubiquitinated cargo proteins, mediate their accumulation, sorting, and deformation of the endosomal membrane. With the exception of ESCRT-0, this machinery is well conserved across eukaryotes, including plants. Here, we discuss the general mechanism of MVB sorting mediated by ESCRT proteins with a special emphasis on what is known about these cellular processes in plants.

Keywords ESCRT · Plant endosomes · Multivesicular bodies

1 Late Endosomal Functions

Endosomal trafficking pathways are emerging as key regulators of plasma membrane (PM) protein abundance and distribution, controlling multiple signalling pathways and developmental processes in eukaryotes.

M. S. Otegui (✉) · R. Buono · F. C. Reyes · H. Roschzttardtz
Botany Department, University of Wisconsin, Lincoln Drive 430,
Madison, WI 53706, USA
e-mail: otegui@wisc.edu

J. Šamaj (ed.), *Endocytosis in Plants*,
DOI: 10.1007/978-3-642-32463-5_13,

Endosomes traffic cargo from both the endocytic and biosynthetic pathways, acting as a central hub for the transport of membrane lipids and proteins in the cell. Plasma membrane (PM) proteins are continuously internalized by endocytosis and moved to endosomes, where they can be sorted either back to the PM (recycling) or to vacuoles/lysosomes for degradation. Endosomal sorting of endocytosed PM proteins, such as signalling receptors and transporters, is a key process that controls PM protein composition and the ability of cells to perceive and respond to extracellular stimuli. Endosomes in animal cells are classified as early, recycling, intermediate, and late endosomes/multivesicular bodies (MVBs). Early and recycling endosomes receive and recycle endocytosed proteins back to the PM and biosynthetic receptors (e.g. vacuolar cargo receptors) back to the *trans*-Golgi network (TGN). Intermediate endosomes and MVBs sort membrane proteins into intraluminal vesicles (ILVs). When mature MVBs fuse with vacuoles/lysosomes, the ILVs are released into the hydrolytic environment of the vacuolar/lysosomal lumen and degraded. In addition, MVBs also carry newly synthesized vacuolar proteins from the post-Golgi compartments such as TGN to lysosomes/vacuoles.

Plants have conserved endosomal functions but they have also evolved specific variations on the organization of their endosomal system and the molecular machineries that control endosomal sorting. Only two endosomal organelles have been clearly indentified and characterized in plants: the TGN that acts as recycling/early endosomes and the MVBs that contain from few to hundreds of ILVs.

Endosomes regulate multiple trafficking pathways by coordinating the recognition, concentration, and packaging of cargo proteins by different protein complexes. Some of these proteins contain BAR (bin-amphiphysin-rvs)- and ENTH (epsin n-terminal homology)-domains and act on vesiculation and membrane tubulation events at endosomes that result in the budding of membranous structures and vesicles into the cytoplasm (Weering et al. 2010; Zimmermann et al. 2010; Weering et al. 2012). On the other hand, the ESCRT (endosomal sorting complex required for transport) machinery mediates the formation of vesicles in the opposite topology, that is, away from the cytoplasm and into the endosomal lumen. In these endosomal sorting pathways, the mechanism of cargo recognition depends on either direct interactions between cargo proteins and sorting complexes or on post-translational modifications of cargo proteins. The ESCRT machinery recognizes ubiquitin moieties on cargo proteins to be sequestered into ILV.

2 The ESCRT Complexes

ESCRT was first identified and characterized as ubiquitin-dependent protein-sorting machinery in yeast based on the pioneer work of Scott Emr and Tom H. Stevens' laboratories in the 1980s and 1990s. These research groups identified different classes of *Saccharomyces cerevisiae* mutants that displayed vacuolar protein sorting (vps) defects (Rothman and Stevens 1986; Banta et al. 1988; Rothman et al. 1989; Robinson et al. 1998). Mutants that exhibited aberrant endosomes were classified as "class E" (Raymond et al. 1992). It was later shown that class E genes are required for the sorting of transmembrane proteins (for example carboxypeptidase S) into the vacuolar lumen via the MVB pathway (Odorizzi et al. 1998; Reggiori and Pelham 2001). Studies performed in yeast and animals indicated that MVB sorting depends on five protein complexes, namely ESCRT-0, I, II, and III and the AAA ATPase SKD1/Vps4p (Suppressor of K^+ Transport Growth Defect 1/Vacuolar protein sorting 4) (Asao et al. 1997; Babst et al. 1997, 1998, 2002a, b; Katzmann et al. 2001, 2003, 2004; Shih et al. 2002; Bache et al. 2003b, 2006; Henne William et al. 2011), which seem to work sequentially in the sorting of membrane cargo proteins and the formation of ILV.

ESCRT proteins are ancient components of the eukaryotic endomembrane machinery and are found in all five major supergroups of eukaryotes: Excavata, Chromalveolata, Archaeplastida, Amoebozoa, and Opisthokonta (Leung et al. 2008). Moreover, genes encoding ESCRT-III- and SKD1/Vps4p-like proteins have been also identified in *Sulfolobus* (Crenarchaea). These archaeal ESCRT components appear to have a fundamental role in cell division (Samson et al. 2008, 2011; Ghazi-Tabatabai et al. 2009). Comparative genomic studies have shown that besides a trend of ESCRT gene conservation across eukaryotes, some ESCRT complexes have undergone expansion of genes encoding for multiple subunits. This is particularly noticeable in plants and multicellular Opisthokonta (fungi and metazoans) (Table 1).

With the exception of ESCRT-0, plants seem to contain orthologs for most of the ESCRT proteins originally identified in metazoans and fungi (Winter and Hauser 2006; Leung et al. 2008) (Table 1) and their interaction networks seem to be also conserved (Spitzer et al. 2006; Haas et al. 2007; Spitzer et al. 2009; Shahriari et al. 2010, 2011; Richardson et al. 2011). However, only a few plant ESCRT proteins have been functionally characterized.

ESCRT-0: Within eukaryotes, only Opisthokonta contain the typical ESCRT-0 subunits Vps27p/Hrs (hepatocyte growth factor-regulated tyrosine kinase substrate) and Hse1p/STAM1/2 (signal transducing adaptor molecule1/2) (Leung et al. 2008). Although both subunits share some structural similarities, they differ in the presence of a FYVE (Fab-1, YGL023, Vps27, and EEA1) zinc finger domain on Vps27p/Hrs (Mao et al. 2000). The FYVE domain binds with high affinity phosphatidylinositol 3-phosphate (PI3P), which is highly enriched in endosomal membranes, and penetrate the lipid bilayer (Stahelin et al. 2002). In addition, both ESCRT-0 subunits contain a VHS (Vps27, Hrs, and STAM) domain that mediates both membrane association and ubiquitin binding (Ren and Hurley 2010; Lange et al. 2012). In fact, recombinant *Caenorhabditis elegans* ESCRT-0 protein complexes are able to bind multiple ubiquitinated cargos simultaneously (Mayers et al. 2011), indicating that ESCRT-0 acts as important recognition module for both endosomal membranes and ubiquitinated cargo during the early stages of MVB sorting. In Metazoans and yeast, ESCRT-0 also binds and recruits ESCRT-I to endosomes (Bache et al. 2003a; Katzmann et al. 2003; Lu et al. 2003).

However, most eukaryotic groups, including plants, do not have these ESCRT-0 components. Which proteins do initially recognize the ubiquitinated cargo on

Table 1 ESCRT and ESCRT-associated proteins in yeast, mammals, and *Arabidopsis thaliana*

Saccharomyces cerevisiae	Mammals	*Arabidopsis thaliana*	Arabidopsis locus name
ESCRT-0			
Vps27	HRS		
Hse1p	STAM1		
	STAM2		
TOM1 and TOM1-like			
	Tom1L1	TOM1A	At2g38410
	Tom1L2-1	TOM1B	At5g01760
	Tom1L2-3	TOM1C	At1g76970
	Tom1L3	TOM1D	At4g32760
	Tom1	TOM1E	At1g06210
		TOM1F	At1g21380
		TOM_L	At5g16880
		TOM1G	At5g63640
		TOM1H	At3g08790
ESCRT I			
Vps23p/Stp22	TSG101/VPS23	ELC/VPS23A	At3g12400
		ELC-like/VPS23B	At5g13860
Vps28p	VPS28	VPS28-1	At4g21560
		VPS28-2	At4g05000
Vps37p	VPS37A	VPS37-1	At3g53120
	VPS37B	VPS37-2	At2g36680
	VPS37C		
	VPS37D		
Mvb12p	MVB12		
ESCRT II			
Vps22p	EAP30	VPS22	At4g27040
Vps25p	EAP25	VPS25	At4g19003
Vps36p	EAP45	VPS36	At5g04920

(continued)

Table 1 (continued)

Saccharomyces cerevisiae	Mammals	*Arabidopsis thaliana*	Arabidopsis locus name
ESCRT III			
Vps2p	CHMP2A	VPS2.1	At2g06530
	CHMP2B	VPS2.2	At5g44560
		VPS2.3	At1g03950
Vps20p	CHMP6	VPS20.1	At5g63880
		VPS20.2	At5g09260
Vps24p	CHMP3	VPS24.1	At5g22950
		VPS24.2	At3g45000
Snf7p/Vps32p	CHMP4A	SNF7.1/VPS32.1	At4g29160
	CHMP4B	SNF7.2/VPS32.2	At2g19830
	CHMP4C		
ESCRT-III-related			
Did2p	CHMP1A	CHMP1A	At1g73030
	CHMP1B	CHMP1B	At1g17730
Vps60p	CHMP5	VPS60.1	At3g10640
		VPS60.2	At5g04850
SKD1/Vps4complex			
Vps4p	VPS4A	SKD1	At2g27600
	VPS4B		
Vta1	LIP5	LIP5	At4g26750

endosomes in the absence of ESCRT-0? TOM1 (target of myb1) proteins, which also contain VHS domains that are able to bind ubiquitin and membranes and are widely distributed in eukaryotes, have been postulated to play the role of an ancestral ESCRT-0 (Blanc et al. 2009; Herman et al. 2011). Thus, Vps27p/Hrs and Hse1p/STAM would be more recent evolutionary acquisitions in Opisthokonta.

There are nine TOM1-like genes in the Arabidopsis genome (Hauser 2006; Richardson et al. 2011; Winter and Herman et al. 2011), but their functional role in MVB sorting has not been explored. However, several ESCRT-I components have been shown to interact with Arabidopsis TOM1-like proteins (Richardson et al. 2011), which is consistent with their putative ESCRT-0 function.

Alternatively, in the absence of a canonical ESCRT-0 complex, plants may rely on a more efficient ESCRT-I endosomal recruitment mechanism, independent of ESCRT-0 function.

ESCRT-I: In Metazoans and yeast, ESCRT-I is a heterotetramer composed of Vps23p/Tsg101 (Tumor Susceptibility Gene 101), Vps28, Vps37, and MVB12 (multivesicular body 12) or the MVB12-like protein UBAP1 (ubiquitin associated protein 1). It assembles as a 1:1:1:1 complex in solution and interacts weakly with acidic phospholipids (Bishop and Woodman 2001; Katzmann et al. 2001; Bache et al. 2004; Stuchell et al. 2004; Eastman et al. 2005; Chu et al. 2006; Curtiss et al. 2007; Gill et al. 2007; Morita et al. 2007). The majority of ESCRT-I is distributed throughout the cytoplasm as a stable complex and is only recruited to endosomes by interactions with the ESCRT-0 complex.

The core yeast ESCRT-I complex consists of an elongated structure of ~20 nm in length, with three of the subunits arranged into a coiled-coil stalk and connected to a globular head group (Kostelansky et al. 2007). Opposite ends of the ESCRT-I complex are able to interact with ESCRT-0 and ESCRT-II. The interaction with ESCRT-0 occurs through the N- terminus of Vps23p whereas the binding to ESCRT-II is mediated by the C-terminal region of Vps28 (Katzmann et al. 2003; Kostelansky et al. 2006). The structural organization of ESCRT-I is thought to be conserved between yeast and metazoans, although mammals present a much greater diversity of ESCRT-I subunits, including four isoforms of VPS37 (VPS37A–D) and two of MVB12 (MVB12A and B) (Bache et al. 2004; Stuchell et al. 2004; Eastman et al. 2005; Morita et al. 2007). Yeast ESCRT-I binds ubiquitin through an N-terminal UEV (ubiquitin E2 variant) domain in Vps23p and the C-terminal domain in Mvb12p, both contributing to the ability of ESCRT-I to sort ubiquitinated cargo proteins in MVBs. Mammalian MVB12A has been shown to bind ubiquitin but together with MVB12B, do not seem to be critical for MVB sorting (Morita et al. 2007; Tsunematsu et al. 2010). Recently, the MVB12-like ESCRT-I subunit UBAP1 has been shown to form part of the mammalian ESCRT-I complex and to play an important role in MVB function (Stefani et al. 2011; Agromayor et al. 2012). The presence of several VPS37 and MVB12/MVB12-like subunits suggest that multiple different ESCRT-I complexes specifically tailored for particular functions/cell types may assemble in mammals (Pashkova and Piper 2012).

Plants contain several isoforms of only three of the ESCRT-I subunits, ELCH (ELC) (homolog of Vps23p/Tsg101) and ELC-like, VPS28-1 and VPS28-2, VPS37-1 and VPS37-2. The ELC protein binds ubiquitin in vitro and forms complexes with other ESCRT-I proteins in vivo (Spitzer et al. 2006). Three studies using yeast two-hybrid, immunoprecipitation, and bifluorescence complementation have shown conserved interactions among multiple Arabidopsis ESCRT-I subunits that are likely important for ESCRT-I complex assembly (Spitzer et al. 2006; Richardson et al. 2011; Shahriari et al. 2011). In addition, ELC, ELC-like and VPS28-1 interact with two ESCRT-II components, VPS36 and VPS22 (Richardson et al. 2011; Shahriari et al. 2011). As in animal cells, mutations in Arabidopsis *ELCH/TSG101* gene are associated with multinucleated cells due to cytokinesis defects (Spitzer et al. 2006). Interestingly, plants seem to be devoid of MVB12 (Leung et al. 2008). However, based on the low similarity between human and yeast MVB12 and its small protein size, it is possible that highly divergent MVB12-like subunits exist in other groups besides yeast and animals.

ESCRT-II: ESCRT-II is a Y-shaped heterotetramer composed of three subunits (Vps22p/EAP30, Vps25p/EAP20, and Vps36p/EAP45) in a 1:2:1 stoichiometry (Babst et al. 2002b). Vps22p and Vps36p form the base whereas two copies of Vps25p forms the arms of the "Y" like structure (Hierro et al. 2004; Teo et al. 2004). Similarly to ESCRT-0, ESCRT-II can bind PI3P with high affinity through the GLUE (gram-like ubiquitin-binding in EAP45) domain of Vps36p/EAP45 (Teo et al. 2006). However, in contrast to ESCRT-0, endogenous constitutively assembled ESCRT-II complexes are mostly found in the cytoplasm. Two NZF (npl4-type zinc finger) domains are inserted into the yeast GLUE domain; the NZF1 binds to the C-terminal domain of Vps28p (ESCRT-I) (Gill et al. 2007), and NZF2 binds ubiquitin (Alam et al. 2004; Teo et al. 2006). Interestingly, neither metazoans nor plants possess these NZF domains (Winter and Hauser 2006). Human VPS36/EAP45 also binds to ubiquitin, but through a non-conserved site on the GLUE domain (Slagsvold et al. 2005; Alam et al. 2006; Hirano et al. 2006). A linker region with a predicted alpha-helix immediately downstream of the human VPS36 GLUE domain has been shown to be necessary but not sufficient, for ESCRT-I binding in vitro (Im and Hurley 2008). ESCRT-II recruits ESCRT-III to the endosomal membranes by a high affinity interaction between Vps25p/EAP20 and Vps20p/CHMP6 (Langelier et al. 2006).

Protein interaction studies on Arabidopsis ESCRT components using yeast two-hybrid assays have shown that VPS22 and VPS36 interact with themselves, with each other, and with VPS25. The binding of VPS36 and VPS25 has been also confirmed by co-immunoprecipitation. As mentioned previously, the interaction between VPS28 (ESCRT-I) and VPS36 is conserved in plants (Richardson et al. 2011; Shahriari et al. 2011). In addition, Arabidopsis VPS25 and VPS36 interact with the ESCRT-III components VPS20A and B and CHMP4A and B, suggesting a conserved mechanism of ESCRT-III recruitment by ESCRT-II in plants.

ESCRT-III: A common feature of the ESCRT-III family members is a distinct distribution of charged amino acids, with two-thirds of the N-terminus carrying a basic amino acid residues and the remaining C-terminal region being acidic.

ESCRT-III proteins, also called CHMPs (charged multivesicular body proteins), consist of 11 isoforms in humans, which, except for CHMP7 (Horii et al. 2006), correspond to one of the six ESCRT-III-like proteins present in yeast. The ESCRT-III complex consists of four core subunits, Vps20p/CHMP6, Snf7p (sucrose non-fermenting7)/CHMP4/Vps32p, Vps24p/CHMP3, and Vps2p/CHMP2 (with several isoforms in animal and plants) and two accessory subunits Did2p/CHMP1, and Vps60p/CHMP5. In addition, IST1 (increased salt tolerance 1) has been shown to act as an ESCRT-III-related protein although its structure is more divergent from the typical ESCRT-III subunits. The core subunits are rather small proteins ranging between 220 and 250 amino acids with similar biochemical and structural properties. Recombinant Vps20p binds membranes weakly but in vivo, it becomes myristoylated, which enhances its ability to associate with endosomal membranes. However, based on studies in yeast, most of the Vps20p pool seems to be soluble in the cytoplasm, just like the other ESCRT-III proteins.

ESCRT-II binds Vps20p which recruits Snf7p. Snf7p oligomerizes and recruits Vps24p and Vps2p, completing the assembly of ESCRT-III (Babst et al. 2002a; Wollert et al. 2009). ESCRT-III oligomers form filaments tightly bound to endosomal membranes that require energy from ATP hydrolysis (through the action of SKD1/Vps4p) to dissociate from membranes and depolymerize. Yeast Vps20p, Snf7p, and Vps24p alone are sufficient for driving membrane scission in vitro (Wollert et al. 2009) whereas Vps2p is required for coupling to the Vps4p complex (Ghazi-Tabatabai et al. 2008).

Unlike ESCRT-I and -II, ESCRT-III subunits do not seem to bind ubiquitin and exist in the cytosol as monomers (Babst et al. 2002a) in an autoinhibited "closed" state. The subunits polymerize into the open active ESCRT-III complex, likely in the form of a fibril (Ghazi-Tabatabai et al. 2008; Hanson et al. 2008) on the endosomal membrane when recruited by ESCRT-II. All ESCRT-III subunits contain a five-helix core (Muziol et al. 2006) with the first two basic helices binding strongly to acidic membranes. The C-terminal region of ESCRT-III subunits has been shown to auto inhibit the assembly of the ESCRT-III complex and to maintain the soluble monomeric pool in closed state (Zamborlini et al. 2006; Shim et al. 2007; Lata et al. 2008a). In this C-terminal domain, various ESCRT-III subunits contain the so called MIT-interacting motifs-1 (MIM1; present in Vps24p/CHMP3, Vps2p/CHMP2, Did2p/CHMP1, IST1) and -2 (MIM2; present in Vps20p/CHMP3, Snf7p/CHMP4, IST1) that can bind MIT (microtubule-interacting and trafficking) domains, like the one present in SKD1/Vps4p (Scott et al. 2005; Kieffer et al. 2008). When ESCRT-III subunits are in active or "open" state, the C-terminal domain becomes available for interaction with other ESCRT-III subunits and with SKD1/Vps4p.

There are three VPS2 isoforms in Arabidopsis, VPS2.1-3, with different functions. VPS2.1 is the most similar to yeast Vps2p and is able to interact with the MIT domain of AMSH3 (associated molecule with the SH3 domain of STAM 3), a deubiquitinating enzyme involved in endosomal sorting (Katsiarimpa et al. 2011). Although Arabidopsis VPS2.1, VPS2.2, and VPS2.3 contain a conserved MIM1 domain, the spacing between two acidic amino acids (Asp-212 and Asp-214 in VPS2.1) is altered in VPS2.2 and VPS2.3 and this difference seems to be crucial for the binding with AMSH3. In addition, VPS2.1 but not VPS2.2 or VPS2.3 localized to aberrant endosomes upon overexpression of a dominant negative form of SKD1. Consistently, a VPS2.1 Arabidopsis mutant is embryo lethal whereas single VPS2.2 and VPS2.3 mutants show only mild root growth defects (Katsiarimpa et al. 2011). These results suggest that the VPS2.1 may perform the canonical VPS2 function in Arabidopsis. However, in a proteomic analysis, VPS2.2-GFP was shown to immunoprecipitate with CHMP1A and B, VPS2.1 and VPS2.3, SNF7.1 and VPS60.1, indicating that VPS2.2 is also part of the ESCRT machinery (Ibl et al. 2011).

The accessory ESCRT-III subunits play a modulatory effect on SKD1/Vps4p function. Whereas Did2p/CHMP1 (Nickerson et al. 2006) and Vps60p/CHMP5 help recruit and activate the SKD1/Vps4p complex, IST1 plays a dual, concentration-dependent function. At low concentration, IST1 stimulate SKD1/VPs4p activity but at high concentration, it is thought to play an inhibitory role (Dimaano et al. 2008).

Genetic studies suggest that Did2p does not play an essential function in yeast but its plant counterparts, the Arabidopsis CHMP1 A and B proteins, are required for plant development (Spitzer et al. 2009). CHMP1 proteins mediate the sorting of several auxin carriers, including PINFORMED1 (PIN1), PIN2, and AUXIN-RESISTANT1 (AUX1) for their degradation inside the vacuole. Double homozygous CHMP1A CHMP1B mutants either fail to complete embryo development and/or die during post-embryogenic seedling development. MVBs form in the double mutant but ILV formation and cargo sorting into ILVs is altered, consistent with a modulatory role of CHMP1 on SKD1 function.

The maize CHMP1 ortholog is called SAL1 (supernumerary aleurone layer 1). The *sal1* mutant develops multiple layers of aleurone (epidermal) cells in the endosperm (Shen et al. 2003) and it has been postulated that SAL1/CHMP1 is critical for the proper endosomal sorting of two plasma membrane proteins, DEFECTIVE KERNEL 1 (DEK1) and CRINKLY 4 (CR4), which are involved in aleurone cell fate specification (Tian et al. 2007).

SKD1/Vps4p complex: The disassembly of ESCRT-III from the endosomal membranes for completion of the ESCRT cycle requires the ATPase activity of the mechanoenzyme SKD1/Vps4p. SKD1/Vps4p consists of an N-terminal MIT domain followed by a linker, the ATPase cassette, a β domain, and a C-terminal helix (Scott et al. 2005; Takasu et al. 2005; Vajjhala et al. 2006, 2008; Xiao et al. 2007; Babst et al. 2011). SKD1/Vps4p is cytoplasmic when ADP-bound and it oligomerizes into most likely a dodecamer of two hexameric rings when it binds ATP (Gonciarz et al. 2008). Whereas ESCRT-III is the principal SKD1/Vps4p substrate, ESCRT-III itself regulates SKD1/Vps4p activity by different mechanisms (Babst et al. 2011). When ESCRT-III subunits are incorporated into fibrils, their MIM domains are assumed to become more accessible to interaction with SKD1/Vps4p, facilitating SKD1/Vps4p recruitment to the ESCRT coat and enhancing its oligomerization through increasing its local concentration. In addition, the interaction with ESCRT-III MIM domains relieves an auto inhibitory effect exerted by the MIT domain of SKD1/Vps4p, leading to an increase in

SKD1/Vps4p ATPase activity (Shim et al. 2007; Lata et al. 2008a; Bajorek et al. 2009; Merrill and Hanson 2010).

SKD1/Vps4p co-assembles onto membrane-bound ESCRT-III with its cofactor LIP5/Vta1p (Shiflett et al. 2004; Ward et al. 2005; Lottridge et al. 2006; Xiao et al. 2008; Shestakova et al. 2010). The C-terminal domain of LIP5/Vta1p binds the SKD1/Vps4p β-domain (Yang and Hurley 2010) and, although it does not stably associate with SKD1/Vps4p in the cytoplasm, it is able to enhance SKD1/Vps4p oligomerization and ATPase activity (Azmi et al. 2006). LIP5/Vta1p contains two MIT domains at its N-terminus (Xiao et al. 2008) that interact with various ESCRT-III proteins, but most strongly with the ESCRT-III accessory protein Vps60p/CHMP5 (Bowers et al. 2004; Azmi et al. 2008; Shim et al. 2008). Vps60p/CHMP5 is believed to function primarily as an adaptor for LIP5/Vta1p to interact with the ESCRT-III complex (Nickerson et al. 2010). The stoichiometry of the yeast Vta1p-Vps4p complex has been reported to be 6:12 (Yu et al. 2008), and therefore, the complex contains 24 MIT domains (12 from Vta1p and 12 from Vps4p) presumably available to interact with ESCRT-III proteins.

Whereas yeast cells can survive without Vps4p function, the constitutive overexpression of an ATPase-deficient SKD1 protein in plants is lethal and its inducible expression in roots results in enlarged late endosomes/MVBs with a reduced number of ILVs (Haas et al. 2007). The specific expression of the same dominant negative version of SKD1 in trichomes leads to vacuole fragmentation and the occasional presence of multiple nuclei (Shahriari et al. 2010).

Arabidopsis SKD1 interacts with LIP5, which also acts as a positive regulator of SKD1 ATPase activity. Interestingly, the knockout LIP5 mutant is viable suggesting that additional regulatory mechanisms controlling SKD1 ATPase activity may exist in plants (Haas et al. 2007).

3 The Mechanism of ILV Formation

In an in vitro study using giant unilamellar vesicles (GUVs), recombinant ESCRT-I and ESCRT-II were localized to the neck of forming vesicles on the GUV surface (Wollert and Hurley 2010). Although the GUVs used in this study were between 50 and 100 times larger than a typical MVB (250–500 nm in diameter) and the resulting vesicles were approximately 2 μm (50 times larger than typical ILVs), this study suggested that either the ESCRT-I and II supercomplex can directly bend membranes (Wollert and Hurley 2010) or alternatively, it could play an indirect role by stabilizing local membrane deformations that occur transiently and spontaneously in GUVs (Mayers and Audhya 2012).

How do ESCRT-I and II bend endosomal membranes? In general, membrane-associated proteins can induce membrane curvature by three mechanisms: (1) inserting part of themselves into membrane monolayers generating an asymmetric organization of the membrane bilayer; (2) altering the lipid composition of the two monolayers; (3) applying mechanical forces (scaffolds) on membranes (Graham

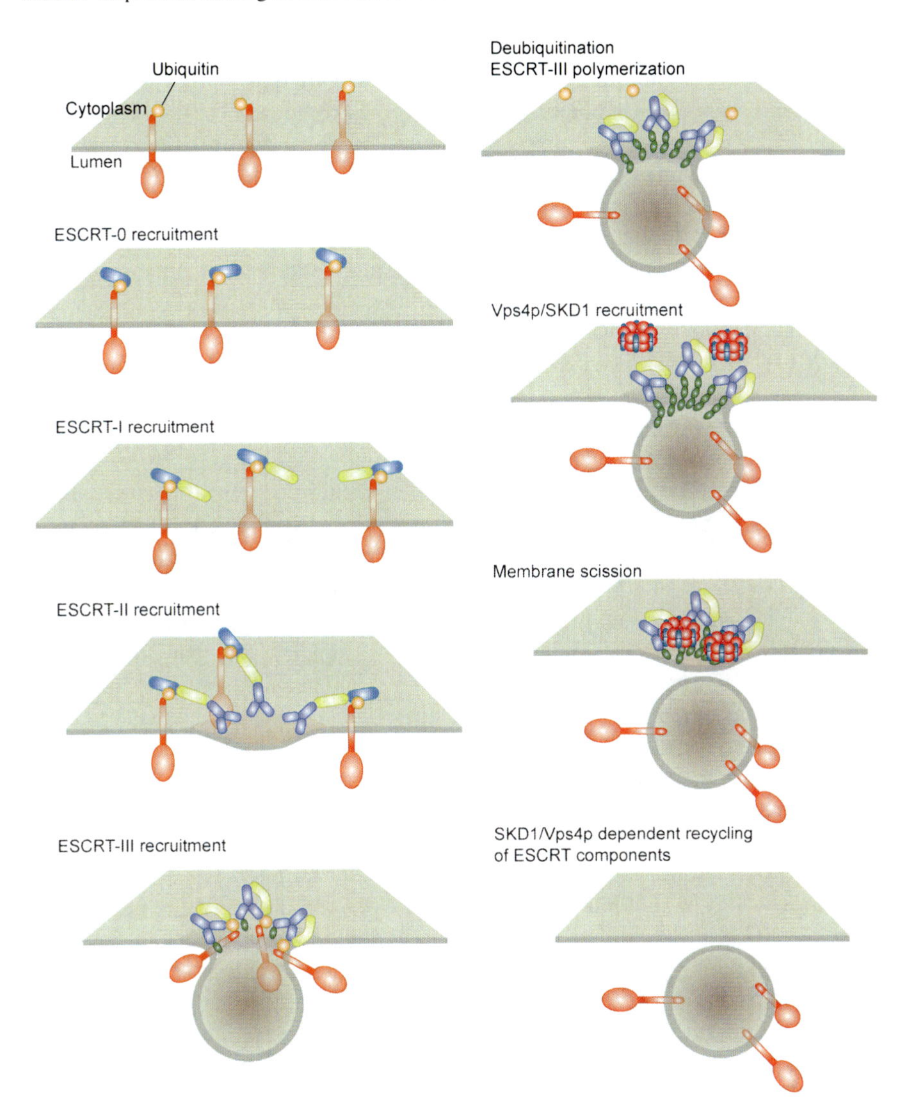

Fig. 1 Hypothetical model of ESCRT-mediated ILV formation on MVBs. Ubiquitinated plasma membrane proteins are recognized by the ESCRT-0 complex. ESCRT-0 recruits ESCRT-I and then ESCRT-II. The crescent shape of the ESCRT-I–II supercomplex would induce/stabilize membrane deformations on the endosomal membrane. Further conformational changes in ESCRT-I–II would facilitate the movement of the ubiquitinated cargo to the vesicle neck. ESCRT-II recruits ESCRT-III to the highly curved vesicle neck. ESCRT-III filaments polymerize and specific deubiquitinating enzymes remove ubiquitin moieties from the cargo proteins. ESCRT-III filaments constrict the vesicle neck and release the ILV into the lumen. SKD1/Vps4p ATPase complex binds ESCRT-III components and disassemble and recycle the ESCRT coat back to the cytoplasm. Based on Boura et al. (2012)

and Kozlov 2010; Mayers and Audhya 2012). If the first mechanism applies to ESCRT-mediated membrane deformation, one would expect that amphipathic domains able to penetrate membrane monolayers become exposed when ESCRT-I and/or ESCRT-II co-assemble on membranes. However, regions of ESCRT-I and ESCRT-II capable of membrane insertion have not been identified. On the other hand, ESCRT-0 to III can associate with acidic phospholipid with different affinities, and therefore they could cause local asymmetric distribution of phospholipids in the two monolayers, potentially reducing the energy requirement for invagination. Interestingly, PLDα1 (phospholipase D alpha 1) was shown to co-immunoprecipitate with VPS2.2 (ESCRT-III) in Arabidopsis cell extracts (Ibl et al. 2011). PLDs produce phosphatidic acid from the hydrolysis of phosphatidylcholine. The cone-shaped geometry of phosphatidic acid that accumulate in the monolayer facing the cytoplasm could induce negative curvature in the endosomal membranes. However, to what extent phospholipid modifications happen during ESCRT-I–II assembly and how much they contribute to endosomal membrane bending is unclear. If the third mechanism is correct, the ESCRT-I–II scaffold should have an intrinsic curvature able to force the lipid bilayer to bend and the energy of the ESCRT complex–phospholipid association should exceed the energy required for membrane bending. Recently, the structure of the yeast ESCRT-I–II supercomplex has been resolved. The ESCRT-I–II supercomplex can adopt closed conformation in which the two complexes are folded back on one another or extended conformations with crescent shapes, which is often characteristic of proteins that induce membrane curvature (Boura et al. 2012). Interestingly, when ESCRT-II binds the ESCRT-III subunit Vps20, the resulting complex also exhibit intrinsic curvature and its membrane binding energy increases with membrane bending (Fyfe et al. 2011). According to the "spoke" model proposed by James Hurley's group (Boura et al. 2012), the concave side of 6–10 ESCRT-I–II super complexes would be in contact with the positive curved profile of the neck of the nascent vesicles, following a radial distribution around the pore rim. In this case, ESCRT-0 would be positioned outside the pore whereas the region of ESCRT-II nucleating ESCRT-III would be facing the center of the forming vesicle. A transition from open to closed state in ESCRT-I would drag the ubiquitinated cargo across the neck of the forming vesicle (Boura et al. 2012) (Fig. 1).

ESCRT-III, which is able to assemble into linear filaments, seems to be responsible for vesicle neck constriction, membrane scission, and ILV release into the endosomal lumen (Fig. 1). As previously mentioned, ESCRT-III subunits are recruited from the cytoplasm to the endosomal membrane following a conformational change, from a closed auto inhibited state to an open active state. The crystal structure of human CHMP3 revealed that four of the five core helices typical of ESCRT-III subunits form a flat lattice that exposes a highly basic interface that binds membrane with high affinity (Muziol 2006; Lata et al. 2009).

Recombinant ESCRT-III proteins form polymers of different architecture in vitro. Recombinant yeast Vps24p forms 15-nm wide filaments (Ghazi-Tabatabai et al. 2008) whereas recombinant yeast Snf7p polymerizes into variables configurations such as sheets, rings, and filaments with 3 nm in diameter. Recent observations using atomic force microscopy on recombinant *C. elegans* Vps32 nucleated by ESCRT-II and Vps20 on supported lipid bilayers demonstrated the presence of long filaments (13.4 nm in diameter) bound specifically to highly curved membranes. These polymers were able to promote membrane remodeling in a manner similar to membrane fission (Fyfe et al. 2011). Consistently, yeast Snf7p has been shown to be the most abundant ESCRT-III component that functions during MVB biogenesis (Teis et al. 2008). Vps24p/CHMP3 and Vps2p/CHMP2 form heterodimers (Babst et al. 2002a) that assemble into tubular structures, some of which exhibit nearly hemispherical dome-like end-caps (Lata et al. 2008b).

How do the ESCRT-III filaments mediate membrane fission? Scission of membrane necks requires the opposing membrane to come as close as 3 nm from each other (Kozlovsky and Kozlov 2003). Although several mechanisms for ESCRT-III-mediated membrane scission have been proposed, they all support a common model where the ESCRT-III spiral/ring/whorls filaments induce membrane constriction from inside of the neck of the nascent vesicles. In one model, the polymerization of the ESCRT-III filament itself generates constriction force by inward curvature during filament outgrowth. According to this model, the ESCRT-III subunits have distinct roles in membrane constriction: Vps20p mediates polymer nucleation; Snf7p polymer outgrowth, and Vps24p together with Vps2p mediate filament capping. SKD1/Vps4p would be only needed to disassemble the ESCRT-III polymers but would not be necessary for membrane constriction (Fig. 1). In the second model, Vps4p plays an active role in membrane constriction by mediating the removal of ESCRT-III subunits assembled on the vesicle neck and remodeling of the ESCRT-III polymers. In the third model, ESCRT-III does not actively deform membranes but could stabilize spontaneous high negative membrane curvature domains on the endosomal membranes. The fact that recombinant human Vps24 and C-terminally truncated Vps2 can assemble in dome-shaped cap structures (Lata et al. 2008b; Bodon et al. 2011) promotes hypothesis that these dome-like configurations are important for the final stages of vesicle neck constriction. Biophysical-based mathematical modeling indicates that this model is energetically feasible since the binding affinity between ESCRT-III and membranes can exceed the deformation energy required to deform the membrane around the dome (Fabrikant et al. 2009).

4 The Role of Ubiquitination in Plant MVB Sorting

Ubiquitination of plasma membrane proteins acts as a signal for both endocytosis (Hicke and Dunn 2003) and sorting into endosomal ILVs by the ESCRT machinery. Addition of a single ubiquitin molecule to some PM proteins such as the yeast Gap1 permease is enough to trigger internalization by endocytosis (Lauwers et al. 2009); however, many receptors require poly-ubiquitination or poly-monoubiquitination for internalization. Endocytosed ubiquitinated PM proteins can either be sorted for degradation or deubiquitinated and recycled back to the PM. Lys63-linked poly-ubiquitination also acts as a degradative sorting signal for many studied MVB cargo proteins in yeast and animals (Lauwers et al. 2009).

Several subunits of ESCRT-0, I, and II (but not III) are able to bind ubiquitinated cargo and these interactions are essential for efficient sorting of protein cargo into ILVs. Nevertheless, it is not clear how all these ubiquitin-binding domains in multiple ESCRT subunits act in a coordinated manner to mediate cargo recruitment, concentration and sorting into ILVs.

In plants, ubiquitination also acts as a signal for endocytic and endosomal sorting of PM protein cargo. For example, mono-ubiquitination of Lys residues in the cytosolic loop of Arabidopsis IRT1 (iron regulated transporter 1) triggers its endocytosis and promotes its degradative endosomal sorting (Barberon et al. 2011). In the case of BOR1 (boron transporter 1), mono or di-ubiquitination of Lys residues (Kasai et al. 2011) together with a Tyr-motif in a cytoplasmic loop (Takano et al. 2010) are necessary for the degradative endosomal sorting of the transporter in the presence of high concentration of boron. The PM auxin efflux carrier PIN2, which is sorted by ESCRT machinery into ILVs (Spitzer et al. 2009), also becomes ubiquitinated (Abas et al. 2006) and its degradative sorting depends on Lys63-linked poly-ubiquitin chains (Leitner et al. 2012).

Ubiquitination is also an important sorting signal for PM receptors in plants. The receptor-like kinase FLAGELLIN-SENSING 2 (FLS2) in Arabidopsis binds bacterial flagellin (and also the flagellin derivative flg22) mediating pathogen-induced responses. Upon flg22 stimulation, FLS2 binds the co-receptor BAK1 and undergoes ligand-mediated endocytosis (Robatzek et al. 2006). BAK1 phosphorylates the U-box (PUB) E3 ubiquitin ligases, PUB12 and 13, leading to the ubiquitination and degradation of FLS2 (Lu et al. 2011).

Recently, it has been shown that the translational fusion of a C-terminal ubiquitin moiety to the PM-localized protein PMA-EGFP (plasma membrane Proton ATPase-enhanced green fluorescent protein) is enough to induce endocytosis and vacuolar delivery likely through the MVB pathway (Herberth et al. 2012).

In metazoans and fungi, ubiquitination and deubiquitination is a dynamic process that occurs at various points along the endocytic pathway. Interestingly, the function of some ESCRT components seems to be also regulated by their reversible ubiquitination. The deubiquitination of both plasma membrane protein cargo and ESCRT subunits depends on deubiquitinating enzymes (DUBs) that enable dynamic regulation of the turnover rate of plasma membrane proteins (Wright et al. 2011). Two human DUBs, such as AMSH and USP8 (ubiquitin-specific protease 8), have been shown to associate directly with both ESCRT-0 complex and with ESCRT-III (Wright et al. 2011). In Arabidopsis, AMSH1, AMSH2, and AMSH3, have significant sequence similarity to human AMSH. AMSH3 is able to hydrolyze Lys48- and Lys63-linked ubiquitin chains in vitro and in vivo. Although its precise role in plant endosomal trafficking is currently unknown, AMSH3 is essential for proper vacuole formation and vacuolar trafficking of both biosynthetic and endocytic cargo (Isono et al. 2010).

5 Biogenesis of MVBs in Plants

The formation of MVBs in Arabidopsis embryo cells has been studied by electron tomography of high-pressure frozen/freeze-substituted samples. In embryo cells, the secretion of storage proteins to the vacuole is very active and electron-dense aggregate of storage protein precursors can be detected in Golgi marginal buds, TGN, dense vesicles, and MVBs. Based on this distribution and progression of processing of storage protein precursors, it has been postulated that Golgi- and TGN-derived vesicles fuse to give rise to larger compartments that start to invaginate ILVs when they reach between 0.05 and 0.1 μm^2 in surface area (Otegui et al. 2006).

In a different study in Arabidopsis roots, tubular connection between MVBs and TGN were reported and the ESCRT-I component VPS28 was localized to Golgi and TGN but not to MVBs, supporting the idea that cargo sorting and ILV formation could start at the TGN (Scheuring et al. 2011). However, in two different studies, a different ESCRT-I component, ELC/VPS23 has been shown to co-localize with MVB markers (Spitzer et al. 2006; Richardson et al. 2011).

These discrepancies on the mechanism of MVB biogenesis and localization of ESCRT components could be due to the different imaging techniques used in these studies and the inherent complexity and diversity of the endosomal system in different plant cell types.

6 Concluding Remarks

Many trafficking and signalling pathways are regulated by endosomal sorting events in eukaryotes. In the last decade, the cell biology field has witnessed a dramatic progress on the elucidation of structural and mechanistic aspects of ESCRT-mediated endosomal sorting. However, there are still many unanswered questions, either about general sorting mechanisms or evolutionary variations in the function and composition of the ESCRT machinery. For instance, how does MVB sorting proceed in the absence of a canonical ESCRT-0 in plants and other eukaryotes? What is the functional significance of the diversification of ESCRT subunits (i.e. VPS2 in plants)? What kinds of lipid modifications occur during membrane bending in MVBs? How do the different ubiquitin-binding domains in ESCRT subunits coordinate cargo concentration, recruitment, and sorting? How do the different aspects of MVB sorting affect the development of multicellular organisms? Current and future studies using combinations of imaging, biochemical, and structural biology approaches will likely answer many of these questions in the coming years.

Acknowledgments We would like to thank Christoph Spitzer for the critical reading of this chapter. Research on Plant ESCRT complexes in the Otegui lab is supported by NSF MCB 1157824 grant to M.S.O.

References

Abas L, Benjamins R, Malenica N, Paciorek T, Wisniewska J, Moulinier-Anzola JC, Sieberer T, Friml J, Luschnig C (2006) Intracellular trafficking and proteolysis of the Arabidopsis auxin-efflux facilitator PIN2 are involved in root gravitropism. Nat Cell Biol 8(3):249–256

Agromayor M, Soler N, Caballe A, Kueck T, Freund SM, Allen MD, Bycroft M, Perisic O, Ye Y, McDonald B, Scheel H, Hofmann K, Neil SJD, Martin-Serrano J, Williams RL (2012) The UBAP1 subunit of ESCRT-I interacts with ubiquitin via a SOUBA domain. Structure 20(3):414–428

Alam SL, Langelier C, Whitby FG, Koirala S, Robinson H, Hill CP, Sundquist WI (2006) Structural basis for ubiquitin recognition by the human ESCRT-II EAP45 GLUE domain. Nat Struct Mol Biol 13 (11):1029–1030. doi:http://www.nature.com/nsmb/journal/v13/n11/suppinfo/nsmb1160_S1.html

Alam SL, Sun J, Payne M, Welch BD, Blake BK, Davis DR, Meyer HH, Emr SD, Sundquist WI (2004) Ubiquitin interactions of NZF zinc fingers. EMBO J 23(7):1411–1421

Asao H, Sasaki Y, Arita T, Tanaka N, Endo K, Kasai H, Takeshita T, Endo Y, Fujita T, Sugamura K (1997) Hrs is Associated with STAM, a signal-transducing adaptor molecule. J Biol Chem 272(52):32785–32791. doi:10.1074/jbc.272.52.32785

Azmi I, Davies B, Dimaano C, Payne J, Eckert D, Babst M, Katzmann DJ (2006) Recycling of ESCRTs by the AAA–ATPase Vps4 is regulated by a conserved VSL region in Vta1. J Cell Biol 172(5):705–717. doi:10.1083/jcb.200508166

Azmi IF, Davies BA, Xiao J, Babst M, Xu Z, Katzmann DJ (2008) ESCRT-III family members stimulate Vps4 ATPase activity directly or via Vta1. Dev Cell 14(1):50–61

Babst M, Davies BA, Katzmann DJ (2011) Regulation of Vps4 during MVB sorting and cytokinesis. Traffic:in press. doi:10.1111/j.1600-0854.2011.01230.x

Babst M, Katzmann DJ, Estepa-Sabal EJ, Meerloo T, Emr SD (2002a) ESCRT-III: an endosome-associated heterooligomeric protein complex required for MVB sorting. Dev Cell 3(2):271–282

Babst M, Katzmann DJ, Snyder WB, Wendland B, Emr SD (2002b) Endosome-associated complex, ESCRT-II, recruits transport machinery for protein sorting at the multivesicular body. Dev Cell 3:283–289

Babst M, Sato TK, Banta LM, Emr SD (1997) Endosomal transport function in yeast requires a novel AAA-type ATPase, Vps4p. EMBO J 16:1820–1831

Babst M, Wendland B, Estepa EJ, Emr SD (1998) The Vps4p AAA ATPase regulates membrane association of a Vps protein complex required for normal endosome function. EMBO J 17(11):2982–2993

Bache KG, Brech A, Mehlum A, Stenmark H (2003a) Hrs regulates multivesicular body formation via ESCRT recruitment to endosomes. J Cell Biol 162(3):435–442

Bache KG, Raiborg C, Mehlum A, Stenmark H (2003b) STAM and hrs are subunits of a multivalent ubiquitin-binding complex on early endosomes. J Biol Chem 278(14):12513–12521

Bache KG, Slagsvold T, Cabezas A, Rosendal KR, Raiborg C, Stenmark H (2004) The growth-regulatory protein HCRP1/hVps37A is a subunit of mammalian ESCRT-I and mediates receptor down-regulation. Mol Biol Cell 15(9):4337–4346

Bache KG, Stuffers S, Malerod L, Slagsvold T, Raiborg C, Lechardeur D, Walchli S, Lukacs GL, Brech A, Stenmark H (2006) The ESCRT-III subunit hVps24 is required for degradation but not silencing of the epidermal growth factor receptor. Mol Biol Cell 17(6):2513–2523

Bajorek M, Schubert HL, McCullough J, Langelier C, Eckert DM, Stubblefield WMB, Uter NT, Myszka DG, Hill CP, Sundquist WI (2009) Structural basis for ESCRT-III protein autoinhibition. Nat Struct Mol Biol 16(7):754–762

Banta LM, Robinson JS, Klionsky DJ, Emr SD (1988) Organelle assembly in yeast: characterization of yeast mutants defective in vacuolar biogenesis and protein sorting. J Cell Biol 107(4):1369–1383. doi:10.1083/jcb.107.4.1369

Barberon M, Zelazny E, Robert S, Conéjéro G, Curie C, Friml J, Vert G (2011) Monoubiquitin-dependent endocytosis of the iron-regulated transporter 1 (IRT1) transporter controls iron uptake in plants. Proc Natl Acad Sci U S A. doi:10.1073/pnas.1100659108

Bishop N, Woodman P (2001) TSG101/mammalian VPS23 and mammalian VPS28 interact directly and are recruited to VPS4-induced endosomes. J Biol Chem 276(15):11735–11742

Blanc C, Charette SJ, Mattei S, Aubry L, Smith EW, Cosson P, Letourneur F (2009) *Dictyostelium* Tom1 participates to an ancestral ESCRT-0 complex. Traffic 10(2):161–171

Bodon G, Chassefeyre R, Pernet-Gallay K, Martinelli N, Effantin G, Hulsik DL, Belly A, Goldberg Y, Chatellard-Causse C, Blot B, Schoehn G, Weissenhorn W, Sadoul R (2011) Charged Multivesicular Body Protein 2B (CHMP2B) of the Endosomal Sorting Complex Required for Transport-III (ESCRT-III) polymerizes into helical structures deforming the plasma membrane. J Biol Chem 286(46):40276–40286. doi:10.1074/jbc.M111.283671

Boura E, Różycki B, Chung Hoi S, Herrick Dawn Z, Canagarajah B, Cafiso David S, Eaton William A, Hummer G, Hurley James H (2012) Solution structure of the ESCRT-I and -II supercomplex: implications for membrane budding and scission. Structure 20(5):874–886

Bowers K, Lottridge J, Helliwell SB, Goldthwaite LM, Luzio JP, Stevens TH (2004) Protein-protein interactions of ESCRT complexes in the yeast *Saccharomyces cerevisiae*. Traffic 5(3):194–210

Chu T, Sun J, Saksena S, Emr SD (2006) New component of ESCRT-I regulates endosomal sorting complex assembly. J Cell Biol 175(5):815–823

Curtiss M, Jones C, Babst M (2007) Efficient cargo sorting by ESCRT-I and the subsequent release of ESCRT-I from multivesicular bodies requires the subunit Mvb12. Mol Biol Cell 18(2):636–645

Dimaano C, Jones CB, Hanono A, Curtiss M, Babst M (2008) Ist1 regulates Vps4 localization and assembly. Mol Biol Cell 19(2):465–474

Eastman SW, Martin-Serrano J, Chung W, Zang T, Bieniasz PD (2005) Identification of human VPS37C, a component of endosomal sorting complex required for transport-I important for viral budding. J Biol Chem 280(1):628–636

Fabrikant G, Lata S, Riches JD, Briggs JAG, Weissenhorn W, Kozlov MM (2009) Computational model of membrane fission catalyzed by ESCRT-III. PLoS Comput Biol 5(11):e1000575

Fyfe I, Schuh AL, Edwardson JM, Audhya A (2011) Association of ESCRT-II with VPS20 generates a curvature sensitive protein complex capable of nucleating filaments of ESCRT-III. J Biol Chem. doi:10.1074/jbc.M111.266411 (in press)

Ghazi-Tabatabai S, Obita T, Pobbati AV, Perisic O, Samson RY, Bell SD, Williams RL (2009) Evolution and assembly of ESCRTs. Biochem Soc Trans 37(Pt 1):151–155

Ghazi-Tabatabai S, Saksena S, Short JM, Pobbati AV, Veprintsev DB, Crowther RA, Emr SD, Egelman EH, Williams RL (2008) Structure and disassembly of filaments formed by the ESCRT-III subunit Vps24. Structure 16:1345–1356

Gill DJ, Teo H, Sun J, Perisic O, Veprintsev DB, Emr SD, Williams RL (2007) Structural insight into the ESCRT-I/-II link and its role in MVB trafficking. EMBO J 26(2):600–612

Gonciarz MD, Whitby FG, Eckert DM, Kieffer, Heroux A, Sundquist WI, Hill CP (2008) Biochemical and structural studies of yeast Vps4 oligomerization. J Mol Biol 384:878–895

Graham TR, Kozlov MM (2010) Interplay of proteins and lipids in generating membrane curvature. Curr Opin Cell Biol 22(4):430–436

Haas TJ, Sliwinski MK, DE M, Preuss M, Ebine K, Ueda T, Nielsen E, Odorizzi G, Otegui MS (2007) The Arabidopsis AAA ATPase SKD1 is involved in multivesicular endosome function and interacts with its positive regulator LYST-INTERACTING PROTEIN5. Plant Cell 19:1295–1312

Hanson PI, Roth R, Lin Y, Heuser JE (2008) Plasma membrane deformation by circular arrays of ESCRT-III protein filaments. J Cell Biol 180(2):389–402

Henne William M, Buchkovich Nicholas J, Emr Scott D (2011) The ESCRT Pathway. Dev Cell 21(1):77–91

Herberth S, Shahriari M, Bruderek M, Hessner F, Müller B, Hülskamp M, Schellmann S (2012) Artificial ubiquitylation is sufficient for sorting of a plasma membrane ATPase to the vacuolar lumen of Arabidopsis cells. Planta. doi: 10.1007/s00425-012-1587-0 (in press)

Herman EK, Walker G, van der Giezen M, Dacks JB (2011) Multivesicular bodies in the enigmatic amoeboflagellate *Breviata anathema* and the evolution of ESCRT 0. J Cell Sci 124(Pt 4):613–621 doi:jcs.078436[pii]

Hicke L, Dunn R (2003) Regulation of membrane protein transport by ubiquitin and ubiquitin-binding proteins. Ann Rev Cell Dev Biol 19:141–172

Hierro A, Sun J, Rusnak AS, Kim J, Prag G, Emr SD, Hurley JH (2004) Structure of the ESCRT-II endosomal trafficking complex. Nature 431:221–225

Hirano S, Suzuki N, Slagsvold T, Kawasaki M, Trambaiolo D, Kato R, Stenmark H, Wakatsuki S (2006) Structural basis of ubiquitin recognition by mammalian Eap45 GLUE domain. Nat Struct Mol Biol 13(11):1031–1032

Horii M, Shibata H, Kobayashi R, Katoh K, Yorikawa C, Yasuda J, Maki M (2006) CHMP7, a novel ESCRT-III-related protein, associates with CHMP4b and functions in the endosomal sorting pathway. Biochem J 400(1):23–32

Ibl V, Csaszar E, Schlager N, Neubert S, Spitzer C, Hauser MT (2011) Interactome of the plant-specific ESCRT-III component AtVPS2.2 in *Arabidopsis thaliana*. J Proteome Res 11(1):397–411. doi:10.1021/pr200845n

Im YJ, Hurley JH (2008) Integrated structural model and membrane targeting mechanism of the human ESCRT-II complex. Dev Cell 14(6):902–913

Isono E, Katsiarimpa A, Müller IK, Anzenberger F, Stierhof YD, Geldner N, Chory J, Schwechheimer C (2010) The deubiquitinating enzyme AMSH3 is required for intracellular trafficking and vacuole biogenesis in *Arabidopsis thaliana*. Plant Cell 22(6):1826–1837. doi:10.1105/tpc.110.075952

Kasai K, Takano J, Miwa K, Toyoda A, Fujiwara T (2011) High boron-induced ubiquitination regulates vacuolar sorting of the BOR1 borate transporter in *Arabidopsis thaliana*. J Biol Chem 286(8):6175–6183. doi:10.1074/jbc.M110.184929

Katsiarimpa A, Anzenberger F, Schlager N, Neubert S, Hauser MT, Schwechheimer C, Isono E (2011) The Arabidopsis deubiquitinating enzyme AMSH3 Interacts with ESCRT-III subunits and regulates their localization. Plant Cell 23:3026–3040

Katzmann DJ, Babst M, Emr SD (2001) Ubiquitin-dependent sorting into the multivesicular body pathway requires the function of a conserved endosomal protein sorting complex, ESCRT-I. Cell 106(2):145–155

Katzmann DJ, Sarkar S, Chu T, Audhya A, Emr SD (2004) Multivesicular body sorting: ubiquitin ligase Rsp5 is required for the modification and sorting of carboxypeptidase S. Mol Biol Cell 15(2):468–480

Katzmann DJ, Stefan CJ, Babst M, Emr SD (2003) Vps27 recruits ESCRT machinery to endosomes during MVB sorting. J Cell Biol 162:413–423

Kieffer C, Skalicky JJ, Morita E, De Domenico I, Ward DM, Kaplan J, Sundquist WI (2008) Two distinct modes of ESCRT-III recognition are required for VPS4 functions in lysosomal protein targeting and HIV-1 budding. Dev Cell 15(1):62–73

Kostelansky MS, Schluter C, Tam YY, Lee S, Ghirlando R, Beach B, Conibear E, Hurley JH (2007) Molecular architecture and functional model of the complete yeast ESCRT-I heterotetramer. Cell 129(3):485–498

Kostelansky MS, Sun J, Lee S, Kim J, Ghirlando R, Hierro A, Emr SD, Hurley JH (2006) Structural and functional organization of the ESCRT-I trafficking complex. Cell 125(1):113–126

Kozlovsky Y, Kozlov MM (2003) Membrane fission: model for intermediate structures. Biophys J 85(1):85–96

Lange A, Castaneda C, Hoeller D, Lancelin JM, Fushman D, Walker O (2012) Evidence for cooperative and domain-specific binding of the signal transducing adaptor molecule 2 (STAM2) to Lys63-linked diubiquitin. J Biol Chem. doi:10.1074/jbc.M111.324954

Langelier C, von Schwedler UK, Fisher RD, De Domenico I, White PL, Hill CP, Kaplan J, Ward D, Sundquist WI (2006) Human ESCRT-II complex and its role in human immunodeficiency virus type 1 release. J Virol 80(19):9465–9480

Lata S, Roessle M, Solomons J, Jamin M, Gottlinger HG, Svergun DI, Weissenhorn W (2008a) Structural basis for autoinhibition of ESCRT-III CHMP3. J Mol Biol 378(4):816–825

Lata S, Schoehn G, Jain A, Pires R, Piehler J, Gottlinger HG, Weissenhorn W (2008b) Helical structures of ESCRT-III are disassembled by VPS4. Science 321(5894):1354–1357

Lata S, Schoehn G, Solomons J, Pires R, Gottlinger HG, Weissenhorn W (2009) Structure and function of ESCRT-III. Biochem Soc Trans 37(Pt 1):156–160

Lauwers E, Jacob C, Andre B (2009) K63-linked ubiquitin chains as a specific signal for protein sorting into the multivesicular body pathway. J Cell Biol 185(3):493–502. doi:10.1083/jcb.200810114

Leitner J, Petrášek J, Tomanov K, Retzer K, Pařezová M, Korbei B, Bachmair A, Zažímalová E, Luschnig C (2012) Lysine63-linked ubiquitylation of PIN2 auxin carrier protein governs hormonally controlled adaptation of Arabidopsis root growth. Proc Natl Acad Sci U S A 109(21):8322–8327. doi:10.1073/pnas.1200824109

Leung KF, Dacks JB, Field MC (2008) Evolution of the multivesicular body ESCRT machinery; retention across the eukaryotic lineage. Traffic 9(10):1698–1716

Lottridge JM, Flannery AR, Vincelli JL, Stevens TH (2006) Vta1p and Vps46p regulate the membrane association and ATPase activity of Vps4p at the yeast multivesicular body. Proc Natl Acad Sci U S A 103(16):6202–6207. doi:10.1073/pnas.0601712103

Lu D, Lin W, Gao X, Wu S, Cheng C, Avila J, Heese A, Devarenne TP, He P, Shan L (2011) Direct ubiquitination of pattern recognition receptor FLS2 attenuates plant innate immunity. Science 332(6036):1439–1442. doi:10.1126/science.1204903

Lu Q, Hope LW, Brasch M, Reinhard C, Cohen SN (2003) TSG101 interaction with HRS mediates endosomal trafficking and receptor down-regulation. Proc Natl Acad Sci U S A 100(13):7626–7631

Mao Y, Nickitenko A, Duan X, Lloyd TE, Wu MN, Bellen H, Quiocho FA (2000) Crystal structure of the VHS and FYVE tandem domains of Hrs, a protein involved in membrane trafficking and signal transduction. Cell 100(4):447–456

Mayers JR, Audhya A (2012) Vesicle formation within endosomes: An ESCRT marks the spot. Commun Integr Biol 5:5–56

Mayers JR, Fyfe I, Schuh AL, Chapman ER, Edwardson JM, Audhya A (2011) ESCRT-0 assembles as a heterotetrameric complex on membranes and binds multiple ubiquitinylated cargoes simultaneously. J Biol Chem 286(11):9636–9645. doi:10.1074/jbc.M110.185363

Merrill SA, Hanson PI (2010) Activation of human VPS4A by ESCRT-III proteins reveals ability of substrates to relieve enzyme autoinhibition. J Biol Chem 285(46):35428–35438. doi:10.1074/jbc.M110.126318

Morita E, Sandrin V, Alam SL, Eckert DM, Gygi SP, Sundquist WI (2007) Identification of human MVB12 proteins as ESCRT-I subunits that function in HIV budding. Cell Host Microbe 2(1):41–53

Muziol T (2006) Structural basis for budding by the ESCRT-III factor CHMP3. Dev Cell 10:821–830

Muziol T, Pineda-Molina E, Ravelli RB, Zamborlini A, Usami Y, Gottlinger H, Weissenhorn W (2006) Structural basis for budding by the ESCRT-III factor CHMP3. Dev Cell 10(6):821–830

Nickerson DP, West M, Henry R, Odorizzi G (2010) Regulators of Vps4 ATPase activity at endosomes differentially influence the size and rate of formation of intraluminal vesicles. Mol Biol Cell 21(6):1023–1032. doi:10.1091/mbc.E09-09-0776

Nickerson DP, West M, Odorizzi G (2006) Did2 coordinates Vps4-mediated dissociation of ESCRT-III from endosomes. J Cell Biol 175:715–720

Odorizzi G, Babst M, Emr SD (1998) Fab1p PtdIns(3)P 5-kinase function essential for protein sorting in the multivesicular body. Cell 95(6):847–858

Otegui MS, Herder R, Schulze J, Jung R, Staehelin LA (2006) The proteolytic processing of seed storage proteins in Arabidopsis embryo cells starts in the multivesicular bodies. Plant Cell 18(10):2567–2581. doi:10.1105/tpc.106.040931

Pashkova N, Piper RC (2012) UBAP1: A new ESCRT member joins the cl_Ub. Structure 20(3):383–385

Raymond CK, Howald-Stevenson I, Vater CA, Stevens TH (1992) Morphological classification of the yeast vacuolar protein sorting mutants: evidence for a prevacuolar compartment in class E vps mutants. Mol Biol Cell 3(12):1389–1402

Reggiori F, Pelham HRB (2001) Sorting of proteins into multivesicular bodies: ubiquitin-dependent and ubiquitin-independent targeting. EMBO J 20(18):5176–5186

Ren X, Hurley JH (2010) VHS domains of ESCRT-0 cooperate in high-avidity binding to polyubiquitinated cargo. EMBO J 29(6):1045–1054. doi:10.1038/emboj20106

Richardson LGL, Howard ASM, Khuu N, Gidda SK, McCartney A, Morphy BJ, Mullen RT (2011) Protein-protein interaction network and subcellular localization of the *Arabidopsis thaliana* ESCRT machinery. Frontiers Plant Sci 2: article 20

Robatzek S, Chinchilla D, Boller T (2006) Ligand-induced endocytosis of the pattern recognition receptor FLS2 in Arabidopsis. Genes Dev 20(5):537–542. doi:10.1101/gad.366506

Robinson DG, Hinz G, Holstein SE (1998) The molecular characterization of transport vesicles. Plant Mol Biol 38(1–2):49–76

Rothman JH, Howald I, Stevens TH (1989) Characterization of genes required for protein sorting and vacuolar function in the yeast *Saccharomyces cerevisiae*. EMBO J 8:2057–2065

Rothman JH, Stevens TH (1986) Protein sorting in yeast: Mutants defective in vacuole biogenesis mislocalize vacuolar proteins into the late secretory pathway. Cell 47(6):1041–1051

Samson RY, Obita T, Freund SM, Williams RL, Bell SD (2008) A role for the ESCRT system in cell division in archaea. Science 322(5908):1710–1713

Samson RY, Obita T, Hodgson B, Shaw MK, Chong PLG, Williams RL, Bell SD (2011) Molecular and structural basis of ESCRT-III recruitment to membranes during archaeal cell division. Mol Cell 41(2):186–196

Scheuring D, Viotti C, Kruger F, Kunzl F, Sturm S, Bubeck J, Hillmer S, Frigerio L, Robinson DG, Pimpl P, Schumacher K (2011) Multivesicular bodies mature from the trans-Golgi network/early endosome in Arabidopsis. Plant Cell 23(9):3463–3481. doi:10.1105/tpc.111.086918

Scott A, Chung HY, Gonciarz-Swiatek M, Hill GC, Whitby FG, Gaspar J, Holton JM, Viswanathan R, Ghaffarian S, Hill CP, Sundquist WI (2005) Structural and mechanistic studies of VPS4 proteins. EMBO J 24(20):3658–3669

Shahriari M, Keshavaiah C, Scheuring D, Sabovljevic A, Pimpl P, Häusler RE, Hülskamp M, Schellmann S (2010) The AAA-type ATPase AtSKD1 contributes to vacuolar maintenance of *Arabidopsis thaliana*. Plant J 64(1):71–85. doi:10.1111/j.1365-313X.2010.04310.x

Shahriari M, Richter K, Keshavaiah C, Sabovljevic A, Huelskamp M, Schellmann S (2011) The *Arabidopsis* ESCRT protein–protein interaction network. Plant Mol Biol 76(1–2):85–96. doi:10.1007/s11103-011-9770-4

Shen B, Li C, Min Z, Meeley RB, Tarczynski MC, Olsen O-A (2003) Sal1 determines the number of aleurone cell layers in maize endosperm and encodes a class E vacuolar sorting protein. Proc Natl Acad Sci U S A 100(11):6552–6557

Shestakova A, Hanono A, Drosner S, Curtiss M, Davies BA, Katzmann DJ, Babst M (2010) Assembly of the AAA ATPase Vps4 on ESCRT-III. Mol Biol Cell 21(6):1059–1071. doi:10.1091/mbc.E09-07-0572

Shiflett SL, Ward DM, Huynh D, Vaughn MB, Simmons JC, Kaplan J (2004) Characterization of Vta1p, a class E Vps protein in *Saccharomyces cerevisiae*. J Biol Chem 279(12):10982–10990. doi:10.1074/jbc.M312669200

Shih SC, Katzmann DJ, Schnell JD, Sutanto M, Emr SD, Hicke L (2002) Epsins and Vps27p/Hrs contain ubiquitin-binding domains that function in receptor endocytosis. Nat Cell Biol 4(5):389–393

Shim S, Kimpler LA, Hanson PI (2007) Structure/function analysis of four core ESCRT-III proteins reveals common regulatory role for extreme C-terminal domain. Traffic 8(8):1068–1079

Shim S, Merrill SA, Hanson PI (2008) Novel interactions of ESCRT-III with LIP5 and VPS4 and their implications for ESCRT-III disassembly. Mol Biol Cell 19(6):2661–2672. doi:10.1091/mbc.E07-12-1263

Slagsvold T, Aasland R, Hirano S, Bache KG, Raiborg C, Trambaiolo D, Wakatsuki S, Stenmark H (2005) Eap45 in mammalian ESCRT-II binds ubiquitin via a phosphoinositide-interacting GLUE domain. J Biol Chem 280(20):19600–19606

Spitzer C, Reyes FC, Buono R, Sliwinski MK, Haas TJ, Otegui MS (2009) The ESCRT-related CHMP1A and B proteins mediate multivesicular body sorting of auxin carriers in Arabidopsis and are required for plant development. Plant Cell 21(3):749–766

Spitzer C, Schellmann S, Sabovljevic A, Shahriari M, Keshavaiah C, Bechtold N, Herzog M, Muller S, Hanisch FG, Hulskamp M (2006) The Arabidopsis elch mutant reveals functions of an ESCRT component in cytokinesis. Development 133(23):4679–4689

Stahelin RV, Long F, Diraviyam K, Bruzik KS, Murray D, Cho W (2002) Phosphatidylinositol 3-phosphate induces the membrane penetration of the FYVE domains of Vps27p and Hrs. J Biol Chem 277(29):26379–26388. doi:10.1074/jbc.M201106200

Stefani F, Zhang L, Taylor S, Donovan J, Rollinson S, Doyotte A, Brownhill K, Bennion J, Pickering-Brown S, Woodman P (2011) UBAP1 is a component of an endosome-specific ESCRT-I complex that is essential for MVB sorting. Curr Biol 21(14):1245–1250

Stuchell MD, Garrus JE, Muller B, Stray KM, Ghaffarian S, McKinnon R, Krausslich HG, Morham SG, Sundquist WI (2004) The human endosomal sorting complex required for transport (ESCRT-I) and its role in HIV-1 budding. J Biol Chem 279(34):36059–36071

Takano J, Tanaka M, Toyoda A, Miwa K, Kasai K, Fuji K, Onouchi H, Naito S, Fujiwara T (2010) Polar localization and degradation of Arabidopsis boron transporters through distinct trafficking pathways. Proc Nat Acad Sci 107(11):5220–5225. doi:10.1073/pnas.0910744107

Takasu H, Jee JG, Ohno A, Goda N, Fujiwara K, Tochio H, Shirakawa M, Hiroaki H (2005) Structural characterization of the MIT domain from human Vps4b. Biochem Biophys Res Commun 334(2):460–465

Teis D, Saksena S, Emr SD (2008) Ordered assembly of the ESCRT-III complex on endosomes is required to sequester cargo during MVB formation. Dev Cell 15(4):578–589

Teo H, Gill DJ, Sun J, Perisic O, Veprintsev DB, Vallis Y, Emr SD, Williams RL (2006) ESCRT-I Core and ESCRT-II GLUE Domain Structures Reveal Role for GLUE in Linking to ESCRT-I and Membranes. Cell 125(1):99–111

Teo H, Perisic O, Gonzalez B, Williams RL (2004) ESCRT-II, an Endosome-Associated Complex Required for Protein Sorting: Crystal Structure and Interactions with ESCRT-III and Membranes. Dev Cell 7(4):559–569

Tian Q, Olsen L, Sun B, Lid SE, Brown RC, Lemmon BE, Fosnes K, Gruis D, Opsahl-Sorteberg HG, Otegui MS, Olsen O-A (2007) Subcellular localization and functional domain studies of DEFECTIVE KERNEL1 in maize and Arabidopsis suggest a model for aleurone cell fate specification involving CRINKLY4 and SUPERNUMERARY ALEURONE LAYER1. Plant Cell 19(10):3127–3145. doi:10.1105/tpc.106.048868

Tsunematsu T, Yamauchi E, Shibata H, Maki M, Ohta T, Konishi H (2010) Distinct functions of human MVB12A and MVB12B in the ESCRT-I dependent on their posttranslational modifications. Biochem Biophys Res Commun 399(2):232–237

Vajjhala PR, Nguyen CH, Landsberg MJ, Kistler C, Gan AL, King GF, Hankamer B, Munn AL (2008) The Vps4 C-terminal helix is a critical determinant for assembly and ATPase activity and has elements conserved in other members of the meiotic clade of AAA ATPases. FEBS J 275(7):1427–1449

Vajjhala PR, Wong JS, To HY, Munn AL (2006) The beta domain is required for Vps4p oligomerization into a functionally active ATPase. FEBS J 273(11):2357–2373

van Weering JR, Verkade P, Cullen PJ (2010) SNX–BAR proteins in phosphoinositide-mediated, tubular-based endosomal sorting. Semin Cell Dev Biol 21:371–380 van Weering JRT, Verkade P, Cullen PJ (2012) SNX–BAR-mediated endosome tubulation is co-ordinated with endosome maturation. Traffic 13(1):94–107. doi:10.1111/j.1600-0854.2011.01297.x

Ward DM, Vaughn MB, Shiflett SL, White PL, Pollock AL, Hill J, Schnegelberger R, Sundquist WI, Kaplan J (2005) The role of LIP5 and CHMP5 in multivesicular body formation and HIV-1 budding in mammalian cells. J Biol Chem 280(11):10548–10555. doi:10.1074/jbc.M413734200

Winter V, Hauser MT (2006) Exploring the ESCRTing machinery in eukaryotes. Trends Plant Sci 11(3):115–123

Wollert T, Hurley JH (2010) Molecular mechanism of multivesicular body biogenesis by ESCRT complexes. Nature 464 (7290):864–869. doi:http://www.nature.com/nature/journal/v464/n7290/suppinfo/nature08849_S1.html

Wollert T, Wunder C, Lippincott-Schwartz J, Hurley JH (2009) Membrane scission by the ESCRT-III complex. Nature 458:172–177

Wright M, Berlin I, Nash P (2011) Regulation of endocytic sorting by ESCRT–DUB-mediated deubiquitination. Cell Biochem Biophys 60(1):39–46. doi:10.1007/s12013-011-9181-9

Xiao J, Xia H, Yoshino-Koh K, Zhou J, Xu Z (2007) Structural characterization of the ATPase reaction cycle of endosomal AAA protein Vps4. J Mol Biol 374(3):655–670

Xiao J, Xia H, Zhou J, Azmi IF, Davies BA, Katzmann DJ, Xu Z (2008) Structural basis of Vta1 function in the multivesicular body sorting pathway. Dev Cell 14(1):37–49

Yang D, Hurley JH (2010) Structural role of the Vps4-Vta1 interface in ESCRT-III recycling. Structure 18(8):976–984

Yu Z, Gonciarz MD, Sundquist WI, Hill CP, Jensen GJ (2008) Cryo-EM structure of dodecameric Vps4p and its 2:1 complex with Vta1p. J Mol Biol 377(2):364–377

Zamborlini A, Usami Y, Radoshitzky SR, Popova E, Palu G, Gottlinger H (2006) Release of autoinhibition converts ESCRT-III components into potent inhibitors of HIV-1 budding. Proc Natl Acad Sci U S A 103(50):19140–19145

Zimmermann J, Chidambaram S, Fischer von Mollard G (2010) Dissecting Ent3p: the ENTH domain binds different SNAREs via distinct amino acid residues while the C-terminus is sufficient for retrograde transport from endosomes. Biochem J 431:123–134

Endocytic Accommodation of Microbes in Plants

Rik Huisman, Evgenia Ovchinnikova, Ton Bisseling and Erik Limpens

Abstract Plants host many different microbes within their cells. These endosymbiotic relationships are characterized by the formation of new specialized membrane compartments inside the plant cells in which the microbes live and where nutrients and signals are efficiently exchanged. Such symbiotic interfaces include arbuscules produced by arbuscular mycorrhiza (AM), organelle-like symbiosomes formed during the rhizobium-legume symbiosis, and haustoria produced by biotrophic fungi and oomycetes. The formation and maintenance of such new membrane compartments require a major reorganization of the host endomembrane system. In the last decade, much progress has been made in understanding how arbuscules, symbiosomes, and haustoria are formed. In this chapter, we will summarize the recent developments in each field, with a major focus on the AM and rhizobial endosymbiosis. It has become clear that rhizobia have co-opted a signalling pathway as well as a cellular mechanism to make the interface membrane compartment from the ancient and most successful AM symbiosis. Both AM symbiosis and rhizobium symbiosis depend on the secretion of lipo-chito-oligosaccharides that trigger a symbiotic signalling cascade, which is required for both arbuscule and symbiosome formation. In both interactions a shared specific exocytosis pathway is recruited to facilitate the formation of the symbiotic interface resulting in a membrane compartment with distinct protein composition. Given the structural similarity of haustoria to arbuscules, similar mechanisms are envisioned to be involved in the formation of a haustorium.

R. Huisman · E. Ovchinnikova · T. Bisseling (✉) · E. Limpens
Laboratory of Molecular Biology,Department of Plant Sciences, Wageningen University, Wageningen, The Netherlands
e-mail: ton.bisseling@wur.nl

J. Šamaj (ed.), *Endocytosis in Plants*,
DOI: 10.1007/978-3-642-32463-5_14,

1 Introduction

Plants form a habitat for a wide array of microbes, which live around or inside the plant in relations that range from mutualistic to parasitic. Perhaps the most intriguing of these biotrophic interactions are endosymbiotic relationships, where the microbes are hosted inside living plant cells. Such extremely intimate interactions are characterized by the formation of a specialized membrane compartment by the host, in which the microbe is accommodated and which forms a symbiotic interface where nutrients and signals are efficiently exchanged (Parniske 2000). The formation of such an interface compartment requires a massive reorganization of the host endomembrane system. Currently, the best studied examples with respect to the intracellular accommodation of microbes in plants are the mutualistic endosymbiosis between plants and arbuscular mycorrhizal (AM) fungi and between Rhizobium bacteria and legumes. During the AM symbiosis, fungal hyphae invade the root and a symbiotic interface compartment is formed inside root cortex cells where highly branched hyphae are contained in a specialized host membrane (the periarbuscular membrane) and together are called arbuscules (Parniske 2008). In the Rhizobium-legume symbiosis rhizobium bacteria are hosted as novel nitrogen-fixing organelles called symbiosomes, inside the cells of a newly formed organ, the root nodule. Recent genetic and cell biological studies, especially in the model legume *Lotus japonicus* (Lotus) and *Medicago truncatula* (Medicago), have started to give insight into the molecular mechanisms by which arbuscules and symbiosomes are made. Although endocytosis likely plays an important role at various stages of the interactions, these studies highlight a major role for regulated exocytosis and have revealed a common molecular basis underlying the formation of the symbiotic interface in both symbioses. Here, we review the current progress in understanding the molecular basis of arbuscule and symbiosome formation and discuss the roles of endo- and exocytosis. Furthermore, we compare the strategy used by to AM fungi and rhizobia to the intracellular accommodation of pathogenic fungi, which form comparable intracellular interface compartments/feeding structures called haustoria.

2 Rhizobial Infection

Most legumes have the unique ability to establish an endosymbiosis with Gram-negative soil bacteria, collectively called rhizobia, which are able to fix nitrogen (Masson-Boivin et al. 2009). This symbiosis is thought to have originated approximately 58 million years ago (Sprent 2007). During the symbiosis a novel organ is formed on the root of the plant, called the root nodule. The formation of this organ starts by dedifferentiation of root cortical cells that enter the cell cycle and become reprogrammed to form a nodule primordium. At the same time, rhizobia invade the root and the developing nodule primordium via cell wall-bound infection structures, called infection threads (Brewin 2004; Gage 2004; Jones et al. 2007). Most commonly, such infection threads originate in root hairs that curl around attached bacteria through a continued reorientation of the growth direction of the root hair by which rhizobia become entrapped in a closed cavity and form a microcolony (Esseling et al. 2003). However, alternative infection mechanisms such as crack entry between cells can occur in certain legumes (Sprent 2007; Held et al. 2010). Inside the curl, the plant cell wall is locally degraded and the plasma membrane invaginates. Through the focal delivery of vesicles new membrane and cell wall material is deposited by which a tubular infection thread is formed that guides the dividing bacteria to the base of the root hair (Fournier et al. 2008). Numerous smooth and coated vesicles have been observed especially at the tip of the infection thread (Robertson and Lyttleton 1982). Such vesicle populations likely represent both endocytic and exocytic vesicles, which need to be balanced to allow the polar growth of the infection thread, in analogy to growing root hairs or pollen tubes (see also by Ovečka et al. in this volume). Proper infection thread growth is further thought to involve a balance between insertion of new membrane and solidification of the infection thread matrix by peroxidases and deposition of cell wall material (Brewin 2004; Gage 2004). Infection thread formation requires a major reorganization of the host cytoskeleton (Timmers et al. 1999). The nucleus moves toward the center of the root hair curl, where a microtubule array forms between the nucleus and the tip of the infection thread, which maintains a fixed distance as the infection thread grows toward the base of the root hair. Recently, two genes *Nap1* and *Pir1* were identified that encode subunits of the SCAR/WAVE complex, which regulates actin polymerization (Yokota et al. 2009; Miyahara et al. 2010). Mutations in these genes significantly impaired infection thread formation and invasion of the nodule primordium.

As infection threads reach the base of the root hair, the bacteria are released into the apoplast at the boundary with the underlying cortical cell. There, a new transcellular infection thread can be initiated and this process continues to guide the bacteria to the nodule primordium cells. The path followed by the infection threads is marked by activated cortical cells (arrested at the G2 phase) that form so-called pre-infection threads. These form cytoplasmic bridges that traverse the central vacuole and contain bundles of microtubules that run the length of the cytoplasmic bridge (Van Brussel et al. 1992; Yang et al. 1994; Timmers et al. 1998). It has been hypothesized that this manipulation of the mitotic cycle is used to weaken the plant cell wall to facilitate the entry of the infection threads. When the infection threads reach the cells of the nodule primordium, the bacteria are released from the infection threads and are taken up into the cells through an "endocytosis-like" process by which symbiosomes are formed. After the infection threads invade the nodule primordium, a meristem is established that continues to add cells to the developing nodule (Timmers et al. 1998). In indeterminate nodules, such as formed in pea or Medicago, this meristem persists at the apex of the nodule by which an elongated nodule is formed and rhizobial infection of meristem-derived cells and symbiosome formation continue to occur in the so-called infection zone (Vasse et al. 1990). In determinate nodules of Lotus, the meristem is active only for a short period and infected cells that already contain symbiosomes can divide resulting in round nodules (Brewin 1991).

Symbiosome formation starts with the formation of a so-called unwalled infection droplet, a region on the infection thread where the cell wall is locally degraded (or not built) and the plasma membrane invaginates. This allows the bacteria to come into close contact with the host plasma membrane and subsequently the bacteria are individually taken up into the cytoplasm by which they become surrounded by a plant-derived membrane, the symbiosome membrane. The uptake of the bacteria correlates with a change in the outer surface of the bacteria (Fraysse et al. 2003), which suggests a physical interaction in analogy to phagocytosis of bacteria by animal cells (Alonso and García-del Portillo 2004; Cossart and Sansonetti 2004). Next, the bacterium (now called bacteroid) and the surrounding symbiosome membrane (also called peribacteroid membrane) divide and start filling the cells. Depending on the plant a symbiosome typically contains only one (in case of Medicago) or several bacteroides (in case of Lotus). Subsequently, the symbiosomes differentiate into their nitrogen-fixing form, which is facilitated by the low free oxygen concentration in the infected nodules cells and correlates with the induction of the bacterial nitrogen fixation (*Nif*) genes (Yang et al. 1991; Soupene et al. 1995; Ott et al. 2005). The cells that form symbiosomes typically endoreduplicate their DNA, which facilitates symbiosome formation and maintenance (Cebolla et al. 1999; Mergaert et al. 2006). Interestingly, host nuclear polyploidy is a common theme in plant endosymbiosis, both symbiotic and pathogenic, where microbes are hosted inside living plant cells (Wildermuth 2010). The induction of endoreduplication cycles in nodule cells of Medicago was shown to be controlled by the cell cycle anaphase-promoting complex activator CCS52A (Cebolla et al. 1999; Vinardell et al. 2003). Knockdown of the *CCS52A* gene specifically in the nodule reduced cell size and ploidy levels and impaired symbiosome formation and maintenance (Vinardell et al. 2003). Some cells in the central tissue of the nodule are never infected by the bacteria and these occur as relatively small uninfected cells in between the large infected cells. Uninfected cells are thought to play a role in metabolite transport to and from infected cells (White et al. 2007). Eventually, as the nodule ages, the symbiosis starts to break down and senescence of symbiosomes and host cells occurs in the so-called senescent zone (Vasse et al. 1990). During this senescence process the symbiosomes fuse to form lytic compartments, where the bacteria are degraded as the host tries to retrieve much of its invested resources (see below).

In several basal legume species, the rhizobia are not released from the infection threads, but instead they fix nitrogen within so-called fixation threads. Such fixation threads are highly branched intracellular thread-like structures that form the symbiotic interface compartment and morphologically resemble somewhat the structure of an arbuscule. Similar fixation threads are formed in *Parasponia* nodules; the only non-legume genus able to establish a rhizobial symbiosis (Op den Camp et al. 2011). Furthermore, in the Caesalpinioidae legume genus *Chamaecrista,* a range of different intracellular structures can be observed, ranging from fixation threads to individual symbiosomes (Naisbitt et al. 1992). Therefore, fixation threads are thought to represent a more primitive stage from which symbiosomes evolved (Sprent 2007). This fits with the observed plasma membrane identity of the symbiosome membrane (see membrane identity below).

3 Arbuscular Mycorrhizal Infection

AM fungi of the order Glomeromycota arguably form the most successful endosymbiosis with plants. This ancient symbiosis originated >450 million years ago and has since then persisted and occurs in over 80 % of current land plants (Redecker et al. 2000; Smith and Read 2008). AM fungi are obligate biotrophs that require a plant host to complete their life cycle. Spores of AM fungi can germinate without the presence of a suitable host plant but then hyphal growth is limited. The plant signals its proximity by secreting strigolactones, which induce enhanced and continuous hyphal growth and branching of the fungus (Akiyama et al. 2005; Besserer et al. 2006). After contact with the host root, the fungal hyphae form hyphopodia on the root surface. Depending on the host plant, fungal hyphae enter the epidermal cell or proceed intercellularly (Smith and Read 2008). Intracellular invasion of the root cells involves the formation of a prepenetration apparatus (PPA), a subcellular structure that facilitates the fungal hyphae to pass through, with similarities to infection thread formation (Genre et al. 2005). First the nucleus moves to the fungal contact side. Here, the cytoplasm accumulates, which is accompanied by substantial cytoskeletal reorganization and prominent proliferation of ER as well as accumulation of Golgi stacks and vesicles (Genre et al. 2005, 2008). Next, the nucleus moves away from the contact site to form a cytoplasmic cytoskeletal/ER column that defines the path for hyphal penetration. By studying various GFP-tagged exocytotic marker proteins, including exocytotic v-SNARE proteins, it was shown that exocytosis occurs at the tips of the invading hypha, marking the sites of perifungal membrane delivery (Genre et al. 2012; see also below). This highlights the importance of focal exocytosis for the intracellular accommodation of the fungi. Such extensive secretory vesicle traffic is likely compensated/balanced by endocytic retrieval of excess membrane to allow the proper growth of the invading hypha, as indicated by the presence of numerous multivesicular bodies around the tips of the hypha (Genre et al. 2008). In the cortex, the fungus grows mostly in between the cells and invades cells in the inner cortex where it forms highly branched hyphal structures called arbuscules (or hyphal coils in certain plant species). These fine hyphal branches are surrounded by an extension of the plant plasma membrane, called the periarbuscular membrane, and together they form the symbiotic interface where nutrient exchange takes place. Via this interface the fungus delivers scarce minerals, especially phosphate, to the plant in return for carbohydrates (Smith and Read 2008). Both invasion of cortical cells and arbuscule development are associated with the formation of complex PPAs (Genre et al. 2008). The differentiation/specialization of the periarbuscular membrane from the plasma membrane with which it is continuous is demonstrated by the altered (primary wall-like) deposition of cell wall material at the interface (Bonfante and Perotto 1995). So, similar to unwalled infection droplets and symbiosomes, arbuscules lack a structured cell wall. Furthermore, specific Medicago phosphate transporters and a H+ ATPase have been shown to localize exclusively to the periarbuscular membrane but not to the plasma membrane (Gianinazzi-Pearson et al. 2000; Harrison et al. 2002). This indicates the acquisition of a specific membrane identity (see below). Arbuscules are typically maintained only for 8–14 days, after which they are degraded and the host cell can be reinfected. Arbuscule degradation likely involves the secretion of lytic enzymes at the interface, allowing the plant to control cooperativity (Parniske 2000; Kiers et al. 2011).

4 signalling in Symbiosis

In the last decades, it has become clear that there is a common genetic basis underlying both the rhizobial and AM endosymbiosis (Kouchi et al. 2010; Oldroyd et al. 2011). Both interactions are tightly regulated by a shared signalling cascade (Fig. 1) and involve highly similar signalling molecules (Oldroyd et al. 2011).

The rhizobium legume symbiosis is initiated by specific lipo-chito-oligosaccharides that are secreted by rhizobia and are called Nod factors (NF) (Truchet et al. 1991). NF are produced by the bacteria in response to specific flavonoids secreted by the plant (Oldroyd and Downie 2004). These NF are active at picomolar concentrations and control both nodule organogenesis as well as infection thread formation (Oldroyd and Downie 2004). Depending on the rhizobial species specific decorations can be present, which determine host specificity especially at the level of infection thread formation (Ardourel et al. 1994).

Nod factors are perceived by the LysM domain-containing receptor kinases such as NFP/LYK3/LYK4 in Medicago and NFR5/NFR1 in Lotus (Limpens et al. 2003; Madsen et al. 2003; Radutoiu et al. 2003; Arrighi et al. 2006; Smit et al. 2007). These receptors trigger ion fluxes leading to a depolarization of the plasma membrane, calcium influx at the plasma membrane, and Ca^{2+} oscillations (Ca^{2+} spiking) in and around the nucleus (Oldroyd and Downie 2006; Capoen et al. 2009). The activation of the Ca^{2+} spiking response also requires a plasma membrane LRR receptor kinase MtDMI2/LjSymRK (Ané et al. 2002; Endre et al. 2002; Stracke et al. 2002) as well as components of a nuclear pore complex (Kanamori et al. 2006; Saito et al. 2007; Groth et al. 2010) and a putative cation channel MtDMI1/LjCastor and LjPollux located at the nuclear envelope (Ané et al. 2004; Charpentier et al. 2008). The resulting Ca^{2+} signature is decoded by a nuclear localized Ca^{2+}/calmodulin-dependent kinase MtDMI3/LjCCaMK (Catoira et al. 2000; Lévy et al. 2004; Mitra et al. 2004; Tirichine et al. 2006). CCaMK activation subsequently triggers transcriptional changes leading to cortical cell divisions as well as root hair curling and infection thread formation, which involves the phosphorylation of its interacting protein MtIPD3/LjCYCLOPS (Messinese et al. 2007; Yano et al. 2008). The NF-induced gene expression requires the

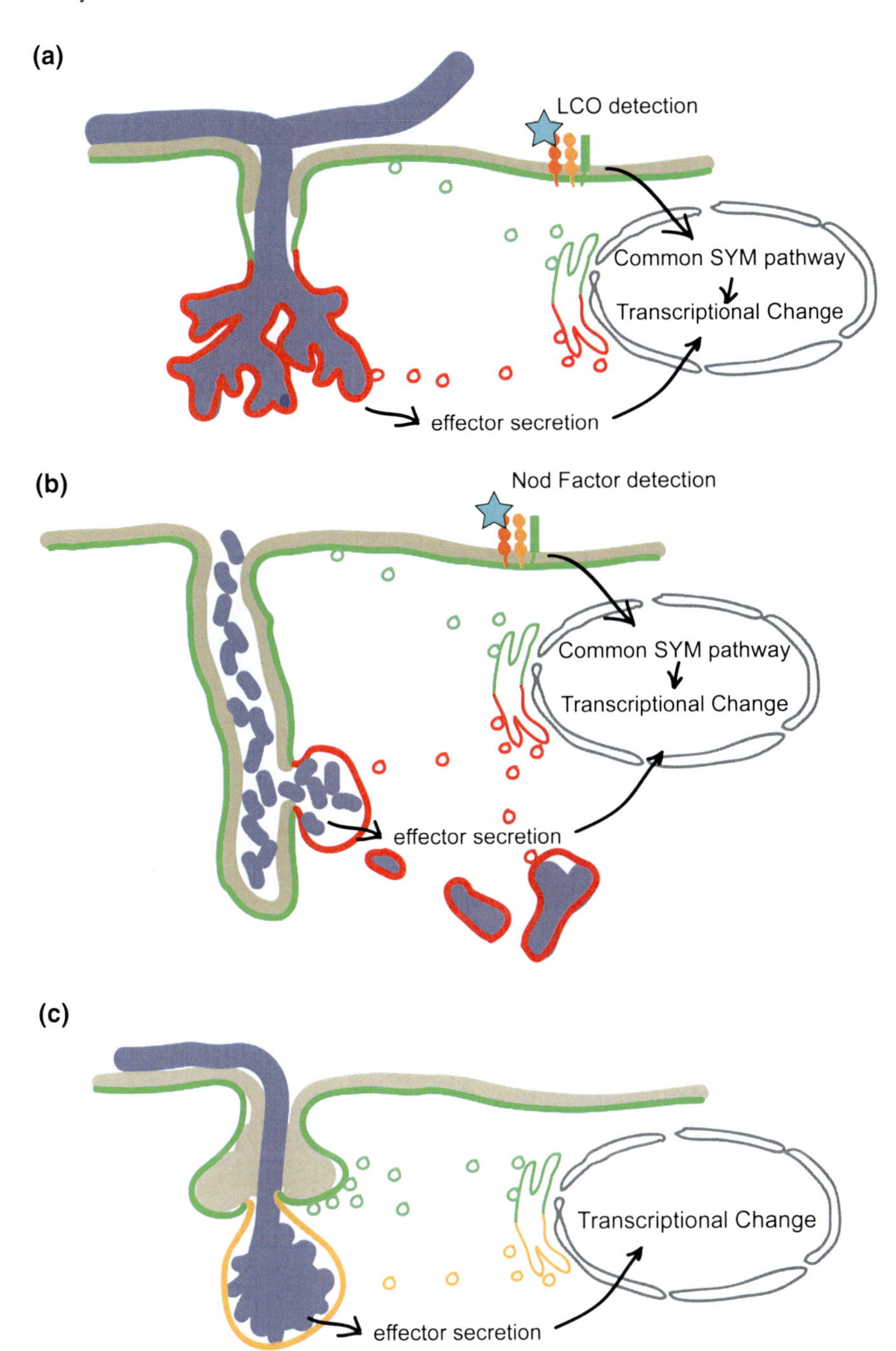
(a)
LCO detection
Common SYM pathway
Transcriptional Change
effector secretion
(b)
Nod Factor detection
Common SYM pathway
Transcriptional Change
effector secretion
(c)
Transcriptional Change
effector secretion

◀ **Fig. 1** Different microbes form biotrophic feeding structures in plants. **a** AM fungi form highly branched arbuscules surrounded by a specialized peri-arbuscular membrane that is devoid of a structured cell wall. **b** Rhizobia are released from unwalled infection droplets that form at local sited on cell wall-bound infection threads that invade the nodule cells. **c** Pathogenic fungi and oomycetes form haustoria in host cells surrounded by a plant-derived extra haustorial membrane. The different interface compartments have in common that their membrane composition is distinct from the surrounding plasma membrane. The distinct interface membrane composition depends on transcriptional changes in the nucleus, as well as a symbiotic membrane identity (*in red or yellow*) that is distinct from the membrane identity of the plasma membrane (*in green*). Different populations of vesicles co-exist, of which one population (*in red or yellow*) is transported to the interface, while the other population (*in green*) is transported to the plasma membrane. Effectors secreted by the endosymbiont induce transcriptional changes in the host. AM and rhizobia predominantly depend on the secretion of LCOs to trigger the common SYM pathway to reprogram their host, whereas pathogens mostly secrete effector proteins to trigger transcriptional changes

transcription factors NSP1, NSP2, and NIN (Schauser et al. 1999; Kalo et al. 2005; Smit et al. 2005; Heckmann et al. 2006; Marsh et al. 2007). Gain-of-function mutations in CCaMK lead to the formation of nodules in the absence of rhizobia. Such gain-of-function constructs are sufficient to allow nodule formation, rhizobial infection, and symbiosome formation when upstream components of the signalling pathway are missing (Hayashi et al. 2010; Madsen et al. 2010). This implies that the primary role of NF perception is the activation of MtDMI3/LjCCaMK. The induction of cortical cell divisions in the Rhizobium-legume symbiosis further requires the activation of the cytokinin receptor, called CRE1 in Medicago (Gonzalez-Rizzo et al. 2006) and LHK1 in Lotus (Tirichine et al. 2007). A *LHK1* knockout mutation blocks cortical cell activation but not infection thread formation in the root hairs (Murray et al. 2007). These infection threads fail to colonize the cortex likely due to absence of pre-infection threads. On the other hand, a gain-of-function mutation in this cytokinin receptor triggers the formation of spontaneous nodules, which depends on the transcription factors NSP1/NSP2 and NIN but it is not sufficient to allow rhizobial infection in the absence of CCaMK/DMI3 activation (Madsen et al. 2010; Plet et al. 2011). These studies suggest that cytokinin is generated upon activation of DMI3 in the epidermis as a mobile signal to coordinate cortical cell activation to allow a successful symbiosis (Oldroyd et al. 2011; Plet et al. 2011).

Several studies indicate that the NF signalling pathway is also active in the nodule to control symbiosome formation. All components are expressed especially in the infection zone of the nodule where symbiosome formation starts (Bersoult et al. 2005; Capoen et al. 2005; Limpens et al. 2005; Mbengue et al. 2010). Nodule-specific knockdown of *SymRK/DMI2* by RNAi or by ectopic expression in a mutant background was shown to cause a block in the release of rhizobia from the cell wall-bound infection threads while nodule formation and infection thread growth still occurred (Capoen et al. 2005; Limpens et al. 2005). Similarly, introduction of a rice *CCaMK/DMI3* ortholog in a Medicago *dmi3* mutant resulted in a partial complementation where nodules contained infection threads; however, bacteria were not released (Godfroy et al. 2006). Further support for the role of

Nod factor signalling in symbiosome formation came from the analysis of *ipd3* mutants in Medicago and pea (Tsyganov et al. 1998; Horvath et al. 2011; Ovchinnikova et al. 2011). Such knock-out mutants are able to form nodules, even though infection thread growth is already hampered; however, the bacteria fail to be released from the cell wall bound infection threads. In addition, attempts to complement *nsp1* mutants with a *Nicotiana benthamiana NSP1* gene resulted in aberrant early senescent symbiosomes (Heckmann et al. 2006). Together, these data highlight a key role for NF signalling in symbiosome formation.

Interestingly, the NF signalling components DMI1, DMI2, DMI3, and IPD3 are also essential to allow a symbiotic association with AM fungi, and therefore this pathway has been called the common symbiotic (SYM) signalling pathway (Parniske 2008). This indicates that rhizobia have co-opted the common SYM pathway from the ancient AM symbiosis. This is further supported by the recent finding that AM fungi produce lipo-chito-oligosaccharide signal molecules that are strikingly similar to the rhizobial Nod factors (Maillet et al. 2011). However, loss-of-function mutations in Nod factor receptors do not affect mycorrhization in model legumes, which suggests that in legumes different receptors activate the common SYM pathway. Strikingly, a recent analysis of Myc factor induced gene expression in Medicago roots showed that NFP is essential for almost all Myc factor induced genes (Czaja et al. 2012). Furthermore, a recent study in *Parasponia*, showed that a single NFP ortholog controls both arbuscule formation and fixation thread formation (Op den Camp et al. 2011). This strongly supports the hypothesis that Nod factor receptors evolved from ancestral Myc receptors. Therefore, it is probable that during evolution rhizobia acquired the ability to produce Nod factors (Sullivan and Ronson 1998) and this allowed them to use the ancient AM machinery to establish an intracellular symbiotic interface. This would imply that Nod factor perception (e.g. Nod factor receptors) in the nodule controls symbiosome formation. Such a role is supported by the observation that the rhizobial *nod* genes are active in the bacteria inside infection threads in a region of the nodule where the Nod factor receptors NFP and LYK3 are also expressed (Sharma and Signer 1990; Schlaman et al. 1991; Marie et al. 1994; Limpens et al. 2005). This involvement is likely not a direct one but probably occurs via the activation of DMI3, as the Nod factor receptors and DMI2 do not appear to be essential for symbiosome formation in the presence of a constitutive-active DMI3/CCaMK version (Madsen et al. 2010). Furthermore, the expression domains of DMI1, DMI2, DMI3 and IPD3 in the nodule appear to be wider than the expression domain of the Nod factor receptors (Limpens et al. 2005; Smit et al. 2005). This suggests that the common SYM pathway may also be activated in the nodule via alternative ways.

In the case of the AM symbiosis in legumes, components of the common SYM pathway upstream of DMI3/CCaMK do not appear to be essential for arbuscule formation in the cortex cells. For example, knockout mutants in lotus *SymRK(DMI2)* gene are still able to form arbuscules in root sectors where the fungus manages to pass the epidermis (Wegel et al. 1998; Demchenko et al. 2004). In contrast, *dmi3* knockout mutants never form arbuscules, even when the fungus manages to pass the epidermis (Demchenko et al. 2004). This suggests that DMI2/SYMRK is not required in the root cortex to allow arbuscule formation. It implies that additional factors lead to the activation of DMI3/CCaMK in the cortex. A similar uncoupling of epidermal and cortical responses was observed in the Lotus *symrk-14* mutant, which contains a mutation in a conserved GDPC sequence in the SYMRK extracellular domain (Kosuta et al. 2011). This indicated that the SYMRK mainly plays a role in the epidermis to control the invasion of the root by the fungus. Loss-of-function mutants of *SYMRK* (as well as other common *SYM* genes) block the formation of the PPA and fungal penetration in the epidermis (Genre et al. 2008).

In Lotus, the CCaMK interacting protein CYCLOPS is essential for arbuscule formation (Yano et al. 2008). However, in Medicago and pea *ipd3* loss-of-function mutants are still able to form arbuscules depending on the genetic background (Horvath et al. 2011; Ovchinnikova et al. 2011). This suggests that depending on the plant additional genetic factors affect the activity (or stability) of DMI3 during the symbioses. Furthermore, CCaMK regulation appears to differ between rhizobial and AM symbioses (Shimoda et al. 2012).

In Medicago, IPD3 is essential for the induction of a nodule-specific remorin (MtSYMREM1), which was shown to interact with DMI2 and the NF receptors LYK3 and NFP (Lefebvre et al. 2010). Interestingly, a knockout mutant of the *MtSYMREM1* gene also causes a block of the release of bacteria from infection threads inside the nodule (Lefebvre et al. 2010). Based on the structural homology of SYMREM1 to caveolins in animal cells, a role for SYMREM1 in the spatial regulation of the signalling receptors was proposed (Lefebvre et al. 2010). Caveolins are scaffolding proteins that localize to a special type of membrane microdomain called caveolae (Williams and Lisanti 2004). Similarly, SYMREM1 has been shown to be highly enriched in detergent-insoluble membrane fractions, indicating a preferred localization to "lipid rafts" (Mongrand et al. 2004). Caveolins in animal cells are involved in compartmentalizing and concentrating signalling molecules at these domains and mark a clathrin-independent endocytic pathway (Pelkmans and Helenius 2002; Conner and Schmid 2003). Therefore, the IPD3-dependent induction of SYMREM1 during nodule formation may offer a feedback regulation on the activity of NF signalling receptors. SYMREM1 localizes not only to the infection thread and infection droplets but also to mature symbiosomes. Therefore, SYMREM1 likely has additional roles during symbiosome development. Interestingly, caveolae typically do not fuse with lysosomes and various pathogens hijack this clathrin-independent endocytic pathway to gain entry into animal cells (Shin and Abraham 2001). It is tempting to speculate that rhizobia use a similar mechanism to gain entry to the plant cells. Whether remorins indeed mark a specific endocytic pathway in plants remains to be demonstrated.

Additional scaffolding proteins that form specific membrane domains and that mark a clathrin-independent endocytic pathway (distinct from caveolae) in animal cells are the flotillins (Glebov et al. 2006; Stuermer 2010). Also, in Arabidopsis, flotillin FLOT1 was shown to be associated with endocytic vesicle formation at membrane microdomains in a clathrin-independent manner (Li et al. 2012).

Recently, two flotillin genes were found to be upregulated during the interaction with rhizobia and were shown to control nodulation (Haney and Long 2010). Both proteins are localized to membrane microdomains. FLOT4 was localized to the infection thread membrane and it was required for proper infection thread growth. Interestingly, it was shown that FLOT4 co-localizes with the Nod factor receptor LYK3 to distinct positionally stable puncta upon rhizobial inoculation in a Nod factor-dependent manner (Haney et al. 2011). Such puncta persisted for several days after inoculation. In uninoculated root hairs LYK3 was shown to be highly mobile in the plasma membrane. This indicates that upon Nod factor perception LYK3 is confined to FLOT4 marked microdomains to regulate the activity of the receptor complex. How these microdomains relate to the SYMREM1 marked membrane domains is currently unknown. Furthermore, the intracellular pool of vesicles containing LYK3 increased 6 h after inoculation with rhizobia, suggesting increased endocytosis. However, the origin of the observed vesicles is unknown. The intracellular LYK3 vesicles were not labeled by FLOT4, which may indicate that they represent secretory compartments transporting newly synthesized LYK3. In soybean it has been shown that a flotillin protein (Nod53b) localizes to the symbiosome membrane (Winzer et al. 1999; Panter et al. 2000). This suggests also a role for these proteins in symbiosome development, possibly in the trafficking of vesicles and their cargo from (or to) the symbiosome membrane (Stuermer 2010). It will be interesting to see whether similar scaffolding proteins control AM fungal colonization or arbuscule formation.

5 Symbiotic Interface Identity

Symbiosome formation involves a major reorganization of the host endomembrane system and requires a massive production of membrane to make the thousands of symbiosome compartments that fill the infected cells. There is approximately 40 times more symbiosome membrane than plasma membrane in an infected cell (Whitehead and Day 1997). It was generally believed that the uptake of the bacteria into the host cells involved an endocytosis process similar to the phagocytic uptake of bacteria in animal cells (Via et al. 1997; Knodler et al. 2001; Brumell and Grinstein 2004; Behnia and Munro 2005). However, a recent systematic analysis of the identity of the symbiosome membrane in the host endomembrane system of Medicago showed that symbiosome formation does not follow the default endocytic pathway (Limpens et al. 2009). By studying the association of membrane identity markers (SNARE proteins and Rab GTPase) for the different endomembrane compartments with the symbiosome membrane, it has been shown that symbiosomes never acquire the early endosomal/*trans*-Golgi network (TGN) marker SYP4 (t-SNARE) or the late endosomal/prevacuolar marker Rab5. Instead, the symbiosome membrane acquires the plasma membrane t-SNARE/SYNTAXIN SYP132 from the start of symbiosome formation throughout its development (Catalano et al. 2007; Limpens et al. 2009).

The presence of the plasma membrane SNARE SYP132 from the start of symbiosome membrane suggests an important role for the delivery of exocytotic vesicles and their associated cargo to allow the uptake of bacteria and subsequent symbiosome development (see Sect. 6 below). This is also supported by the fusion of ER and Golgi-derived vesicles with the infection droplet and symbiosome membranes (Robertson and Lyttleton 1982; Roth and Stacey 1989; Cheon et al. 1994). The involvement of a nodule-specific secretory pathway in symbiosome development was revealed by the identification of a nodule-specific signal peptidase complex. Due to a mutation in the SPC22 subunit of this complex in the Medicago *does not fix nitrogen 1 (dnf1)* mutant symbiosomes fail to differentiate to their nitrogen-fixing form and the bacteroides morphologically resemble free-living rhizobia. Van de Velde and coworkers (Van De Velde et al. 2010) showed that symbiosome differentiation in Medicago is triggered by nodule-specific antimicrobial-like cysteine-rich peptides (NCRs). Failure to remove the signal peptide from these NCRs in the *dnf1* mutants results in the accumulation of these peptides in the ER instead of accumulating at the symbiosomes leading to the block in symbiosome differentiation (Van de Velde et al. 2010; Wang et al. 2010).

After the symbiosomes stop dividing and differentiate to their nitrogen-fixing form, the vacuolar/transitory late endosomal marker Rab7 is additionally found on the symbiosome membrane. Rab7 is required to control the proper development of the symbiosome (Cheon et al. 1993; Limpens et al. 2009). Despite the presence of this key regulator of vacuole formation, symbiosomes do not acquire a vacuolar identity. Only at the onset of senescence, when the symbiosomes fuse and form lytic compartments, do symbiosomes acquire vacuolar SNAREs, such as SYP22 and VTI11. Therefore, symbiosomes appear to acquire a mosaic plasma membrane (SYP132) and late endosomal (Rab7) identity, and the delay in acquiring (lytic) vacuolar identity (e.g. vacuolar SNAREs) most likely ensures their survival and maintenance as individual units (Limpens et al. 2009). The presence of Rab7 on the symbiosome membrane may indicate that symbiosomes can intercept vacuole-targeted proteins, which could explain the reported presence of several vacuolar enzymes in the symbiosome space which are likely transported to the vacuole via late endosomes (Mellor 1989; Whitehead and Day 1997; Jones et al. 2007). How symbiosomes prevent/delay the acquisition of a vacuolar identity is currently unknown.

Similar to symbiosome formation, there is a tremendous expansion of membrane during the formation of an arbuscule, which must be specifically targeted toward the growing periarbuscular membrane (PAM). The central vacuole persists, while ER, Golgi stacks, and peroxisomes relocalize to the close proximity of the developing arbuscule (Pumplin and Harrison 2009). The protein composition of the PM and the membrane surrounding the arbuscular trunk is different from the protein composition of the PAM, suggesting selective targeting of proteins toward the PAM (Pumplin and Harrison 2009). Recently, Pumplin et al. (2012) showed that the targeting of different proteins toward the PAM or the PM is both protein and promoter timing dependent. When the phosphate transporter PT1 is expressed under its native promoter which is only active in uninfected cells, it localizes toward the PM. When PT1 is expressed under a MtBcp1 promoter which switches on after fungal entry, but before arbuscules are produced, it localizes to both the PM and the PAM. When expressed under a MtPT4 promoter which is only active when arbuscules are produced, it localizes exclusively to the PAM. The essential phosphate transporter PT4 is specifically targeted to the PAM, and when expressed in non-infected cells it does not leave the ER. In contrast, the aquaporin AtPIP2a localized to the plasma membrane in uninfected cells but when expressed from the PT4 promoter was stuck in the ER, and under the control of the Bcp1 promoter localized to the ER and PM but not to the PAM. These observations prompted the authors to hypothesize that during the formation of an arbuscule, a default exocytosis pathway is reoriented toward the PAM, and targeting of some proteins is fully dependent on the timing of their expression. This reorientation appeared to be accompanied by a change in the cargo that was allowed to enter the secretory pathway, possibly regulated by specific chaperones in the ER (González et al. 2005; Dharmasiri et al. 2006; Wu et al. 2011). However, not all proteins showed a similar localization when expressed under the same promoter. Since only some proteins such as phosphate and sugar transporters, but not aquaporin AtPIP2a localized to the PAM when expressed under the appropriate promoter (Pumplin et al. 2012), we hypothesize that only a specific exocytosis pathway is reoriented toward the PAM. This implies that there are at least two distinguishable populations of exocytotic vesicles, of which one population is targeted toward the PAM during arbuscule formation, while the other remains directed to the PM.

6 Recruitment of Specific Exocytosis Pathways

Ivanov et al. (2012) showed that two specific exocytotic vesicle-associated membrane proteins (v-SNAREs/VAMPs) are required for the formation of symbiotic membrane interface in both Rhizobium–legume symbiosis and arbuscular mycorrhiza. Co-silencing of these Medicago *VAMP72* genes blocked symbiosome as well as arbuscule formation whereas it did not affect root colonization by the microbes' (Ivanov et al. 2012). Interestingly, orthologs of these "symbiotic *v-SNAREs*" are lacking in the genome of Arabidopsis, which cannot establish a symbiosis with either rhizobia or AM fungi. Identification of these VAMP72 s as common symbiotic regulators in exocytotic vesicle trafficking suggests that the ancient exocytotic pathway forming the periarbuscular membrane compartment also has been co-opted in the Rhizobium–legume symbiosis (Ivanov et al. 2012). Thus, the Rhizobium symbiosis and arbuscular mycorrhiza use not only the same signalling pathway but also the same cellular process, namely a specific exocytotic pathway, to form intracellular symbiotic compartments (Fig. 1).

Because the formation of unwalled infection droplets, which could be regarded as the beginning of a symbiotic interface, was blocked in the *vamp72* knockdown nodules, it can be hypothesized that the VAMP72 marked vesicles deliver cargo to degrade (or prevent the build up) of a cell wall at these sites and to allow the uptake of bacteria. Induction of such cargo may be controlled by the common DMI3/IPD3 module activated via upstream DMI2 and DMI1 signalling (Fig. 1). Recently, a nodule-specific secreted pectin methyl esterase (LjNPL) involved in cell wall remodeling was found to control the infection thread formation (Xie et al. 2012). Knockout mutations of the corresponding gene in Lotus blocked the formation of infection threads after the bacteria formed a microcolony in curled root hairs, a similar phenotype as observed in Lotus *cyclops/ipd3* mutants (Yano et al. 2008). Interestingly, the orthologous *NPL* gene in Medicago is most strongly expressed specifically in the distal infection zone where infection threads invade the nodule cells, and symbiosomes are formed (Limpens unpublished). Therefore, this pectin methyl esterase may be one of the putative cargos of the VAMP72 vesicles that contribute to the formation of an unwalled infection droplet. It has been postulated by Timmers et al. (2005) that cell wall degrading enzymes may lead to the accumulation of osmotically active molecules, leading to the build up of a pressure in the unwalled droplet that drives the entry of bacteria. Alternatively or additionally, the common SYM pathway (Fig. 1) could control the targeting of vesicles to the infection threads by affecting the cytoskeleton or the formation of specific membrane domains. It has been previously suggested that the cytoskeleton rearrangements and targeted vesicle transport play a role in this process as release of bacteria from the infection threads seemed to involve a change in the kind of vesicles and their cargo fusing with the tips of the infection thread (Brewin 1998). A potential regulator may be the recently identified VAPYRIN protein, which is essential for both rhizobial infection as well as AM fungal colonization and arbuscule formation and is induced by the common SYM pathway (Feddermann et al. 2010; Pumplin et al. 2010; Murray et al. 2011). VAPYRIN contains an N-terminal VAMP-associated protein (VAP)/major sperm protein (MSP) domain which indicates a role in vesicle transport.

7 The Establishment of a Pathogenic Host–Microbe Interface

The infection of plants by biotrophic pathogens shows striking similarities to the accommodation of mutualistic endosymbionts. A biotrophic lifestyle has evolved several times and is found across the fungal kingdom as well as in oomycetes. Despite this lack of phylogenetic relation, many biotrophic pathogens form a similar biotrophic feeding structure, a haustorium, which is surrounded by a plant-derived membrane, the extra haustorial membrane (EHM). Besides the transfer of nutrients to the microbe, the haustorium has a prominent role in the delivery of effectors into the host to suppress plant defense and to reprogram it to host the pathogen (Kemen et al. 2005). Interestingly, it was recently shown that also AM fungi secrete effector proteins that are translocated in the host cells to facilitate infection (Kloppholz et al. 2011). Such effectors may enter the cell by binding to phosphatidylinositol-3-phosphate (PI3P) located at the outer surface of the PM, in analogy to effector proteins from oomycetes and fungi (Kale et al. 2010). Entry of such effectors appears to involve a "lipid-raft"-mediated endocytic pathway (Kale et al. 2010). Also, some rhizobial strains are known to secrete effectors into the plant using a type III secretion system that is commonly used by pathogenic bacteria for effector secretion (Alfano and Collmer 2004; Bartsev et al. 2004). Some effectors secreted by AM and rhizobia are shown to inhibit the expression of plant defense genes (Bartsev et al. 2004; Kloppholz et al. 2011). Therefore, the main function of AM and rhizobial effectors may be to suppress defense responses. In contrast, pathogens are also believed to rely on their effectors to reprogram their host to form a feeding structure (Caillaud et al. 2012).

Plants sense the presence of microbes by PM localized receptors that recognize typical microbe-associated molecular patterns (MAMPs) or danger-associated molecular patterns (DAMPs), which are plant-derived compounds that indicate the disturbance of plant cell wall integrity by pathogens (Zipfel and Felix 2005). When a pathogen contacts a plant cell, this results in the local deposition of antimicrobial compounds as well as cell wall reinforcement by deposition of callose and lignin (Schmelzer 2002). In order to form a haustorium, adapted pathogens break through this initial line of defense and form a membrane compartment that lacks a structured cell wall. In contrast, cell wall material continues to be transported to the periphery of the haustorium, which results in a cup-shaped collar that may eventually lead to full encasement of the haustorium (Meyer et al. 2009; Fig. 1).

Research on parasitic host–microbe interactions has traditionally focused on defense responses. The initial recognition of the pathogen, followed by the local delivery of defense molecules relies on the re-orientation of exocytotic pathways. It was shown in barley and Arabidopsis that t-SNARE SYP121 (ROR2 in barley) accumulates at the site of attempted penetration by both adapted and non-adapted powdery mildew fungi (Collins et al. 2003; Bhat et al. 2005; Meyer et al. 2009). Arabidopsis SYP121 was shown to form complexes with the t-SNAREs SNAP33 and v-SNAREs VAMP721 and VAMP722. Since a VAMP721/VAMP722 double knockout is not viable, the authors argue that a default exocytosis pathway essential for plant development is recruited for defense (Kwon et al. 2008).

In the last decade, a lot of progress has been made on the biogenesis of haustoria. It was shown for the haustoria of different fungal and oomycete pathogens that PM markers are often excluded from the EHM (Koh et al. 2005; Lu et al. 2012), while some proteins exclusively label the EHM, but not the PM (Micali et al. 2011; Caillaud et al. 2012). Two models explain the formation of an EHM with a distinct composition from the PM (Koh et al. 2005). The first model states that vesicles are delivered to the periphery of the haustorium. Subsequently, the haustorial neckband is selectively allowing membrane and EHM proteins to diffuse to the EHM, while PM proteins are unable to pass the neckband. The second model states that distinct vesicle populations exist, of which one is targeted to the PM, while the other is targeted to the EHM. Consequently, the formation of a haustorium requires protein trafficking dedicated to the formation and maintenance of the EHM. The existence of a vesicle population dedicated to EHM formation and maintenance is supported by the observation that in both fungal and oomycete interactions defense molecules and cell wall reinforcements are continuously transported to the periphery of the haustorium resulting in the encasement of haustoria (Meyer et al. 2009; Caillaud et al. 2012). This implies that pathogens use an exocytotic pathway for haustorium formation that is distinct from the pathway delivering cell wall reinforcements. The observation that fungi are still able to form haustoria in the SYP121 or VAMP721/VAMP722 mutants further supports the involvement of different exocytosis pathways in EHM formation and haustorial encasement or PM maintenance.

The fact that plants retain the ability to host pathogens suggests that the pathogens hijack processes fundamental for plant development to form haustoria. It is most unlikely that all biotrophic pathogens use similar effectors to reprogram their host, since effectors are under constant pressure to escape recognition by the plant defense system, and are consequently often unique for a pathogen species (Baxter et al. 2010; Spanu et al. 2010; Duplessis et al. 2011). However, there may be a common plant pathway that is recruited by different biotrophic pathogens in order to create a biotrophic feeding structure.

Arguably the most likely genes to hijack in order to establish a pathogenic host–microbe interface are the genes that are responsible for the formation of a mutualistic host–microbe interface. However, the common symbiosis genes were shown to be dispensable for haustorium formation in leaves, since the rust fungus *Uromyces loti* was able form haustoria in all tested common signalling mutants of *Lotus japonicus* (Meleresh and Parniske 2006). Furthermore, model species *Arabidopsis thaliana* is lacking most common symbiosis genes and it is unable to engage a symbiotic relationship with AM fungi. Nevertheless, Arabidopsis can host many biotrophic pathogens. These observations show that biotrophic endosymbionts do probably not directly target the common symbiosis pathway in order to trick the plant into cooperating to form an interface. However, there are some indications that downstream processes of the common symbiosis genes may influence plant responses toward biotrophic fungi. It was shown that upon contact with the compatible biotrophic fungus *Colletotrichum trifolii*, the cytoplasm of *Medicago truncatula* root epidermal cells accumulates at the site of contact. This was associated with the repositioning of the nucleus and resembled the formation of a pre-penetration apparatus in the plant-AM interaction (Genre et al. 2009). This cytoplasmic aggregation was abolished in Medicago *dmi3* mutants, resulting in collapse of the cytoplasm after contact by the pathogen. As argued before, the formation of a host–microbe interface requires both transcriptional changes as well as local changes in membrane identity. It is likely that the initial induction of transcriptional change through effectors by pathogens is fundamentally different from the induction of transcriptional change by mutualistic endosymbionts (Fig. 1). To what extent the microbes use/co-opt a common cellular mechanism(s) to make the interface membranous compartment will be an interesting topic for future studies.

8 Conclusions and Future Prospects

We are starting to understand that diverse exocytic pathways are involved in the formation of a host–microbe interface and in the maintenance of the PM during the intracellular accommodation of a microbe inside a plant cell. However, it remains unknown how different vesicles and their cargos are specifically targeted to the interface membrane. The interface most likely acquires a distinct membrane identity resulting in a specific membrane composition, but it remains obscure how this membrane identity is initiated. We consider it likely that local signalling between the endosymbiont and host plays a key role, together with the formation of specific membrane domains that recruit specific exocytotic vesicles or may trigger endocytosis of specific membrane proteins. It is clear that in evolution rhizobia acquired the ability to infect legumes by producing NFs that resemble the LCOs secreted by AM. However, it is questionable whether NF signalling is directly required at the host-microbe interface, or whether only NF-induced gene expression is sufficient. Understanding how the common SYM pathway connects with the recruitment of a specific exocytic pathway will be a major task for the future.

The structural similarity of haustoria to symbiotic host–microbe interfaces is striking. Now that we start to unravel the plant proteins that mark and control the symbiotic interfaces, it should be tested whether these also mark/control pathogenic interfaces in order to establish whether pathogens hijack the process of symbiotic interface formation. If so, this will be a great step forward in understanding how it is possible that plants have maintained the intrinsic ability to host enemies. Furthermore, understanding how rhizobia and AM fungi control the switch to a lytic compartment could have major implications to manipulate such mechanism to control pathogenic interactions. Studying plant–microbe interactions will continue to give valuable insight into the regulation of the host endomembrane system.

References

Akiyama K, Matsuzaki K, Hayashi H (2005) Plant sesquiterpenes induce hyphal branching in arbuscular mycorrhizal fungi. Nature 435:824–827

Alfano JR, Collmer A (2004) Type III secretion system effector proteins: double agents in bacterial disease and plant defense. Annu Rev Phytopathol 42:385–414

Alonso A, García-del Portillo FG (2004) Hijacking of eukaryotic functions by intracellular bacterial pathogens. Int Microbiol 7:181–191

Ane JM, Kiss GB, Riely BK, Penmetsa RV, Oldroyd GED, Ayax C, Levy J, Debelle F, Baek JM, Kalo P, Rosenberg C, Roe BA, Long SR, Denarie J, Cook DR (2004) *Medicago truncatula* DMI1 required for bacterial and fungal symbioses in legumes. Science 303:1364–1367

Ane JM, Levy J, Thoquet P, Kulikova O, de Billy F, Penmetsa V, Kim DJ, Debelle F, Rosenberg C, Cook DR, Bisseling T, Huguet T, Denarie J (2002) Genetic and cytogenetic mapping of

DMI1, DMI2, and DMI3 genes of *Medicago truncatula* involved in nod factor transduction, nodulation, and mycorrhization. Mol Plant-Microbe Interact 15:1108–1118
Ardourel M, Demont N, Debelle FD, Maillet F, Debilly F, Prome JC, Denarie J, Truchet G (1994) *Rhizobium meliloti* lipooligosaccharide nodulation factors—different structural requirements for bacterial entry into target root hair-cells and induction of plant symbiotic developmental responses. Plant Cell 6:1357–1374
Arrighi JF, Barre A, Ben Amor B, Bersoult A, Soriano LC, Mirabella R, de Carvalho-Niebel F, Journet EP, Gherardi M, Huguet T, Geurts R, Denarie J, Rouge P, Gough C (2006) The *Medicago truncatula* lysine motif-receptor-like kinase gene family includes NFP and new nodule-expressed genes. Plant Physiol 142:265–279
Bartsev AV, Deakin WJ, Boukli NM, McAlvin CB, Stacey G, Malnoe P, Broughton WJ, Staehelin C (2004) NopL, an effector protein of Rhizobium sp NGR234, thwarts activation of plant defense reactions. Plant Physiol 134:871–879
Baxter L et al (2010) Signatures of adaptation to obligate biotrophy in the Hyaloperonospora arabidopsidis genome. Science 330:1549–1551
Behnia R, Munro S (2005) Organelle identity and the signposts for membrane traffic. Nature 438:597–604
Bersoult A, Camut S, Perhald A, Kereszt A, Kiss GB, Cullimore JV (2005) Expression of the *Medicago truncatula* DMI2 gene suggests roles of the symbiotic nodulation receptor kinase in nodules and during early nodule development. Mol Plant-Microbe Interact 18:869–876
Besserer A, Puech-Pages V, Kiefer P, Gomez-Roldan V, Jauneau A, Roy S, Portais JC, Roux C, Becard G, Sejalon-Delmas N (2006) Strigolactones stimulate arbuscular mycorrhizal fungi by activating mitochondria. PLoS Biol 4:1239–1247
Bhat RA, Miklis M, Schmelzer E, Schulze-Lefert P, Panstruga R (2005) Recruitment and interaction dynamics of plant penetration resistance components in a plasma membrane microdomain. Proc Natl Acad Sci U S A 102:3135–3140
Bonfante P, Perotto S (1995) Strategies of arbuscular mycorrhizal fungi when infecting host plants. New Phytol 130:3–21
Brewin NJ (1991) Development of the legume root nodule. Annu Rev Cell Biol 7:191–226
Brewin NJ (1998) Tissue and cell invasion by *Rhizobium*: the structure and development of infection threads and symbiosomes. In: Spaink HP, Kondorosi A, Hooykaas PJJ (eds) The Rhizobiaceae. Kluwer Academic Publishers, Dordrecht
Brewin NJ (2004) Plant cell wall remodelling in the rhizobium-legume symbiosis. Crit Rev Plant Sci 23:293–316
Brumell JH, Grinstein S (2004) Salmonella redirects phagosomal maturation. Curr Opin Microbiol 7:78–84
Caillaud MC, Piquerez SJM, Fabro G, Steinbrenner J, Ishaque N, Beynon J, Jones JDG (2012) Subcellular localization of the Hpa RxLR effector repertoire identifies a tonoplast-associated protein HaRxL17 that confers enhanced plant susceptibility. Plant J 69:252–265
Capoen W, Den Herder J, Sun JH, Verplancke C, De Keyser A, De Rycke R, Goormachtig S, Oldroyd G, Holsters M (2009) Calcium Spiking patterns and the role of the calcium/calmodulin-dependent kinase CCaMK in lateral root base nodulation of *Sesbania rostrata*. Plant Cell. 21:1526–1540
Capoen W, Goormachtig S, De Rycke R, Schroeyers K, Holsters M (2005) SrSymRK, a plant receptor essential for symbiosome formation. Proc Natl Acad Sci U S A 102:10369–10374
Catalano CM, Czymmek KJ, Gann JG, Sherrier DJ (2007) *Medicago truncatula* syntaxin SYP132 defines the symbiosome membrane and infection droplet membrane in root nodules. Planta 225:541–550
Catoira R, Galera C, de Billy F, Penmetsa RV, Journet EP, Maillet F, Rosenberg C, Cook D, Gough C, Denarie J (2000) Four genes of *Medicago truncatula* cone symbiosome membrane and infection dansduction pathway. Plant Cell 12:1647–1665
Cebolla A, Vinardell JM, Kiss E, Olah B, Roudier F, Kondorosi A, Kondorosi E (1999) The mitotic inhibitor ccs52 is required for endoreduplication and ploidy-dependent cell enlargement in plants. EMBO J 18:4476–4484

Charpentier M, Bredemeier R, Wanner G, Takeda N, Schleiff E, Parniske M (2008) Lotus japonicus CASTOR and POLLUX are ion channels essential for perinuclear calcium spiking in legume root endosymbiosis. Plant Cell 20:3467–3479

Cheon CI, Hong Z, Verma DPS (1994) Nodulin-24 follows a novel pathway for integration into the peribacteroid membrane in soybean root nodules. J Biol Chem 269:6598–6602

Cheon CI, Lee NG, Siddique ABM, Bal AK, Verma DPS (1993) Roles of plant homologs of Rab1p and Rab7p in the biogenesis of the peribacteroid membrane, a subcellular compartment formed de novo during root nodule symbiosis. EMBO J 12:4125–4135

Collins NC, Thordal-Christensen H, Lipka V, Bau S, Kombrink E, Qiu JL, Hückelhoven R, Steins M, Freialdenhoven A, Somerville SC, Schulze-Lefert P (2003) SNARE-protein-mediated disease resistance at the plant cell wall. Nature 425:973–977

Conner SD, Schmid SL (2003) Regulated portals of entry into the cell. Nature 422:37–44

Cossart P, Sansonetti PJ (2004) Bacterial invasion: the paradigms of enteroinvasive pathogens. Science 304:242–248

Czaja LF, Hogekamp C, Lamm P, Maillet F, Andres Martinez E, Samain E, Dénarié J, Küster H, Hohnjec N. (2012) Transcriptional responses towards diffusible signals from symbiotic microbes reveal MtNFP- and MtDMI3-dependent reprogramming of host gene expression by AM fungal LCOs. Plant Physiol. First Published on May 31; doi: 10.1104/pp.112.195990, PMID:22652128

Demchenko K, Winzer T, Stougaard J, Parniske M, Pawlowski K (2004) Distinct roles of *Lotus japonicus* SYMRK and SYM15 in root colonization and arbuscule formation. New Phytol 163:381–392

Dharmasiri S, Swarup R, Mockaitis K, Dharmasiri N, Singh SK, Kowalchyk M, Marchant A, Mills S, Sandberg G, Bennett MJ, Estelle M (2006) AXR4 is required for localization of the auxin influx facilitator AUX1. Science 312:1218–1220

Duplessis S et al (2011) Obligate biotrophy features unraveled by the genomic analysis of rust fungi. Proc Natl Acad Sci U S A 108:9166–9171

Endre G, Kereszt A, Kevei Z, Mihacea S, Kalo P, Kiss GB (2002) A receptor kinase gene regulating symbiotic nodule development. Nature 417:962–966

Esseling JJ, Lhuissier FGP, Emons AMC (2003) Nod factor–induced root hair curling: Continuous polar growth towards the point of nod factor application. Plant Physiol 132:1982–1988

Feddermann N, Muni RRD, Zeier T, Stuurman J, Ercolin F, Schorderet M, Reinhardt D (2010) The PAM1 gene of petunia, required for intracellular accommodation and morphogenesis of arbuscular mycorrhizal fungi, encodes a homologue of VAPYRIN. Plant J 64:470–481

Fournier J, Timmers ACJ, Sieberer BJ, Jauneau A, Chabaud M, Barker DG (2008) Mechanism of infection thread elongation in root hairs of *Medicago truncatula* and dynamic interplay with associated rhizobial colonization. Plant Physiol 148:1985–1995

Fraysse N, Couderc F, Poinsot V (2003) Surface polysaccharide involvement in establishing the rhizobium–legume symbiosis. Eur J Biochem 270:1365–1380

Gage DJ (2004) Infection and invasion of roots by symbiotic, nitrogen–fixing rhizobia during nodulation of temperate legumes. Microbiol Mol Biol Rev 68:280

Genre A, Bonfante P (2005) Building a mycorrhizal cell: how to reach compatibility between plants and arbuscular mycorrhizal fungi. J Plant Interact 1:3–13

Genre A, Chabaud M, Faccio A, Barker DG, Bonfante P (2008) Prepenetration apparatus assembly precedes and predicts the colonization patterns of arbuscular mycorrhizal fungi within the root cortex of both *Medicago truncatula* and *Daucus carota*. Plant Cell 20:1407–1420

Genre A, Ivanov S, Fendrych M, Faccio A, Žárský V, Bisseling T, Bonfante P (2012) Multiple exocytotic markers accumulate at the sites of perifungal membrane biogenesis in arbuscular mycorrhizas. Plant Cell Physiol 53:244–255

Genre A, Ortu G, Bertoldo C, Martino E, Bonfante P (2009) Biotic and abiotic stimulation of root epidermal cells reveals common and specific responses to arbuscular mycorrhizal fungi. Plant Physiol 149:1424–1434

Gianinazzi-Pearson V, Arnould C, Oufattole M, Arango M, Gianinazzi S (2000) Differential activation of H+-ATPase genes by an arbuscular mycorrhizal fungus in root cells of transgenic tobacco. Planta 211:609–613

Glebov OO, Bright NA, Nichols BJ (2006) Flotillin-1 defines a clathrin-independent endocytic pathway in mammalian cells. Nat Cell Biol 8:46–54

Godfroy O, Debelle F, Timmers T, Rosenberg C (2006) A rice calcium- and calmodulin-dependent protein kinase restores nodulation to a legume mutant. Mol Plant-Microbe Interact 19:495–501

Gonzalez E, Solano R, Rubio V, Leyva A, Paz-Ares J (2005) PHOSPHATE TRANSPORTER TRAFFIC FACILITATOR1 is a plant-specific SEC12-related protein that enables the endoplasmic reticulum exit of a high-affinity phosphate transporter in Arabidopsis. Plant Cell 17:3500–3512

Gonzalez-Rizzo S, Crespi M, Frugier F (2006) The *Medicago truncatula* CRE1 cytokinin receptor regulates lateral root development and early symbiotic interaction with *Sinorhizobium meliloti*. Plant Cell 18:2680–2693

Groth M, Takeda N, Perry J, Uchida H, Draxl S, Brachmann A, Sato S, Tabata S, Kawaguchi M, Wang TL, Parniske M (2010) NENA, a Lotus japonicus Homolog of Sec13, is required for rhizodermal infection by arbuscular mycorrhiza fungi and rhizobia but dispensable for cortical endosymbiotic development. Plant Cell 22:2509–2526

Haney CH, Long SR (2010) Plant flotillins are required for infection by nitrogen-fixing bacteria. Proc Natl Acad Sci U S A 107:478–483

Haney CH, Riely BK, Tricoli DM, Cook DR, Ehrhardt DW, Long SR (2011) Symbiotic rhizobia bacteria trigger a change in localization and dynamics of the *Medicago truncatula* receptor kinase LYK3. Plant Cell 23:2774–2787

Harrison MJ, Dewbre GR, Liu JY (2002) A phosphate transporter from *Medicago truncatula* involved in the acquisiton of phosphate released by arbuscular mycorrhizal fungi. Plant Cell 14:2413–2429

Hayashi T, Banba M, Shimoda Y, Kouchi H, Hayashi M, Imaizumi-Anraku H (2010) A dominant function of CCaMK in intracellular accommodation of bacterial and fungal endosymbionts. Plant J 63:141–154

Heckmann AB, Lombardo F, Miwa H, Perry JA, Bunnewell S, Parniske M, Wang TL, Downie JA (2006) Lotus japonicus nodulation requires two GRAS domain regulators, one of which is functionally conserved in a non-legume. Plant Physiol 142:1739–1750

Held M, Hossain MS, Yokota K, Bonfante P, Stougaard J, Szczyglowski K (2010) Common and not so common symbiotic entry. Trends Plant Sci 15:540–545

Horvath B, Yeun LH, Domonkos A, Halasz G, Gobbato E, Ayaydin F, Miro K, Hirsch S, Sun JH, Tadege M, Ratet P, Mysore KS, Ane JM, Oldroyd GED, Kalo P (2011) *Medicago truncatula* IPD3 Is a member of the common symbiotic signaling pathway required for rhizobial and mycorrhizal symbioses. Mol Plant-Microbe Interact 24:1345–1358

Ivanov S, Fedorova EE, Limpens E, De Mita S, Genre A, Bonfante P, Bisseling T (2012) Rhizobium-legume symbiosis shares an exocytotic pathway required for arbuscule formation. Proc Natl Acad Sci U S A 109:8316–8321

Jones KM, Kobayashi H, Davies BW, Taga ME, Walker GC (2007) How rhizobial symbionts invade plants: the Sinorhizobium-Medicago model. Nat Rev Microbiol 5:619–633

Kale SD, Gu B, Capelluto DGS, Dou D, Feldman E, Rumore A, Arredondo FD, Hanlon R, Fudal I, Rouxel T, Lawrence CB, Shan W, Tyler BM (2010) External lipid PI3P mediates entry of eukaryotic pathogen effectors into plant and animal host cells. Cell 142:284–295

Kalo P, Gleason C, Edwards A, Marsh J, Mitra RM, Hirsch S, Jakab J, Sims S, Long SR, Rogers J, Kiss GB, Downie JA, Oldroyd GED (2005) Nodulation signaling in legumes requires NSP2, a member of the GRAS family of transcriptional regulators. Science 308:1786–1789

Kanamori N, Madsen LH, Radutoiu S, Frantescu M, Quistgaard EMH, Miwa H, Downie JA, James EK, Felle HH, Haaning LL, Jensen TH, Sato S, Nakamura Y, Tabata S, Sandal N, Stougaard J (2006) A nucleoporin is required for induction of Ca2+ spiking in legume nodule development and essential for rhizobial and fungal symbiosis. Proc Natl. Acad. Sci. U S A 103:359–364

Kemen E, Kemen AC, Rafiqi M, Hempel U, Mendgen K, Hahn M, Voegele RT (2005) Identification of a protein from rust fungi transferred from haustoria into infected plant cells. Mol Plant-Microbe Interact 18:1130–1139

Kiers ET, Duhamel M, Beesetty Y, Mensah JA, Franken O, Verbruggen E, Fellbaum CR, Kowalchuk GA, Hart MM, Bago A, Palmer TM, West SA, Vandenkoornhuyse P, Jansa J, Bücking H (2011) Reciprocal rewards stabilize cooperation in the mycorrhizal symbiosis. Science 333:880–882

Kloppholz S, Kuhn H, Requena N (2011) A secreted fungal effector of glomus intraradices promotes symbiotic biotrophy. Curr Biol 21:1204–1209

Knodler LA, Celli J, Finlay BB (2001) Pathogenesis trickery: deception of host cell processes. Nat Rev Mol Cell Biol 2:578–588

Koh S, André A, Edwards H, Ehrhardt D, Somerville S (2005) Arabidopsis thaliana subcellular responses to compatible *Erysiphe cichoracearum* infections. Plant J 44:516–529

Kosuta S, Held M, Hossain MS, Morieri G, MacGillivary A, Johansen C, Antolin-Llovera M, Parniske M, Oldroyd GED, Downie AJ, Karas B, Szczyglowski K (2011) Lotus japonicus symRK-14 uncouples the cortical and epidermal symbiotic program. Plant J 67:929–940

Kouchi H, Imaizumi-Anraku H, Hayashi M, Hakoyama T, Nakagawa T, Umehara Y, Suganuma N, Kawaguchi M (2010) How many peas in a pod? legume genes responsible for mutualistic symbioses underground. Plant Cell Physiol 51:1381–1397

Kwon C, Neu C, Pajonk S, Yun HS, Lipka U, Humphry M, Bau S, Straus M, Kwaaitaal M, Rampelt H, Kasmi FE, Jürgens G, Parker J, Panstruga R, Lipka V, Schulze-Lefert P (2008) Co-option of a default secretory pathway for plant immune responses. Nature 451:835–840

Lefebvre B, Timmers T, Mbengue M, Moreau S, Hervé C, Tóth K, Bittencourt-Silvestre J, Klaus D, Deslandes L, Godiard L, Murray JD, Udvardi MK, Raffaele S, Mongrand S, Cullimore J, Gamas P, Niebel A, Ott T (2010) A remorin protein interacts with symbiotic receptors and regulates bacterial infection. Proc Natl Acad Sci U S A 107:2343–2348

Levy J, Bres C, Geurts R, Chalhoub B, Kulikova O, Duc G, Journet EP, Ane JM, Lauber E, Bisseling T, Denarie J, Rosenberg C, Debelle F (2004) A putative Ca2+ and calmodulin-dependent protein kinase required for bacterial and fungal symbioses. Science 303:1361–1364

Li R, Liu P, Wan Y, Chen T, Wang Q, Mettbach U, Baluška F, Šamaj J, Fang X, Lucas WJ, Lin J (2012) A membrane microdomain-associated protein, AtFlot1, is involved in a clathrin-independent endocytic pathway and is required for seedling development. Plant Cell 24:2105–2122

Limpens E, Franken C, Smit P, Willemse J, Bisseling T, Geurts R (2003) LysM domain receptor kinases regulating rhizobial nod factor-induced infection. Science 302:630–633

Limpens E, Ivanov S, van Esse W, Voets G, Fedorova E, Bisseling T (2009) Medicago N(2)-fixing symbiosomes acquire the endocytic identity marker Rab7 but delay the acquisition of vacuolar identity. Plant Cell 21:2811–2828

Limpens E, Mirabella R, Fedorova E, Franken C, Franssen H, Bisseling T, Geurts R (2005) Formation of organelle-like N-2-fixing symbiosomes in legume root nodules is controlled by DMI2. Proc Natl Acad Sci U S A 102:10375–10380

Lu YJ, Schornack S, Spallek T, Geldner N, Chory J, Schellmann S, Schumacher K, Kamoun S, Robatzek S (2012) Patterns of plant subcellular responses to successful oomycete infections reveal differences in host cell reprogramming and endocytic trafficking. Cell Microbiol 14:682–697

Madsen EB, Madsen LH, Radutoiu S, Olbryt M, Rakwalska M, Szczyglowski K, Sato S, Kaneko T, Tabata S, Sandal N, Stougaard J (2003) A receptor kinase gene of the LysM type is involved in legume perception of rhizobial signals. Nature 425:637–640

Madsen LH, Tirichine L, Jurkiewicz A, Sullivan JT, Heckmann AB, Bek AS, Ronson CW, James EK, Stougaard J (2010) The molecular network governing nodule organogenesis and infection in the model legume *Lotus japonicus*. Nat Commun 1:10

Maillet F, Poinsot V, Andre O, Puech-Pages V, Haouy A, Gueunier M, Cromer L, Giraudet D, Formey D, Niebel A, Martinez EA, Driguez H, Becard G, Denarie J (2011) Fungal lipochitooligosaccharide symbiotic signals in arbuscular mycorrhiza. Nature 469:58 (U1501)

Marie C, Plaskitt KA, Downie JA (1994) Abnormal bacteroid development in nodules induced by glucosamine synthase mutant of *Rhizobium leguminosarum*. Mol Plant-Microbe Interact 7:482–487

Marsh JF, Rakocevic A, Mitra RM, Brocard L, Sun J, Eschstruth A, Long SR, Schultze M, Ratet P, Oldroyd GED (2007) *Medicago truncatula* NIN is essential for rhizobial-independent nodule organogenesis induced by autoactive calcium/calmodulin-dependent protein kinase. Plant Physiol 144:324–335

Masson-Boivin C, Giraud E, Perret X, Batut J (2009) Establishing nitrogen-fixing symbiosis with legumes: how many rhizobium recipes? Trends Microbiol 17:458–466

Mbengue M, Camut S, de Carvalho-Niebel F, Deslandes L, Froidure S, Klaus-Heisen D, Moreau S, Rivas S, Timmers T, Hervé C, Cullimore J, Lefebvre B (2010) The *Medicago truncatula* E3 ubiquitin ligase PUB1 interacts with the LYK3 symbiotic receptor and negatively regulates infection and nodulation. Plant Cell 22:3474–3488

Meleresh D, Parniske M (2006) Symbiosis genes of *Lotus japonicus* are not required for intracellular accomodation of the rust fungus Uromyces loti. New Phytol 170:641–644

Mellor RB (1989) Bacteroids in the rhizobium-legume symbiosis inhabit a plant internal lytic compartment: implications for other microbial endosymbioses. J Exp Bot 40:831–839

Mergaert P, Uchiumi T, Alunni B, Evanno G, Cheron A, Catrice O, Mausset AE, Barloy-Hubler F, Galibert F, Kondorosi A, Kondorosi E (2006) Eukaryotic control on bacterial cell cycle and differentiation in the rhizobium-legume symbiosis. Proc Natl Acad Sci. U S A 103:5230–5235

Messinese E, Mun JH, Yeun LH, Jayaraman D, Rouge P, Barre A, Lougnon G, Schornack S, Bono JJ, Cook DR, Ane JM (2007) A novel nuclear protein interacts with the symbiotic DMI3 calcium- and calmodulin-dependent protein kinase of *Medicago truncatula*. Mol Plant-Microbe Interact 20:912–921

Meyer D, Pajonk S, Micali C, O'Connell R, Schulze-Lefert P (2009) Extracellular transport and integration of plant secretory proteins into pathogen-induced cell wall compartments. Plant J 57:986–999

Micali CO, Neumann U, Grunewald D, Panstruga R, O'Connell R (2011) Biogenesis of a specialized plant-fungal interface during host cell internalization of Golovinomyces orontii haustoria. Cell Microbiol 13:210–226

Mitra RM, Gleason CA, Edwards A, Hadfield J, Downie JA, Oldroyd GED, Long SR (2004) A Ca^{2+}/calmodulin-dependent protein kinase required for symbiotic nodule development: gene identification by transcript-based cloning. Proc Natl Acad Sci U S A 101:4701–4705

Miyahara A, Richens J, Starker C, Morieri G, Smith L, Long S, Downie JA, Oldroyd GED (2010) Conservation in function of a SCAR/WAVE component during infection thread and root hair growth in *Medicago truncatula*. Mol Plant-Microbe Interact 23:1553–1562

Mongrand S, Morel J, Laroche J, Claverol S, Carde JP, Hartmann MA, Bonneu M, Simon-Plas F, Lessire R, Bessoule JJ (2004) Lipid rafts in higher plant cells: purification and characterization of triton X-100-insoluble microdomains from tobacco plasma membrane. J Biol Chem 279:36277–36286

Murray JD, Karas BJ, Sato S, Tabata S, Amyot L, Szczyglowski K (2007) A cytokinin perception mutant colonized by rhizobium in the absence of nodule organogenesis. Science 315:101–104

Murray JD, Muni RRD, Torres-Jerez I, Tang YH, Allen S, Andriankaja M, Li GM, Laxmi A, Cheng XF, Wen JQ, Vaughan D, Schultze M, Sun J, Charpentier M, Oldroyd G, Tadege M, Ratet P, Mysore KS, Chen RJ, Udvardi MK (2011) Vapyrin, a gene essential for intracellular progression of arbuscular mycorrhizal symbiosis, is also essential for infection by rhizobia in the nodule symbiosis of *Medicago truncatula*. Plant J 65:244–252

Naisbitt T, James EK, Sprent JI (1992) The evolutionary significance of the legume genus Chamaecrista, as determined by nodule structure. New Phytol 122:487–492

Oldroyd GED, Downie JA (2004) Calcium, kinases and nodulation signalling in legumes. Nat Rev Mol Cell Biol 5:566–576

Oldroyd GED, Downie JA (2006) Nuclear calcium changes at the core of symbiosis signalling. Curr Opin Plant Biol 9:351–357

Oldroyd GED, Murray JD, Poole PS, Downie JA (2011) The rules of engagement in the legume-rhizobial symbiosis. Annu Rev Genet 45:119–144

Op den Camp R, Streng A, De Mita S, Cao QQ, Polone E, Liu W, Ammiraju JSS, Kudrna D, Wing R, Untergasser A, Bisseling T, Geurts R (2011) LysM-type mycorrhizal receptor recruited for rhizobium symbiosis in nonlegume parasponia. Science 331:909–912

Ott T, van Dongen JT, Gunther C, Krusell L, Desbrosses G, Vigeolas H, Bock V, Czechowski T, Geigenberger P, Udvardi MK (2005) Symbiotic leghemoglobins are crucial for nitrogen fixation in legume root nodules but not for general plant growth and development. Curr Biol 15:531–535

Ovchinnikova E, Journet EP, Chabaud M, Cosson V, Ratet P, Duc G, Fedorova E, Liu W, den Camp RO, Zhukov V, Tikhonovich I, Borisov A, Bisseling T, Limpens E (2011) IPD3 controls the formation of nitrogen-fixing symbiosomes in pea and Medicago Spp. Mol Plant-Microbe Interact 24:1333–1344

Panter S, Thomson R, De Bruxelles G, Laver D, Trevaskis B, Udvardi M (2000) Identification with proteomics of novel proteins associated with the peribacteroid membrane of soybean root nodules. Mol Plant-Microbe Interact 13:325–333

Parniske M (2000) Intracellular accommodation of microbes by plants: a common developmental program for symbiosis and disease? Curr Opin Plant Biol 3:320–328

Parniske M (2008) Arbuscular mycorrhiza: the mother of plant root endosymbioses. Nat Rev Microbiol 6:763–775

Pelkmans L, Helenius A (2002) Endocytosis via caveolae. Traffic 3:311–320

Plet J, Wasson A, Ariel F, Le Signor C, Baker D, Mathesius U, Crespi M, Frugier F (2011) MtCRE1-dependent cytokinin signaling integrates bacterial and plant cues to coordinate symbiotic nodule organogenesis in *Medicago truncatula*. Plant J 65:622–633

Pumplin N, Harrison MJ (2009) Live-cell imaging reveals periarbuscular membrane domains and organelle location in *Medicago truncatula* roots during arbuscular mycorrhizal symbiosis. Plant Physiol 151:809–819

Pumplin N, Mondo SJ, Topp S, Starker CG, Gantt JS, Harrison MJ (2010) *Medicago truncatula* Vapyrin is a novel protein required for arbuscular mycorrhizal symbiosis. Plant J 61:482–494

Pumplin N, Zhang XC, Noar RD, Harrison MJ (2012) Polar localization of a symbiosis-specific phosphate transporter is mediated by a transient reorientation of secretion. Proc Natl Acad Sci U S A 109:E665–E672

Radutoiu S, Madsen LH, Madsen EB, Felle HH, Umehara Y, Gronlund M, Sato S, Nakamura Y, Tabata S, Sandal N, Stougaard J (2003) Plant recognition of symbiotic bacteria requires two LysM receptor-like kinases. Nature 425:585–592

Redecker D, Kodner R, Graham LE (2000) Glomalean fungi from the ordovician. Science 289:1920–1921

Robertson JG, Lyttleton P (1982) Coated and smooth vesicles in the biogenesis of cell-walls, plasma-membrane, infection threads and peribacteroid membranes in root hairs an nodules of white clover. J Cell Sci 58:63–78

Roth LE, Stacey G (1989) Bacterium release into host cells of nitrogen-fixing soybean nodules: the symbiosome membrane comes from three sources. Eur J Cell Biol 49:13–23

Saito K, Yoshikawa M, Yano K, Miwa H, Uchida H, Asamizu E, Sato S, Tabata S, Imaizumi-Anraku H, Umehara Y, Kouchi H, Murooka Y, Szczyglowski K, Downie JA, Parniske M, Hayashi M, Kawaguchi M (2007) NUCLEOPORIN85 is required for calcium spiking, fungal and bacterial symbioses, and seed production in *Lotus japonicus*. Plant Cell 19:610–624

Schauser L, Roussis A, Stiller J, Stougaard J (1999) A plant regulator controlling development of symbiotic root nodules. Nature 402:191–195

Schlaman HRM, Horvath B, Vijgenboom E, Okker RJH, Lugtenberg BJJ (1991) Suppression of nodulation gene expression in bacteroids of *Rhizobium leguminosarum* biovar *viciae*. J Bacteriol 173:4277–4287

Schmelzer E (2002) Cell polarization, a crucial process in fungal defence. Trends Plant Sci 7:411–415

Sharma SB, Signer ER (1990) Temporal and spatial regulation of the symbiotic genes of *Rhizobium meliloti* in planta revealed by transposon Tn5-gusA. Gene Dev 4:344–356

Shimoda Y, Han L, Yamazaki T, Suzuki R, Hayashi M, Imaizumi-Anraku H (2012) Rhizobial and fungal symbioses show different requirements for calmodulin binding to calcium calmodulin-dependent protein kinase in *Lotus japonicus*. Plant Cell 24:304–321

Shin JS, Abraham SN (2001) Co-option of endocytic functions of cellular caveolae by pathogens. Immunology 102:2–7

Smit P, Limpens E, Geurts R, Fedorova E, Dolgikh E, Gough C, Bisseling T (2007) Medicago LYK3, an entry receptor in rhizobial nodulation factor signaling. Plant Physiol 145:183–191

Smit P, Raedts J, Portyanko V, Debelle F, Gough C, Bisseling T, Geurts R (2005) NSP1 of the GRAS protein family is essential for rhizobial Nod factor-induced transcription. Science 308:1789–1791

Smith S, Read D. (2008) Mycorrhizal Symbiosis, 3rd edn. Academic Press, San Diego. ISBN 0123705266.

Soupene E, Foussard M, Boistard P, Truchet G, Batut J (1995) Oxygen as a key developmental regulator of Rhizobium-meliloti N2-fixation gene-expression whitin the alfalfa root-nodule. Proc Natl Acad Sci U S A 92:3759–3763

Spanu PD et al (2010) Genome expansion and gene loss in powdery mildew fungi reveal tradeoffs in extreme parasitism. Science 330:1543–1546

Sprent JI (2007) Evolving ideas of legume evolution and diversity: a taxonomic perspective on the occurrence of nodulation. New Phytol 174:11–25

Stracke S, Kistner C, Yoshida S, Mulder L, Sato S, Kaneko T, Tabata S, Sandal N, Stougaard J, Szczyglowski K, Parniske M (2002) A plant receptor-like kinase required for both bacterial and fungal symbiosis. Nature 417:959–962

Sullivan JT, Ronson CW (1998) Evolution of rhizobia by acquisition of a 500-kb symbiosis island that integrates into a phe-tRNA gene. Proc Natl Acad Sci U S A 95:5145–5149

Stuermer CA (2010) The reggie/flotillin connection to growth. Trends Cell Biol 20:6–13

Timmers ACJ, Auriac MC, de Billy F, Truchet G (1998) Nod factor internalization and microtubular cytoskeleton changes occur concomitantly during nodule differentiation in alfalfa. Development 125:339–349

Timmers ACJ, Auriac MC, Truchet G (1999) Refined analysis of early symbiotic steps of the Rhizobium-Medicago interaction in relationship with microtubular cytoskeleton rearrangements. Development 126:3617–3628

Timmers ACJ, Holsters M, Goormachtig S (2005) Endocytosis and endosymbiosis. In: Šamaj J, Baluška F, Menzel D (eds) Plant endocytosis. Springer, Berlin

Tirichine L, Imaizumi-Anraku H, Yoshida S, Murakami Y, Madsen LH, Miwa H, Nakagawa T, Sandal N, Albrektsen AS, Kawaguchi M, Downie A, Sato S, Tabata S, Kouchi H, Parniske M, Kawasaki S, Stougaard J (2006) Deregulation of a Ca2+/calmodulin-dependent kinase leads to spontaneous nodule development. Nature 441:1153–1156

Tirichine L, Sandal N, Madsen LH, Radutoiu S, Albrektsen AS, Sato S, Asamizu E, Tabata S, Stougaard J (2007) A gain-of-function mutation in a cytokinin receptor triggers spontaneous root nodule organogenesis. Science 315:104–107

Truchet G, Roche P, Lerouge P, Vasse J, Camut S, Debilly F, Prome JC, Denarie J (1991) Sulfated lipo-oligosaccharide signals of Rhizobium-meliloti elicit root nodule organogenesis in alfalfa. Nature 351:670–673

Tsyganov VE, Morzhina EV, Stefanov SY, Borisov AY, Lebsky VK, Tikhonovich IA (1998) The pea (Pisum sativum L.) genes sym33 and sym40 control infection thread formation and root nodule function. Mol Gen Genet 259:491–503

Van Brussel AAN, Bakhuizen R, Van Spronsen PC, Spaink HP, Tak T, Lugtenberg BJJ, Kijne JW (1992) Induction of preinfection thread structures in the leguminous host plant by mitogenic lipooligosaccharides of rhizobium. Science 257:70–72

Van De Velde W, Zehirov G, Szatmari A, Debreczeny M, Ishihara H, Kevei Z, Farkas A, Mikulass K, Nagy A, Tiricz H, Satiat-Jeunemaître B, Alunni B, Bourge M, Kucho KI, Abe M, Kereszt A, Maroti G, Uchiumi T, Kondorosi E, Mergaert P (2010) Plant peptides govern terminal differentiation of bacteria in symbiosis. Science 327:1122–1126

Vasse J, Debilly F, Camut S, Truchet G (1990) Correlation between ultrastructural differentiation of bacteroids and nitrogen-fixation in alfalfa nodules. J Bacteriol 172:4295–4306

Via LE, Deretic D, Ulmer RJ, Hibler NS, Huber LA, Deretic V (1997) Arrest of mycobacterial phagosome maturation is caused by a block in vesicle fusion between stages controlled by rab5 and rab7. J Biol Chem 272:13326–13331

Vinardell JM, Fedorova E, Cebolla A, Kevei Z, Horvath G, Kelemen Z, Tarayre S, Roudier F, Mergaert P, Kondorosi A, Kondorosi E (2003) Endoreduplication mediated by the anaphase-promoting complex activator CCS52A is required for symbiotic cell differentiation in *Medicago truncatula* nodules. Plant Cell 15:2093–2105

Wang D, Griffitts J, Starker C, Fedorova E, Limpens E, Ivanov S, Bisseling T, Long S (2010) A nodule-specific protein secretory pathway required for nitrogen-fixing symbiosis. Science 327:1126–1129

Wegel E, Schauser L, Sandal N, Stougaard J, Parniske M (1998) Mycorrhiza mutants of *Lotus japonicus* define genetically independent steps during symbiotic infection. Mol Plant-Microbe Interact 11:933–936

White J, Prell J, James EK, Poole P (2007) Nutrient sharing between symbionts. Plant Physiol 144:604–614

Whitehead LF, Day DA (1997) The peribacteroid membrane. Physiol Plant 100:30–44

Wildermuth MC (2010) Modulation of host nuclear ploidy: a common plant biotroph mechanism. Curr Opin Plant Biol 13:449–458

Williams TM, Lisanti MP (2004) The caveolin proteins. Genome Biol 5:214–222

Winzer T, Bairl A, Linder M, Linder D, Werner D, Muller P (1999) A novel 53-kDa nodulin of the symbiosome membrane of soybean nodules, controlled by *Bradyrhizobium japonicum*. Mol Plant-Microbe Interact 12:218–226

Wu GS, Otegui MS, Spalding EP (2010) The ER-localized TWD1 immunophilin is necessary for localization of multidrug resistance-like proteins required for polar auxin transport in Arabidopsis roots. Plant Cell 22:3295–3304

Yang WC, Deblank C, Meskiene I, Hirt H, Bakker J, Van Kammen A, Franssen H, Bisseling T (1994) Rhizobium nod Factors reactivate the cell-cycle during infection and nodule primordium formation, but the cycle is only completed in primordium formation. Plant Cell 6:1415–1426

Yang WC, Horvath B, Hontelez J, Van Kammen A, Bisseling T (1991) In situ localization of rhizobium messenger-RNAs in pa root-ndules—NifA and NifH localization. Mol Plant-Microbe Interact 4:464–468

Yano K, Yoshida S, Müller J, Singh S, Banba M, Vickers K, Markmann K, White C, Schuller B, Sato S, Asamizu E, Tabata S, Murooka Y, Perry J, Wang TL, Kawaguchi M, Imaizumi-Anraku H, Hayashi M, Parniske M (2008) CYCLOPS, a mediator of symbiotic intracellular accommodation. Proc Natl Acad Sci U S A 05:20540–20545

Yokota K, Fukai E, Madsen LH, Jurkiewicz A, Rueda P, Radutoiu S, Held M, Hossain MS, Szczyglowski K, Morieri G, Oldroyd GED, Downie JA, Nielsen MW, Rusek AM, Sato S, Tabata S, James EK, Oyaizu H, Sandal N, Stougaard J (2009) Rearrangement of actin cytoskeleton mediates invasion of *Lotus japonicus* roots by *Mesorhizobium loti*. Plant Cell 21:267–284

Zipfel C, Felix G (2005) Plants and animals: a different taste for microbes? Curr Opin Plant Biol 8:353–360

Endocytosis of LeEix and EHD Proteins During Plant Defense Signalling

Maya Bar and Adi Avni

Abstract Plants are exposed to pathogenic microorganisms in their environment, and have developed various defense mechanisms to avoid disease and death. Active defense reactions can also be triggered by treatment with microbial compounds called elicitors or microbial-associated molecular patterns (MAMPs), which may be characteristic of a whole group of organisms or limited to specific strains of a microbial species. Endocytosis has been demonstrated to be involved in plant immunity. EH domain-containing proteins (EHDs) are involved in various aspects of the endocytic process, primarily via protein–protein interactions. Here, we characterize endocytosis and signalling occurring during plant defense responses induced by elicitors, focusing particularly on EIX (ethylene inducing xylanase) and the involvement of EHD proteins in these processes.

1 Plant Defense Responses

The gene-for-gene (Flor 1947) model of plant–pathogen interactions proposes that each resistance gene (*R*-gene) confers resistance only to pathogens carrying the corresponding avirulence gene.

Gene-for-gene resistance responses have been observed in interaction of plants with a wide variety of pathogens, including fungi, bacteria, and viruses (Dewit

M. Bar
The Robert H. Smith Institute of Plant Sciences and Genetics in Agriculture, Hebrew University, 76100 Rehovot, Israel

A. Avni (✉)
Department of Molecular Biology and Ecology of Plants, Tel Aviv University, 69978 Tel Aviv, Israel
e-mail: lpavni@post.tau.ac.il

J. Šamaj (ed.), *Endocytosis in Plants*,
DOI: 10.1007/978-3-642-32463-5_15,

et al. 1985; Blein et al. 1991; Fluhr et al. 1991; Basse et al. 1992). A simple molecular explanation for gene-for-gene resistance is that an avirulence gene encodes a ligand that binds to a receptor encoded by the plant *R*-gene (Ebel and Cosio 1994; Bent 1996). Ligand binding triggers activation of a signal transduction cascade culminating in expression of defense responses that inhibit the pathogen and confer the resistance (Glazebrook et al. 1997).

The defense mechanisms are triggered when an organic compound (termed elicitor) is recognized by a plant *R*-gene. The list of elicitors is long and includes glucans, pectic fragments, and proteins with and without carbohydrate side chains. Elicitors may be of pathogen or non-pathogen origin (Dewit et al. 1985; Blein et al. 1991; Basse et al. 1992; Furman-Matarasso et al. 1999).

Recognition between R-proteins and their corresponding elicitors is likely to activate a signal transduction cascade which involves various responses including cell wall fortification (Hammond-Kosack and Jones 1996), production of reactive oxygen species (ROS), induction of pathogenesis related (PR) genes (Hammond-Kosack and Jones 1996), Ethylene biosynthesis (Boller 1991), and the hypersensitive response (HR) (Yu et al. 1998; Elbaz et al. 2002).

In some cases, the "R" gene is a receptor which contains an endocytic motif, and endocytosis has been shown to be a crucial step in the recognition between the "R" gene and the elicitor in a few cases, including the LeEix system which will be described in detail below.

1.1 Plant Defense Receptors

Leucine-rich-repeat receptor kinase (LRR-RLKs) and LRR-RLPs are involved in signalling and defense responses in plants (Becraft 2002). The most intensively studied LRR-RLK in the context of plant defense responses is FLS2, which recognizes bacterial flagellin and the flagellin-derived peptide flg22 (Felix et al. 1999; Gomez-Gomez et al. 1999; Gomez-Gomez and Boller 2000). FLS2 recognition of flg22 leads to induction of defense responses (Felix et al. 1999; Asai et al. 2002; Zipfel et al. 2004). Mutations in FLS2 compromised the ability of the plant to mount an efficient defense against bacterial pathogens (Zipfel et al. 2004; Robatzek et al. 2006).

LRR-RLPs have also been implicated in responses to pathogens. The tomato Cf genes which mediate resistance to *Cladosporium fulvum* encode LRR-RLPs, the LRR domain of which was shown to be important for avirulence (Avr) gene recognition (van der Hoorn et al. 2005). Additional LRR-RLPs include the tomato Ve-resistant proteins (Kawchuk et al. 2001; Fradin et al. 2009) and the LeEix proteins (Ron and Avni 2004).

*LeEix*1 and *LeEix*2 are responsible for the plant response to the fungal elicitor ethylene-inducing xylanase (EIX). The *LeEix* genes show homology to *R*-genes of transmembrane proteins which contain an extracellular LRR, like the *Cf* gene family and *Ve R*-genes. The LeEix proteins are transmembrane proteins and contain a signal peptide within the N-terminus, an extracellular domain, a transmembrane domain, and a short cytoplasmic tail in the C-terminus. (Ron and Avni 2004).

1.2 Elicitors

Elicitors [microbial-associated molecular patterns (MAMPs)] that trigger plant defense responses have been isolated from a variety of phytopathogenic and non-pathogenic microorganisms (Fuchs et al. 1989; Ebel and Cosio 1994; Felix et al. 1999). In soybean cell culture, the *Verticillium* elicitor was shown to enter the cell via an endocytic process (Horn et al. 1989). Flg 22 stimulates endocytosis of FLS2, in a process which requires kinase activity (Robatzek et al. 2006). In tobacco, the cryptogein elicitor was reported to induce endocytosis in correlation with its defense response activation (Leborgne-Castel et al. 2008).

1.2.1 Ethylene-Inducing Xylanase

The plant hormone ethylene is involved in modulating a broad spectrum of physiological processes including pathogenesis, senescence, and fruit ripening (Goeschi et al. 1966).

A 22 kDa fungal β-1-4-endoxylanase protein referred to as EIX was isolated from xylan-induced *Trichoderma viride* cultures (Dean et al. 1989; Fuchs et al. 1989). Similar xylanases have been identified in xylan-induced filtrates of plant pathogenic fungi (Dean et al. 1989; Wu et al. 1997). Injection of EIX into the leaf mesophyll intercellular spaces induces ethylene production and HR, as well as other plant defense responses in *Nicotiana tabacum* cv. Xanthi (Fuchs et al. 1989; Lotan and Fluhr 1990; Bailey et al. 1990, 1992; Dean and Anderson 1991), *Solanum lycopersicon* (tomato) leaf tissue (Ron et al. 2000), *Nicotiana tabacum* cell suspensions (Yano et al. 1998), and in other plant species. These responses are characteristic of plants responding to exogenously applied elicitors (Blein et al. 1991; Felix et al. 1993).

EIX induces defense responses in specific plant species and/or varieties (Bailey et al. 1990, 1992; Ron et al. 2000; Elbaz et al. 2002), and was shown to specifically bind to the plasma membrane of both tomato and tobacco EIX-responding cultivars (Hanania and Avni 1997).

2 Endocytosis in Plant Defense Responses

Many apparent roles for regulated endocytosis in plant development and immunity have emerged (Robatzek 2007). The tomato Ve2, Cf9, Cf4, and LeEix proteins (Jones et al. 1994; Takken et al. 1998; Kawchuk et al. 2001; Ron and Avni 2004) contain the conserved endocytosis signal Yxxϕ within the short cytoplasmic domain. However, the extensively studied flagellin defense response receptor FLS2 does not contain a Yxxϕ motif, but it was reported to contain a PEST-like motif which has also been implicated in endocytosis (Robatzek et al. 2006). Mutation in the endocytic motif of LeEix2 resulted in abolishment of HR induction in response to EIX, suggesting endocytosis plays a key role in mediating the signal generated by EIX (Ron and Avni 2004). Similarly, it has been reported that impairing the PEST-like motif in FLS2 may compromise FLS2 internalization and abolish some elements of the flg22-triggered defense response (Robatzek et al. 2006).

Receptor-mediated endocytosis has also been reported to be important for the response to pathogens in mammalian systems, such as in the case of the toll-like receptors (TLR), which also contain extra-cellular LRR domains (Husebye et al. 2006).

3 Epidermal Growth Factor Receptor Substrate-15 (EPS-15) Homology Domain Containing Proteins in Plant Cells

Endocytosis involves many protein–protein interactions. One module which mediates such interactions is the EH domain-containing protein (EHD) (EPS15 homology domain) first identified in EPS15 (Wong et al. 1995; Carbone et al. 1997). Sequence analysis and preliminary functional characterization of plant EHDs suggest a high level of functional homology with mammalian EHDs. Orthologs of EHD proteins exist in Arabidopsis and the entire *Solanaceae* family as well as in rice, maize, and other plant species. The Arabidopsis EHD proteins share high homology with their mammalian counterparts (Mintz et al. 1999; Pohl et al. 2000; Galperin et al. 2002). Structurally, AtEHD1, AtEHD2, and its spliced variant termed AtEHD2-2 contain the same domains as the mammalian proteins (Pohl et al. 2000; Galperin et al. 2002; Naslavsky and Caplan 2005). The EH domain is present at the N-terminus of the plant proteins, with the center domain harboring the nucleotide binding site and the DxxG motif that is completely conserved in evolution (Rotem-Yehudar et al. 2001; Galperin et al. 2002). A major difference between the plant EHD proteins and the mammalian, Drosophila, and *Caenorhabditis elegans* proteins is the location of the EH domain. While many N-terminal EH domain-containing proteins exist in mammals and other organisms (Naslavsky and Caplan 2005), the AtEHDs described herein bear the most resemblance to the mammalian EHDs (EHD1-4). The conservation of EHD proteins throughout the plant kingdom and their relatively high homology to the mammalian EHDs indicate their relative importance in plant systems. Despite the difference in domain arrangement, we have demonstrated that the plant EHD proteins have similar functions as their mammalian counterparts. Available microarray data (Zimmermann et al. 2004) indicate that the Arabidopsis *EHD* genes are expressed in all plant tissues. This correlates with the data obtained for the mammalian EHD proteins (Mintz et al. 1999; Pohl et al. 2000; Rapaport et al. 2006; George et al. 2007), and may also indicate the importance of EHDs in ubiquitous functions occurring in all types of cells. Given the role of EHDs in endocytosis, it is likely that these conserved proteins serve integral roles in signalling in a variety of cell types in diverse species (Polo et al. 2003).

AtEHD1 and AtEHD2 are localized to endosomes and colocalize with endocytic markers in both plant and mammalian systems (Bar et al. 2008). The fact that both proteins colocalize with FM4-64 shows that they are localized to membranous endocytic organelles.

AtEHD1 fully colocalizes with hEHD1 and hEHD3 but does not colocalize with hEHD2 or hEHD4 (Bar et al. 2008). This could indicate that AtEHD1 and hEHD1/3 share similar functions, which is consistent with reported phenotypes for hEHD1 in knock-out mice.

Arabidopsis plants silenced in the *AtEHD*1 gene demonstrated a delay in internalization of the fluorescent dye FM4-64 (Bar et al. 2008). As AtEHD1 colocalizes with hEHD1, it is possible that this delay may be a similar phenomenon to the delayed recycling observed in hEHD1 knock-out mice (Rapaport et al. 2006). Silenced *AtEHD*1 plants did not show any distinctive phenotype. By contrast, *AtEHD*2 overexpression suppresses endocytosis in both plant and mammalian cells (Bar et al. 2008), as does hEHD2 in mammalian cells (Guilherme et al. 2004a). Thus, it is possible that AtEHD1 and AtEHD2 have coevolved in plants to exert opposite effects; one may act to stimulate endocytosis under certain conditions, while the other one can suppress endocytosis under certain conditions. The rate of endocytosis depends on a multitude of factors, and many of them remain unknown. However, these parameters could be influenced by the expression level (or other regulatory elements) of one or both AtEHDs. One could envisage a decrease of active AtEHD1 or an increase of active AtEHD2 (or vice versa) in a situation when endocytosis must be precisely regulated.

Considering the inhibitory effect of AtEHD2 overexpression on endocytosis, it is possible that AtEHD2 is involved in a particular rate-limiting step of the endocytic process. In such case, overexpression of AtEHD2 may cause the endocytic process to become "stuck" in this particular step, thus inhibiting faster entry of typically endocytosed material into the cell. This could also explain why the expression level of *AtEHD*2 is normally very low in wild-type cells compared to the expression of *AtEHD*1 (Bar et al. 2008).

3.1 EHD1 and Recycling

We have recently demonstrated that knock-down of *EHD*1 causes a delayed recycling phenotype and reduces brefeldin A sensitivity in Arabidopsis seedlings (Bar and Avni, unpublished results). Interestingly, internalization of LeEix2 depends primarily on recycling as opposed to de novo protein synthesis (Bar and Avni 2009a). The EH domain of EHD1 was found to be crucial for the localization of EHD1 to endosomal structures. Mutant EHD1 lacking the EH domain did not localize to endosomes and showed a phenotype similar to that of *EHD*1 knock-down seedlings. Mutants lacking the coiled-coil domain, however, showed a phenotype similar to wild-type or *EHD*1 overexpressing seedlings. Interestingly, transgenic plants overexpressing *EHD*1 possess enhanced tolerance to salt stress, a property which also requires an intact EH domain.

3.2 EHD2 and Inhibition of Endocytosis

Notably, mammalian EHD2 appears to be the most unique mammalian EHD protein, and the same is true for plant EHD2 (Bar et al. 2008). This is interesting, given that EHD2 is not usually endosomal and it was the only EHD protein found to inhibit endocytosis both in mammals and Arabidopsis.

AtEHD2 does not significantly colocalize with any of the mammalian EHD proteins (Bar et al. 2008). Though plant proteins are by no means guaranteed to localize properly in mammalian cells, it would seem that AtEHD2 shares similar function with hEHD2 based on the inhibitory effect on endocytosis. This effect was observed both in plant cells and in mammalian cells using plant AtEHD2, indicating that AtEHD2 is able to exert at least some of its native biological activity in mammalian cells. This shows the high level of functional homology between plant and mammalian endocytosis (Ortiz-Zapater et al. 2006; Lam et al. 2007a).

AtEHD1 was found to be colocalized with ARA6 and FYVE (Bar et al. 2008) as well as with RabA1e and RabD2b (Bar and Avni, unpublished results), indicating that it resides partly on early endosomes (Ueda et al. 2001; Ŝamaj et al. 2005; Voigt et al. 2005; Golomb et al. 2008), from which recycling back to the plasma membrane occurs in plant cells (Ueda et al. 2001). Mammalian EHD1 was found to reside primarily in the endocytic recycling compartment (ERC) (Mintz et al. 1999; Grant et al. 2001), as well as on early endosomes (Naslavsky et al. 2004) and vesicular/tubular structures (Caplan et al. 2002). Though evidence of recycling endosomes exists in plants (Jaillais and Gaude 2007; Jaillais et al. 2008), such endosomes have not been well characterized. Previous work in the field of plant endocytosis has shown that materials are recycled to the TGN and back to the plasma membrane from early endosomes (Lam et al. 2007a, b). Thus, AtEHD1 resides on early endosomes and recycling endosomes, which may partially overlap. From EHD1 positive endosomes, some of the PM receptors are recycled back to the cell surface. Mammalian and plant EHD1 clearly function together with other endocytic/recycling proteins, as knock-down of *EHD*1 results only in a mild recycling phenotype, in both plants and mammals (Rapaport et al. 2006; Bar et al. 2008).

EHD2 resides primarily at the plasma membrane. Though the expression level of *AtEHD*2 is very low under normal conditions, upon overexpression it acts to diminish internalization of such "classical" endocytic cargos as FM4-64 and transferrin in plant and mammalian cells, respectively (Guilherme et al. 2004a; Bar et al. 2008). EHD2 most likely inhibits the clathrin-dependent endocytic pathway (Dhonukshe et al. 2007; Bar et al. 2008), though it could possibly affect other pathways as well.

Plant EHD2 inhibits endocytosis of other receptors upon overexpression, including LRR-receptor-like proteins in plant cells (Bar and Avni 2009a, b). We have previously shown (Ron and Avni 2004) that signalling of the tomato LeEix2 receptor requires the endocytic process. EHD2 controls LeEix2 signalling via modulation of its endocytosis, thereby limiting the level of the plant response. Plant EHD2 may also serve to modulate other signalling processes in which endocytosis is involved. For example, plant EHD2 may regulate auxin signalling via regulation of PIN (auxin efflux transporter) endocytosis which is clathrin dependent (Dhonukshe et al. 2007). This could indicate that plant EHD2 is part of a more general ubiquitous control mechanism associated with receptor-mediated endocytosis (RME, see also Chap. 7 by Di Rubbo and Russinova in this volume).

Both EHD1 and EHD2 were found to be coupled to the actin cytoskeleton in mammals (Guilherme et al. 2004a; Braun et al. 2005). Our results demonstrate that this is the case in plants as well (Bar et al. 2008, 2009). It is not clear at this point how EHD2 exerts its function in RME under native conditions, but one clue could be that it shows plasma membrane and not endosomal localization in both mammalian (Benjamin et al. 2011) and plant cells (Guilherme et al. 2004a; Bar et al. 2008). Perhaps fluctuations in local concentration of EHD2 at micro domains within the plasma membrane can regulate the level of endocytosis at different locations throughout the cell. Soluble TLR was shown to inhibit the signalling of membrane TLR by binding to the TLR-specific ligand, thus serving as decay receptors. Similarly, the expression level of *EHD2* may modulate the endocytic process and provide negative regulation when required.

EHD2 appears to be an essential component in the endocytosis of the LeEix2, Cf4, and Cf9 receptors, causing inhibition of HR and ethylene biosynthesis upon its overexpression, while it does not seem to be involved in the FLS2 system (Bar and Avni 2009a, b). EHD2 may affect endocytosis directly, though it is also possible that EHD2 modulates LeEix2 internalization through an effect on the plasma membrane.

3.3 Functional Analysis of EHD2 Domains

We conducted an analysis of the importance of various domains within EHD2 for the protein function. The coiled-coil domain of EHD2 is crucial for the ability of EHD2 to inhibit endocytosis in plants (Bar et al. 2009). This domain was also required for binding of EHD2 to the LeEix2 receptor. Therefore, we suggest that binding of EHD2 to the LeEix2 receptor is required for inhibition of LeEix2 internalization. The P-loop of EHD2 is important for EHD2 to function properly (Bar et al. 2009), as evidenced by the loss of the ability to inhibit EIX-induced HR and to bind LeEix2 in the *AtEHD2_G221R P-loop* mutant. Our observations together with the published importance of the P-loop in mammalian EHDs suggest the possibility that the P-loop is required for proper membrane localization of AtEHD2, while the coiled-coil in fact mediates the binding to "target" proteins, thereby enabling the inhibitory function on endocytosis. Neither the P-loop mutant (G221R) nor a coiled-coil deletion (ΔCC) was able to bind the LeEix2 receptor, and both mutants lost the ability to inhibit HR.

Interestingly, the EH domain of AtEHD2 does not appear to be involved in the inhibition of endocytosis. Both a point mutation in the EH domain (G37R) and a complete deletion of this domain (ΔEH) did not affect the inhibition of endocytosis, as these mutants retained wild-type level activity. Further, swapping the EH domain between AtEHD1 (which does not inhibit endocytosis) and AtEHD2 had no effect. As mentioned above, EHD2 is localized primarily to the plasma membrane in both mammals and plants. Interestingly, a truncation mutant of mammalian EHD2 lacking the EH domain was also localized to the plasma membrane (Blume et al. 2007). Additionally, this mutant was able to inhibit internalization of transferrin in a manner similar to that of wild-type EHD2 (Guilherme et al. 2004a, b). It seems that although the EHDs share a high level of homology and similar structure/domains, in mammals and plants, the fact that each EHD possesses different functions could be related to the different domains present in the protein, whereby each function is exerted primarily through a different domain. Thus, different domains might have varying importance in different EHD proteins. The EH domain, which appears to be very important in EHD1, may not be crucial for the function of EHD2.

We suggest that upon EIX binding, μ-adaptin binds to the YXXϕ motif within the cytoplasmic domain of the LeEix2 receptor. The AP-2 complex is assembled, and AtEHD2 binds the ó-subunit of AP-2 and/or the LeEix2 receptor directly via the coiled-coil domain (Bar et al. 2009). However, the involvement of additional proteins in this complex cannot be excluded. Tethering of this complex to the actin cytoskeleton via additional proteins, as was reported for EHD2 in mammals (Guilherme et al. 2004a), may take part in the inhibition of endocytosis, particularly given the actin reorganization phenotype caused by *EHD*2 overexpression (Bar et al. 2009). The binding of AtEHD2 to AP-2 and/or LeEix2 needs to be examined further in order to elucidate the activity of different protein complexes in LeEix2 internalization and function.

4 LeEix2/EIX Endocytosis

4.1 Parameters of LeEix2 Endocytosis

Endocytosis is a crucial step in the defense response triggered in plants by EIX and additional MAMPs. Similar to previous work done with FLS2 (Robatzek et al. 2006), we were able to show (Bar and Avni 2009b) that signalling of the fungal elicitor EIX is dependent on internalization of its receptor LeEix2 via endocytosis, in a process which requires components of the cytoskeleton. After EIX application (15–20 min), LeEix2 is internalized into highly motile endosomes, in a swift endocytic process which follows a similar time course to that described for flg22-induced FLS2 (Robatzek et al. 2006), and similar to the time-frame of mammalian endocytosis (Gruenberg and Howell 1987).

LeEix2, Cf9, and Cf4 are LRR-RLPs showing structural similarities. They all possess extra-cellular LRR repeats and short cytoplasmic domains containing the Yxxϕ endocytic motif (Jones et al. 1994; Ron and Avni 2004). FLS2 is a receptor-like kinase (LRR-RLK) and has an intra-cellular kinase domain which does not contain theYxxϕ motif but instead it contains a non-classical PEST-like endocytic motif (Gomez-Gomez and Boller 2000; Robatzek et al. 2006). Another difference between LeEix2 and FLS2 is that FLS2 appears to be degraded and synthesized de novo after flg22-induced internalization, while LeEix2 is probably returned (at least in part) to the plasma membrane by recycling vesicles, given that cycloheximide does not affect its presence in the membrane (Bar and Avni 2009b). The recycling of LeEix2 does not require protein synthesis but may be amplified by the synthesis of certain proteins involved.

Further evidence that EHD2 is specific to the endocytic pathway of LeEix2 but not FLS2 comes from the fact that EIX but not flg22 can induce the expression of *NtEHD*2 (Bar and Avni 2009b). Thus, EIX application triggers *NtEHD*2 expression, and NtEHD2 acts to inhibit the defense response in the short term. Longer exposure to the MAMP leads to a "full-blown" defense response including HR which is free of the EHD2 inhibitory influence, suggesting that a control mechanism based on the interplay of different proteins may be at work. The kinase activity of RLKs such as FLS2 may be required for receptor internalization and signalling (Robatzek et al. 2006) and may provide the specificity which is not possible in the case of LeEix2 or Cf9. Concerning LeEix2 and Cf9, one could envisage a mechanism in which the MAMP triggers expression of the endocytosis inhibitory protein in order to more tightly regulate the HR.

4.2 Characterization of EIX Endocytosis

Internalization of defense receptors is required for proper defense signalling in several cases (Ron and Avni 2004; Robatzek et al. 2006; Bar and Avni 2009a). As we have previously reported, internalization of both EIX (Hanania and Avni 1997; Rotblat et al. 2002) and the LeEix2 receptor (Ron and Avni 2004; Bar and Avni 2009a) are required for the plant to mount a proper response to EIX. Prevention of internalization of endocytic vesicles by inhibition of dynamin further demonstrated the necessity of EIX/LeEix2 internalization for signalling and defense responses. The same is true for inhibition of the actin cytoskeleton using the F-actin depolymerizing drug latrunculinB (Sharfman et al. 2011).

Characterization of endosomal movement and content following EIX treatment allowed us to demonstrate that EIX causes a sub-population of endosomes to move faster and in a more directional manner (Sharfman et al. 2011). Further, this subpopulation appears to contain smaller endosomes or endosomes with a lower PI-3-P content (as revealed by FYVE, a PI-3-P binding molecular marker) (Voigt et al. 2005; Sharfman et al. 2011). We have also shown that at least some of the endosomes containing LeEix show increased speed and directionality of movement (Sharfman et al. 2011). Given the requirement of an intact cytoskeleton for EIX signalling, it seems probable that directional movement following EIX treatment occurs on actin filaments.

In addition, components of membrane lipid synthesis and signalling are involved in the plant response to EIX. Thus, manipulating PLD or PLC activity impairs membrane functions and ultimately leads to inability of the cell to form endosomes, which in turn prevents endocytosis. Moreover, certain products of the membrane lipid synthesis such as phosphatidic acid may themselves serve as secondary messengers and be involved to a certain extent in EIX/LeEix2 signalling.

In addition to endocytosis of both the receptor and elicitor, proper EIX signalling also requires tyrosine kinase activity (Sharfman et al. 2011). The LeEix2 receptor is devoid of kinase activity, unlike some plant defense receptors such as FLS2 (Gomez-Gomez and Boller 2000; Robatzek et al. 2006), but may be phosphorylated by another protein. If the LeEix2 receptor is phosphorylated, such phosphorylation may be required for receptor internalization and for proper signalling. Alternatively, tyrosine kinase activity may be involved in further downstream signalling events ensuring proper plant response to EIX.

5 Conclusions and Future Prospects

Results obtained in connection with LeEix2 internalization indicate that recycling might be involved in some aspects of EIX signalling, as inhibition of protein synthesis did not result in receptor degradation following internalization, but did cause LeEix2 to persist on endosomes (Bar and Avni 2009a). Additional results suggest that recycling of the EIX receptor is perhaps responsible for the amplitude of the response to EIX, since treatment of cells with BFA caused attenuation of EIX signalling in several examined parameters (Sharfman et al. 2011).

signalling endosomes have been previously reported in plants (Geldner et al. 2007). Considering data obtained upon temporal separation of the internalization event (by several inhibitors) from the endosome trafficking (by treatment with endosidin1 (Robert et al. 2008), it seems that much of the EIX-LeEix2 signalling occurs from endosomal compartments (Sharfman et al. 2011). Further exploration of endosomal signalling in the EIX/LeEix2 system as well as in plants in general is required.

Endocytosis is involved in both biotic and abiotic stresses in plant cells and the EHD proteins as regulators of endocytosis participate in plant responses to these stresses. The elucidation of various components of EIX/LeEix2 endocytosis and signalling provides framework for future research aiming to further clarify plant defense signalling mechanisms. Based on previous and recent work presented here (Hanania et al. 1999; Rotblat et al. 2002; Ron and Avni 2004; Bar et al. 2008), we propose the model depicted in Fig. 1.

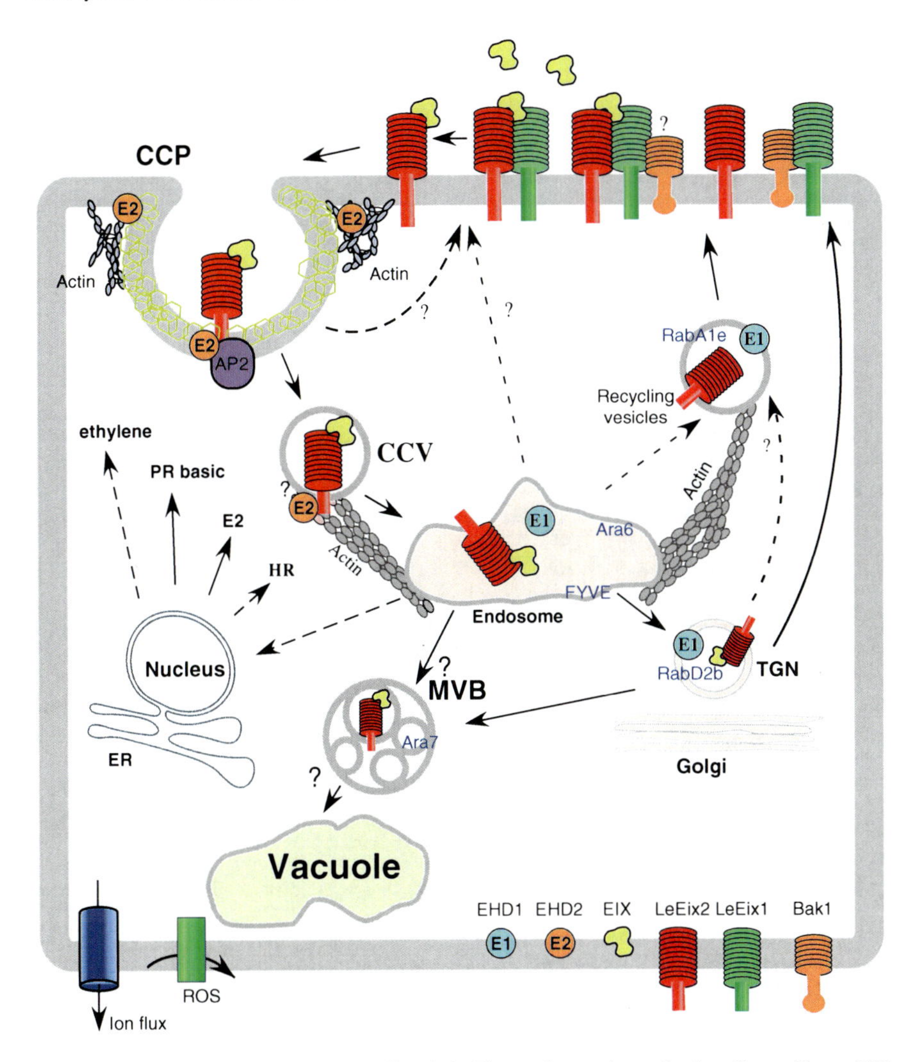

Fig. 1 Proposed model for EIX-mediated LeEix endocytosis and signalling. Upon EIX application, it binds the LeEix2 receptor on the outside of the plasma membrane (Hanania and Avni 1997; Ron and Avni 2004). The expression of both LeEix receptors causes attenuation of EIX endocytosis and signalling, in a BAK1 dependent manner (Bar et al. 2011). The EIX-LeEix2 complex probably binds an endocytic protein complex using the Yxxϕ motif present within the cytoplasmic tail of LeEix2. Two subunits of AP-2 bind to the Yxxϕ motif of LeEix2 and to EHD2, respectively (Bar and Avni 2009a; Bar et al. 2009). LeEix2 and EIX are internalized into FYVE endosomes, which probably also contain EHD1 (Bar et al. 2008). LeEix2 is probably recycled back to the plasma membrane on recycling vesicles which contain RabA1e and also EHD1, as well as TGN/early endosomes, which may overlap with recycling vesicles, and contain RabD2b and EHD1 (Bar and Avni, unpublished results) The internalization of LeEix2 and EIX is required for induction of defense responses, (Bailey et al. 1990, 1992; Laxalt et al. 2007). EIX application also triggers EHD2 and LeEix1 expression, which act to inhibit the defense response in the short term. Longer exposure to EIX (or other elicitors) leads to a "full-blown" defense response. LeEix2 and possibly also EIX are likely also partially degraded via the multivesicular body (MVB) pathway

References

Asai T, Tena G, Plotnikova J, Willmann MR, Chiu WL, Gomez-Gomez L, Boller T, Ausubel FM, Sheen J (2002) MAP kinase signalling cascade in Arabidopsis innate immunity. Nature 415:977–983

Bailey BA, Dean JFD, Anderson JD (1990) An ethylene biosynthesis-inducing endoxylanase elicits electrolyte leakage and necrosis in *Nicotiana tabacum* Cv xanthi leaves. Plant Physiol 94:1849–1854

Bailey BA, Korcak RF, Anderson JD (1992) Alterations in *Nicotiana tabacum-L* Cv xanthi cell-membrane function following treatment with an ethylene biosynthesis-inducing endoxylanase. Plant Physiol 100:749–755

Bar M, Aharon M, Benjamin S, Rotblat B, Horowitz M, Avni A (2008) AtEHDs, novel Arabidopsis EH-domain containing proteins involved in endocytosis. Plant J

Bar M, Avni A (2009a) EHD2 inhibits ligand-induced endocytosis and signaling of the leucine-rich repeat receptor-like protein LeEix2. Plant J 59:600–611

Bar M, Avni A (2009b) EHD2 inhibits signaling of leucine rich repeat receptor-like proteins. Plant Signal Behav 4:682–684

Bar M, Sharfman M, Avni A (2011) LeEix1 functions as a decoy receptor to attenuate LeEix2 signaling. Plant Signal Behav 6:455–457

Bar M, Sharfman M, Schuster S, Avni A (2009) The coiled-coil domain of EHD2 mediates inhibition of LeEix2 endocytosis and signaling. PLoS One 4:e7973

Basse CW, Bock K, Boller T (1992) Elicitors and suppressors of the defense response in tomato cells—purification and characterization of glycopeptide elicitors and glycan suppressors generated by enzymatic cleavage of yeast invertase. J Biol Chem 267:10258–10265

Becraft PW (2002) Receptor kinase signaling in plant development. Annu Rev Cell Dev Biol 18:163–192

Benjamin S, Weidberg H, Rapaport D, Pekar O, Nudelman M, Segal D, Hirschberg K, Katzav S, Ehrlich M, Horowitz M (2011) EHD2 mediates trafficking from the plasma membrane by modulating Rac1 activity. Biochem J 439:433–442

Bent AF (1996) Plant disease resistance genes: function meets structure. Plant Cell 8:1757–1771

Blein JP, Milat ML, Ricci P (1991) Responses of cultured tobacco cells to cryptogein, a proteinaceous elicitor from *Phytophthora cryptogea*—possible plasmalemma involvement. Plant Physiol 95:486–491

Blume JJ, Halbach A, Behrendt D, Paulsson M, Plomann M (2007) EHD proteins are associated with tubular and vesicular compartments and interact with specific phospholipids. Exp Cell Res 313:219–231

Boller T (1991) Ethylene in pathogenesis and disease resistance. In: Mattoo AK, Suttle JC (eds) The plant hormon ethylen. CRC press, Boca Raton, pp 293–314

Braun A, Pinyol R, Dahlhaus R, Koch D, Fonarev P, Grant BD, Kessels MM, Qualmann B (2005) EHD proteins associate with syndapin I and II and such interactions play a crucial role in endosomal recycling. Mol Biol Cell 16:3642–3658

Caplan S, Naslavsky N, Hartnell LM, Lodge R, Polishchuk RS, Donaldson JG, Bonifacino JS (2002) A tubular EHD1-containing compartment involved in the recycling of major histocompatibility complex class I molecules to the plasma membrane. EMBO J 21:2557–2567

Carbone R, Fre S, Iannolo G, Belleudi F, Mancini P, Pelicci PG, Torrisi MR, Di Fiore PP (1997) EPS15 and EPS15R are essential components of the endocytic pathway. Cancer Res 57:5498–5504

Dean JFD, Anderson JD (1991) Ethylene biosysnthesis-inducing endoxylanase. II Purification and physical characterization of the enzyme produced by *Trichoderma viride*. Plant Physiol 95:316–323

Dean JFD, Gamble HR, Anderson JD (1989) The ethylene biosynthesis inducing xylanase: Its induction in *Trichoderma viride* and certain plant pathogens. Phytopathology 79:1071–1078

Dewit PJGM, Hofman AE, Velthuis GCM, Kuc JA (1985) Isolation and characterization of an elicitor of necrosis isolated from intercellular fluids of compatible interactions of *Cladosporium fulvum* (Syn Fulvia-Fulva) and tomato. Plant Physiol 77:642–647

Dhonukshe P, Aniento F, Hwang I, Robinson DG, Mravec J, Stierhof YD, Friml J (2007) Clathrin-mediated constitutive endocytosis of PIN auxin efflux carriers in Arabidopsis. Curr Biol 17:520–527

Ebel J, Cosio EG (1994) Elicitors of plant defense responses. Int Rev Cytol—Surv Cell Biol 148:1–36

Elbaz M, Avni A, Weil M (2002) Constitutive caspase-like machinery executes programmed cell death in plant cells. Cell Death Differ 9:726–733

Felix G, Duran JD, Volko S, Boller T (1999) Plants have a sensitive perception system for the most conserved domain of bacterial flagellin. Plant J 18:265–276

Felix G, Regenass M, Boller T (1993) Specific preception of subnanomolar concentrations of chitin fragments by tomato cells: induction of extracellular alkalization, changes in protein phosphorylation, and establishment of a refractory state. Plant J 4:307–316

Flor HH (1947) Host-parasite interaction in flax rust-it's genetics and other implications. Phytopathology 45:680–685

Fluhr R, Sessa G, Sharon A, Ori N, Lotan T (1991) Pathogenesis-related proteins exhibit both pathogenesis-induced and developmental regulation. Kluwer, Dordrecht

Fradin EF, Zhang Z, Juarez Ayala JC, Castroverde CD, Nazar RN, Robb J, Liu CM, Thomma BP (2009) Genetic dissection of Verticillium wilt resistance mediated by tomato Ve1. Plant Physiol 150:320–332

Fuchs Y, Saxena A, Gamble HR, Anderson JD (1989) Ethylene biosynthesis-inducing protein from cellulysin is an endoxylanase. Plant Physiol 89:138–143

Furman-Matarasso N, Cohen E, Du Q, Chejanovsky N, Hanania U, Avni A (1999) A point mutation in the ethylene-inducing xylanase elicitor inhibits the beta-1-4-endoxylanase activity but not the elicitation activity. Plant Physiol 121:345–352

Galperin E, Benjamin S, Rapaport D, Rotem-Yehudar R, Tolchinsky S, Horowitz M (2002) EHD3: a protein that resides in recycling tubular and vesicular membrane structures and interacts with EHD1. Traffic 3:575–589

Geldner N, Hyman DL, Wang X, Schumacher K, Chory J (2007) Endosomal signaling of plant steroid receptor kinase BRI1. Genes Dev 21:1598–1602

George M, Ying GG, Rainey MA, Solomon A, Parikh PT, Gao QS, Band V, Band H (2007) Shared as well as distinct roles of EHD proteins revealed by biochemical and functional comparisons in mammalian cells and C-elegans. BMC Cell Biol 8:3

Glazebrook J, Rogers EE, Ausubel FM (1997) Use of Arabidopsis for genetic dissection of plant defense responses. Annu Rev Genet 31:547–569

Goeschi JD, Rappaport L, Pratt HK (1966) Ethylene as a factor regulating the growth of pea epicotyls subjected to physical stress. Plant Physiol 42:877–884

Golomb L, Abu-Abied M, Belausov E, Sadot E (2008) Different subcellular localizations and functions of Arabidopsis myosin VIII. BMC Plant Biol 8:3

Gomez-Gomez L, Boller T (2000) FLS2: an LRR receptor-like kinase involved in the perception of the bacterial elicitor flagellin in Arabidopsis. Mol Cell 5:1003–1011

Gomez-Gomez L, Felix G, Boller T (1999) A single locus determines sensitivity to bacterial flagellin in *Arabidopsis thaliana*. Plant J 18:277–284

Grant B, Zhang Y, Paupard MC, Lin SX, Hall DH, Hirsh D (2001) Evidence that RME-1, a conserved C. elegans EH-domain protein, functions in endocytic recycling. Nat Cell Biol 3:573–579

Gruenberg J, Howell KE (1987) An internalized transmembrane protein resides in a fusion-competent endosome for less than 5 minutes. Proc Natl Acad Sci U S A 84:5758–5762

Guilherme A, Soriano NA, Bose S, Holik J, Bose A, Pomerleau DP, Furcinitti P, Leszyk J, Corvera S, Czech MP (2004a) EHD2 and the novel EH domain binding protein EHBP1 couple endocytosis to the actin cytoskeleton. J Biol Chem 279:10593–10605

Guilherme A, Soriano NA, Furcinitti PS, Czech MP (2004b) Role of EHD1 and EHBP1 in perinuclear sorting and insulin-regulated GLUT4 recycling in 3T3-L1 adipocytes. J Biol Chem 279:40062–40075

Hammond-Kosack KE, Jones JD (1996) Resistance gene-dependent plant defense responses. Plant Cell 8:1773–1791

Hanania U, Avni A (1997) High-affinity binding site for ethylene-inducing xylanase elicitor on *Nicotiana tabacum* membranes. Plant J 12:113–120

Hanania U, Furman-Matarasso N, Ron M, Avni A (1999) Isolation of a novel SUMO protein from tomato that suppresses EIX-induced cell death. Plant J 19:533–541

Horn MA, Heinstein PF, Low PS (1989) Receptor-mediated endocytosis in plant cells. Plant Cell 1:1003–1009

Husebye H, Halaas O, Stenmark H, Tunheim G, Sandanger O, Bogen B, Brech A, Latz E, Espevik T (2006) Endocytic pathways regulate toll-like receptor 4 signaling and link innate and adaptive immunity. EMBO J 25:683–692

Jaillais Y, Fobis-Loisy I, Miege C, Gaude T (2008) Evidence for a sorting endosome in Arabidopsis root cells. Plant J 53:237–247

Jaillais Y, Gaude T (2007) Sorting out the sorting functions of endosomes in Arabidopsis. Plant Signal Behav 2:556–558

Jones DA, Thomas CM, Hammond-Kosack KE, Balint-Kurti PJ, Jones JD (1994) Isolation of the tomato Cf-9 gene for resistance to *Cladosporium fulvum* by transposon tagging. Science 266:789–793

Kawchuk LM, Hachey J, Lynch DR, Kulcsar F, van Rooijen G, Waterer DR, Robertson A, Kokko E, Byers R, Howard RJ, Fischer R, Prufer D (2001) Tomato Ve disease resistance genes encode cell surface-like receptors. Proc Natl Acad Sci U S A 98:6511–6515

Lam SK, Siu CL, Hillmer S, Jang S, An G, Robinson DG, Jiang L (2007a) Rice SCAMP1 defines clathrin-coated, trans-golgi-located tubular-vesicular structures as an early endosome in tobacco BY-2 cells. Plant Cell 19:296–319

Lam SK, Tse YC, Robinson DG, Jiang L (2007b) Tracking down the elusive early endosome. Trends Plant Sci 12:497–505

Laxalt AM, Raho N, ten Have A, Lamattina L (2007) Nitric oxide is critical for inducing phosphatidic acid accumulation in xylanase-elicited tomato cells. J Biol Chem 282:21160–21168

Leborgne-Castel N, Lherminier J, Der C, Fromentin J, Houot V, Simon-Plas F (2008) The plant defense elicitor cryptogein stimulates clathrin-mediated endocytosis correlated with reactive oxygen species production in bright yellow-2 tobacco cells. Plant Physiol 146:1255–1266

Lotan T, Fluhr R (1990) Xyalanase, a novel elicitor of pathogenesis-related proteins in tobacco, use a non-ethylene pathway for induction. Plant Physiol 93:811–817

Mintz L, Galperin E, Pasmanik-Chor M, Tulzinsky S, Bromberg Y, Kozak CA, Joyner A, Fein A, Horowitz M (1999) EHD1–an EH-domain-containing protein with a specific expression pattern. Genomics 59:66–76

Naslavsky N, Boehm M, Backlund PS Jr, Caplan S (2004) Rabenosyn-5 and EHD1 interact and sequentially regulate protein recycling to the plasma membrane. Mol Biol Cell 15:2410–2422

Naslavsky N, Caplan S (2005) C-terminal EH-domain-containing proteins: consensus for a role in endocytic trafficking, EH? J Cell Sci 118:4093–4101

Ortiz-Zapater E, Soriano-Ortega E, Marcote MJ, Ortiz-Masia D, Aniento F (2006) Trafficking of the human transferrin receptor in plant cells: effects of tyrphostin A23 and brefeldin A. Plant J 48:757–770

Pohl U, Smith JS, Tachibana I, Ueki K, Lee HK, Ramaswamy S, Wu Q, Mohrenweiser HW, Jenkins RB, Louis DN (2000) EHD2, EHD3, and EHD4 encode novel members of a highly conserved family of EH domain-containing proteins. Genomics 63:255–262

Polo S, Confalonieri S, Salcini AE, Di Fiore PP (2003) EH and UIM: endocytosis and more. Sci STKE, re17

Rapaport D, Auerbach W, Naslavsky N, Pasmanik-Chor M, Galperin E, Fein A, Caplan S, Joyner AL, Horowitz M (2006) Recycling to the plasma membrane is delayed in EHD1 knockout mice. Traffic 7:52–60

Robatzek S (2007) Vesicle trafficking in plant immune responses. Cell Microbiol 9:1–8

Robatzek S, Chinchilla D, Boller T (2006) Ligand-induced endocytosis of the pattern recognition receptor FLS2 in Arabidopsis. Gene Dev 20:537–542

Robert S, Chary SN, Drakakaki G, Li S, Yang Z, Raikhel NV, Hicks GR (2008) Endosidin1 defines a compartment involved in endocytosis of the brassinosteroid receptor BRI1 and the auxin transporters PIN2 and AUX1. Proc Natl Acad Sci U S A 105:8464–8469

Ron M, Avni A (2004) The receptor for the fungal elicitor ethylene-inducing xylanase is a member of a resistance-like gene family in tomato. Plant Cell 16:1604–1615

Ron M, Kantety R, Martin GB, Avidan N, Eshed Y, Zamir D, Avni A (2000) High-resolution linkage analysis and physical characterization of the EIX-responding locus in tomato. Theor Appl Genet 100:184–189

Rotblat B, Enshell-Seijffers D, Gershoni JM, Schuster S, Avni A (2002) Identification of an essential component of the elicitation active site of the EIX protein elicitor. Plant J 32:1049–1055

Rotem-Yehudar R, Galperin E, Horowitz M (2001) Association of insulin-like growth factor 1 receptor with EHD1 and SNAP29. J Biol Chem 276:33054–33060

Ŝamaj J, Read ND, Volkmann D, Menzel D, Baluška F (2005) The endocytic network in plants. Trends Cell Biol 15:425–433

Sharfman M, Bar M, Ehrlich M, Schuster S, Melech-Bonfil S, Ezer R, Sessa G, Avni A (2011) Endosomal signaling of the tomato leucine-rich repeat receptor-like protein LeEix2. Plant J 68:413–423

Takken FLW, Schipper D, Nijkamp HJJ, Hille J (1998) Identification and Ds-tagged isolation of a new gene at the Cf-4 locus of tomato involved in disease resistance to *Cladosporium fulvum* race 5. Plant J 14:401–411

Ueda T, Yamaguchi M, Uchimiya H, Nakano A (2001) Ara6, a plant-unique novel type Rab GTPase, functions in the endocytic pathway of *Arabidopsis thaliana*. EMBO J 20:4730–4741

van der Hoorn RAL, Wulff BBH, Rivas S, Durrant MC, van der Ploeg A, de Wit PJGM, Jones JDG (2005) Structure-function analysis of Cf-9, a receptor-like protein with extracytoplasmic leucine-rich repeats. Plant Cell 17:1000–1015

Voigt B, Timmers AC, Ŝamaj J, Hlavačka A, Ueda T, Preuss M, Nielsen E, Mathur J, Emans N, Stenmark H, Nakano A, Baluŝka F, Menzel D (2005) Actin-based motility of endosomes is linked to the polar tip growth of root hairs. Eur J Cell Biol 84:609–621

Wong WT, Schumacher C, Salcini AE, Romano A, Castagnino P, Pelicci PG, Di Fiore P (1995) A protein-binding domain, EH, identified in the receptor tyrosine kinase substrate EPS15 and conserved in evolution. Proc Natl Acad Sci U S A 92:9530–9534

Wu S-C, Ham K-S, Darvill AG, Albersheim P (1997) Deletion of two Endo-beta-1,4-Xylanase genes reveals additional isozymes secreted by the rice blast fungus. Mol Plant Microbe Interact 10:700–708

Yano A, Suzuki K, Uchimiya H, Shinshi H (1998) Induction of hypersensitive cell death by fungal protein in cultures of tobacco cells. Mol Plant Microbe Interact 11:115–123

Yu I, Parker J, Bent A (1998) Gene-for-gene disease resistance without the hypersensitive response in Arabidopsis *dnd1* mutant. Proc Natl Acad Sci 95:7819–7824

Zimmermann P, Hirsch-Hoffmann M, Hennig L, Gruissem W (2004) GENEVESTIGATOR. Arabidopsis microarray database and analysis toolbox. Plant Physiol 136:2621–2632

Zipfel C, Robatzek S, Navarro L, Oakeley EJ, Jones JDG, Felix G, Boller T (2004) Bacterial disease resistance in Arabidopsis through flagellin perception. Nature 428:764–767

Endocytosis and Cytoskeleton: Dynamic Encounters Shaping the Portals of Cell Entry

Anirban Baral and Pankaj Dhonukshe

Abstract Genetic and pharmacological studies coupled with live imaging have portrayed the crosstalk between cytoskeleton and endocytic pathways of yeast, animals, and plants. Localized actin nucleation at endocytic foci seems to be the driving force for endocytic vesicle formation in yeasts and animals. Actin microfilaments also serve as tracks for intracellular transport of internalized endocytic vesicles. In addition, microtubules serve as the tracks for long range transport of endosomes in mammalian cells. Distinct actin and microtubule associated motor proteins facilitate this transport processes. Depolymerization of cortical actin in plants does not block entry of cargo in cells. However, subsequent trafficking processes are affected indicating a major role of actin in long range transport of endocytic vesicles. In plants involvement of microtubules in endocytic processes specializes both in non-dividing and dividing cells. In interphase cells, cortical microtubules co-align with pinching endocytic vesicles while endoplasmic microtubules direct the trajectories of endocytic materials during mitosis. Microtubules play key roles in delivering secreted and endocytic cargos to the newly assembling cell plate. Thus, with some conserved features of cytoskeletal involvement in endocytosis from yeast and animals, plants shape a unique dialogue between the cytoskeleton and membrane trafficking in order to meet plant-specific needs.

A. Baral
National Centre for Biological Sciences (TIFR), GKVK Campus,
Bellary Road 560065 Bangalore, India

P. Dhonukshe (✉)
Department of Biology, Utrecht University, Padualaan 8
3584 CH Utrecht, The Netherlands
e-mail: P.B.Dhonukshe@uu.nl

J. Šamaj (ed.), *Endocytosis in Plants*,
DOI: 10.1007/978-3-642-32463-5_16, © Springer-Verlag Berlin Heidelberg 2012

1 Introduction

Endocytosis is the process by which cells internalize plasma membrane (PM) along with resident proteins and extracellular milieu to regulate membrane homeostasis, to down-regulate PM receptors and to trigger signalling cascades vital for cell survival. Endocytosis has diversified during evolution across different kingdoms ranging from unicellular yeast and protozoa to multicellular animals and plants. Different pathways of endocytosis exist that vary in terms of the cargo they internalize and the molecular players they exploit. Based on the cell type, one endocytic pathway may dominate or several distinct endocytic pathways may operate in parallel. However, despite the diversity of endocytic pathways, all forms of endocytosis critically depend on the cytoskeletal elements that provide structural scaffold right from the pinching off an endocytic vesicle at the PM up to its intracellular destination deep inside the cell. In eukaryotic systems three kinds of cytoskeletal elements are found, namely actin filaments, microtubules and intermediate filaments. Intermediate filaments are mostly involved in conferring rigidity to cellular structures and have no known endocytic functions. Moreover, they are not found in all eukaryotes. Actin and microtubule cytoskeletons are ubiquitous in all eukaryotes and their structures have been conserved during evolution. Actin cytoskeleton plays predominant roles in controlling endocytosis in unicellular yeasts. Whereas both actin and microtubule cytoskeleton along with a plethora of accessory proteins play distinct roles in endocytic processes in plants and in animals. In the following sections we discuss roles of actin filaments and microtubules in endocytosis, with a focus on plant endocytosis.

2 Actin Cytoskeleton and Endocytosis

The first indication of actin involvement in endocytosis came from studies in budding yeast *Saccharomyces cerevisiae*. In yeast it was demonstrated that mutations in actin, in actin cross linking protein fimbrin (Sac 6P) and in actin associated myosin motors block the receptor mediated and fluid phase endocytosis (Geli and Riezman 1996). Subsequent studies using compounds depolymerizing or stabilizing actin filaments such as latrunculin or jasplakinolide supported the credibility of dynamic actin in endocytosis (Ayscough 2000). Time lapse imaging of live yeast cells expressing GFP-tagged actin revealed two kinds of structures, the cable like actin filaments and transient punctate structures that assemble and disassemble at the PM very rapidly, in the timescale of seconds. Multicolor real-time fluorescence microscopy revealed that these actin patches colocalize with a set of endocytic regulatory proteins and represent zones of very rapid actin polymerization and endocytic internalization (Kaksonen et al. 2003; Huckaba et al. 2004). Inhibiting functions of actin cytoskeleton components by genetic or chemical perturbation coupled with live cell imaging approaches in animal cells

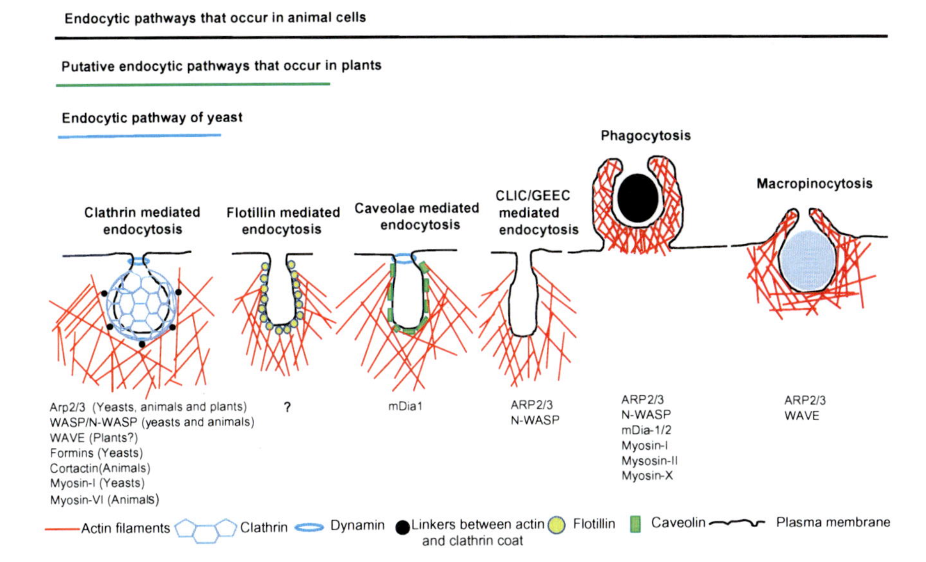

Fig. 1 Involvement of actin in major endocytic pathways. Clathrin-dependent endocytic pathway is consensus among yeast, animals, and plants. Plants have a flotillin-dependent endocytic pathway in addition. Other pathways of endocytosis like caveolar pathway, CLIC/GEEC pathway, macropinocytosis, and phagocytosis are reported to occur exclusively in animal cells till now. Actin is involved in all the reported endocytic pathways. However, the proteins involved in the control of actin dynamics in case of different pathways vary. Arp2/3 complex is a common player in many pathways that initiate actin nucleation at the onset of endocytosis. Activity of ARP2/3 complex is regulated by nucleation promoting factors like WASP, N-WASP, cortactin, WAVE complex and their homologous. Notably, plants do not have homologous for WASP complex but they have homologous for component of WAVE complex. Other than ARP2/3, the formins like mDia-1 and mDia-2 nucleate actin polymerization during uptake of caveolae and during phagocytosis. Plants have several formin isoforms which might play similar role in the endocytic uptake. In case of clathrin-dependent endocytosis, actin is linked to the clathrin coat components via adaptor proteins like ankyrins, amphiphysin, HIP1, and HIP1R. Plant SH3 group proteins might serve similar functions

have uncovered critical roles of actin machinery in different endocytic processes. Here, we present a brief overview of the major endocytic pathways discovered in yeast and animal systems including roles of actin in their functionality (also, see Fig. 1).

2.1 Function of Actin in Diverse Endocytic Pathways

2.1.1 Clathrin-dependent Endocytosis

Clathrin-and dynamin-dependent endocytosis is the most well-characterized endocytic pathway in terms of molecular details. This is also the most conserved endocytic route that operates across diverse eukaryotic systems like yeasts, animals, and plants. In this endocytic process the coat protein clathrin is recruited to the membrane for mediating membrane invagination by forming basketlike self-assembly that is followed by scission of the clathrin coated pit by the GTPase dynamin. Clathrin-dependent endocytosis is responsible for the uptake of a variety of transmembrane receptors from the PM. Evanescent wave Microscopy (also known as Total Internal Reflection of Fluorescence Microscopy or TIRFM) enabled visualization of molecular events at or immediately underneath the PM. Using this technique the dynamics of clathrin and actin at the PM was investigated. In yeast, clathrin is always recruited to the nascent endocytic foci. Intriguingly, abolishing clathrin function causes a significant reduction in the number of endocytic foci at the PM but it does not block endocytosis completely. This suggests that although clathrin plays an important role in recruitment of actin machinery to the membrane, it is dispensable for membrane invagination and vesicle formation (Kaksonen et al. 2005). Moreover, dynamin is not required for scission of endocytic vesicles in yeast. Although the endocytic pathways in mammals are more complex and varied, the function of actin cytoskeleton in endocytic processes is similar to that observed in yeast. In mammalian cells, depolymerization or stabilization of actin filaments blocks internalization of clathrin-dependent cargo, similar to that in yeast cells. Such interference with actin dynamics not only blocks internalization of clathrin coated vesicles but also their lateral mobility and the fission and fusion of already formed vesicles. Biochemical and microscopic analyzes revealed actin dependency of vesicle constriction and scission. Using electron microscopy and TIRF imaging, it was found that when actin dynamics is perturbed, the number of invaginated coated pits is higher as compared to that in control cells (Merrifield et al. 2005; Yarar et al. 2005). This indicates that clathrin coated invagination may proceed without actin in mammalian cells which is in contrast to the yeast cells where membrane invagination requires intact actin cytoskeleton. The homologous of yeast proteins that regulate actin polymerization during vesicle assembly, scission and transport have also been identified in mammalian cells. Sequential recruitment and stepwise action of these proteins regulates each step of endocytosis (Smythe and Ayscough 2006).

2.1.2 Caveolae-Mediated Endocytosis

Caveolae are flask-shaped membrane invaginations that form from sterol and sphingolipid enriched membrane domains containing caveolin. Caveolins are integral membrane proteins that bind directly to membrane cholesterol. Caveolae dependent endocytosis is a major portal for clathrin independent entry in the cell and this pathway has been implicated in uptake of cargo such as modified albumins, membrane components such as modified glycosphingolipids and crosslinked GPI anchored proteins as well as SV-40 virus particles (Nabi and Le 2003). Caveolae mediated cargo uptake is dynamin dependent. The role of actin in caveolae dependent endocytosis was revealed by latrunculin and jasplakinolide treatments that inhibited SV-40 virus internalization. Further analysis revealed that SV-40 particles localize to caveolin rich domains and nucleate actin polymerization that facilitates virus internalization (Pelkmans et al. 2002).

2.1.3 Flotillin-Mediated Endocytosis

Flotillins are lipid associated protein that form microdomains within the PM which are distinct from caveolae. It has been shown that apart from PM, flotillins can be found in intracellular compartments that contain endocytosis cell surface materials. Reduction in flotillin-1 by RNA interference reduces the uptake of cell surface GPI anchored protein CD-59 and Cholera Toxin subunit B in cultured animal cells (Glebov et al. 2006). The role of dynamin in flotillin-dependent endocytic uptake is not clear. Overexpression of flotillins generates actin-rich filopodia-like protrusions in various epithelial cells. Flotillins are also hypothesized to have a role in differential actin dynamics during leukocyte chemotaxis (Affentranger et al. 2011). Such actin modulating properties might be important for endocytosis of flotillin-dependent cargo as well.

2.1.4 CLIC/GEEC-Dependent Endocytosis

A unique form of endocytosis is reported in animal cells and it is independent of clathrin, caveolin, and dynamin. This pathway takes up GPI-anchored proteins from PM as well as fluid phase endocytic markers like dextrans (Kalia et al. 2006). Uptake of cargo through this pathway is believed to be dependent on hitherto unidentified clathrin independent carriers (CLIC) and involves GEECs (GPI-AP enriched early endosomal compartments) as intermediates. This pathway is critically dependent on actin cytoskeleton. Mild perturbation of actin using low concentration of latrunculinB specifically blocks this pathway without significantly affecting clathrin-dependent endocytosis (Chadda et al. 2007).

2.1.5 Macro Scale Endocytosis (Phagocytosis and Macropinocytosis)

Phagocytosis and macropinocytosis are the endocytic processes by which cell ingests large particles (more than 500 nm) and bulk amounts of fluid respectively. These endocytic vesicles are much larger than those known in other endocytic pathways. Another important difference is that unlike other endocytic pathways where membrane invagination and scission generates endocytic vesicles, in macro scale endocytosis protrusions from the PM encapsulates the cargo and then seals off to generate large enclosed vesicles which can measure up to several micrometers. Actin machinery is shown to be a key component for this pathway (Swanson 2008).

2.1.6 Endocytic Pathways in Plants

In plants clathrin-dependent endocytic pathway seems to be the predominant one. Caveolin homologous are absent in the sequenced plant genomes. Due to the presence of cell wall and high turgor pressure phagocytosis is not a very common event seen in plants except for the special cases such as the phagocytic uptake of Rhizobia during root nodule formation in leguminous plants (Jones et al. 2007). Flotillin homologous have been reported in Arabidopsis which form prominent foci on PM (Li et al. 2011). It is reported very recently that flotilin-1 mediates a clathrin-independent endocytic pathway in plants which is important for uptake of plasma membrane sterols and has important implications in plant development (Li et al. 2012).

In plants, the role of actin is investigated in context of clathrin-dependent uptake and subsequent intracellular trafficking for a number of transmembrane proteins. The role of actin in the earliest steps of clathrin-dependent endocytosis was investigated by evanescent wave microscopy (Konopka et al. 2008). Endocytic foci formation at the PM of Arabidopsis root epidermal cells was monitored with the aid of fluorescently tagged clathrin light chain (CLC) and a plant specific dynamin isoform DRP1C. It was found that the response of clathrin and dynamin foci to actin depolymerization by latrunculin B is dose dependent. Low concentration of latrunculin B, which depolymerizes cortical actin but does not prevent cytoplasmic streaming, had no effect on dynamics of DRP1C-GFP foci and slightly increased their lifetime at the PM. On the other hand, much higher concentration of the inhibitor that stops cytoplasmic streaming blocked internalization of both clathrin and dynamin foci. This indicates that actin driven cytoplasmic streaming might have some functions in endocytic foci formation and their scission. Actin perturbations impacts flotillin-dependent endocytosis as well (Li et al. 2012). Treatment with low concentration of latrunculin-B shortened the trajectories of flotillin-GFP punctae at PM of Arabidopsis root epidermal cells and caused almost six fold decrease in the diffusion coefficient of such punctae. Thus it appears that the cortical actin has important regulatory role near the plasma membrane at the onset of different types of endocytic events.

The role of actin in subsequent intracellular trafficking of vesicles is particularly well studied in the context of polarized growth in plants such as elongation of root hairs and pollen tubes which require directed delivery of PM and cell wall materials toward the tip (see also chapter by Ovečka et al. in this volume). This secretory event is counterbalanced by rapid endocytosis that retrieves excess membranous materials. Dynamic actin cytoskeleton is essential for the polarized transport of secretory vesicles as well as endosomes, as evident by the effect of perturbation of actin dynamics which abruptly blocks vesicle dynamics and ceases polarized growth (Wang et al. 2006). Live cell imaging also revealed that at the tip of growing root hairs highly motile endosomes localize with dynamic actin patches (Voigt et al. 2005). The movement of endosomes is largely dependent on actin polymerization and not so much on myosin motor activity. In subapical regions the endosomes align with actin cables and move toward the basal regions of the root

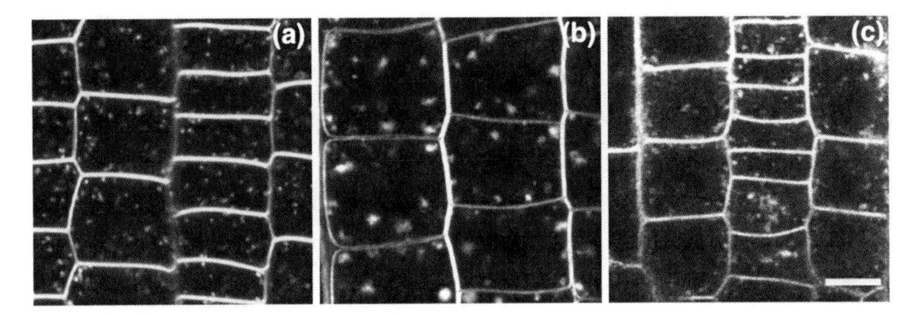

Fig. 2 Uptake of endocytic tracer dye FM4-64 in Arabidopsis root tip epidermal cells in control plant (**a**) and in the presence of cytoskeleton perturbing agents (**b**) and (**c**). Plants are treated with 10 μM latrunculin-B (**b**) or 20 μM oryzalin (**c**) for one hour followed by 30 min incubation with 5 μm FM4–64 in the presence of the inhibitors. Scale bar: 10 μM

hair and pollen tube along the actin tracks (Voigt et al. 2005; Zhang et al. 2010). The role of actin in endocytosis has been probed in non polarized cells as well. Actin patches are found in plasmodesmata and pit fields of maize root cortical cells and depolymerization of actin by latrunculin treatment hinders fluid phase uptake of the dye Lucifer Yellow in these cells (Baluska et al. 2004). Ligand induced endocytosis of flagellin receptor FLS2 in Arabidopsis leaves is also inhibited by actin depolymerization (Robatzek et al. 2006). Depolymerization of actin in Arabidopsis and maize roots by treatment with latrunculin B or cytochalasin D causes accumulation of various endocytic cargos in small intracellular aggregates. Although the precise molecular identity of such compartments is unknown, they accumulate endocytosed sterols, endocytic tracer dye FM4-64 (Fig. 2b) and endocytosed plasma membrane proteins like Lti6A and PIN1 (Grebe et al. 2003; Boutte et al. 2006). Probably, these bodies are formed by aggregation of endosomes with each other as the actin tracks carrying them collapse. On the other hand, stabilization of actin cytoskeleton by treatment with the drug jasplakinolide impairs endocytic uptake of FM4-64 in Arabidopsis root cells (Dhonukshe et al. 2008). In a recent study it is shown that stabilization of cortical F-actin by overexpression of the GTPase ROP2 or its effector RIC-4 inhibits endocytosis in pavement cells of Arabidopsis leaves. Depolymerization of cortical actin by mild dosage of latrunculin-B restores the normal level of endocytosis in these cells (Nagawa et al. 2012). Thus, actin stabilization has an inhibitory role on endocytosis in yeasts, animals and in plants but actin depolymerization does not inhibit the entry of endocytosed cargo into plant cells. However, subsequent downstream trafficking processes are severely affected when actin cytoskeleton is perturbed. For example, trafficking of endocytosed auxin transporter PIN2 toward vacuole (Kleine-Vehn et al. 2008) or redistribution of endocytosed cargo from BFA bodies (Grebe et al. 2003) is blocked by actin depolymerization and by a mutation in the *Actin2* gene in the latter case. This indicates that the meshwork of cortical filamentous actin might act as a inhibitory barrier to endocytosis but downstream trafficking is dependent on thicker cytoplasmic actin cables (Nagawa et al. 2012).

2.1.7 Myosins and Endocytosis

Myosins are actin filament associated motor proteins that travel along actin tracks and carry associated membrane enclosed vesicles. Studies in yeast and animal systems suggested distinct roles of myosins in early stage of endocytic foci nucleation at the PM. Based on sequence variation, myosins have been grouped into eighteen subfamilies of which only subfamily VI myosins are minus end directed and the rest are plus end directed (Berg et al. 2001). Budding yeast has five myosin genes which belong to three subfamilies. Two subfamily I members, Myo-3, and Myo-5 have been implicated to have endocytic functions and mutation in these genes cause severe defects in receptor mediated endocytosis (Geli and Riezman 1996). Live imaging revealed that these myosins are localized to nascent endocytic foci during actin nucleation phase and myosin activity is needed for vesicle invagination and scission (Sun et al. 2006).

In animal cells, different endocytic pathways employ different myosins. In case of clathrin-dependent endocytosis myosin VI is utilized. This myosin localizes to the newly forming clathrin foci at the PM presumably by binding to the adaptor protein AP2 through the protein disabled-2 (Morris et al. 2002). It is not known if presence of myosin is essential for binding of actin to the coat proteins but the motors are believed to be responsible for transport of a newly formed endocytic vesicle inside the cell. In clathrin-independent micro scale endocytosis, involvement of particular myosins is not yet reported. However, the phagocytic cup formation is mediated by myosin-Ic, myosin-X, and myosin-II (Soldati and Schliwa 2006).

Arabidopsis has seventeen myosins which are divided into two groups, myosin VIII group which has four members (ATM1, ATM2, myosin VIIIA, and myosin VIIIB) and myosin XI group which has 13 members (myosins XIA,B,C,D,E,F,-G,H,I,J, and-K, MYA1and MYA2) (Lee and Liu 2004). Myosin XI localizes to various organelles like peroxisomes, Golgi bodies, mitochondria, chloroplasts, and endoplasmic reticulum (Lee and Liu 2004). Class XI myosins are functionally redundant. Multiple mutants of class XI myosins show defects in F-actin assembly as well as defects in the cell shape and polarized growth of root hairs (Peremyslov et al. 2010). Myosin-VIII localizes to the plasmodesmata and cell plate of dividing cell (Reichelt et al. 1999) and is proposed to be involved in endocytosis (Golomb et al. 2008; Sattarzadeh et al. 2008). Truncated forms of Arabidopsis class VIII myosins ATM1 and ATM2 are found to localize to endocytic compartments marked with endosomal marker FYVE, endocytic tracer FM4-64, and also with endosomal Rab proteins (Golomb et al. 2008; Sattarzadeh et al. 2008). Inhibition of myosin ATPase activity by 2,3-butanedionemonoxime (BDM) blocks fluid phase endocytosis in inner cortex cells of maize (Baluska et al. 2004) and inhibits endocytic foci formation in Arabidopsis epidermal cells (Konopka et al. 2008). Taken together, these observations implicate myosins (especially group VIII myosins) in endocytic processes in plants.

2.2 Actin Regulatory Proteins and Adaptors: The Controllers and the Connectors

2.2.1 Actin Regulatory Proteins

As evident from studies in animal and yeast systems, precise regulation of actin cytoskeleton proteins during each steps of endocytosis require a battery of proteins that regulate the actin cytoskeleton through different modes of action. Some of the actin regulators nucleate actin for its polymerization at the endocytic foci. The most important one is the multiprotein ARP2/3 complex which can initiate actin nucleation de novo or at the branch points of actin filaments. ARP2/3 complex and it's activators like cortactin and WASP complex are also recruited to nascent endocytic foci during clathrin mediated endocytosis in yeast and animal cells while actin nucleation by these players is suggested to be the major driving force behind clathrin coated vesicle formation (Kaksonen et al. 2006; Sun et al. 2006). ARP2/3 and associated WASP complex are also found be the major players for the uptake of GPI anchored proteins (Chadda et al. 2007) and for actin mediated membrane protrusion and engulfment of cargo during phagocytosis. Whereas, in case of macropinocytosis ARP2/3 complex activity is regulated by another multiprotein complex called SCAR/WAVE complex (Swanson 2008).

Analysis of Arabidopsis genome revealed a number of genes which encode the components of ARP2/3 complex. These genes are *ARP2*, *ARP3*, *ARPC1A*, *ARPC1B*, *ARPC2A*, *ARPC2B*, *ARPC3*, *ARPC4* and *ARPC5* (Li et al. 2003; Szymanski 2005). Two main components (ARPC2 and ARP3) of this complex can complement the cognate mutations in yeast indicating functional conservation. It is also shown that the ARP2/3 complex localizes to PM, suggesting its putative role in endocytosis and membrane trafficking (Kotchoni et al. 2009). Although the homologous of WASP and cortactin are not reported in Arabidopsis, mutant screens have identified positive regulator of the ARP2/3 complex which is a heteromer of five proteins, NAP1, SRA1, ABI1L1, BRIC1 and SCAR2 (Frank et al. 2004). These proteins are close homologous of the components of animal SCAR/WAVE complex. However, in some instances actin nucleation necessary for endocytic vesicle formation is mediated by players other than ARP2/3 complex. Recently, it has been reported that the actin polymerization required for caveolae internalization during adhesion loss of animal cells is mediated by the formin mDia-1 instead of the ARP2/3 complex (Echarri et al. 2012). Arabidopsis has several formin like (FH) proteins. Some of these proteins are PM associated and have actin nucleating activity (Banno and Chua 2000; Cheung and Wu 2004; Favery et al. 2004). Hence, these proteins become probable candidates as actin nucleators during endocytic vesicle formation. In addition to actin nucleators, there are proteins that trigger actin depolymerization, branching, severing, and cross-linking. Arabidopsis contains other actin crosslinking protein fimbrin (Wang et al. 2004), the actin depolymerizing proteins (ADFs) and cofillin (Dong et al. 2001; Hussey et al. 2002; Tian et al. 2009) as well as actin capping proteins (Huang et al.

2003). The counterparts of these proteins are involved in the precise control of different steps of endocytic uptake and membrane trafficking in animal and yeast systems. Notably in yeast, cofillin, capping proteins, and fimbrin are recruited to endocytic foci at the onset of actin nucleation. Bundling and capping of actin filaments by fimbrin and capping proteins is thought to provide the necessary force for internalization of the vesicle (Smythe and Ayscough 2006). In plants, role of these proteins in modulating the actin cytoskeleton and their impact on plant development have been documented. However, there are no reports on the spatiotemporal interaction and the regulatory functions of these players during endocytic vesicle formations in plants. In addition, the mutant phenotypes for these regulators are subtle, questioning their putative role in the endocytosis. For example, mutations or knock down of ARP2/3 complex components cause severe defects in endocytosis, growth arrest, and lethality in yeasts and animals (Winter et al. 1999; Sawa et al. 2003) whereas Arabidopsis mutants of ARP2/3 complex components are viable and show some growth defects (Li et al. 2003; Mathur et al. 2003) which are much milder than the phenotypes reported when clathrin-dependent endocytosis is blocked (Kitakura et al. 2011). So far, evidence for the direct involvement of these actin modulators in the process of endocytosis is missing and detailed studies are required to bridge this gap.

2.2.2 Small GTPase Regulators of Actin Cytoskeleton and Endocytosis

Involvement of some small GTPases that modulate actin cytoskeleton has been investigated in context of endocytosis in plants. In animal system, active forms of Ras superfamily GTPases can initiate actin nucleation by activating the ARP2/3 complex through WASP or SCAR/WAVE complex. For example CDC-42 dependent actin nucleation through ARP2/3–WASP complex is critical for uptake of GPI-anchored proteins and fluid through GEEC pathway (Chadda et al. 2007). In plants, CDC-42 or other RAS superfamily GTPases are not found but a subfamily of Rho like GTPases called ROP GTPases are found in diverse plant species. The role of ROP GTPases in regulating endocytosis through modulation of cytoskeleton is gradually becoming evident especially in terms of the polarized growth of pollen tubes and root hairs. Various ROP GTPases as well as components of ARP2/3 complexes are localized to tip of the growing root hairs and pollen tubes indicating an involvement of ROP GTPases in the process (Molendijk et al. 2001). Mutations in the *ROP2* gene that alter actin dynamics also alter root hair morphology drastically (Jones et al. 2002). Moreover, the relation between ROP GTPase function, actin organization, and endocytosis has been convincingly demonstrated by overexpression of constitutively active forms of the ROP GTPase, AtRac10. AtRac10 overexpression causes a change in the actin organization resulting in reduced endocytosis which ultimately culminates in dramatic changes in root hair morphology (Bloch et al. 2005). The recent study by (Nagawa et al. 2012) also reports similar findings on constitutively active ROP2 overexpression, causing inhibition of endocytosis and alterations in leaf cell morphology.

2.2.3 Linkers Between Clathrin Coat and Actin Cytoskeleton

Another important question is: How actin cytoskeleton interacts with the vesicle coat components? A number of proteins have been identified in yeasts and animal cells through biochemical and microscopy experiments which connect the coat components and the actin cytoskeleton. In animals, some of these linker proteins are ACK1 and ACK2, ankyrin, amphiphysin, HIP1 and HIP1R. Dynamin, the component required for scission of the vesicle also interacts with actin. The interaction of dynamin with actin is probably through SH3 domain containing proteins like syndapin, Abp1, cortactin and intersectin (Smythe and Ayscough 2006). In plants, this kind of linker proteins awaits identification and functional characterization. However, homologous of several yeast and human proteins working at the interface of clathrin machinery and actin cytoskeleton at different stages of endocytosis like epsin, auxilin, synaptojanin, synaptotagmin, amphiphysin, annexin and Eps15 are present in plants (Šamaj et al. 2004). Eps15 and other epsin homology domain (EHD) proteins are important accessory proteins for assembling clathrin coats and they also have the ability to interact with actin cytoskeleton (Duncan et al. 2001; Guilherme et al. 2004). EHD proteins from Arabidopsis have been localized to endosomal structures and shown to have regulatory roles in endocytosis (Bar et al. 2008; Bar and Avni 2009, see also chapter by Bar and Avni in this volume). Although, the interaction of these proteins with actin and their regulatory roles in control of actin assembly have not yet been demonstrated.

Till date, only Arabidopsis SH3 domain containing proteins (AtSH3Ps) have been shown to interact with both clathrin and actin. It is revealed by electron microscopy that AtSH3P1 colocalizes with clathrin in newly budding endocytic vesicles. Actin association of this protein is also shown by biochemical experiments. AtSH3P1 is also found to associate with auxilin like proteins which might facilitate binding of AtSH3P1 to clathrin coats and their uncoating (Lam et al. 2001). However, much detailed investigation is required to gain comparable understating in plants similar to that available in animal and yeast fields.

3 Microtubules and Endocytosis

Microtubules play key roles during mitotic spindle formation and chromosomal segregation during cell division in all eukaryotes. During interphase, microtubules are responsible for important functions in animal cells such as intracellular movement and clustering of various membrane bound organelles (de Forges et al. 2012). From an endocytosis perspective, the role of microtubules in clathrin coated vesicle formation and internalization is not that prominent in mammals. However, there are few reports that provide indication of the microtubule involvement in clathrin-dependent endocytosis. Depolymerization of microtubules inhibits uptake of clathrin-dependent cargo in some suspension cultured animal cells (Gekle et al. 1997; Subtil and Dautry-Varsat 1997). Involvement of microtubules in clathrin-independent endocytic routes is poorly understood. However, microtubule depolymerization is reported to increase the number of invaginated caveolae at the PM and to inhibit internalization and intracellular transport those caveolae in animal cells (Mundy et al. 2002). Microtubules play important roles in later stages of endocytic trafficking, especially, in long range transport of endocytic vesicles in animal cells. This is a divergence from the endocytic machinery of yeasts, where the intracellular trafficking is dependent on the actin cytoskeleton (Girao et al. 2008). In animal cells, movement of endocytosed cargo from early endosomes to lysosomes through late endosomes is found to be dependent on microtubules (Aniento et al. 1993; Apodaca et al. 1994). Thus, the earliest events of endocytic uptake in animal cells usually rely primarily on actin cytoskeleton, whereas the later steps of endosome trafficking are mediated by the microtubules. The mechanism of transfer of endosomes from actin to microtubule tracks is an area of active research. It is believed to be regulated by the interaction of endosomes with actin- and microtubule-associated motors (Ross et al. 2008). In interphase plant cells depolymerization of microtubules by oryzalin treatment does not affect uptake of transmembrane proteins like the auxin efflux carrier PIN1 (Geldner et al. 2001), sterols (Grebe et al. 2003) and pectins (Baluska et al. 2003). It appears that oryzalin treatment mildly reduces FM4-64 uptake (Fig. 2c) but it does not clump FM4-64 labeled endosomes as in the case of actin depolymerization in Arabidopsis roots. This indicates that microtubules are not much involved in intracellular transport of endocytosed cargo in interphase cells and this process is mostly regulated by the actin cytoskeleton (Voigt et al. 2005). However, microtubule depolymerization by oryzalin treatment is found to have an effect on endocytic foci formation in Arabidopsis epidermal cells (Konopka et al. 2008). Oryzalin treatment increases the lifetime of clathrin and dynamin foci on the PM and it also increases the lateral mobility of such foci suggesting that microtubules can limit the lateral mobility nascent clathrin coated vesicles on plasma membrane by forming the restricted corridors. Associations between microtubules and clathrin have been demonstrated by electron microscopic studies in Arabidopsis pollen grains (Lam et al. 2001). Moreover, ligand induced uptake of flagellin receptor FLS2 in Arabidopsis leaves is blocked in response to oryzalin treatment as it is blocked by actin perturbation (Robatzek et al. 2006). Microtubules also seem to play a prominent role in flotillin-dependent endocytosis as treatment with oryzalin severely restricts the motility of flotillin vesicles at plasma membrane of Arabidopsis root epidermal cells (Li et al. 2012). In dividing plant cells, however, involvement of microtubules in endocytosis is more prominent. Cell division is tightly coupled with endocytosis in plants. PM components are rapidly endocytosed in dividing plant cells and these endocytosed materials are directed to the nascent cell plate and serve as a building blocks as the cell plate expands (Dhonukshe et al. 2006). The site for the formation of cell plate is determined even before the cell division commences by a narrow band of microtubules and cortical actin, known as preprophase band (PPB). The PPB assembles at the site of future cell plate docking to the maternal side wall during G2 phase and attains a compact mature form during late prophase. The PPB does not remain in place till cell plate is formed and disassembles at the onset of metaphase. It is still a matter of speculation how PPB specifies the site of cell plate formation in its absence. However, PPB marks the endocytic hot spots (Fig. 3b). Live cell imaging indicated the link between endocytosis and PPB (Dhonukshe et al. 2005). In interphase cells, FM4-64 labeled endocytic vesicles are randomly distributed in the cytoplasm but at the onset of mitosis, these vesicles align with endoplasmic microtubules (EMTs) that act as internalization tracks. FM4-64 labeled endocytic vesicles travels along these MT tracks and are aligned in the PPB region to form an endosomal belt. This internalization is mostly dependent on microtubules as oryzalin treatment prevents it. Cryofixation and electron microscopic studies revealed the colocalization of clathrin coated vesicles at the PM and in the cytoplasmic areas adjacent to the PPB (Karahara et al. 2009).

The precise significance of this endocytosis and PPB formation is not known. However, PPB formation generates a cytoplasmic region that traverses the central vacuole and persists till cell plate formation. This cytoplasmic region known as actin depleted zone (ADZ) is depleted of cortical actin. Actin depletion in this region is found to be critical for proper orientation of cell division plane (Hoshino et al. 2003). It may be hypothesized that either endocytosis selectively removes actin nucleating molecules from PM during PPB formation or microtubules and actin filaments act as reciprocating entities to limit presence of each other. Later, during telophase, microtubules and actin filaments build phragmoplast at the site of future cell plate formation and both Golgi derived secretory material and endocytosed material from cell surface supplies the building blocks to this cell plate. Microtubules seem to play a critical role in delivery of material to the cell plate. Oryzalin treatment inhibits delivery of the syntaxin KNOLLE to the cell plate (Jurgens 2005). Endocytic delivery of PM-resident proteins such as auxin efflux carrier PIN1 to the cell plate is perturbed in response to oryzalin treatment (Geldner et al. 2001). This led the authors to propose that in dividing cells, PM derived materials might be endocytosed by a microtubule-dependent pathway for their delivery to the cell plate in addition to the actin dependent trafficking mechanism that operates mainly in interphase cells (Fig. 3c). In dividing plant cells, the PPB region is decorated by RanGAP (activator protein for the small GTPase Ran) and this localization is not disrupted even after PPB is disassembled (Xu et al. 2008). Considering the fact that activated Ran GTPases are promoters of plus end directed microtubular trafficking and microtubule arrays of the phragmoplast are oriented with the plus ends toward the growing cell plate it is postulated that perhaps Ran GTPases control plus end directed traffic of endocytosed material from cell periphery to the newly forming cell plate (Xu et al. 2008).

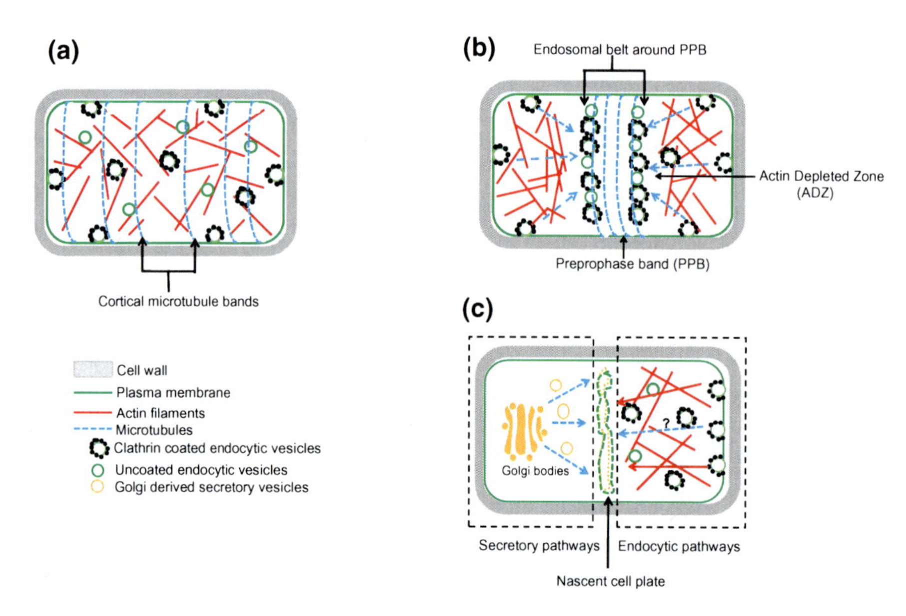

Fig. 3 Overview of spatial organization of the cytoskeleton and endocytosis in interphase and dividing cells. **a** In interphase cells microtubules are arranged in cortical arrays around cell periphery (*blue dotted lines*). Internalized coated and non-coated endocytic vesicles are distributed throughout the cytoplasm by using transport on actin tracks (*red lines*). **b** During early prophase, microtubules accumulate at the site of future cell plate formation to form the preprophase band (PPB). The cytoplasmic area near PPB is depleted in actin filaments and is known as actin depleted zone (ADZ) Endoplasmic microtubules (*blue dashed arrows*) align form the cell periphery to the PPB and act as internalization track for endocytic vesicles. Internalized endosomes accumulate around the PPB to form endosomal belt. **c** During cytokinesis, microtubules play key role in transport of newly derived secretory vesicles to assembling cell plate. Endocytosed materials are also deposited to the cell plate by actin filaments (*red arrows*) and probably also by microtubules (*blue dashed arrow*)

3.1 Microtubule-Associated Motors and Their Roles in Endocytic Processes

There are two major types of motor proteins associated with microtubules. Fourteen families of kinesin motors have been reported in mammals, all of which move cargo toward plus end of microtubules, except for kinesin-14 which is minus end directed. The other type of motors, dyneins, move exclusively toward minus end of microtubule tracks (Soldati and Schliwa 2006). The minus end directed trafficking that carry endosomes from cell periphery toward the cell interior is mediated mostly by the dynein motors or the unique group of minus end directed kinesins. On the other hand, recycling of endosomes to the cell surface is dependent on plus end directed motor activity of kinesins (Caviston and Holzbaur 2006). In plants, the role of microtubule motors in endocytosis is not well understood. Dyneins or homologous proteins have not been detected in

Arabidopsis genome although four dynein heavy chain like proteins have been found in genome of rice (*Oryza sativa*) (King 2002). Arabidopsis genome has sequences for 61 kinesin motors. Twenty one of them are predicted to be minus end directed and others are predicted to be plus end directed (Lee and Liu 2004). Some of the kinesins like AtPAKRP2 associate with Golgi derived secretory vesicles and they are implicated in the trafficking of these vesicles to the developing cell plate by plus end directed motor activity (Lee et al. 2001). Some other kinesins are hypothesized to deliver the secreted materials to the cell wall (Lu et al. 2005). The roles of plant kinesins in endocytosis are not explored in detail. The kinesin KCA-1 is localized to PM and it is depleted from the PM near PPB during early prophase till the end of cytokinesis. However, KCA-1 remains localized in the rest of the PM and at the endoplasmic microtubules while it also localizes to the newly forming cell plate during cytokinesis (Vanstraelen et al. 2004, 2006). It is possible that the minus end directed motor activity of KCA-1 drives transport of endocytic vesicles from PM to the cell plate along microtubule tracks. Detailed studies using genetic perturbations are required to present a clear picture of the involvement of these motor proteins in the regulation of microtubule mediated endocytosis.

4 Conclusions and Future Prospects

In-depth studies spanning more than a decade have elucidated the role of cytoskeletal elements in endocytic processes in yeasts and animals to appreciable details. In plants, similar studies are relatively recent and the picture depicting the cognate scenario in plants is just emerging. The current evidences demonstrate that the basic design of the cytoskeletal elements is conserved between plants and other systems. Homologous of many players that modulate/interact with actin and microtubule cytoskeletons and may regulate endocytosis have been found in the sequenced plant genomes. However, there are some unique features that set plants apart from yeast and animal models in terms of cytoskeleton and endocytosis. Due to considerable turgor pressure the sizes of clathrin coated endocytic vesicles are smaller as compared to those in non-plant systems. Further, microtubule dependent trafficking processes that are initiated in early prophase in dividing cells appear to be an exclusive feature of plant system. In addition, there are likely subtle differences in the mechanism of endocytic vesicle formation between plant and other systems even in case of interphase cells. Actin nucleation with positive contribution of actin at the onset of endocytic foci formation is a conserved feature in yeasts and animals because depolymerization of actin prevents endocytic foci formation in these systems. On the contrary, in plants depolymerization of cortical actin by mild chemical perturbation does not inhibit endocytic foci formation or entry of endocytosed cargo inside cells. At the same time, stabilization of cortical actin by either pharmacological or genetic means inhibits endocytosis in plants. These observations imply that cortical actin might act as an inhibitory barrier to endocytosis in plants. Interestingly, microtubule depolymerization has a more prominent effect on dynamics of endocytic foci. This observation and the fact that the PPB acts as a localized hub for enhanced clathrin mediated endocytosis hints for a positive regulatory role of microtubules in endocytic vesicle formation. During development of Arabidopsis leaf pavement cells, actin and microtubule localizes to mutually exclusive domains beneath the PM and this provided a model to study the roles of actin and microtubules in endocytosis simultaneously. Signalling by ROP-2/RIC-4 initiates actin nucleation at the lobe regions which in turn suppress clathrin mediated endocytosis selectively in these regions. On the other hand, ROP6/RIC1 mediates microtubule organization at the cell indentations where endocytosis takes place (Xu et al. 2010). Concerning intracellular endosome trafficking, plants seem to deviate from animal systems. Microtubules mediate long range transport of endocytosed cargo in animal cells whereas this function is mediated by oriented cables of filamentous actin in plant system as demonstrated in the case of polar growth of root hairs and pollen tubes. Considering above findings, plants seem to employ microtubules for cargo internalization events and the actin cytoskeleton for cargo transport in interphase plant cells. During cell division, plants seem to employ both cytoskeletal elements for cargo delivery.

Despite the knowledge gathered so far, the finer details of the regulation of endocytosis by cytoskeleton are much less understood in plants as compared to the yeast and animal models. The spatio-temporal interactions of cytoskeleton and endocytic foci at higher resolution should be employed in plant endocytic research. Cytoskeletal regulators and motor proteins play important roles in different stages of endocytosis in yeast and animals as revealed by genetic perturbations coupled with live imaging. Although plants possess homologous of these proteins, their roles in endocytosis have not yet been probed. In the near future combined genetic and cell biological studies involving high resolution real-time live imaging will be instrumental to provide a clear understanding of the involvement of cytoskeleton in the process of endocytosis in plant cells.

References

Affentranger S, Martinelli S, Hahn J, Rossy J, Niggli V (2011) Dynamic reorganization of flotillins in chemokine-stimulated human T-lymphocytes. BMC Cell Biol 12:28

Aniento F, Emans N, Griffiths G, Gruenberg J (1993) Cytoplasmic dynein-dependent vesicular transport from early to late endosomes. J Cell Biol 123:1373–1387

Apodaca G, Katz LA, Mostov KE (1994) Receptor-mediated transcytosis of IgA in MDCK cells is via apical recycling endosomes. J Cell Biol 125:67–86

Ayscough KR (2000) Endocytosis and the development of cell polarity in yeast require a dynamic F-actin cytoskeleton. Curr Biol 10:1587–1590

Baluška F, Šamaj J, Hlavačka A, Kendrick-Jones J, Volkmann D (2004) Actin-dependent fluid-phase endocytosis in inner cortex cells of maize root apices. J Exp Bot 55:463–473

Baluška F, Šamaj J, Wojtaszek P, Volkmann D, Menzel D (2003) Cytoskeleton-plasma membrane-cell wall continuum in plants. Emerging links revisited. Plant Physiol 133:482–491

Banno H, Chua NH (2000) Characterization of the arabidopsis formin-like protein AFH1 and its interacting protein. Plant Cell Physiol 41:617–626
Bar M, Aharon M, Benjamin S, Rotblat B, Horowitz M, Avni A (2008) AtEHDs, novel Arabidopsis EH-domain-containing proteins involved in endocytosis. Plant J 55:1025–1038
Bar M, Avni A (2009) EHD2 inhibits ligand-induced endocytosis and signaling of the leucine-rich repeat receptor-like protein LeEix2. Plant J 59:600–611
Berg JS, Powell BC, Cheney RE (2001) A millennial myosin census. Mol Biol Cell 12:780–794
Bloch D, Lavy M, Efrat Y, Efroni I, Bracha-Drori K, Abu-Abied M, Sadot E, Yalovsky S (2005) Ectopic expression of an activated RAC in Arabidopsis disrupts membrane cycling. Mol Biol Cell 16:1913–1927
Boutte Y, Crosnier MT, Carraro N, Traas J, Satiat-Jeunemaitre B (2006) The plasma membrane recycling pathway and cell polarity in plants: studies on PIN proteins. J Cell Sci 119:1255–1265
Caviston JP, Holzbaur EL (2006) Microtubule motors at the intersection of trafficking and transport. Trends Cell Biol 16:530–537
Chadda R, Howes MT, Plowman SJ, Hancock JF, Parton RG, Mayor S (2007) Cholesterol-sensitive Cdc42 activation regulates actin polymerization for endocytosis via the GEEC pathway. Traffic 8:702–717
Cheung AY, Wu HM (2004) Overexpression of an Arabidopsis formin stimulates supernumerary actin cable formation from pollen tube cell membrane. Plant Cell 16:257–269
de Forges H, Bouissou A, Perez F (2012) Interplay between microtubule dynamics and intracellular organization. Int J Biochem Cell Biol 44:266–274
Dhonukshe P, Baluška F, Schlicht M, Hlavačka A, Šamaj J, Friml J, Gadella TW Jr (2006) Endocytosis of cell surface material mediates cell plate formation during plant cytokinesis. Dev Cell 10:137–150
Dhonukshe P, Grigoriev I, Fischer R, Tominaga M, Robinson DG, Hasek J, Paciorek T, Petrášek J, Seifertová D, Tejos R, Meisel LA, Zažímalová E, Gadella TW Jr, Stierhof YD, Ueda T, Oiwa K, Akhmanova A, Brock R, Spang A, Friml J (2008) Auxin transport inhibitors impair vesicle motility and actin cytoskeleton dynamics in diverse eukaryotes. Proc Natl Acad Sci U S A 105:4489–4494
Dhonukshe P, Mathur J, Hulskamp M, Gadella TW Jr (2005) Microtubule plus-ends reveal essential links between intracellular polarization and localized modulation of endocytosis during division-plane establishment in plant cells. BMC Biol 3:11
Dong CH, Xia GX, Hong Y, Ramachandran S, Kost B, Chua NH (2001) ADF proteins are involved in the control of flowering and regulate F-actin organization, cell expansion, and organ growth in Arabidopsis. Plant Cell 13:1333–1346
Duncan MC, Cope MJ, Goode BL, Wendland B, Drubin DG (2001) Yeast Eps15-like endocytic protein, Pan1p, activates the Arp2/3 complex. Nat Cell Biol 3:687–690
Echarri A, Muriel O, Pavon DM, Azegrouz H, Escolar F, Sanchez-Cabo F, Martinez F, Montoya MC, Llorca O, Del Pozo MA (2012) Caveolar domain organization and trafficking is regulated by Abl kinases and mDia1. J Cell Sci. doi:10.1242/jcs.090134
Favery B, Chelysheva LA, Lebris M, Jammes F, Marmagne A, De Almeida-Engler J, Lecomte P, Vaury C, Arkowitz RA, Abad P (2004) Arabidopsis formin AtFH6 is a plasma membrane-associated protein upregulated in giant cells induced by parasitic nematodes. Plant Cell 16:2529–2540
Frank M, Egile C, Dyachok J, Djakovic S, Nolasco M, Li R, Smith LG (2004) Activation of Arp2/3 complex-dependent actin polymerization by plant proteins distantly related to Scar/WAVE. Proc Natl Acad Sci U S A 101:16379–16384
Gekle M, Mildenberger S, Freudinger R, Schwerdt G, Silbernagl S (1997) Albumin endocytosis in OK cells: dependence on actin and microtubules and regulation by protein kinases. Am J Physiol 272:668–677
Geldner N, Friml J, Stierhof YD, Jurgens G, Palme K (2001) Auxin transport inhibitors block PIN1 cycling and vesicle trafficking. Nature 413:425–428

Geli MI, Riezman H (1996) Role of type I myosins in receptor-mediated endocytosis in yeast. Science 272:533–535

Girao H, Geli MI, Idrissi FZ (2008) Actin in the endocytic pathway: from yeast to mammals. FEBS Lett 582:2112–2119

Glebov OO, Bright NA, Nichols BJ (2006) Flotillin-1 defines a clathrin-independent endocytic pathway in mammalian cells. Nat Cell Biol 8:46–54

Golomb L, Abu-Abied M, Belausov E, Sadot E (2008) Different subcellular localizations and functions of Arabidopsis myosin VIII. BMC Plant Biol 8:3

Grebe M, Xu J, Mobius W, Ueda T, Nakano A, Geuze HJ, Rook MB, Scheres B (2003) Arabidopsis sterol endocytosis involves actin-mediated trafficking via ARA6-positive early endosomes. Curr Biol 13:1378–1387

Guilherme A, Soriano NA, Bose S, Holik J, Bose A, Pomerleau DP, Furcinitti P, Leszyk J, Corvera S, Czech MP (2004) EHD2 and the novel EH domain binding protein EHBP1 couple endocytosis to the actin cytoskeleton. J Biol Chem 279:10593–10605

Hoshino H, Yoneda A, Kumagai F, Hasezawa S (2003) Roles of actin-depleted zone and preprophase band in determining the division site of higher-plant cells, a tobacco BY-2 cell line expressing GFP-tubulin. Protoplasma 222:157–165

Huang S, Blanchoin L, Kovar DR, Staiger CJ (2003) Arabidopsis capping protein (AtCP) is a heterodimer that regulates assembly at the barbed ends of actin filaments. J Biol Chem 278:44832–44842

Huckaba TM, Gay AC, Pantalena LF, Yang HC, Pon LA (2004) Live cell imaging of the assembly, disassembly, and actin cable-dependent movement of endosomes and actin patches in the budding yeast, *Saccharomyces cerevisiae*. J Cell Biol 167:519–530

Hussey PJ, Allwood EG, Smertenko AP (2002) Actin-binding proteins in the Arabidopsis genome database: properties of functionally distinct plant actin-depolymerizing factors/cofilins. Philos Trans R Soc Lond B Biol Sci 357:791–798

Jones KM, Kobayashi H, Davies BW, Taga ME, Walker GC (2007) How rhizobial symbionts invade plants: the Sinorhizobium-Medicago model. Nat Rev Microbiol 5:619–633

Jones MA, Shen JJ, Fu Y, Li H, Yang Z, Grierson CS (2002) The Arabidopsis Rop2 GTPase is a positive regulator of both root hair initiation and tip growth. Plant Cell 14:763–776

Jurgens G (2005) Plant cytokinesis: fission by fusion. Trends Cell Biol 15:277–283

Kaksonen M, Sun Y, Drubin DG (2003) A pathway for association of receptors, adaptors, and actin during endocytic internalization. Cell 115:475–487

Kaksonen M, Toret CP, Drubin DG (2005) A modular design for the clathrin- and actin-mediated endocytosis machinery. Cell 123:305–320

Kaksonen M, Toret CP, Drubin DG (2006) Harnessing actin dynamics for clathrin-mediated endocytosis. Nat Rev Mol Cell Biol 7:404–414

Kalia M, Kumari S, Chadda R, Hill MM, Parton RG, Mayor S (2006) Arf6-independent GPI-anchored protein-enriched early endosomal compartments fuse with sorting endosomes via a Rab5/phosphatidylinositol-3'-kinase-dependent machinery. Mol Biol Cell 17:3689–3704

Karahara I, Suda J, Tahara H, Yokota E, Shimmen T, Misaki K, Yonemura S, Staehelin LA, Mineyuki Y (2009) The preprophase band is a localized center of clathrin-mediated endocytosis in late prophase cells of the onion cotyledon epidermis. Plant J 57:819–831

King SM (2002) Dyneins motor on in plants. Traffic 3:930–931

Kitakura S, Vanneste S, Robert S, Lofke C, Teichmann T, Tanaka H, Friml J (2011) Clathrin mediates endocytosis and polar distribution of PIN auxin transporters in Arabidopsis. Plant Cell 23:1920–1931

Kleine-Vehn J, Leitner J, Zwiewka M, Sauer M, Abas L, Luschnig C, Friml J (2008) Differential degradation of PIN2 auxin efflux carrier by retromer-dependent vacuolar targeting. Proc Natl Acad Sci U S A 105:17812–17817

Konopka CA, Backues SK, Bednarek SY (2008) Dynamics of Arabidopsis dynamin-related protein 1C and a clathrin light chain at the plasma membrane. Plant Cell 20:1363–1380

Kotchoni SO, Zakharova T, Mallery EL, Le J, El-Assal Sel D, Szymanski DB (2009) The association of the Arabidopsis actin-related protein2/3 complex with cell membranes is linked to its assembly status but not its activation. Plant Physiol 151:2095–2109
Lam BC, Sage TL, Bianchi F, Blumwald E (2001) Role of SH3 domain-containing proteins in clathrin-mediated vesicle trafficking in Arabidopsis. Plant Cell 13:2499–2512
Lee YR, Giang HM, Liu B (2001) A novel plant kinesin-related protein specifically associates with the phragmoplast organelles. Plant Cell 13:2427–2439
Lee YR, Liu B (2004) Cytoskeletal motors in Arabidopsis. Sixty-one kinesins and seventeen myosins. Plant Physiol 136:3877–3883
Li R, Liu P, Wan Y, Chen T, Wang Q, Mettbach U, Baluška F, Šamaj J, Fang X, Lucas WJ, Lin J (2012) A membrane microdomain-associated protein, Arabidopsis Flot1, is involved in a clathrin-independent endocytic pathway and is required for seedling development. Plant Cell. doi:10.1105/tpc.112.095695
Li S, Blanchoin L, Yang Z, Lord EM (2003) The putative Arabidopsis arp2/3 complex controls leaf cell morphogenesis. Plant Physiol 132:2034–2044
Li X, Wang X, Yang Y, Li R, He Q, Fang X, Luu DT, Maurel C, Lin J (2011) Single-molecule analysis of PIP2;1 dynamics and partitioning reveals multiple modes of Arabidopsis plasma membrane aquaporin regulation. Plant Cell 23:3780–3797
Lu L, Lee YR, Pan R, Maloof JN, Liu B (2005) An internal motor kinesin is associated with the Golgi apparatus and plays a role in trichome morphogenesis in Arabidopsis. Mol Biol Cell 16:811–823
Mathur J, Mathur N, Kirik V, Kernebeck B, Srinivas BP, Hulskamp M (2003) Arabidopsis CROOKED encodes for the smallest subunit of the ARP2/3 complex and controls cell shape by region specific fine F-actin formation. Development 130:3137–3146
Merrifield CJ, Perrais D, Zenisek D (2005) Coupling between clathrin-coated-pit invagination, cortactin recruitment, and membrane scission observed in live cells. Cell 121:593–606
Molendijk AJ, Bischoff F, Rajendrakumar CSV, Frim J, Braun M, Gilroy S, Palme K (2001) *Arabidopsis thaliana* Rop GTPases are localized to tips of root hairs and control polar growth. EMBO J 20:2779–2788
Morris SM, Arden SD, Roberts RC, Kendrick-Jones J, Cooper JA, Luzio JP, Buss F (2002) Myosin VI binds to and localises with Dab2, potentially linking receptor-mediated endocytosis and the actin cytoskeleton. Traffic 3:331–341
Mundy DI, Machleidt T, Ying YS, Anderson RG, Bloom GS (2002) Dual control of caveolar membrane traffic by microtubules and the actin cytoskeleton. J Cell Sci 115:4327–4339
Nabi IR, Le PU (2003) Caveolae/raft-dependent endocytosis. J Cell Biol 161:673–677
Nagawa S, Lin D, Dhonukshe P, Zhang X, Friml J, Scheres B, Fu Y, Yang Z (2012) ROP GTPase-dependent actin microfilaments promote PIN1 polarization by localized inhibition of clathrin-dependent endocytosis. PLoS Biol 10:e1001299
Pelkmans L, Puntener D, Helenius A (2002) Local actin polymerization and dynamin recruitment in SV40-induced internalization of caveolae. Science 296:535–539
Peremyslov VV, Prokhnevsky AI, Dolja VV (2010) Class XI myosinss are required for development, cell expansion, and F-Actin organization in Arabidopsis. Plant Cell 22:1883–1897
Reichelt S, Knight AE, Hodge TP, Baluška F, Šamaj J, Volkmann D, Kendrick-Jones J (1999) Characterization of the unconventional myosin VIII in plant cells and its localization at the post-cytokinetic cell wall. Plant J 19:555–567
Robatzek S, Chinchilla D, Boller T (2006) Ligand-induced endocytosis of the pattern recognition receptor FLS2 in Arabidopsis. Genes Dev 20:537–542
Ross JL, Ali MY, Warshaw DM (2008) Cargo transport: molecular motors navigate a complex cytoskeleton. Curr Opin Cell Biol 20:41–47
Šamaj J, Baluška F, Voigt B, Schlicht M, Volkmann D, Menzel D (2004) Endocytosis, actin cytoskeleton, and signaling. Plant Physiol 135:1150–1161
Sattarzadeh A, Franzen R, Schmelzer E (2008) The Arabidopsis class VIII myosin ATM2 is involved in endocytosis. Cell Motil Cytoskeleton 65:457–468

Sawa M, Suetsugu S, Sugimoto A, Miki H, Yamamoto M, Takenawa T (2003) Essential role of the C. Elegans Arp2/3 complex in cell migration during ventral enclosure. J Cell Sci 116:1505–1518

Smythe E, Ayscough KR (2006) Actin regulation in endocytosis. J Cell Sci 119:4589–4598

Soldati T, Schliwa M (2006) Powering membrane traffic in endocytosis and recycling. Nat Rev Mol Cell Biol 7:897–908

Subtil A, Dautry-Varsat A (1997) Microtubule depolymerization inhibits clathrin coated-pit internalization in non-adherent cell lines while interleukin 2 endocytosis is not affected. J Cell Sci 110(19):2441–2447

Sun Y, Martin AC, Drubin DG (2006) Endocytic internalization in budding yeast requires coordinated actin nucleation and myosin motor activity. Dev Cell 11:33–46

Swanson JA (2008) Shaping cups into phagosomes and macropinosomes. Nat Rev Mol Cell Biol 9:639–649

Szymanski DB (2005) Breaking the WAVE complex: the point of Arabidopsis trichomes. Curr Opin Plant Biol 8:103–112

Tian M, Chaudhry F, Ruzicka DR, Meagher RB, Staiger CJ, Day B (2009) Arabidopsis actin-depolymerizing factor AtADF4 mediates defense signal transduction triggered by the *Pseudomonas syringae* effector AvrPphB. Plant Physiol 150:815–824

Vanstraelen M, Torres Acosta JA, De Veylder L, Inze D, Geelen D (2004) A plant-specific subclass of C-terminal kinesins contains a conserved a-type cyclin-dependent kinase site implicated in folding and dimerization. Plant Physiol 135:1417–1429

Vanstraelen M, Van Damme D, De Rycke R, Mylle E, Inze D, Geelen D (2006) Cell cycle-dependent targeting of a kinesin at the plasma membrane demarcates the division site in plant cells. Curr Biol 16:308–314

Voigt B, Timmers AC, Šamaj J, Hlavacka A, Ueda T, Preuss M, Nielsen E, Mathur J, Emans N, Stenmark H, Nakano A, Baluška F, Menzel D (2005) Actin-based motility of endosomes is linked to the polar tip growth of root hairs. Eur J Cell Biol 84:609–621

Wang X, Teng Y, Wang Q, Li X, Sheng X, Zheng M, Šamaj J, Baluška F, Lin J (2006) Imaging of dynamic secretory vesicles in living pollen tubes of *Picea meyeri* using evanescent wave microscopy. Plant Physiol 141:1591–1603

Wang YS, Motes CM, Mohamalawari DR, Blancaflor EB (2004) Green fluorescent protein fusions to Arabidopsis fimbrin 1 for spatio-temporal imaging of F-actin dynamics in roots. Cell Motil Cytoskeleton 59:79–93

Winter DC, Choe EY, Li R (1999) Genetic dissection of the budding yeast Arp2/3 complex: a comparison of the in vivo and structural roles of individual subunits. Proc Natl Acad Sci U S A 96:7288–7293

Xu T, Wen M, Nagawa S, Fu Y, Chen JG, Wu MJ, Perrot-Rechenmann C, Friml J, Jones AM, Yang Z (2010) Cell surface- and rho GTPase-based auxin signaling controls cellular interdigitation in Arabidopsis. Cell 143:99–110

Xu XM, Zhao Q, Rodrigo-Peiris T, Brkljacic J, He CS, Muller S, Meier I (2008) RanGAP1 is a continuous marker of the Arabidopsis cell division plane. Proc Natl Acad Sci U S A 105:18637–18642

Yarar D, Waterman-Storer CM, Schmid SL (2005) A dynamic actin cytoskeleton functions at multiple stages of clathrin-mediated endocytosis. Mol Biol Cell 16:964–975

Zhang Y, He J, Lee D, McCormick S (2010) Interdependence of endomembrane trafficking and actin dynamics during polarized growth of Arabidopsis pollen tubes. Plant Physiol 152:2200–2210

Index

J. Šamaj (ed.), *Endocytosis in Plants*,
DOI: 10.1007/978-3-642-32463-5, © Springer-Verlag Berlin Heidelberg 2012

Printed by Publishers' Graphics LLC